THE MECHANICS OF THE CIRCULATION

SECOND EDITION

Continuing demand for this book confirms that it remains relevant over 30 years after its first publication. The fundamental explanations are largely unchanged, but in the introduction to this second edition the authors are on hand to guide the reader through major advances of the last three decades.

With an emphasis on physical explanation rather than equations, Part I clearly presents the background mechanics. The second part applies mechanical reasoning to the component parts of the circulation: blood, the heart, the systemic arteries, microcirculation, veins and the pulmonary circulation. Each section demonstrates how an understanding of basic mechanics enhances our understanding of the function of the circulation as a whole.

This classic book is of value to students, researchers and practitioners in bioengineering, physiology and human and veterinary medicine, particularly those working in the cardiovascular field, and to engineers and physical scientists with multidisciplinary interests.

'... essential reading for anyone who is interested in the mechanics of the circulation. The normally incomprehensible mechanical laws are explained so clearly that even the non-mathematically minded will have no difficulty, which makes me very sorry that it was not available when I was grappling with these problems.'

DAVID MENDEL, Journal of the Royal Society of Medicine

'Like a good sculpture which leaves no chisel marks on the marble, there are no marks of individual specialization in this book. All is well integrated toward the physiology of circulation ... After reading the book, one would wonder how can circulation physiology be understood without such a study of mechanics. It cannot! I recommend this book to all physiology teachers and students.'

Y. C. FUNG, Journal of Biomechanical Engineering

'Here is a book on the mechanics of the circulation that is equally accessible to those trained in the life sciences and in the mechanical sciences.'

SIR JAMES LIGHTHILL, Journal of Fluid Mechanics

THE MECHANICS OF
THE CIRCULATION

SECOND EDITION

C.G. CARO
Imperial College

T.J. PEDLEY
University of Cambridge

R.C. SCHROTER
Imperial College

W.A. SEED
Imperial College

With the Assistance of

K.H. PARKER
Imperial College

CAMBRIDGE
UNIVERSITY PRESS

CAMBRIDGE
UNIVERSITY PRESS

University Printing House, Cambridge CB2 8BS, United Kingdom

One Liberty Plaza, 20th Floor, New York, NY 10006, USA

477 Williamstown Road, Port Melbourne, VIC 3207, Australia

314-321, 3rd Floor, Plot 3, Splendor Forum, Jasola District Centre, New Delhi - 110025, India

103 Penang Road, #05-06/07, Visioncrest Commercial, Singapore 238467

Cambridge University Press is part of the University of Cambridge.

It furthers the University's mission by disseminating knowledge in the pursuit of
education, learning and research at the highest international levels of excellence.

www.cambridge.org
Information on this title: www.cambridge.org/9780521151771

First published 2012

A catalogue record for this publication is available from the British Library

Library of Congress Cataloging in Publication data
The mechanics of the circulation / C.G. Caro . . .[et al.]. – 2nd ed.
p.   cm.
Includes bibliographical references and index.
ISBN 978-0-521-15177-1 (pbk.)
1. Hemodynamics.   2. Blood – Circulation.   I. Caro, Colin G. (Colin Gerald)   II. Title.
QP105.M4   2011
612.1´1 – dc23   2011027494

ISBN 978-0-521-15177-1 Paperback

Contents

Foreword

When I arrived at the Physiological Flow Studies Unit, Imperial College, in 1971, the writing of *The Mechanics of the Circulation* was already underway. The book had been commissioned by Oxford University Press to be delivered in 1972 and the Tuesday afternoon book meeting was a regular event. From the outset, the purpose of the book was seen as presenting cardiovascular mechanics in a rigorous but accessible way. It was not meant to be a textbook, but an introduction to the subject that would be useful to a wide range of readers from medical students to experts in either mechanics or cardiovascular physiology.

The Mechanics of the Circulation was finally published in 1978 and it was obvious that the authors had succeeded in their purpose. It was a truly interdisciplinary book, its authors having trained in medicine, mathematics and engineering, but there was a continuity of style and content that remains unusual in multidisciplinary, multi-author books. Individual authors wrote the first drafts of the different sections of the book closest to their expertise, but they all had an equal say in the final product which, as evidenced by the time it took to write the book and the heat that was generated in those weekly meetings, was no easy task.

The book had an enormous impact on the emerging field of cardiovascular mechanics and, by extension, on the development of the discipline of bioengineering as an essentially multidisciplinary field of study. It was reprinted and published as a paperback. Then, for reasons known only to the publisher, it was allowed to go out of print. In the years that followed there were occasional discussions about writing a second edition to incorporate the many advances that had taken place in the understanding of the cardiovascular system. But, because of other pressures and activities, the authors never found the time and the book became unavailable (except for the Russian and Chinese translations which continued to be available for several decades).

With the authors all retired, new discussions arose about a second edition and I was very honoured to be asked to be involved. We had many meetings about the changes that were needed and how the book could be made more relevant to the present time. It very quickly became evident, however, that the explosion of knowledge about the physiology and mechanics of the cardiovascular system during the past 30 years made it impossible to embrace the whole subject in a single book. After much discussion, it was decided that the one thing that has remained constant over the years is the basic mechanics, which was the primary subject of the original book. We therefore decided to republish rather than rewrite the original book. This volume, with the addition of this foreword, a new preface, a few minor corrections and a greatly enhanced index, is the result.

Some flavour of the differences in research on the circulation between 1970 and 2010 can be gained simply by considering the way the old and the new versions of the book were produced. The original book was written in longhand and transcribed by a typist. Editing involved handwritten comments in the margins and the index involved annotated filing cards that were sorted by hand. The book was set by hand and the figures were reproduced using photolithography. The current book was prepared by scanning the original into a text file generated by an optical character recognition program. The text files were edited on a computer into a LaTeX file which generated the final format electronically. The output of the LaTeX program was edited via email and the new index was generated using the 'makeindex' function in LaTeX. Finally, a LaTeX-compatible printing press was used to convert the electronic form of the book into the printed hard copy. Every aspect of cardiovascular science has undergone a similar revolution.

In the compilation of a new, greatly expanded index for this volume, I have been struck by two things about *The Mechanics of the Circulation*: its completeness and its cohesiveness.

In the course of introducing the mechanics of the circulation, the book covers the anatomy and physiology of the circulation in considerable detail and even includes some examples of its pathology. There are, inevitably, some omissions. For instance, the extracellular material at the outer surface of endothelial cells receives only the most fleeting of mentions in Chapter 13 and is never named as the glycocalyx. And in the extensive discussion of waves in the circulation there is no mention of the water hammer equation that so conveniently relates changes in pressure and velocity. It is remarkable, however, how rare these omissions are given the breadth of material covered.

Even more impressive is the cohesiveness of the book. The authors have taken great care in the cross-referencing between the different sections. 'Part I: Background mechanics' provides a thorough grounding in basic mechanics, with extensive links to the application of these principles to the cardiovascular system. 'Part II: Mechanics of the circulation' deals with the different parts of the circulation in turn. Here the links are not only to the basic principles, but also to the other parts of the circulation with similar or opposing properties.

From my personal experience and from the experience of other colleagues working on the circulation, *The Mechanics of the Circulation* is a very valuable book. It provides an introduction to mechanics for those trained in physiology, medicine and biology and an introduction to the anatomy and physiology of the circulation for those trained in mechanics, engineering and mathematics. Virtually everyone I know in the field has a well-thumbed copy on their bookshelf and many have used it as a basic text for both undergraduate and graduate courses.

Thirty years after its original publication, I am delighted that this classic book is once again being made available to experts and, most importantly, to students – the experts-to-be.

Kim H. Parker

July 2010

Preface to the First Edition

In 1808 Thomas Young introduced his Croonian lecture to the Royal Society on the function of the heart and arteries with the words:

The mechanical motions, which take place in an animal body, are regulated by the same general laws as the motions of inanimate bodies ... and it is obvious that the inquiry, in what manner and in what degree, the circulation of the blood depends on the muscular and elastic powers of the heart and of the arteries, supposing the nature of those powers to be known, must become simply a question belonging to the most refined departments of the theory of hydraulics.

For Young this was a natural approach to physiology; like many other scientists in the nineteenth century, he paid scant attention to the distinction between biological and physical science. Indeed, during his lifetime he was both a practising physician and a professor of physics; and, although he is remembered today mainly for his work on the wave theory of light and because the elastic modulus of materials is named after him, he also wrote authoritatively about optic mechanisms, colour vision, and the blood circulation, including wave propagation in arteries.

This polymath tradition seems to have been particularly strong among the early students of the circulation, as names like Borelli, Hales, Bernoulli, Euler, Poiseuille, Helmholtz, Fick, and Frank testify; but, as science developed, so did specialization and the study of the cardiovascular system became separated from physical science. This process was not, of course, complete because collaborative work between scientists from different disciplines has always gone on. However, its scale was quite limited, and many medical and physiological workers found it difficult to comprehend because of their inadequate background in mathematics and mechanics, just as physical scientists found the complexity and empiricism of physiological studies, as well as the terminology, forbidding.

The separation caused by specialization has now assumed new importance. Over about the last twenty years physical scientists and engineers have made considerable contributions to the understanding of the mechanics of the circulation. These have strongly stimulated collaborative research, but at the same time have made the field increasingly difficult for those with a limited training in physics and mathematics. Several recent reviews and monographs bear witness to the importance of this inter-disciplinary work, but do little to help the medical reader, since they invariably assume an understanding of mechanics and are often quite mathematical in format.

This book is an attempt to alleviate the problem. It is intended as an introductory text on the mechanics of the circulation which, so far as is practicable, avoids mathematical formulations and presents mechanics in readily comprehensible terms. Our experience in teaching students of physiology and medicine, and cardiological physicians and surgeons, suggests that this approach is helpful, and it is to such a readership that the book is primarily directed. In addition, we think that the book will prove useful to physical scientists, mathematicians, and engineers interested in the field, since it provides the relevant anatomical and physiological background to the mechanics, and gives definitions of terms and numerical data wherever possible.

The book is divided into two parts. The first part, 'Background mechanics', provides a non-mathematical outline of the physical processes and mechanisms which have general importance in the circulation. Thus it forms a physical introduction to the later material, though since it is self-contained and deals in a general rather than a specific way with solid and fluid mechanics and mass transport, it may also prove useful as a background to the study of systems other than the circulation.

The second part, 'The mechanics of the circulation', examines in some detail the physiological events that occur in the circulation and the physical mechanisms that underlie them. It deals first with the relevant properties of blood and then considers the circulation systematically, starting with the heart and moving forward chapter by chapter through the circulation. No attempt is made to deal in detail with active physiological mechanisms such as reflexes, but the resulting changes in the physical properties of the system are studied. In each chapter the relevant anatomical and physiological background is presented first, followed by a discussion of the mechanics. There is extensive cross-referencing to physical processes already examined earlier in the book; more specialized physical processes, relevant to the mechanics of a part of the circulation, are introduced as they arise.

We have attempted to cover all the mechanical features of the circulation which are currently considered important. However, the book is not intended as a research review, and we have therefore largely avoided citing original research references in the text. Instead, we have provided a reading list for each chapter in the second part of the book, chosen to guide readers unfamiliar with the literature to suitable reviews and sources. In addition, we have, wherever possible, taken our illustrations from important sources, in many cases the original research literature, so that the references given in the figure captions supplement the reading list.

A temptation in writing an interdisciplinary book of this kind is to oversimplify, usually at the expense of one of the disciplines; we have tried hard to avoid doing this. We have also tried, wherever possible, to supply numerical data; for convenience the more important measured and derived values, which are referred to repeatedly throughout the book, are collected together in a table reproduced on the end-page. The units used are those of the Système International, though with quantities such as pressure, where confusion might arise, we have added the traditional units. Since physical scale is important to so much of mechanics, and the dog is the only species for which anything like a comprehensive range of reliable measurements is available, we have given values from this animal throughout the book. Even so, we have had to turn to other species in describing the microcirculation, though this is a region where inter-species differences in scale appear to be relatively slight. Finally we have referred specifically to the human circulation wherever the mechanics appears to be different, or where we believe that it has relevance to a circulatory disease process.

July 1977

C.G.C.
T.J.P.
R.C.S.
W.A.S.

Acknowledgements

We owe debts of gratitude to many colleagues for their advice and help. In particular we would like to thank Dr Laurence Smaje, who contributed a major part of the physiological material contained in the chapter on the microcirculation, and without whose guidance we could not have surveyed that topic.

Mr Paul Minton of Imperial College, Dr Giorgio Gabella of University College, and Drs Graham Miller and Derek Gibson of the Brompton Hospital all made available to us original data or material for figures. Dr Michael Sudlow contributed greatly to early discussions on the scope and form of the book, and provided material for the chapter on veins; and Professor Ilsley Ingram and Drs Julien Hoffmann and Michael Hughes made valuable comments and suggestions about the chapters on the blood, heart and pulmonary circulation respectively. In addition, we owe our thanks to all the authors and journals cited in the figure captions for permission to reproduce figures. Every effort has been made to secure necessary permissions to reproduce copyright material in this work, though in some cases it has proved impossible to trace copyright holders. If any omissions are brought to our notice, we will be happy to include appropriate acknowledgements on reprinting.

Finally, we give our special thanks to Miss Evelyn Edwards, whose editorial precision helped (and chastened) us throughout the preparation of the book.

Introduction to the Second Edition

In the Preface to the first edition, we commented on the benefits and drawbacks of interdisciplinary research; the contributions of specialists to advance our understanding and the difficulty for the non-specialist in understanding these advances. We were thinking particularly about the mechanics of the circulation and the contributions that had been made by engineers, physicists and mathematicians working in collaboration with physiologists and medical doctors. Our goal in writing the book was to alleviate the problem of understanding these advances by providing an introductory text on the mechanics of the circulation that was accessible to physiologists and medical practitioners.

The three decades since the book was published have seen an explosive growth in research on the cardiovascular system. In 1978, bioengineering did not exist as a separate academic discipline and the field of cardiovascular mechanics was relatively small, although it had a long and distinguished history extending over more than three centuries. Today, bioengineering is widely recognized as an academic discipline and interdisciplinary research is generally accepted as essential to progress.

Our understanding of the circulation is immeasurably greater today than it was in 1978, but many problems remain unsolved and cardiovascular disease is still the largest single cause of death world-wide. Again, however, these advances have brought increased difficulty in understanding. We believe that the need for an introductory text on the mechanics of the circulation that is accessible to the non-specialist is even greater now than it was when the book was first published. We consider that the book will be valuable not only to circulatory scientists and practitioners, but also to physical scientists working on imaging, cell biologists working on the responses to mechanical stimuli, molecular biologists interested in cell signalling and researchers using computational fluid dynamics to study haemodynamics; to give only a few examples.

When the idea of a new edition of *The Mechanics of the Circulation* was mooted, we briefly considered expanding the book to include the many advances in cardiovascular science. We soon realized, however, that it would be impossible to be comprehensive without compromising rigour because of the sheer volume of new work. Since the basic mechanics, the theme of our introductory text, has not changed over the years, we decided to reissue the book with only minor corrections to the original text. As a result, readers will frequently come across phrases such as 'are incompletely understood', 'has not been explored', 'is still unknown' or 'the data are sparse'. In many cases the more recent developments render these phrases untrue or misleading, although in some cases they are still as true as they were in 1978. Rather than attempt to adjust the text, we have left it unchanged, and hope that this global disclaimer will absolve us from any accusation of egregious ignorance.

In addition to the therapeutic innovations which have transformed clinical practice, most of the advances in cardiovascular science can be categorized into four broad areas: imaging and measurement techniques, computational mechanics, endothelial biology, and molecular biology and genomics. It would be impossible to summarize any one of these fields here, and so we will only mention a few examples and provide some references to books and review articles that will give the interested reader an introduction to modern developments.

Imaging and measurement techniques

In the 1970s, imaging of the arteries was limited to X-ray methods and direct measurement of arterial blood velocity was difficult and highly invasive. The first quantitative measurements of local blood velocity *in vivo* were made using hot-film anemometers. Doppler ultrasound is mentioned in *The Mechanics of the Circulation*, but only to allude to its 'tremendous potential'. This tremendous potential of ultrasound has now been realized and the anatomy and function of the cardiovascular system are routinely imaged and measured non-invasively at the bedside. A number of other imaging modalities followed, such as magnetic resonance imaging (MRI), computer tomography (CT) and positron emission tomography (PET). These techniques have been developed in many different ways, enabling not only anatomical images and maps of blood and tissue velocity, but also dynamic functional imaging that could only be dreamt of 30 years ago. Today, the fine degrees of spatial and temporal resolution of imaging techniques allow the observation and consequently dynamic mechanistic modelling of many fast physiological and physico-chemical processes from the scale of whole organs all the way down to the molecular level.

The emergence of fluorescence microscopy and fast-pulsed laser methods have given us a much better understanding of structure and dynamics at the microscopic and sub-microscopic level. It is now possible to image single molecules with two-photon methods and to probe the nano-environment of proteins using state-sensitive fluorescent markers.

All of these imaging techniques progressed rapidly from the research lab to clinical practice, making the flow patterns shown in Chapter 12 of *The Mechanics of the Circulation* look particularly dated. Figures 12.38, 12.40 and 12.41 show velocity patterns measured in animal experiments and represent the state of the art in the measurement of arterial blood flow in 1978. Now, more detailed and much more accurate measurements are routinely made non-invasively in the clinic or even the GP surgery.

- W. Manning and D. J. Pennell (eds) (2001) *Cardiovascular Magnetic Resonance*, Churchill Livingstone.
- J. B. Pawley (ed.) (2006) *Handbook of Biological Confocal Microscopy*, Springer, ISBN 38725921X 9780387259215.
- G. K. Von Schulthess (2006) *Molecular Anatomic Imaging: PET-CT and SPECT-CT Integrated Modality Imaging.* Lippincott Williams & Wilkins, ISBN 9780781776745.
- G. J. Tearney, S. Waxman, M. Shishkov, B. J. Vakoc, M. J. Suter, M. I. Freilich, A. E. Desjardins, W.-Y. Oh, L. A. Bartlett, M. Rosenberg and B. E. Bouma (2008) Three-dimensional coronary artery microscopy by intracoronary optical frequency domain imaging. *J. Am. Coll. Cardiol. Img.* **1**, 752–761.
- G. T. Herman (2009) *Fundamentals of Computerized Tomography: Image Reconstruction from Projections* (2nd edn). Springer, ISBN 978-1-85233-617-2.
- H. Feigenbaum, W. F. Armstrong and T. Ryan (2010) *Echocardiography* (7th edn). Lippincott Williams & Wilkins, ISBN 0781795575.

Computational mechanics

The impact of computers on every facet of our lives is self-evident. Since 1978, advances in computer technology have enabled innumerable developments in cardiovascular mechanics. For instance, all the developments in imaging depend upon fast, memory-intensive calculations and many of the advances in molecular biology and genetics would not have been possible without mass data storage and analysis. Advances in computer power have also given rise to computational fluid dynamics (CFD) which has greatly increased our understanding of the complex flows that occur in the cardiovascular system. The basic equations of mechanics are well known and have been rigorously tested for centuries (as described in this book). They are, however, very complex and their solution is well established as a 'hard' problem. Powerful computer programs have been developed that enable us to predict the dynamics of the fluid and solid components of the system and, very recently, the interaction between the two. The emerging ability to model these 'fluid–structure interactions' in detail will undoubtedly lead to a far greater understanding of how the behaviour of one influences the other, particularly at the cellular and molecular levels.

CFD has a significant advantage, in that it can predict properties of the cardiovascular system, such as wall shear stress, that cannot be measured in vivo. CFD has also brought better understanding of the influence of complex three-dimensional geometry on mixing and mass transport in the circulation. It is probably fair to say that many of the advances in CFD have been stimulated by haemodynamics and that CFD will be necessary to reveal the aetiology of cardiovascular disease.

However, much remains to be done. Because of the nonlinearity of the basic equations, many potential solutions may be possible for the same conditions; for turbulent flows it seems that there can be an infinity of such solutions for some conditions. Thus, finding *an* answer to a patient-specific flow problem is not necessarily the same as finding *the* answer. Consider, for example, the effort that goes into the prediction of the weather, a fluid dynamics problem comparable in complexity to flow in the circulation, with results that have only recently advanced (somewhat) beyond 'tomorrow's weather will be like today's'.

- C. G. Caro, D. J. Doorly, M. Tarnawski, K. T. Scott, Q. Long and C. L. Dumoulin (1996) Non-planar curvature and branching of arteries and non-planar-type flow. *Proc. R. Soc.: Math. Phys. Eng. Sci.*, **452**, 185–197.
- C. A. Taylor and M. T. Draney (2004) Experimental and computational methods in cardiovascular fluid mechanics. *Annu. Rev. Fluid Mech.*, **36**, 197–231.
- Y. S. Chatzizisis, A. U. Coskun, M. Jonas, E. R. Edelman, C. L. Feldman and P. H. Stone (2007) Role of endothelial shear stress in the natural history of coronary atherosclerosis and vascular remodeling. *J. Am. Coll. Cardiol.* **49**, 2379–2393.
- G. Coppola and C. G. Caro (2009) Arterial geometry, flow pattern, wall shear and mass transport: potential physiological significance. *J. R. Soc. Interface*, **6**, 519–528. doi: 10.1098/rsif.2008.0417.
- D. J. Doorly, D. J. Taylor, A. M. Gambaruto, R. C. Schroter and N. Philos (2008) Nasal architecture: form and flow. *Philos. Trans. R. Soc. A* **366**, 3225–3246.
- L. Formaggia, A. Quarteroni and A. Veneziani (eds) (2009) *Cardiovascular Mathematics: Modeling and Simulation of the Circulatory System.* Springer-Verlag Italia, Milan.
- K. H. Parker (2009) An introduction to wave intensity analysis. *Med. Biol. Eng. Comput.*, **47**, 175–188, ISSN:0140-0118.
- C. A. Taylor and C. A. Figueroa (2009) Patient-specific modeling of cardiovascular mechanics. *Annu. Rev. Biomed. Eng.* **11**, 109–134.

Endothelial biology

Cardiac physiologists were active in the 1970s mapping out the complex control systems that govern the function of the heart, but larger blood vessels were generally considered to be relatively inert. The tone of the smooth muscle in the arteries and arterioles was known to respond to nervous stimulation and to a small number of vasoactive substances circulating in the blood. EDRF (endothelium-derived relaxant factor) was discovered at the end of the decade and NO (nitric oxide) was not identified as the principle EDRF until the early 1980s. In the last three decades, this picture has changed beyond recognition. The endothelium is now recognized as the largest organ in the body and endothelial cells are known to transduce and respond via normal or disturbed biological changes to subtle alterations in the pattern of blood flow over them.

The NO story provides a prime example of progress in this area of cardiovascular science. The Nobel Prize in Physiology or Medicine was awarded in 1998 to three scientists for their work between 1977 and 1986 on the detection and characterization of the role of NO in cardiovascular signalling. The idea that a short-lived gas which could rapidly diffuse through membranes and tissue could be produced by a cell and used to regulate the function of other cells was a radical departure which had profound effects in virtually every area of physiology and medicine. NO is now recognized as a primary signalling pathway throughout the body and the endothelium is its major producer. NO is also a neurotransmitter, and vascular-derived NO has been shown to regulate neural activity, providing another level of complexity to the cardiovascular control mechanisms. The production, regulation and response to NO is now one of the most studied, and still incompletely understood, topics in physiology and medicine.

- M. H. Friedman, C. B. Bargeron, O. J. Deters, G. M. Hutchins and F. F. Mark (1987) Correlation between wall shear and intimal thickness at a coronary artery branch. *Atherosclerosis* **68**, 27–33.
- A. M. Malek, S. L. Alper and S. Izumo (1999) Hemodynamic shear stress and its role in atherosclerosis. *JAMA*, **282**, 2035–2042.
- J. Loscalzo and J. A. Vita (eds) (2000) *Nitric Oxide and the Cardiovascular System.* Humana Press, ISBN: 9780896036208.
- J. M. Tarbell (2003) Mass transport in arteries and the localization of atherosclerosis. *Annu. Rev. Biomed. Eng.* **5**, 79–118.
- C. D. Searles (2006) Transcriptional and posttranscriptional regulation of endothelial nitric oxide synthase expression. *Am. J. Physiol. Cell Physiol.* **291**, 803–816.
- P. F. Davies (2007) Hemodynamics in the determination of the endothelial phenotype and flow mechanotransduction. In *Endothelial Biomedicine; A Comprehensive Treatise* (ed. W. C. Aird), Cambridge University Press, pp. 230–245.

Molecular biology and genomics
The separation of molecular and genetic biology from cell biology is artificial, since one subsumes the other. However, specialists in these areas are making very rapid advances and it is not always clear how they affect particular cells and organs; general texts on molecular cell biology usually omit the cardiovascular system entirely. The complexity of the molecular pathways that govern cell signalling and function is beautiful but bewildering. This is perhaps the field that is currently advancing most rapidly, many of the advances being expressed in a language of acronyms that can seem impenetrable to the newcomer.

The list of molecules involved in the complex transduction and signalling pathways within the endothelial cell now numbers in the thousands and the number of genes known to be involved increases almost daily. Indeed, the mechanical environment of a cell is now generally recognized to be as important as its chemical environment in determining its function and development. This is particularly true of progenitor cells, and so genetic medicine and tissue engineering are reliant on a good understanding of their micro-mechanical environment. Decoding these influences is one of the most active areas of cardiovascular science and it is difficult to see how this research will progress, other than to say that it is certain to be important.

- S. Chien (1990) *Molecular Biology of the Cardiovascular System.* Lippincott Williams & Wilkins, ISBN-13: 9780812113129.
- K. R. Chien (ed.) (2004) *Molecular Basis of Cardiovascular Disease: A Companion to Braunwald's Heart Disease* W. R. Saunders Company.
- H. Morita, J. Seidman and C. E. Seidman (2005) Genetic causes of human heart failure. *J. Clin. Invest.* **115**, 518–526.
- R. A. Walsh (ed.) (2005) *Molecular Mechanisms of Cardiac Hypertrophy and Failure*, Informa Healthcare/Taylor & Francis, ISBN 1-84214-248-8.
- J. A. Hill and E. N. Olson (2008) Cardiac plasticity. *N. Engl. J. Med.* **358**, 1370–1380.
- R. Passier, L. W. van Laake and C. L. Mummery (2008) Stem-cell-based therapy and lessons from the heart. *Nature* **453**, 322–329.
- E. M. Small, R. J. Frost and E. N. Olson (2010) MicroRNAs add a new dimension to cardiovascular disease. *Circulation* **121**, 1022–1032.

There have been, as we note, major advances in the 30 years between the publication of the first and second editions of this book, both in the understanding of circulatory mechanics and in the technologies applied to its study. But much remains to be done. Recent work has revealed complicated and highly important interactions between cellular pathways and mechanical events that make the need for a better understanding of the mechanics of the circulation ever more apparent. Future developments are certain to be as exciting and unpredictable as those of the past three decades, but, whatever they are, they must adhere to the basic principles of mechanics. This book aims to provide a detailed yet accessible account of these principles and we are delighted that *The Mechanics of the Circulation* is being reissued.

CGC

TJP

RCS

WAS

August 2010

Part I

Background mechanics

1

Particles and continuous materials

The science of mechanics comprises the study of motion (or equilibrium) and the forces which cause it. The blood moves in the blood vessels, driven by the pumping action of the heart; the vessel walls, being elastic, also move; the blood and the walls exert forces on each other, which influence their respective motions. Thus, in order to study the mechanics of the circulation, we must first understand the basic principles of the mechanics of fluids (e.g. blood), and of elastic solids (e.g. vessel walls), and the nature of the forces exerted between two moving substances (e.g. blood and vessel walls) in contact.

As well as studying the relatively large-scale behaviour of blood and vessel walls as a whole, we can apply the laws of mechanics to motions right down to the molecular level. Thus, 'mechanics' is taken here to include all factors affecting the transport of material, including both diffusion and bulk motion.

The study of mechanics began in the time of the ancient Greeks, with the formulation of 'laws' governing the motion of isolated solid bodies. The Greeks believed that, for a body to be in motion, a force of some sort had to be acting upon it all the time; the physical nature of this force, exerted for example on an arrow in flight, was mysterious. The need for such a force was related to one of the paradoxes of the Greek philosopher Zeno: that the arrow occupies a given position during one instant, yet is simultaneously moving to occupy a different position at a subsequent instant.

These matters were not fully resolved until the seventeenth century when Isaac Newton formulated his three *laws of motion*, which form the basis of all the mechanics described in this book. The laws refer to the motion of individual particles, which are defined as objects with *mass* (so that, for example, the Earth exerts a gravitational pull on them), but which occupy single points (that is, they have no size). Of course, every real body, even one as small as an atom or an electron, has a finite size, but the laws of particle mechanics can be directly applied both to real bodies in isolation (like the arrow of Zeno's paradox, or the Earth in its motion round the Sun, or an individual red blood cell) and to extensive regions of continuous matter which can be deformed into different configurations. Examples of such deformable materials

3

include all elastic solids, like steel, rubber and blood vessel walls, and all fluids, like water, treacle, blood plasma and air. Both liquids and gases are described here as fluids, since the laws of motion are applied in exactly the same way to each.

Newton's laws can be applied to bodies of finite size because it can be proved that a body will move as if all of its mass and all the external forces acting on it were concentrated at one point. This point is called the *centre of mass*.[1] Thus, the flight of the centre of mass of Zeno's arrow is the same as that of a particle of the same mass, acted on by the same forces of gravity and air resistance. Similarly, the motion through space of the Moon, or the Earth, or another planet, can be described by particle mechanics. So can the motion of the centre of mass of a blood cell, as long as the forces exerted on it by the surrounding plasma are known. However, the tumbling of a red cell, or the rotation of the Earth about its axis, or any other motion of a body relative to its centre of mass, depends on the detailed shape of the body and cannot be described as if the body were a particle.

The application of Newton's laws to the motion of continuous deformable materials is more difficult to justify. It is bound up with the implicit assumption that the fluids and solids we are interested in are continuous materials. In fact, physicists have long known that all matter is made up of molecules, bound together in various configurations by forces of various strengths,[2] and consisting of numbers of atoms. These in turn consist of central nuclei, surrounded by clouds of electrons, moving in orbits whose diameters are large compared with those of the nuclei. The motion of electrons round a nucleus is analogous to that of the planets round the sun, and like the solar system, most of an atom (and hence most matter) consists of empty space. Some typical dimensions are given in **Table 1.1.** It might be supposed that each nucleus, and each electron, or each atom, or even each molecule could be regarded as a particle, and its motion under the influence of the intermolecular forces deduced from Newton's laws. However, in air at normal temperature and pressure, for example, there are roughly 10^{20} molecules per cubic centimetre, and the position of each one would have to be specified precisely. Such a task is virtually impossible. The fact that the spacing between molecules is usually very small compared with the dimensions of the natural or experimental regions of fluid whose motion we wish to describe (see **Table 1.1**) indicates how we can overcome the difficulty. We may suppose the material to be divided up into a large number of elements whose dimensions are very small

---

[1] The centre of mass of a body is the same as its centre of gravity: if in the region of the Earth's surface the body is suspended by a string successively attached to various parts of it, there is one point in the body through which the straight line formed by extending the line of the string downwards always passes. This point is the *centre of gravity*.

[2] In a *solid*, the intermolecular forces are very strong and the molecules vary their relative positions only slightly; the spacing between molecules is comparable to their size. In a *liquid*, the intermolecular forces are less strong; molecules can move about readily (although their spacing is still comparable to their size) and they undergo frequent collisions. In a *gas*, the intermolecular forces are weak and the spacing is large compared with molecular dimensions, although it is still a very small distance (approximately 3×10^{-9} m (3 nm) for air at normal temperature and pressure).

Table 1.1. *Typical dimensions*

	Dimension (m)
Diameter of:	
an atomic nucleus	2×10^{-15}
an atom or gas molecule	6×10^{-10}
a polymer molecule	$\sim 10^{-8}$
Spacing of gas molecules	3×10^{-9}
Diameter of:	
a red blood cell	8×10^{-6}
a capillary	$4\text{--}10 \times 10^{-6}$
an artery	10^{-2}
the Earth	1.2×10^{7}
the Sun	1.4×10^{9}
the solar system	1.2×10^{13}
a galaxy	10^{20}
Spacing between galaxies	10^{22}

compared with those of the region of interest, but which still contain a very large number of molecules. With regard to the experiment, such an element effectively occupies a point, and can therefore be considered as a particle; with regard to molecular motions, however, it is very large, and its overall properties, like its velocity, or the density of the material in it, can be obtained by averaging over all the molecules which comprise it. We are thus able to ignore the random nature of molecular motion and treat materials as continuous. Newton's laws can now be applied to each element of the material (called a *fluid element*, or *fluid particle*, when the material is a fluid), and a precise and useful description of the motion as a whole will emerge.

In blood there are some very large molecules (e.g. lipoproteins, diameter about 3–5×10^{-8} m), and it flows in some very narrow tubes (some capillaries have a diameter as low as 4×10^{-6} m); but even so, the tube diameter is large compared with molecular dimensions. Thus blood plasma, for example, can be treated as a continuous fluid in the manner outlined above. Whole blood, however, cannot always be so treated, since it consists not only of plasma, but also of large numbers of cells which amount to about 45% of volume in normal man, and consist primarily of red blood cells (see Chapter 10). It would be convenient if the cells were small and numerous enough for their separate identity to be ignored, and their effect on the motion of whole blood, regarded as a continuous fluid, to be described in an average way. This is the case in large arteries (the diameter of the aorta, for example, is roughly 2000 times that of a red cell), but the diameter of a capillary is comparable to that of a red cell, and a description of flow in such small vessels must treat plasma and cells separately. To sum up, then, whole blood is effectively continuous in large vessels, but is not so in the microcirculation; plasma is continuous in both.

In Part I of this book, we shall develop the fundamental mechanics of continuous fluids and solids, although we must first outline Newton's laws of particle mechanics.

The mathematical symbols which appear are used solely as a form of shorthand, facilitating the precise expression of mechanical laws. They are all explained in words wherever they first appear, and a reader who knows some calculus will find much of the notation familiar.

2

Particle mechanics

Position

In order to describe the motion of a particle we must be able to describe accurately its position in space, which changes as the particle moves. To do this we suppose three straight lines to be drawn and fixed in space, all passing through a given point O, and each one perpendicular to the other two. The lines of intersection of two walls and the floor of a room are examples, with O in the corner of the room. If a fly were walking on the wall of the room (**Fig. 2.1**a), we could specify its position at any instant by recording its distance (say z) from the floor and its distance (say x) from a perpendicular wall. Similarly, if the fly were flying in the room, its position could be specified by recording its perpendicular distances from the three mutually perpendicular planes (the floor and two walls). And so it is with any point P whose position we wish to specify. Suppose that lines are drawn through P which intersect the three original lines at right angles at the points X, Y, Z (**Fig. 2.1**b). The

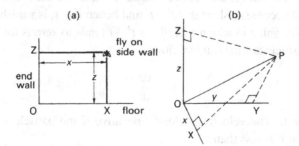

Fig. 2.1. (a) The position of a fly on the side wall of a room, specified by its distance x from the end wall and its distance z from the floor; x and z are its coordinates relative the the axes formed by the lines OX, OZ. (b) The position of a point P in three dimensions (a fly flying in a room) can be specified by its distances (x, y, z) from three mutually perpendicular planes (two walls and the floor). The coordinates of P are (x, y, z). The corner of the room O is the origin of coordinates.

7

three lengths OX (say x), OY(y), and OZ(z) then uniquely specify the position of P. These lengths are called the *coordinates* of P with respect to the three *axes* through O. The lines OX, OY and OZ are usually called the x-axis, y-axis and z-axis respectively. The total distance of P from O can be shown from Pythagoras' theorem to be equal to $\sqrt{x^2 + y^2 + z^2}$; this quantity is independent of the directions of the axes.

It is essential to remember two things implicit in this description. First, although the choice of the point O and the three axes is arbitrary, once it has been made it must be adhered to consistently. For instance, it would be hopeless to try to discuss the interaction between two particles if their positions were specified in relation to different corners of the room. We usually choose axes in the most convenient way – for example, if a particle is moving about on a flat plane (the fly on the wall), it is sensible to take one of the axes (say OY) perpendicular to that plane, so that y always remains constant and only two lengths, x and z, need be specified. Second, the units of length by which x, y and z are measured must be specified explicitly, and always used consistently. A length is not just a number, it is a quantity with dimension, and units are required to measure it. In this book, we shall usually use metres (m), centimetres ($1\,cm = 10^{-2}\,m$) or micrometres ($1\,\mu m = 10^{-6}\,m$); the whole question of units will be fully discussed later (Chapter 3).

Velocity

Another quantity of importance in describing the motion of a particle is its velocity, or the rate at which its position changes. Consider a particle moving along a straight line, OX (**Fig. 2.2**a), so that its position is specified by one coordinate, its distance x from O. If its coordinate at time t is x and its coordinate a little time later (t') is x', then the average velocity or speed of the particle in the interval of time from t to t' is $v = (x' - x)/(t' - t)$. This is well defined however short the time interval is; even if we let the interval become so short that $t' - t$, and hence $x' - x$, is vanishingly small, v is still defined. The value to which v tends as $t' - t$ tends to zero is the instantaneous velocity of the particle at time t; it is written

$$v = \frac{dx}{dt} \tag{2.1}$$

evaluated at time t. The velocity is clearly negative if the particle is moving back towards O, so that x' is less than x.

The definition of dx/dt can be understood graphically from (**Fig. 2.2**b). The upper graph shows how x varies with time t. Representative points P(x,t) and P'(x',t') are marked. The quantity $(x' - x)/(t' - t)$ is the tangent of the angle ϕ between the line joining the two points and the time axis. This quantity is called the slope of that line. As $t' - t$ tends to zero, the point P' moves towards the point P, and the line joining them approaches the tangent to the curve at P (shown in **Fig. 2.2**b as a dash–dot line).

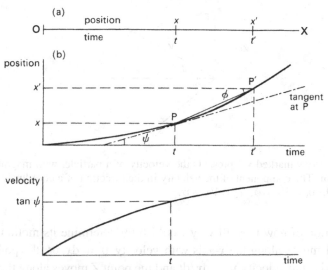

Fig. 2.2. (a) A particle moving along a line OX is at distance x from a fixed point O at time t, and at distance x' at a later time t'. The quantity $(x'-x)/(t'-t)$ is the average velocity of the particle during the time interval from t to t'. As this interval is made shorter, so that $t'-t$ tends to zero, $x'-x$ also becomes shorter, but the average velocity tends to a well-defined limit, v. This is the velocity of the particle at time t. (b) The upper graph shows the distance x plotted against time (for a particular motion of the particle). The quantity $(x'-x)/(t'-t)$ is the slope of the line PP' (and is equal to $\tan\phi$). As $t'-t$ tends to zero, this line becomes the tangent to the curve at the point P (broken line), whose slope is equal to v ($=\tan\psi$), the velocity of the particle at time t. The lower graph shows the corresponding plot of v against t.

The quantity dx/dt, i.e. v, is thus seen to be the slope of the tangent to the curve at P, and takes the value $\tan\psi$. The corresponding graph showing how v varies with t is also presented in **Fig. 2.2**b.

The resolution of Zeno's paradox lies in the performance of this limiting procedure; without it there is no way of defining the instantaneous velocity of a particle in terms of its position at successive times. The procedure was in fact not thought of until the seventeenth century, when the calculus was first developed by Newton and Leibniz. In the notation of the calculus, the symbol d/dt represents the rate of change of a quantity with time; in this example, all we mean by dx/dt is the rate at which x changes with time t. The units of velocity must be taken to be consistent with the units chosen for distance and time. If distance is measured in metres and time in seconds, then velocity must be measured in metres per second ($\mathrm{m\,s^{-1}}$).

The above definition of velocity can readily be extended to situations where the particle is moving in three dimensions. If the fly already referred to were to fly from one corner of the room to the opposite corner, all its coordinates would change with time.

2. Particle mechanics

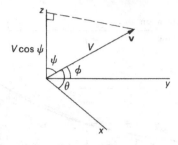

Fig. 2.3. The arrow marked **v** represents the velocity of a particle, with magnitude V and a certain direction. The component of the velocity in the direction of a coordinate axis is equal to $v_x = V \cos \theta$, $v_y = V \cos \phi$, $v_z = V \cos \psi$.

The specification of how they all vary would fully determine its motion. The point X of **Fig. 2.1** moves along the x-axis with velocity $v_x = \mathrm{d}x/\mathrm{d}t$, the point Y moves along the y-axis with velocity $v_y = \mathrm{d}y/\mathrm{d}t$ and the point Z moves along the z-axis with velocity $v_z = \mathrm{d}z/\mathrm{d}t$. The (three-dimensional) velocity of P is thus fully determined by the three quantities (v_x, v_y, v_z), which are called the velocity components of P, in the x, y and z directions respectively. They clearly depend on the directions of the coordinate axes, but are independent of the position of the origin O. The total speed at which P is travelling, i.e. the component of its velocity along a line instantaneously parallel to the direction of motion, cannot depend on the directions of the axes. The speed, sometimes called the magnitude of the velocity, can be shown to be equal to $\sqrt{v_x^2 + v_y^2 + v_z^2}$, which is a positive quantity even if some or all of v_x, v_y, v_z are negative.

We can specify the velocity of a particle in three dimensions just as precisely by giving both its magnitude and its direction relative to any two of the coordinate axes (for example, if we know the direction of motion of the fly, and its total speed, then its velocity is fully determined). If the magnitude of the velocity is V, and the angles it makes with the x- and y-axes respectively are θ and ϕ (**Fig. 2.3**), then the components v_x and v_y are given by[1]

$$v_x = V \cos \theta, \qquad v_y = V \cos \phi.$$

The third component, v_z, is then given by

$$v_z = \sqrt{V^2 - v_x^2 - v_y^2} = V \sqrt{1 - \cos^2 \theta - \cos^2 \phi},$$

which is also equal to $V \cos \psi$, where ψ is the angle between the direction of the velocity and the z-axis.

The velocity of a particle is an example of a physical quantity which has a certain magnitude and a certain direction. It exists independently of how we choose to

[1] The reader is assumed to be familiar with the elementary properties of sines and cosines.

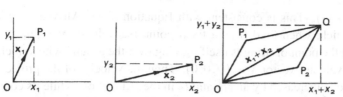

Fig. 2.4. These diagrams show the geometrical interpretation of Equation (2.3) for the addition of vectors, illustrated in two dimensions. The vectors x_1 and x_2, represented by the lines OP_1 and OP_2, are added to form the vector $x_1 + x_2$, represented by the line OQ. The points O and Q are the opposite corners of a parallelogram two sides of which are OP_1 and OP_2.

measure it, although the value of the magnitude depends on the units used to measure it, and the specification of the direction depends on the orientation of the chosen coordinate axes. Such a quantity is called a *vector*, and vectors will be represented in this book by symbols in bold type. The velocity of a particle, for example, can then be written by the single symbol **v**. The quantities (v_x, v_y, v_z) are the components of the vector **v**, and **v** can be regarded as equivalent to its three components taken together. We therefore often write $\mathbf{v} = (v_x, v_y, v_z)$.

Another example of a vector quantity is the position of the particle P (**Fig. 2.1b**), which has magnitude equal to the length of OP ($\sqrt{x^2 + y^2 + z^2}$) and direction given by the cosines of the angles between OP and any two of OX, OY, OZ; alternatively, its components are the coordinates (x, y, z) themselves. This *position vector*, say **x**, is in fact a special type of vector, in that it does depend on the position of the origin O; all other vectors describing physical quantities, like velocity, are independent of the position of the origin. Vector notation, like the use of d/dt for 'rate of change of', is just a convenient form of shorthand. From the definitions of v_x, v_y, v_z as the rates of change of x, y, z (i.e. $v_x = dx/dt$, etc.), we can combine the two shorthand notations in an obvious way as follows:

$$\mathbf{v} = \frac{d\mathbf{x}}{dt}. \tag{2.2}$$

Velocity (a vector) is the rate of change of position (also a vector).

Vectors representing two quantities of the same type (for example, two velocities, or two position vectors) are added together by adding their components. Let $\mathbf{x}_1 = (x_1, y_1, z_1)$ and $\mathbf{x}_2 = (x_2, y_2, z_2)$ be two such vectors; then

$$\mathbf{x}_1 + \mathbf{x}_2 = (x_1 + x_2, y_1 + y_2, z_1 + z_2). \tag{2.3}$$

To see this, consider the situation in two dimensions (**Fig. 2.4**).

We add the vector $\mathbf{x}_1 = (x_1, y_1)$, representing the point P_1, to the vector $\mathbf{x}_2 = (x_2, y_2)$, representing the point P_2. For the geometric interpretation to remain consistent, the resulting vector should represent the point Q, whose coordinates are

$(x_1 + x_2, y_1 + y_2)$. This is consistent with Equation (2.3). Alternatively, consider a projectile which, when fired from a fixed point, has velocity $\mathbf{v} = (v_x, v_y, v_z)$. Now suppose that the point of firing is itself moving over the ground with velocity U in the x-direction, velocity vector $\mathbf{V} = (U, 0, 0)$. Then the velocity of the projectile relative to the ground is increased by an amount U in the x-direction, while its components in the y- and z-directions are unchanged, i.e. $\mathbf{v} + \mathbf{V} = (v_x + U, v_y, v_z)$.

Acceleration

In the same way as the velocity of a particle is defined as the rate of change of position, so the acceleration of the particle, defined as the rate of change of velocity, can also be written down. For motion along a line, the acceleration is dv/dt, which is the same as the slope of the tangent of the graph of v against t (**Fig. 2.2**b). It too has three components, the rates of change of the three velocity components, and is also a vector, say $\mathbf{a}$:

$$\mathbf{a} = \frac{d\mathbf{v}}{dt} = \left(\frac{dv_x}{dt}, \frac{dv_y}{dt}, \frac{dv_z}{dt} \right). \tag{2.4}$$

In the notation of calculus, if $u = dx/dt$, then du/dt can be written d^2x/dt^2, a useful shorthand for the rate of change of the rate of change of x. Thus we can write

$$\mathbf{a} = \frac{d^2\mathbf{x}}{dt^2} = \left(\frac{d^2x}{dt^2}, \frac{d^2y}{dt^2}, \frac{d^2z}{dt^2} \right). \tag{2.5}$$

The units of acceleration must, for consistency, be metres per second squared ($m\,s^{-2}$).

It is perhaps a little difficult to grasp the precise definition of acceleration as a three-dimensional quantity. In one dimension, with the particle moving on a straight line OX, it is fairly easy: if the velocity v is increasing at a given moment, then the acceleration $a = dv/dt$ is positive; if v is decreasing a is negative. If the particle is moving back towards O with a positive value of x, then v is negative, but if it is at the same time slowing down, the acceleration a is positive. To make it clearer, **Fig. 2.5** shows graphs of x, v and a against time t for a particle which starts from rest at O, accelerates up to a uniform speed which is maintained for some time, then decreases speed with constant negative acceleration until it has changed direction and is returning to O with the same uniform speed. Finally, it is slowed down and stopped at O again by the application of a positive acceleration. The direction of the acceleration (the sign of a) is independent of the direction of motion (the sign of u).

In two or three dimensions the direction of the acceleration is also independent of the direction of the velocity. Whenever the velocity is changing, either in magnitude or in direction, the particle experiences an acceleration. For example, suppose that a particle is travelling in a circle with constant speed, like a ball twirled on the end of a string or a satellite in its orbit round the Earth. In this case the *magnitude* of the

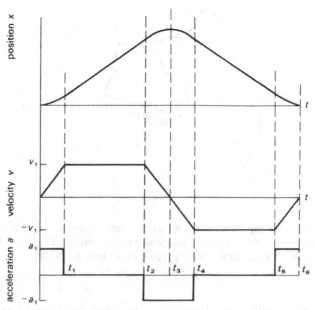

Fig. 2.5. Graphs of distance x travelled along a line, velocity v and acceleration a against time t, for a particle which starts from rest at the origin ($x = 0$) at time $t = 0$. Until time t_1 it accelerates with uniform acceleration a; until it has velocity v_1. This is maintained until time t_2, when the particle acquires a constant negative acceleration $-a$, until it has changed direction and is returning towards the origin with the same uniform speed (velocity $-v_1$) at time $t = t_4$. At time t_3, the velocity is zero and the particle is at its farthest distance from the origin. The uniform velocity $-v_1$ is maintained until time t_5, and finally the particle is slowed down with uniform acceleration a until it comes to rest at the origin at time t_6.

velocity is constant, but the particle experiences an acceleration because the *direction* of the velocity is changing (**Fig. 2.6**). Since the speed is constant, this acceleration must be in a direction perpendicular to the direction of motion (otherwise there would be a component in the direction of motion, and the total speed would change), and therefore the direction of the acceleration passes through the centre of the circle. If at time t the velocity (magnitude v) is in a certain direction, then a little time later (t') it will have acquired a component in the perpendicular direction because of the acceleration, and the velocity vector will point in a different direction. The component of velocity acquired in the time $t' - t$ is equal to v times twice the sine of $\frac{1}{2}\theta$, which, when θ is very small, is equal to the distance traversed by the particle in that time, $v(t' - t)$, divided by the radius of the circle r. Thus, the perpendicular component of velocity is $v^2(t' - t)/r$, and its rate of increase, when $t' - t$ tends to zero, is v^2/r, which is the magnitude of the acceleration towards the centre. A planet travelling

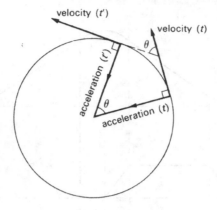

Fig. 2.6. A particle moving in a circle of radius r with constant speed u experiences an acceleration of magnitude v^2/r towards the centre of the circle. The angle θ is the angle swept out by the line from the centre to the particle in the time from t to t'.

round the Sun, or a satellite travelling round the Earth, experiences an acceleration towards the sun or the earth respectively. Similarly, a particle of blood streaming round the curved arch of the aorta must experience an acceleration towards the centre of curvature.

Newton's laws of motion: mass and force

We are now able to describe the position, velocity and acceleration of a particle as it moves on a given path in space. But what causes particles or bodies to move or stop moving? The answer is *forces*. It is unfortunately difficult to give a precise definition of force, although we all have a qualitative idea of its nature from everyday experience. From a logical point of view, the argument we shall give is a circular one, since we say that force causes motion, but is defined by the motion it causes. Nevertheless, the argument is scientifically respectable because it incorporates the results of simple experiments which remove all ambiguity.

If we lift a heavy weight, we use more force than if we lift a light weight. If we wish to prevent a car rolling downhill we have to exert a force in the uphill direction. We can stop a ball rolling along a table by applying a force to it, and if no force is applied it goes on rolling. We do not need to involve the mysterious force which the Greeks thought was required to maintain a body in motion, since we can recognize that a moving body will remain in motion with a constant velocity unless a force is exerted to change that velocity. A much more satisfactory explanation of everyday experience can be obtained, as Newton showed, if forces are associated with changes in velocity, or accelerations, not with velocity itself. Zero force is associated with zero acceleration, i.e. with constant velocity (which is itself zero if the body is at rest).

What, then, is the relationship between force and acceleration? If two balls, one heavier than the other, are rolling with the same velocity on a flat horizontal surface, and if we wish to stop them over the same period of time (i.e. give them the same negative acceleration), we have to exert a larger force on the heavier ball. The quality possessed by the heavier ball which makes it more difficult to stop it or give it positive acceleration is called inertia, and in this book we shall use this word in a purely qualitative way. The *quantity* which represents the *inertia* of a body is its *mass*, which is a measure of the total amount of matter in the body. A body with greater mass is more difficult to accelerate (or decelerate); similarly, it is its greater mass which causes one object to be heavier, i.e. have a greater weight, than another. The unit of mass is independent of the units of length and time, and the three together form a fundamental system of units in terms of which all other mechanical quantities are measured. We shall use the kilogram (kg) as the fundamental unit of mass, although some quantities will be given in grams ($1\,\mathrm{g} = 10^{-3}\,\mathrm{kg}$) (see Chapter 3).

The precise relationship between force, mass and acceleration, from which the definition of force will become clearer, is given by Newton's laws of motion, which are set out below.

(1) *First law.* Every particle continues in a state of rest or of uniform motion in a straight line unless acted on by some external force or forces.

In other words, the velocity **v** remains constant, maybe zero, if no net force (see below) is acting.

(2) *Second law.* When a particle of mass m is acted on by a force or forces so that it experiences an acceleration **a**, the net force acting on it is equal to the mass multiplied by the acceleration.

In other words, the net force is a vector, which we may call **F**, given by

$$\mathbf{F} = m\mathbf{a}. \tag{2.6}$$

This equation is often called the *equation of motion* of the particle.

By net force we mean the sum of all forces acting on the particle, which may be exerted in different ways. For example, if we push an object along a horizontal table (**Fig. 2.7**), the forces on it include the gravitational attraction of the Earth (i.e. the weight of the object, **W**), the force **R** exerted by the table on the body (which will in general include a component holding it up, N, and a frictional force T in the direction opposite to the direction of motion) and the forward force **P** (which we are exerting directly). If the forward acceleration is **a**, then

$$\mathbf{P} + \mathbf{R} + \mathbf{W} = m\mathbf{a}. \tag{2.7}$$

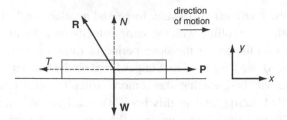

Fig. 2.7. Diagram representing all the forces acting on a body being pulled horizontally along a flat table to the right. **P** is the force pulling it, **W** is its weight and **R** is the force exerted on the body by the table. The force **R** has two components, one vertically upwards (N), which would be present even if the body were just resting there, and a frictional component T in the direction opposite to the direction of motion.

Addition of forces is carried out by the addition of their components as in Equation (2.3). Thus, if we choose x- and y-axes fixed relative to the table as shown (supposing the z-direction to be out of the paper) and if $\mathbf{P} = (P, 0, 0)$, $\mathbf{R} = (-T, N, 0)$, $\mathbf{W} = (0, -W, 0)$ and $\mathbf{a} = (a, 0, 0)$, then Equation (2.7) becomes

$$(P - T, N - W, 0) = (ma, 0, 0).$$

Clearly, two vectors can be equal only if each component is equal; we deduce that $N - W = 0$ (i.e. the upward force exerted by the table on the object is equal to its weight, so that there is no vertical acceleration), and $P - T = ma$, which relates the horizontal component of force to the acceleration of the body.

From Equation (2.6) we deduce the unit of force: since mass is measured in kg and acceleration in $\mathrm{m\,s^{-2}}$, force must have the unit $\mathrm{kg\,m\,s^{-2}}$ (kilogram metre per second per second), which is appropriately called the newton (the weight of an average apple is about one newton (1 N)). As we shall discuss in more detail later (Chapters 3 and 6), it is important that the quantities used in any equation, such as Equation (2.6), should be given the correct units. Indeed, a useful way of testing the consistency of an equation is to verify that the terms on either side of it are measured by the same units. For example, a length can never be equal to an area, nor a velocity to an acceleration, nor a force to a mass; an equation in which such equalities were stated would not only be wrong, but meaningless too.

(3) *Third law.* To every action there is an equal and opposite reaction.

That is, if one body exerts a force ($\mathbf{F}$) on another body, the second body must exert an equal and opposite force ($-\mathbf{F}$) on the first body. So in the example of **Fig. 2.7**, the fact that the table exerts an upward component of force N and a retarding component of force $-T$ on the body means that the body exerts a downward force $-N$ and a forward force T on the table. If you catch a ball by applying a force to it, you are aware that the ball applies an opposite force to you. When you hold a heavy weight, you have to exert an upward force on it because it is exerting a downward force on you; you also exert a downward force on the ground, which again reciprocates.

The types of force which are present in the example of the object on the horizontal plane surface (**Fig. 2.7**) are of two kinds: long range and short range.

First there is the weight of the body, which is an example of the gravitational attraction exerted between any two bodies with mass. In this case one body is the Earth, and the long-range force exerted by it on a particle of mass m near its surface (the weight of the particle, **W**) is directed approximately towards the centre of the Earth (i.e. vertically downwards) and is proportional to m. Its magnitude varies only slightly with position on the Earth, and can be regarded as constant in any local experiment. So we can write

$$\mathbf{W} = m\mathbf{g}, \tag{2.8}$$

where **g** is a vector whose magnitude is a constant and which is directed vertically downwards. A comparison of Equations (2.8) and (2.6) shows that **g** has the dimensions of acceleration, and use of Equation (2.6) shows that a body falling freely, when the only force acting is its weight, will experience the downward acceleration **g**. We call **g** the acceleration due to gravity; the magnitude g is approximately $9.8\,\mathrm{m\,s^{-2}}$. The approximate constancy of **g** gives rise to a convenient way of measuring the mass of a body, that is by comparing its weight with that of a given standard body, which itself need not be measured at the same time or place.

Gravitational attraction is a long-range force, in that it is exerted on any given particle by all other particles, whether they are in contact with it or not. Another example of a long-range force is the force a moving charged particle experiences in an electric or a magnetic field.

The other forces acting on the body in **Fig. 2.7** are short-range forces, which can be exerted by two bodies only when they are in contact: the force **P** is applied by us by direct pushing or pulling; the force **R**, exerted by the table, would not exist if the bodies were not in contact. Its normal component N exists because the body and table are pressing together, and a similar force is present wherever two bodies are in contact (when you touch something, you feel it because it exerts a normal force on you). The tangential component T of **R** is a frictional force, which would be absent if the bodies were perfectly smooth. For two given surfaces in relative motion, T is usually proportional to N (i.e. $T = CN$), the constant of proportionality C being called the coefficient of friction. If the applied force P is less than CN, the body will not move ($a = 0$), and the actual frictional force T is just enough to prevent movement, i.e. is equal to P. We shall see below that the forces exerted on the 'particles' of a continuous *deformable* material can also be divided into long-range and short-range forces of a similar kind.

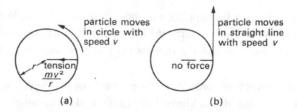

Fig. 2.8. (a) A particle is attached to a string of fixed length and moves in a circle of radius r with speed v. The tension in the string must be equal to the mass of the particle (m) times the acceleration which it experiences towards the centre of the circle (v^2/r). (b) When the string breaks, no force acts on the particle, and it moves off in a straight line with speed v.

The whole of mechanics follows from Newton's three laws. Before we develop their application to continuous fluids or solids, it might be helpful here to see a couple of examples of their use in the motion of particles (or of the centres of mass of large bodies, which are equivalent).

(1) If a man is driving a car, total mass M, at velocity $\mathbf{V}$, and wishes to stop it, he usually causes the brakes to exert a fairly uniform retarding force $\mathbf{F}$ in the opposite direction, so that his acceleration is $\mathbf{F}/M$ (Newton's second law, Equation (2.6)); the component of $\mathbf{F}$ in the direction of motion is of course negative, causing deceleration. The driver (mass m) remains at rest relative to the car, because he exerts a small forward force on it (through his legs and the pedals, his arms and the steering wheel, etc.), causing it to exert a backward force of the same magnitude on him (Newton's third law). This magnitude must be m/M times the magnitude of $\mathbf{F}$ in order that the driver should experience the same deceleration as the car. If the retarding force $\mathbf{F}$ is large and applied very suddenly (e.g. if the car hits a tree), the driver may be unable to exert a large enough force on the car in the usual way to cause it to give him the same deceleration as the car. He will continue forward, relative to the car, until stopped by something which can supply the necessary force (e.g. his chest against the steering wheel, or seat belt).

(2) Consider a particle (mass m) travelling in a circle with constant speed (**Fig. 2.8**a). We have seen that it experiences an acceleration towards the centre of the circle. Therefore, by Newton's second law, there must be a force acting on it in the same direction. If the particle represents a ball on a string, the force is supplied by the string, and is called the *tension* in the string.[2] If it represents a planet moving round the Sun, the force is gravitational. The magnitude of the force is equal to m

---

[2] This tension is the total force exerted by the string on the ball, and in mechanics the word always represents a force. It should not be confused with the tensile stress experienced by the material of which the string is made, since this is the force per unit cross-sectional area (see Chapter 7). Nor should it be mistaken for the surface tension in a film or membrane, which is the force exerted by such a surface per unit length of its boundary. Also, the use of the word 'tension' to describe the partial pressure of a gas in solution is misleading and should be abandoned.

times the magnitude of the acceleration, i.e. is mv^2/r, where v is the speed of the particle and r the radius of the circle.

If the force suddenly stops (e.g. if the string breaks) the particle can no longer experience any acceleration and, therefore, flies off in a straight line at a tangent to the circle, with constant velocity, until acted on by another force (**Fig. 2.8**b). This flying off has sometimes led people to suppose that a particle moving in a circle is acted on by an outward force, called the centrifugal force, which is balanced during circular motion by the tension in the string; and that when the string breaks, this force causes the particle to leave its circular orbit. Such a view is misleading and should be avoided: a broken string exerts no force.

Another example of bodies moving in circles is supplied by manned artificial satellites. Here, both the satellite and the men in it are travelling in the same circular orbit, which we may take to have radius r, with the same speed (say, v). Thus, they each experience the same acceleration, v^2/r, towards the centre of the earth, and each object must experience a force in that direction equal to v^2/r times its mass. This force is supplied by the Earth's gravitational field, which exerts a force on any body proportional to its mass and inversely proportional to the square of its distance from the centre of the Earth (i.e. $F = km/r^2$, where k is a constant and m the mass of the body). Thus, Newton's second law can be applied to any object in the given orbit, to give

$$\frac{km}{r^2} = \frac{mv^2}{r}.$$

So, as long as $k = v^2 r$, any object can travel in that orbit without any additional force being supplied. Thus, no force is applied between a man inside the capsule and the capsule itself, as long as the man is not moving relative to the capsule. He experiences weightlessness, in that as he floats about inside the capsule he experiences no force tending to bring him in contact with the 'floor' of the capsule. The same goes for the blood in his circulation. For any particle of blood, there is a balance between the earthward force and the earthward acceleration resulting from its circular motion. This causes considerable physiological problems, because humans have evolved in an environment where gravity is always experienced. A man on a space walk outside the capsule is in the same circumstance as a man inside. No extra force need be applied to him as long as he remains at rest relative to the capsule. Nor is there any air resistance or other force tending to slow him down relative to the capsule, so he does not require a counteracting tension in the lines holding him to it. These lines, therefore, do not become stretched out as they would within the Earth's atmosphere, where the drag of the air is important.

There are one or two more mechanical terms which should be defined before we outline the application of Newton's laws to continuous materials.

Momentum

The second law states that force equals mass times acceleration, i.e. mass times 'rate of change of velocity'. This can just as easily be written as the rate of change of 'mass times velocity', as long as the mass is constant; in symbols:

$$\mathbf{F} = m\frac{d\mathbf{v}}{dt} = \frac{d}{dt}(m\mathbf{v}).\tag{2.9}$$

It is this form of the equation which is in fact the more general, because it can also be applied to the motion of bodies whose mass is changing, like a rocket which burns fuel and shoots it out behind. The quantity $m\mathbf{v}$ is called the *momentum* of a particle, and Newton's law can be expressed in the form 'force equals rate of change of momentum'. In the absence of external forces, the momentum of a particle, or of a body or system of particles, remains constant, or is *conserved*. For example, if two particles (masses m_1, m_2; velocities $\mathbf{v}_1$, $\mathbf{v}_2$) collide and coalesce, the combined body, mass $(m_1 + m_2)$, must have the same momentum as the two original bodies put together, so its velocity must be $(m_1\mathbf{v}_1 + m_2\mathbf{v}_2)/(m_1 + m_2)$. If the collision were instantaneous, this would be true immediately afterwards even if external forces (e.g. gravity) were acting, because there would be no time for the total momentum to be changed during the collision. If a car of mass M is travelling at speed V in a certain direction, say the x-direction, so that its velocity vector is $(V,0,0)$, and a lorry of mass $10M$ is travelling in the opposite direction with speed $(-V/10)$, velocity vector $(-V/10,0,0)$, the total momentum of the two is zero and their centre of mass is at rest. It would remain at rest if the two vehicles collided, although the motion of the two vehicles relative to their combined centre of mass would change dramatically.

Work and energy

When a force of magnitude F is applied to a particle while it travels a distance d in the direction of action of the force, then the force is said to do *work*, and the amount of work done is equal to Fd. This result also holds when the force and the direction of motion are not parallel (**Fig. 2.9**); d is then the projection of the distance travelled on the direction of the force. Alternatively, the work done is equal to the component of force in the direction of motion ($F\cos\theta$ in **Fig. 2.9**) times the distance travelled d_1; this is the same as Fd, since $d = d_1\cos\theta$. If $\mathbf{F} = (F_1, F_2, F_3)$ is the force and the position of the end of the path relative to the beginning is $\mathbf{x} = (x_1, x_2, x_3)$, the work done can be shown to be $(F_1x_1 + F_2x_2 + F_3x_3)$, which is commonly written $\mathbf{F}\cdot\mathbf{x}$. If $\mathbf{F}$ is not constant, or if the direction of motion is variable, then the path of the particle has to be split into small straight segments over which $\mathbf{F}$ is constant, and the total work done is obtained by adding together the work done over each segment.

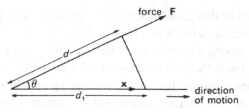

Fig. 2.9. The work done by a force of magnitude F acting on a particle which moves a distance d_1 in a straight line is equal to $Fd_1 \cos \theta$, where θ is the angle between the direction of the force and the direction of motion. This is the same as Fd, where d $(= d \cos \theta)$ is the projection of d_1 on the direction of action of the force. It is often written $\mathbf{F} \cdot \mathbf{x}$, where $\mathbf{F}$ is the vector force, and $\mathbf{x}$ is the position vector of the end of the path travelled, relative to the beginning.

For a body moving in a circle with constant speed, like a ball on the end of a string (see p. 18), the force acting on it is directed towards the centre of the circle and is always at right angles to the direction of motion. It therefore does no work.

It can be shown, from Newton's second law, that the total work done on a particle, over a period of time, by all the forces acting on it, is equal to the change in the quantity $\frac{1}{2}mv^2$, where m is the mass of the particle and v is its speed. This quantity is called the *kinetic energy* of the particle. If we apply the quoted result over a very short time, we obtain the additional result that the rate at which forces do work on the particle is equal to the rate of increase of its kinetic energy at any instant.

When the particle moves in the Earth's gravitational field, the gravitational force on it is mg vertically downwards; if the z-axis is taken vertically upwards, this force has components $(0, 0, -mg)$. We can now define a quantity mgz, called the *potential energy* of the particle. The work done per unit time by the gravitational force is $(-mg)dz/dt$, which is equal to $-d/dt(mgz)$, i.e. minus the rate of change of potential energy. In the absence of other forces this must be equal to the rate of change of kinetic energy. In other words, the rate of change of total energy (kinetic plus potential) is zero; this result is called the *principle of conservation of energy*.

$$\text{Kinetic energy} + \text{potential energy} = \tfrac{1}{2}mv^2 + mgz = E, \qquad \text{a constant.} \qquad (2.10)$$

The value of E is arbitrary, since it depends on the origin of coordinates (the level at which z is taken to be zero), but once that has been chosen, E remains fixed. A body possesses kinetic energy by virtue of its motion, while its potential energy is determined by its *position*. Suppose that during motion under the action of no force but gravity a body loses height, i.e. that its potential energy is reduced. The principle of conservation of energy then shows that the kinetic energy of the body, and hence its speed, must increase. Similarly, if the body rises, it must lose speed.

Potential energy can be thought of as stored energy, which may be transformed into kinetic energy (i.e. motion) when it is released. There are numerous ways of storing energy other than by raising a body in a gravitational field. The most important from the point of view of cardiovascular mechanics is by stretching (or compressing) an elastic material. When a particle is attached to a spring which is stretched and then released the stored potential energy is converted into kinetic energy and the particle moves. Similarly, if a balloon is blown up to a high pressure and the nozzle is then released, the air rushes out; i.e. the potential energy stored in the stretched rubber is converted into kinetic energy of the air. The same is true of blood vessels, which become distended when the local blood pressure is high, but contract again, contributing to the motion of the blood, when the pressure falls again (see Chapter 12).

Elastic forces and gravity have the property that when a body is caused to move against them, then the work done is stored as potential energy, and is subsequently recoverable as kinetic energy. Then the principle of conservation of energy holds, and the forces are called *conservative forces*. Not all forces are conservative. In particular, forces of a frictional nature, like friction between two solids sliding on each other (as in the example on p. 15), or air resistance, are such that any work done against them is lost as mechanical energy. Mechanical energy is not conserved but is dissipated (in fact it is converted into heat), and these forces are known as *dissipative forces*. When dissipative forces are acting, the principle of conservation of mechanical energy does not hold. However, if the mechanical energy lost by dissipation can be shown to be small compared with the otherwise conservative energy changes expected in a given motion, then dissipation can be neglected in a calculation with little loss of accuracy in the results.

As an example of the use of the energy principle, consider a ball thrown up into the air and suppose that the only force acting is its weight (i.e. neglect air resistance). If the ball is released with upwards speed V, from a certain level which we may take to be $z = 0$, then its subsequent speed v is related to its height z above that level by Equation (2.10). The values of the constant E are given by the conditions at release, where $z = 0$ and $v = V$; thus, $E = \frac{1}{2}mV^2$ and Equation (2.10) give

$$gz + \tfrac{1}{2}v^2 = \tfrac{1}{2}V^2. \tag{2.11}$$

Note that there is no horizontal component of the motion, because the horizontal component of velocity is zero initially and there is no horizontal force. From Equation (2.11) we can calculate the maximum height attained by the ball, since at that height ($z = h$, say) its velocity must be zero ($v = 0$). Hence

$$h = V^2/2g. \tag{2.12}$$

Similarly, the downward velocity of the ball when it returns to its point of release ($z = 0$) again has magnitude V: on the way up, the velocity vector had components $(0, 0, V)$, while on the way down it has components $(0, 0, -V)$.

It can readily be verified that Equation (2.12) is dimensionally correct, since V is a velocity $(m\,s^{-1})$ and g is an acceleration $(m\,s^{-2})$, so that the units of $V^2/2g$ are

$$\left(\frac{m}{s}\right)^2 \times \frac{s^2}{m} = m.$$

Thus $V^2/2g$ has the units of length. The unit of energy is that of a mass times the square of a velocity, i.e. $kg\,m^2\,s^{-2}$; this is of course the same as the unit of work (force times distance). The unit of momentum (mass times velocity) is $kg\,m\,s^{-1}$.

3

Units

It soon becomes clear to any student of physiology that there are many systems of units and forms of terminology. For example, respiratory physiologists measure pressures in centimetres of water and cardiovascular physiologists use millimetres of mercury. As the study of any single branch of physiology becomes increasingly sophisticated, more and more use is made of other disciplines in science. As a result, the range of units has increased to such an extent that conversion between systems takes time and can easily cause confusion and mistakes.

We see also frequent misuse of terminology which can only confuse; for example, the partial pressure of oxygen in blood is often referred to as the 'oxygen tension', when in reality tension means a tensile force and is hardly the appropriate word to use.

In order to combat a situation which is deteriorating, considerable effort is being made to reorganize and unify the systems of nomenclature and units as employed in physiology. For any agreed procedure to be of value, it must be self-consistent and widely applicable. Therefore, it has to be based upon a proper understanding of mathematical principles and the laws of physics.

The system of units which has been adopted throughout the world and is now in use in most branches of science is known as the *Système International* or SI (see p. 28). It is a coherent system of units based on the metric system of units of the kilogram mass, metre and second, and it provides a suitable basis for the unification of systems currently employed in the various branches of physiology. Conversion from the older c.g.s. system is, moreover, quite simple.

The difference between units and dimensions

The need to state *units* when specifying a physical *quantity* is recognized in everyday life as well as in the field of scientific research. Thus, distances are commonly measured in kilometres, centimetres, ångströms, etc., and velocities are measured in miles per hour or centimetres per second. The number, which indicates the magnitude or amount of a given quantity in a particular set of units, is conveniently called the measure and is inversely proportional to the size of the units used. Thus, 1 km is measured as 10^3 m or 10^5 cm.

The word *dimension*, however, is used rather differently in physical science from the way in which it is commonly employed. In everyday usage dimensions indicate the physical size of an object; for example, the size of a piece of paper is $10 \text{ in} \times 8 \text{ in}$, or the volume of a box has the dimensions $a \times b \times c \, \text{cm}^3$. Implicit in these descriptions are the concepts of area and volume respectively, which are formed as the square and the cube of the unit of length.

This leads us to the specific scientific concept of the word 'dimension', in which the dimensions of area are those of (length)^2 and those of volume are (length)^3. It is important to realize that we are here concerned solely with the *nature* of the quantity and not with its *measure* in any particular set of units. Convention has established the notation $[L^2]$ and $[L^3]$ for these two quantities. Similarly, the dimensions of velocity are $[L]/[T]$, or $[LT^{-1}]$, and those of density $[ML^{-3}]$.

The dimensions of the quantities so far considered are all self-evident from the nature of the quantity or follow at once from its definition. Often, however, the dimensions of a quantity can be related to those of another quantity only by inference from a physical law.

An example of this is the derivation of the dimensions of force from those of mass, length and time. The law involved is Newton's second law (p. 15). Thus:

$$\text{force} = \text{mass} \times \text{acceleration}$$

or

$$F = ma,$$

where a denotes acceleration whose dimensions are $[LT^{-2}]$. Hence we have

$$[F] = [ma] = [MLT^{-2}]$$

or, conversely, we may define the dimensions of mass as

$$[M] \equiv [FL^{-1}T^2].$$

The consequence of this is that the dimensions of all quantities in mechanics can be expressed in terms of $[M]$, $[L]$ and $[T]$ or in terms of $[F]$, $[L]$ and $[T]$. These alternative methods of deriving dimensional expressions are widely known as the 'mass-based' and 'force-based' systems, and each has its corresponding system of units.

Mass, length and time as fundamental units

Thus, it can be seen that the dimensions of all physical quantities not involving temperature can be described in terms of the fundamental dimensions of mass, length and time. In science we also generally use mass, length and time as the fundamental or *primary* units. These are treated as being independent of one another and the units are defined on the basis of arbitrary accepted standards.

For example, we base our measurement of length on the metre; the metre standard used to be a metal bar kept in Paris, and the distance between two marks on the bar (under specified conditions of temperature) was defined as the reference standard. Today the standard is the 'optical metre'; it is defined as 1 650 763.73 vacuum wavelengths of orange light from a krypton-86 discharge lamp! Although this may seem a bizarre and difficult way of defining a metre, it has two important advantages. First, the new standard is virtually the same length as the old one and, second any competent laboratory can set up its own reference standard without the need to travel to Paris.

From the fundamental units of mass, length and time, other units in mechanics can be derived. Thus, in the c.g.s. system the unit of area is the square centimetre or cm^2; in SI it is the square metre or m^2. The units of density in SI become kilogram per cubic metre ($kg\,m^{-3}$).

The unit of force is defined in Newton's second law as that which produces unit acceleration when acting on unit mass. The unit of force is then given a special name depending upon the system of units in which it is derived: in the c.g.s. system it is the *dyne* (1 dyne = $1\,g\,cm\,s^{-2}$); in SI it is called the *newton* ($1\,N = 1\,kg\,m\,s^{-2}$), which is 10^5 dynes. Hence, the units of stress or pressure (see Chapter 4, p. 31), which is force per unit area, become, in SI, $N\,m^{-2}$ or $kg\,m^{-1}\,s^{-2}$.

The use of mass, length and time as the fundamental units is not the only possible choice; different independent units would be equally permissible and indeed occasionally provide convenient solutions to particular problems. Thus, engineering systems often use force, length and time as the fundamental units; the unit of mass then becomes a derived unit. However, the choice of mass, length and time is practically universal in pure science and is both convenient and well founded.

The inconvenience of force as a fundamental unit

In the force-based system of units, the units of force were originally defined as the *weights* of unit mass. Now the weight of a body results from the action of the force of gravity upon it; gravitational acceleration has been introduced and this varies slightly from place to place on the Earth – and is considerably less on the Moon. Thus, the force-based system suffers from a severe disadvantage because of the variability of gravitational attraction with location.

The mass of a body is simply an indication of the amount of matter within it. We may measure this by measuring its resistance to a change of motion with the help of Newton's second law:

$$\text{force} = \text{mass} \times \text{acceleration}.$$

Thus, if we subject a body to a known acceleration we may quantify its mass by measuring the force required to maintain that acceleration. Alternatively, we may compare the mass of two bodies by comparing the forces required to maintain the same acceleration in both of them.

The force-based units have become known as gravitational units in contrast with the *absolute* units of mass, length and time. The implication of gravity in the definition and the casual use of the words 'weight' and 'mass' have led to inconsistencies in calculations on many occasions. One of the important aspects of SI is that it is based on mass, not force, and problems associated with the use of the weight of a material are overcome.

Energy and heat

In mechanics it will be seen that we are continuously concerned with the energy content of a system; the dimensions of energy are $[ML^2T^{-2}]$, and in SI we measure energy in the units of joules. In physiology, as in other branches of science, we are often concerned with the measurement of heat, which in fact is a form of energy and so has the above dimensions and is measured in the same units. Temperature, which is related to heat, is also considered as a fundamental quantity and is given the dimensions $[\theta]$, the unit used being the kelvin.

The concept of substance

When we consider the mass of a body we are unable to make any statements about its molecular content. However, when we are interested in studying processes of chemical change we want to know something about the number of molecules we are dealing with. We want to know the *amount* of *substance*, not its mass. The amount is expressed as the number of *moles* of the material, and in turn a mole is defined as the molecular weight, usually expressed in grams.

Dimensional homogeneity and consistency of units

As we have already noted, physically meaningful equations express relationships between quantities of the same physical character, and are therefore dimensionally homogeneous. This is known as the *principle of dimensional homogeneity*, and it has been tacitly assumed above in deducing derived dimensions and units. It is by the application of this principle that we can guarantee the consistency of derived units.

The use of volume and flow rate in physiology

In the past, the concepts of volume and flow rate have been used in a loose manner which is unacceptable in precise scientific descriptions. Therefore, it is worthwhile to consider these quantities in further detail.

Volume indicates a region of space whose dimensions are $[L^3]$, so that in SI units it is measured in cubic metres (m^3). That volume of space may contain a certain amount of some substance or it may be completely empty. It is a mistake to consider

Table 3.1. *The fundamental SI units*

Quantity	Name of unit	Symbol
Length	metre	m
Mass	kilogram	kg
Time	second	s
Thermodynamic temperature	kelvin	K
Electric current	ampere	A
Luminous intensity	candela	cd
Substance	mole	mol

Table 3.2. *Common derived SI units*

Physical quantity	SI unit	Symbol
Force, tension	newton	$N = kg\,m\,s^{-2}$
Work, energy, quantity of heat	joule	$J = N\,m$
Power	watt	$W = J\,s^{-1}$
Frequency	hertz	$Hz = s^{-1}$
Area	metre^2	m^2
Volume	metre^3	m^3
Density (mass density)	kilogram/metre^3	$kg\,m^{-3}$
Velocity	metre/second	$m\,s^{-1}$
Angular velocity	radian/second	$rad\,s^{-1}$
Acceleration	metre/second^2	$m\,s^{-2}$
Pressure, stress	newton/metre^2	$N\,m^{-2}$
	(also called pascal)	
Surface tension	newton/metre	$N\,m^{-1}$
Dynamic viscosity	newton second/metre^2	$N\,s\,m^{-2}$
Kinematic viscosity	metre^2/second	$m^2\,s^{-1}$
Diffusion coefficient	metre^2/second	$m^2\,s^{-1}$

that 'volume' implies anything about the amount of material within it. Of course, if an independent statement is made about the density of the material in that space, then we can specify its mass.

Flow rate implies the rate of transport of a given amount of material from one region of space to another. We are thus concerned with the movement of a given mass of material in a given time – not the movement of a volume. The dimensions of flow rate are mass per unit time $[MT^{-1}]$ and its units in the SI are $kg\,s^{-1}$; the dimensions are not $[L^3T^{-1}]$, as is so often assumed (see Chapter 4).

Système International (SI)

SI is a rationalized selection of units from the metric system, so the individual units are not new. There are seven fundamental units (**Table 3.1**) and several derived units, some of which have special names (**Table 3.2**). Derived units are merely for our convenience and they can all be expressed in terms of the fundamental units.

Table 3.3. *Conversion factors for some common units to SI units*

Quantity	Common unit	SI
Volume	$1\,ft^3$	$0.02832\,m^3$
Mass	$1\,lb$	$0.4536\,kg$
Force	$1\,dyne$	$10^{-5}\,N$
Work	$1\,erg$	$10^{-7}\,Nm$
Pressure	$1\,cm\,H_2O$	$98.1\,Nm^{-2}$
	$1\,mm\,Hg$	$133.3\,Nm^{-2}$
Viscosity	$1\,poise$	$0.1\,Nsm^{-2}$

Table 3.4. *Multiplicative factors in SI units*

Factor by which unit is multiplied	Prefix	Symbol	Example
10^{12}	tera	T	
10^{9}	giga	G	
10^{6}	mega	M	megawatt (MW)
10^{3}	kilo	k	kilometre (km)
10^{2}	hecto	h	
10^{1}	deca	da	decagram (dag)
10^{-1}	deci	d	decimetre (dm)
10^{-2}	centi	c	centimetre (cm)
10^{-3}	milli	m	milligram (mg)
10^{-6}	micro	μ	microsecond (μs)
10^{-9}	nano	n	nanometre (nm)
10^{-12}	pico	p	picogram (pg)

Although SI is simply a development of the existing metre–kilogram–second system, it is superior because it is coherent. This means that the product of unit quantities yields a unit resultant quantity. For example:

$$1\,N \times 1\,m = 1\,J$$

and

$$1\,kg \times 1\,m \div 1\,s = 1\,N.$$

No numerical factors are involved, and this makes calculation much more straightforward and eliminates the tedious problem of applying conversion factors. **Table 3.3** lists a number of conversion factors for common units to the appropriate SI values.

The system also possesses a clearly defined organization for describing multiples of the basic and derived units. The way in which it works can be seen from **Table 3.4**. Great care should be taken in the use of these prefixes. The prefix should always be written immediately adjacent to the unit to be qualified, e.g. meganewton (MN), kilojoule (kJ), microsecond (μs). Only one prefix can be applied to a given unit at any one time, e.g. one thousand kilograms is one megagram (Mg), not one kilo-kilogram.

The symbol m stands for the fundamental unit 'metre' and also for the prefix 'milli' so to avoid confusion it has to be used very carefully in certain circumstances. For example, mN stands for millinewton, whereas m N denotes the metre–newton or unit of work. However, the subtle use of spacing between the letters can lead to confusion, so it is better to write the metre as the second unit, i.e. newton–metre (Nm).

Another important point to note is that when a multiple of a fundamental unit is raised to a power, the power applies to the whole multiple and not the fundamental unit alone, e.g. $1\,km^2$ means $1\,(km)^2 = 10^6\,m^2$, not $1\,k(m)^2 = 10^3\,m^2$.

4

Basic ideas in fluid mechanics

We saw in Chapter 1 how real materials, in particular fluids, can be regarded as continuous if the distances over which their gross properties (like density) change is much larger than the molecular spacing. They can then be split up into small elements, to each of which the laws of particle mechanics can be applied. We have also set down those laws. Before applying them, however, we must know what forces act on such an element. As with the body sliding along the table (**Fig. 2.7**), the forces experienced by a representative fluid element are of two kinds: long-range and short-range.

The forces which act at long range, the *body forces*, are experienced by all fluid elements; the two most common examples are gravitational and electromagnetic in origin. The electromagnetic force on an element depends on quantities like its electrical charge, but the gravitational force, i.e. the weight of the element, depends only on its mass; this is the only example of body force to be considered from now on. If a fluid element P which occupies the point $\mathbf{x}$ at a certain time t has volume V and if the fluid in the neighbourhood of $\mathbf{x}$ at that time has density ρ,[1] then the gravitational force on the element is $\rho V \mathbf{g}$.

Stress

Short-range forces are exerted on the element P by those other elements with which it is in contact, and by no other. They consist of all the intermolecular forces exerted by molecules just outside the surface of P on the molecules just inside. If we consider a small portion of the surface of the element and add up all those forces exerted *by* molecules on side I, *on* molecules on side II, such that the line of action of the force intersects the portion of surface, the result will be a certain force $\mathbf{F}$ (a vector quantity, with a certain magnitude and direction). By Newton's third law, the force exerted *by* the molecules on side II *on* those on side I is equal to $-\mathbf{F}$. The magnitude of $\mathbf{F}$ is proportional to the area A of the portion of surface; a quantity independent of A

---

[1] Density is mass (kg) per unit volume (m^3) and is measured in units $kg\,m^{-3}$; the use of the Greek letter ρ to represent density is standard notation.

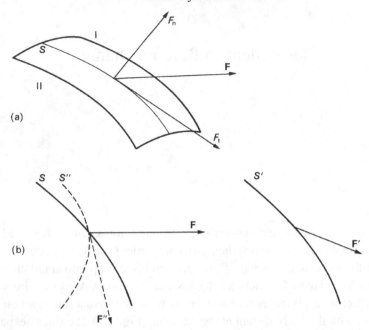

Fig. 4.1. (a) A portion S of the surface of a fluid element, which we imagine to be drawn in the interior of the fluid. The sum of those forces exerted by molecules on side I of S on molecules on side II (such that their line of action crosses S) is $\mathbf{F}$; this quantity has components tangential to S (namely F_t) and normal to it (namely F_n). If S has area A, the stress exerted across it is $\mathbf{F}/A$. (b) Side views of portions of surface such as S. The stress $\mathbf{F}'/A'$ exerted across a parallel portion S' (area A') at a different position in the fluid will normally be different from $\mathbf{F}/A$. So will the stress $\mathbf{F}''/A''$ exerted across another portion S'' (area A'') at the same location as S, but with a different orientation.

is the force per unit area, $\mathbf{F}/A$, which is called the *stress*. The stress usually varies with both the position and the orientation of the portion of surface being considered. Consider, for example, a portion of surface S', with area A', parallel to the surface S, but centred on a different point of space (**Fig. 4.1**b); the stress $\mathbf{F}'/A'$ acting across S' is usually different from the stress $\mathbf{F}/A$ which acts across S. Furthermore, the stress acting across a portion of surface S'' (area A'') centred on the same point of space as S, but with a different orientation (the broken curve of **Fig. 4.1**b), is also usually different from F/A. We may note that, like the force exerted by a flat table on a body sliding over it (see **Fig. 2.7**), the (stress) force $\mathbf{F}$ in general has two components: a normal component F_n and a tangential component F_t (see **Fig. 4.1**a).

In most of the flows which we shall consider, the important contribution to the normal stress is the *pressure*, which acts equally in all directions and is *independent* of the orientation of the surface portion considered. The tangential stress is analogous to the frictional force T (**Fig. 2.7**). It resists the relative motion of neighbouring layers

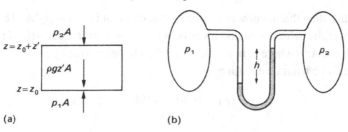

Fig. 4.2. (a) A small rectangular block of fluid, of vertical depth z' and horizontal area A, in equilibrium. The pressure forces on its sides must cancel out, so that pressure is independent of horizontal position. The vertical forces on the block (marked by arrows) must also balance, which leads to Equation (4.1). (b) The pressure difference $p_1 - p_2$ between the two regions shown is measured by the difference in height h between the two arms of a U-tube containing a fluid of density ρ. In fact $p_1 - p_2 = g\rho h$.

of fluid, so if fluid is flowing over a plane surface (see **Fig. 4.4**), with fluid further from the wall flowing more rapidly than that nearby, a tangential stress is exerted between neighbouring layers, tending to hold back the faster moving fluid and speed up the slower fluid. This stress owes its presence to a property of the fluid which Newton called 'defect of slipperiness', and which is now called *viscosity*.

Hydrostatic pressure

Let us first examine what stresses are present in a fluid which is completely at rest, such as still water in a fixed tank. Suppose that instruments which measure the stresses acting on a small portion of surface are put into the fluid. They show that no tangential stresses act anywhere in the fluid and that the normal stresses measured at a point are independent of the orientation of the portion of surface. In other words, the only stress present in a stationary fluid is the pressure. Measurements show further that the pressure does not vary in any horizontal direction, but does depend on the depth of the measuring point below the free surface of the tank. The difference between the pressure measured at a depth z and that at the free surface is directly proportional to z.

This result can also be derived theoretically, from Newton's second law (Equation (2.6)). Consider the balance of forces on a small rectangular block of fluid (**Fig. 4.2**a). The body force is entirely vertical, so the horizontal components of pressure force must balance out, which implies that the pressure is independent of horizontal position, as observed. Pressure must, however, vary with height (the coordinate z) because there is no motion, and the weight of the element must therefore be balanced by an upward pressure force on the bottom which is greater than the downward one on the top. If A is the horizontal cross-sectional area of the element, and if p_1 and p_2 are the pressures on surfaces 1 (the bottom) and 2 (the top) respectively (see **Fig. 4.2**a), then the net upward force on the element is $(p_1 - p_2)A$. This must be equal to its weight, which acts downwards, and is equal to the density of the fluid ρ times its volume Az' (where z' is the depth of the element) times g.

Hence the force balance equation is

$$(p_1 - p_2)A = g\rho Az'$$

or

$$p_1 - p_2 = g\rho z'. \tag{4.1}$$

Thus, as the height increases by an amount z', the pressure decreases by an amount $g\rho z'$. If we take the pressure to be atmospheric (p_A) at the surface of the tank and choose the level $z = 0$ to be at that surface (so that z is negative in the fluid), the pressure everywhere in the tank is given by

$$p = p_A - g\rho z. \tag{4.2}$$

As the depth below the surface increases, and z becomes more negative, so the pressure increases. The pressure given by Equation (4.2) is called the *hydrostatic pressure*, since it was derived for a fluid at rest. It can readily be checked that the units of pressure (force per unit area, or $kg\,m^{-1}s^{-2}$) are the same as those of $g\rho z$. If the density ρ itself varies with z, as in the sea, which becomes colder and saltier (and hence denser) with increasing depth, or in the atmosphere, which becomes more rarefied with increasing height, then Equation (4.2) would not be correct although Equation (4.1) would be. To calculate the pressure at some level, given its value at another level, we would have to add up the contributions $g\rho z'$ from each slice of thickness z' over which ρ was effectively constant.[2] Equation (4.2) gives rise to the most commonly used method of measuring pressure. If two regions, containing gas, say, in which the pressures are different, are connected to the two arms of a U-tube containing a liquid of known density (**Fig. 4.2**b), the levels of the liquid in the two arms will differ, by a height h. The pressure difference between the two regions is then immediately given by Equation (4.2) to be $g\rho h$, where ρ is the liquid density. Two of the liquids most commonly used for pressure measurements are water (density $0.001\,kg\,m^{-3}$) and mercury (density $0.013\,kg\,m^{-3}$). This has led to pressures being quoted in units of $cm\,H_2O$ or $mm\,Hg$, referring to the equivalent height of a water or mercury column. However, since a mixture of units makes comparison of different pressures extremely complicated, we use the internationally accepted unit of newton

---

[2] If we divide Equation (4.1) by z' and take the limit in which z' tends to zero, we obtain

$$dp/dz = -g\rho, \tag{4.1a}$$

where the left-hand side is the rate of increase of pressure with height. Compare the symbol for 'rate of change with time' defined in Equation (2.1).

per square metre, or $kg\,m^{-1}\,s^{-2}$. In terms of this:

$$1\,cm\,H_2O = 98.1\,N\,m^{-2}$$

and

$$1\,mm\,Hg = 133.3\,N\,m^{-2}$$

(see **Table 3.3**).

A body immersed in fluid at rest in the Earth's gravitational field experiences inwardly directed pressure forces over all of its surface. Because of the increase of pressure with depth, these forces are greater on the lower portions of surface than on the upper, and the net effect is an upward force called the *upthrust*. If the body is a cylinder with vertical sides and horizontal top and bottom surfaces of equal area A, then the upthrust is equal to the pressure difference between top and bottom, times A. If the height of the cylinder is h, we therefore have

$$\text{Upthrust} = g\rho hA. \tag{4.3}$$

But hA is the volume of the cylinder and ρhA is the mass of fluid displaced by it, so the right-hand side of Equation (4.3) is equal to the weight of fluid displaced. That is, upthrust equals weight of fluid displaced, which can be shown to be true for bodies of general shape, and is a law first propounded by the Greek philosopher Archimedes. If the body is made of a material of uniform density ρ', the net force on it due to gravity (weight plus upthrust) is equal to the volume of the body times g times the density difference $(\rho - \rho')$. It is directed upwards if ρ is greater than ρ', and downwards if ρ is less than ρ'. Only when ρ equals ρ' does the body remain at rest with no net vertical force acting.

Stress in a moving fluid: viscosity

Let us now turn our attention to the stresses present in moving fluids. For this purpose it is convenient to consider the forces experienced by a small cubic element of fluid and to suppose further that there is one direction in which the components of all the stress forces acting on the element always cancel out. The situation is therefore two-dimensional, and all forces of interest act in the plane perpendicular to this direction: this is the plane of the diagrams in **Fig. 4.3**. If the element were at rest in the absence of body forces like gravity, it would experience equal, inwardly directed normal stresses on each face, because pressure is independent of the orientation of the surface over which it acts (**Fig. 4.3a**). These stresses would be in equilibrium and would not tend to deform the cube in any way, apart from a uniform compression if the fluid were compressible; from now on let us suppose it to be effectively incompressible (see p. 41 for conditions in which this is permitted). A pressure, applied uniformly round the surface of a small fluid element in this way, can neither cause nor

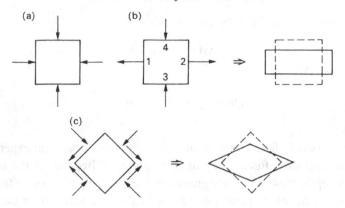

Fig. 4.3. Two-dimensional diagrams of the deformations produced in an incompressible cubic fluid element by stress systems of the type indicated by the arrows. (a) No deformation associated with pressure on its own. (b) Equal and opposite tensile stresses on faces 1 and 2, together with equal and opposite compressive stresses on faces 3 and 4, result in the deformation shown on the right. (c) If the cube of (b) is tilted through 45°, the stress system then consists of a uniform pressure superimposed on tangential stresses in the directions indicated; these result in deformation as shown.

resist a change in shape of the element. However, if the normal stresses on the faces of the cube are not equal, they do result in deformation.

Now suppose that there are equal, outward, tensile stresses on a pair of opposite faces of the cube (faces 1 and 2 in **Fig. 4.3**b), and equal, inward, compressive stresses on another pair (faces 3 and 4). Since the forces on opposite faces are equal, the element as a whole is not given an acceleration and, therefore, its centre of mass remains at rest. However, a stress system of this nature would tend to deform the cube in the manner indicated in **Fig. 4.3**b. A similar deformation would result even if the stress were inwards on all faces, but of smaller magnitude on faces 1 and 2 than on faces 3 and 4; this would be equivalent to a uniform pressure with a system like that of **Fig. 4.3**b superimposed on it. If the cube were tilted through 45°, as in **Fig. 4.3**c, it would tend to be distorted in the manner shown there, and the stress system then would consist of equal *tangential stresses*, superimposed on the uniform pressure. Thus, either unequal normal stresses or any tangential stresses on the faces of a fluid element are associated with deformation of the element, and hence with motion. In a general motion, the element will be moved about bodily and rotated bodily as well as being deformed, but it is clearly only the deformations which can be associated with stress systems other than pressure.

As already indicated, all stresses other than pressure owe their presence to viscosity. We have seen on p. 17 that when two solid surfaces slide over each other, a resisting frictional force is set up. In the same way, when a body of fluid is set in motion in such

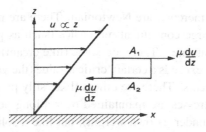

Fig. 4.4. Fluid flowing in the x-direction over a plane wall ($z = 0$), with velocity u directly proportional to the distance z from the wall; i.e. the velocity gradient du/dz is uniform. The top surface A_1 of the fluid particle shown experiences a forward tangential stress $\mu du/dz$; the bottom surface A_2 suffers backward tangential stress of equal magnitude. The (forward) stress exerted on the wall also has the same magnitude.

a way that local deformation of fluid elements takes place, viscous stresses which tend to resist the deformation are set up. The magnitudes of the viscous stresses depend upon the rate of deformation. For example, when a body is moved rapidly through a fluid, it causes more rapid deformation of fluid elements than one moving more slowly. Therefore, the resistive viscous stresses which are set up are also greater, and the faster moving body experiences a larger drag force. In most everyday fluids, the stresses are directly proportional to the rate of deformation. A complete mathematical description of the relationship for general flows would be out of place here, but there is one type of flow, depicted in **Fig. 4.4**, for which the relationship is very simple. In this flow, all fluid elements travel in the same direction (parallel to the x-axis) with a velocity u which is proportional to the coordinate z in one of the perpendicular directions. In this case the stress on a surface element aligned in a plane of constant z (A_1 or A_2 in **Fig. 4.4**) is tangential, in the x-direction, and has a magnitude proportional to the velocity gradient (rate of change of u with z) at that point. In other words, the stress on A_1 in the x-direction is equal to

$$S = \mu \frac{du}{dz} \tag{4.4}$$

where μ, a constant, is called the coefficient of viscosity (or often just the viscosity) of the fluid. The stress on A_2 is in the opposite direction, and of equal magnitude. Since stress has the units of force per unit area, and the velocity gradient has units of inverse time (s^{-1}), Equation (4.4) shows that the viscosity μ has units $kg\,m^{-1}\,s^{-1}$. A tangential stress of this kind is often called a *shearing*, or *shear stress*, since it is associated with a shearing motion in which neighbouring layers of fluid slide over each other. The velocity gradient is sometimes called the *rate of shear*, or shear rate.

Fluids in which the stress is always directly proportional to the local rate of shear, i.e. have uniform viscosity, are called Newtonian fluids; many everyday fluids, like air, water, glycerine and mercury, are Newtonian. There are many liquids, however, particularly those with large constituent molecules (such as proteins), in which the viscosity varies with shear rate. There are some (like cream) which do not begin to flow until the shear stress exceeds a certain critical value (the yield stress) and behave like solids for smaller shears. There are others (like 'silly putty') which behave like a fluid if low shearing stresses are maintained over a long period of time (as when it rests on a flat surface under its own weight), but like a solid for large stresses of short duration (a ball of this material bounces, and will even shatter if struck by a hammer). These are all examples of *non-Newtonian fluids*; we shall see in Chapter 10 that, over a certain range of stresses and shear rates, blood is a non-Newtonian fluid, exhibiting both a yield stress and shear-dependent viscosity. In large arteries, blood is effectively Newtonian, so that considerable simplification is possible in describing its fluid dynamics; this is unfortunately not the case in the microcirculation.

In addition to resisting the relative motion of neighbouring layers of fluid, viscosity also prevents a fluid from slipping over any solid boundary with which it is in contact. A fluid element which is in contact with a solid boundary adheres to it, and thus has the same velocity as the boundary. This requirement, known as the *no-slip condition*, is an empirical law. No adequate proof of it, based on the physical laws governing molecular interactions, is available; on the other hand, no experiment has been done which contradicts it. The no-slip condition is a constraint to which any theoretical deductions about particular flow patterns must conform, and it has important consequences in the circulation, as we shall see. It implies, for example, that fluid flowing past a stationary solid boundary always exerts a shear stress on that boundary (as in **Fig. 4.4**). This is because the fluid a little way from the boundary must be travelling faster than the fluid at rest on the boundary. Thus, there is a non-zero shear rate at the wall, and hence, from Equation (4.4), a non-zero shear stress.

The equation of motion of a fluid

Let us now consider the application of Newton's laws to fluid elements in motion. The second law states that the acceleration of a fluid element multiplied by its mass is equal to the sum of the long-range (body) force and short-range (stress) force acting on it. Consider, for example, a fluid element travelling in the x-direction in a unidirectional flow whose velocity varies with z (directed vertically upwards) but not with x, y, or time, as in **Fig. 4.5**. Gravity acts downwards (in the negative z-direction) and there is no variation of pressure or velocity in the third (y-) direction The balance of forces in the vertical direction shows that the pressure at a particular value of x varies with height as in a fluid at rest, and is given by Equation (4.1). Now consider the horizontal force balance (per unit length in the y-direction). If there were no motion, there would be no viscous stresses, and the pressure on the left- and right-hand faces of the element (at $x = x_0$ and $x = x_0 + x'$ respectively) would be equal. When there is

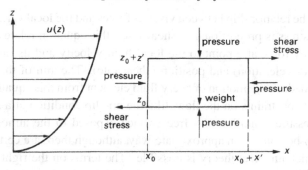

Fig. 4.5. The forces acting on a rectangular fluid element in a flow where the velocity $u(z)$ is in the x-direction and varies only with z. The vertical force balance shows that the pressure at a particular value of x varies with height (z) as in a fluid at rest (Equation (4.1)). The forward shear stress ($\mu du/dz$) is smaller on the upper surface ($z = z_0 + z'$) than the backward shear stress on the lower surface ($z = z_0$) because of the shape of the graph of u against z. Therefore, the pressure on the left-hand face ($x = x_0$) must exceed that on the right-hand face ($x = x_0 + x'$), i.e. a negative (favourable) pressure gradient is acting on the fluid.

motion, viscous forces do act; in the case shown, they tend to slow the element down, because the backward shear stress on the bottom surface of the element (at $z = z_0$) is greater (because the shear rate du/dz is greater) than the forward shear stress on the top surface (at $z = z_0 + z'$). If the pressures on the vertical faces were equal, the fluid element would therefore decelerate. The only way for the motion to be maintained, let alone accelerated, is for the pressure on the left-hand face to exceed that on the right-hand face. That is:

$$p(x_0) - p(x_0 + x') > 0.$$

In other words, there is a negative gradient of pressure in the x-direction;[3] a *negative* pressure gradient is favourable to flow, whereas a *positive* one would be adverse. Unless there are body forces driving the motion, a favourable pressure gradient is required to accelerate a fluid or maintain its motion against the action of viscosity. For example, fluid can flow steadily along a horizontal pipe, in which the component of body force (gravity) in the direction of motion is zero, only if the pressure at one end is greater than that at the other.

For general flows, then, the equation of motion of any fluid element (i.e. Newton's second law) may be written in the form

$$\text{mass} \times \text{acceleration} = \text{body force} + \text{pressure gradient force} + \text{viscous force.} \quad (4.5)$$

[3] The pressure gradient is defined as the slope of the graph of p against x, and so is the limit of the quantity $[p(x_0 + x') - p(x_0)]/x'$ as x' becomes very small. In other words, it is dp/dx (see Chapter 2, p. 8, for a discussion of this limiting process).

Since we know the relationship between viscous forces and the local variations in fluid velocity (shear stress is proportional to shear rate), this equation relates the pressure gradient and body force at a point to the local fluid velocity and its rates of change with both time (acceleration) and position (shear rate). The aim of theoretical fluid mechanics is to deduce the motion of every fluid element from this equation, and from the other known constraints on the flow, like the no-slip condition at a solid wall, or the constant pressure condition at a free surface exposed to the atmosphere. This can usually only be done in an approximate way, although there are certain restricted situations in which an exact theory is possible. The terms on the right-hand side of Equation (4.5) all represent forces, whereas the left-hand side consists of the mass × acceleration term, representing the inertia of the fluid elements. If this term were taken over to the other side of the equation, with opposite sign, it could be thought of as another force, and in fact it is common in fluid mechanics to speak of it as the '*inertia force* term'. However, it is important, as with 'centrifugal force', to remember that it represents mass × acceleration, not force.

Convective and local acceleration

If we wish to measure velocity components or pressure in a flowing fluid, we normally put our measuring instrument into the flow at a certain point, leave it stationary while the fluid flows past and record how its output varies with time during the experiment. This procedure can be repeated with the device in different positions, and the way the flow changes from one region to another, where conditions may be different, can be observed. This is a different approach from that used in the mechanics of isolated particles (Chapter 2), in which individual particles are followed as they move from point to point. If the latter approach were used for fluid elements, continuously being deformed, the relationship between stress and shear rate would be very complicated. It is therefore much simpler in theoretical work, as well as corresponding more closely with experimental technique, to use the former approach. There is, however, one disadvantage, which is that it is more difficult to write down the acceleration of a fluid element because the rate of change of velocity at a point represents the difference in velocity of two elements which successively occupy that point. The acceleration of a single element must also depend both on how the velocity at a point varies with time and on how the velocity varies from point to point, since the same element successively occupies different points. Particles can have acceleration even when the velocity at every point is constant with time (and the flow is said to be *steady*), since they can move between points of different velocity. A good example of this is in the flow of water over a weir: fluid elements well upstream of the weir move extremely slowly, but they are accelerated as they approach it, and shoot over it very quickly. Nevertheless, the flow is still steady, because the fluid velocity at any point is constant. This contribution to the acceleration of a particle is called the *convective acceleration* (experienced as the particle is convected from point to point by the flow); the contribution from the time variation of the velocity at a given point is called the *local acceleration*. An example of an *unsteady* flow, in which the local acceleration is the more important contribution, is to be found in the flow of blood in large arteries (see Chapter 12). It is the presence of the convective acceleration term which makes the equation of fluid motion (Equation (4.5)) very difficult to solve in general.

Conservation of mass

Before we discuss the detailed nature of the flow of fluid in given situations, in particular that of blood in the circulation, there are two further general principles which must be considered. The first states that mass cannot be destroyed or created (except where nuclear reactions take place, in which case mass can be converted into energy and vice versa), and is known as the principle of conservation of mass. It is physically very obvious, but it exercises an important constraint on the types of motion which can exist in any given flow situation. It implies, for example, that the same mass of fluid must flow out of a system of tubes as flows into it, unless either the density of the fluid inside increases (so that a greater mass can occupy the same volume), or the volume of the system of tubes increases through the expansion of flexible walls. Now, all real fluids are compressible (can change their density) to some extent, but liquids such as water and blood are far less so than gases, and it is often a very good approximation to treat them as incompressible. The approximation breaks down when the fluid speed approaches the speed of sound in the fluid, which is the speed at which small pressure changes are propagated through the fluid (not through the walls of the vessel containing the fluid; propagation through the walls is important in the circulation, as shown in Chapter 12). In the circulation the maximum blood velocities observed are well below 1% of the sound speed, so from now on we shall treat blood as an incompressible fluid and shall ignore any phenomena associated with its compressibility. Thus, if blood flows in a system of rigid tubes, what goes in must come out; in particular, if a single tube has a cross-sectional area A which varies along its length (**Fig. 4.6**), the speed of the flow at any time will also vary along its length. The product of the area A with the average longitudinal velocity u at the same cross-section is a constant, independent of which cross-section is chosen. This constant is the *volume flow rate* Q through the tube: $uA = Q$.

Thus, if a wide tube becomes narrower, the velocity of the fluid increases; if it becomes wider, the fluid slows down. If the tube were not rigid, the volume flow rate would not necessarily be spatially uniform, because the cross-sectional area could be variable in time. For example, if the area of a segment of tube increases with time, more fluid must flow in than flows out, so that the values of uA at the two ends must be different (**Fig. 4.6**). Only in steady flow would the volume flow rate in a flexible tube necessarily be uniform.

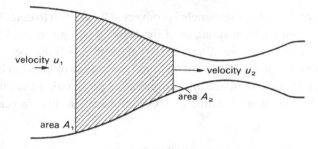

Fig. 4.6. Diagram of a tube, containing incompressible fluid, whose cross-sectional area A varies along its length; in consequence, the longitudinal velocity u (averaged over the cross-section) also varies along its length. In a rigid tube the volume flow rate $Q\ (= uA)$ is uniform at any given time, even if it varies with time; that is, $u_1A_1 = u_2A_2$. In a flexible tube A can vary with time, and at any time $(u_1A_1 - u_2A_2)$ is equal to the rate of increase of the shaded volume.

Bernoulli's theorem

The second general principle concerns the conservation of energy. In the absence of nuclear reactions, the total energy of a system is made up of mechanical energy (kinetic and potential – see Chapter 2, p. 21 – and of thermal energy (heat)). The laws of thermodynamics state that the rate of change of the total energy of a system is equal to the rate at which applied forces do work on it. When there are no applied forces, the total energy is constant, i.e. energy is conserved, although in general it is possible for mechanical energy to be converted into heat (and vice versa), for example by the action of viscous forces (see p. 38). The principle of conservation of energy can be used to derive an important result concerning the steady flow of fluids in which the effects of viscosity are locally extremely small, and may be neglected. This is usually true far from solid boundaries. The absence of viscosity means that, if there are no applied forces, mechanical energy is conserved and is not converted into heat.

It is convenient here to introduce the idea of a *streamline*; this is an imaginary curve drawn in space at a given instant so that at every point it is parallel to the direction of motion (the velocity vector) of the fluid at that point. In a steady flow, where the velocity at a point does not change with time, the streamlines coincide with the paths followed by individual fluid particles. That is not true in unsteady flow because although a fluid particle always travels in the direction of the local velocity vector, that direction may change at a later time, and the streamline passing through the particle may not pass through the previous location of the particle. Consider a steady flow and imagine a narrow tube drawn in the fluid in such a way that its boundary consists everywhere of streamlines. There is thus no component of velocity across the boundary, and the fluid flowing in the tube is constrained always to remain in it (**Fig. 4.7**). This 'streamline tube' then resembles a rigid tube of non-uniform

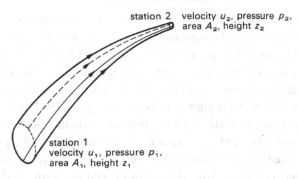

Fig. 4.7. Diagram of a stream tube, in a fluid in steady flow, whose long boundary consists everywhere of streamlines; no fluid crosses that boundary. The cross-sectional area A, height z, fluid velocity u, and pressure p are different at the two ends. The principle of conservation of energy shows that, when viscosity is unimportant, the quantity H (given by Equation (4.6)) is uniform (Bernoulli's theorem).

cross-sectional area, with the difference that the walls are not solid, and the viscous no-slip condition need not be applied. Now suppose that the two ends of a finite length of the tube are at different levels, and the fluid is flowing up from level $z = z_1$ at one end (where its velocity is u_1, cross-sectional area A_1) to level $z = z_2$ at the other (velocity u_2, area A_2). The principle of conservation of (mechanical) energy can be applied to the fluid instantaneously occupying this finite length of tube, as it flows upwards for a short time. In words it says that the gain in kinetic energy (which is positive in **Fig. 4.7** because the fluid is flowing towards a narrower piece of streamtube where, by the conservation of mass, the velocity is greater) plus the gain in potential energy (again positive in the case drawn because flow is upwards) is equal to the work done by the pressure forces on the ends (pressure $p = p_1$ on end 1, p_2 on end 2). The result of applying this principle is, in mathematical terms, that the quantity H_1, given by

$$H_1 = p_1 + \tfrac{1}{2}\rho U_1^2 + g\rho z_1$$

(where ρ is the fluid density), is equal to the quantity H_2, given by the same equation with suffix 2 replacing suffix 1. If we define a quantity H, called the *total head*, by the equation

$$H = p + \tfrac{1}{2}\rho u^2 + g\rho z, \tag{4.6}$$

where p, u and z are the pressure, velocity and height at an arbitrary point in the fluid, then the above result shows that H is constant along a streamline in steady flow (as long as viscous effects are negligible). The constant value of H in general varies from one streamline to another. The above result, known as *Bernoulli's theorem*, could be deduced directly from the mathematical form of the equation of motion, Equation (4.5), in the same way as the equation of conservation of energy of a particle can be deduced from its equation of motion (p. 21). The consequences of Bernoulli's theorem in the circulation will be explored in detail later; however, as long as viscous effects are small (as they often are in large arteries; see Chapter 12), and if the flow is horizontal (i.e. z is a constant), then when the magnitude of the velocity u increases along a streamline, the pressure p correspondingly falls, and vice versa. The lowest pressures are associated with the highest velocities at a given level, in the same way as the lowest pressures are associated with the highest levels in a fluid at rest. Because changes in the terms $\frac{1}{2}\rho u^2$ and $g\rho z$ cause reciprocal changes in the pressure, these terms can be regarded as equivalent to pressures. The term $g\rho z$ is the hydrostatic pressure equivalent (or just *hydrostatic pressure*), and the term $\frac{1}{2}\rho u^2$ is called the dynamic pressure equivalent (or just *dynamic pressure*), since it is associated with motion.

In reality, all fluids have viscosity, which can cause mechanical energy to be irrecoverably converted into heat, or *dissipated*. The rate of increase of the mechanical energy of a fluid element is then equal to the rate at which work is done on it by the applied forces, *less* the rate at which mechanical energy is dissipated and becomes thermal energy. Far from solid boundaries, the effects of viscosity are small, and the total head H is approximately constant over short lengths of streamline. However, energy dissipation does occur, and a fluid element loses mechanical energy as it flows, experiencing a progressive (but gradual) decrease in the value of H with distance along the streamline. Over large distances this 'loss of head' can become appreciable. Mechanical energy is dissipated in all motions where frictional or viscous forces operate; another example is in the motion of a body on a rough surface (p. 22), where the dissipation occurs as the body does work against the frictional force, and the mechanical energy lost cannot be recovered. The heat produced by viscous dissipation in a fluid can itself have an effect on the fluid motion, by raising the temperature, and hence lowering the density, of some fluid elements, causing them to rise relative to other fluid elements. This phenomenon, called *free convection*, is not important in the circulation, because the temperature variations are not large enough (indeed, they cannot be measured).

5

Flow in pipes and around objects

Poiseuille flow in a tube

In order that we can understand the flow properties of biological fluids such as blood which may exhibit non-Newtonian properties, it is first necessary to discuss the behaviour of simple or Newtonian fluids. Let us look at the flow properties of a simple liquid like water in a very long horizontal pipe. Imagine that this pipe is circular in cross-section and is d units in diameter (**Fig. 5.1**). Its entrance and exit are connected to large reservoirs so that the pressure drop between the ends of the tube may be maintained constant and a steady flow of water through the pipe achieved. Small side hole, or lateral, pressure tappings are made in the pipe at frequent intervals along its length and these tappings are connected to a series of manometers. It is thus possible to measure the pressure drop per unit length or pressure gradient along the pipe.

If the pressure at the inlet to the pipe is p_I and that at the outlet p_0, then we shall observe that, as $p_I - p_0$ (or Δp)[1] is increased by raising the level in the upstream reservoir, so is the flow rate Q through the pipe. However, if we look at the distribution of the pressure drop along the pipe we see that initially the pressure falls very rapidly with distance from the entrance, but after a long distance, which may be as much as

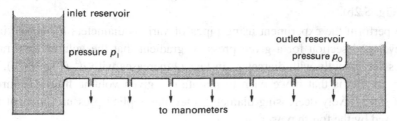

Fig. 5.1. Schematic drawing of a long, straight, horizontal tube with steady flow showing inlet and outlet reservoirs and the regular location of lateral pressure tappings.

---

[1] The symbol Δp means the difference in pressure between any two stations 1 and 2; thus, it is a form of shorthand for $p_1 - p_2$. In general, ΔM means the difference between the value of the property M at stations 1 and 2.

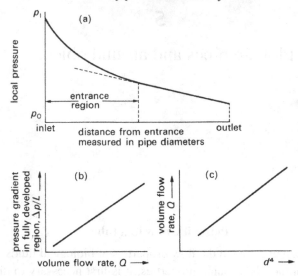

Fig. 5.2. (a) The variation in local lateral pressure with distance down the tube. Initially, in the entrance region, the pressure falls rapidly with distance but far from the entrance the pressure falls linearly with distance, (b) The linear variation of the pressure gradient ($\Delta p/L$) in the region remote from the entrance as a function of the volume flow rate Q through the tube. (c) The linear increase in flow rate Q with increasing pipe diameter to the fourth power d^4 at a constant applied pressure gradient in the region remote from the entrance.

$100d$, the pressure falls linearly with distance (**Fig. 5.2**a). The region in which the pressure is falling relatively rapidly is called the *entrance region* of the pipe. The flow remote from the entrance is said to be *fully developed* or *established*. In the fully developed section of the pipe the pressure falls linearly with distance, i.e. the pressure gradient $\Delta p/L$ (pressure difference between two stations distance L apart divided by L) is constant. In this region also the pressure gradient increases linearly with flow rate Q (**Fig. 5.2**b).

If we perform the experiment using pipes of various diameters d, we see in the fully developed section for a given pressure gradient that the volume flow rate Q increases very rapidly with diameter. In fact Q increases with d^4 (**Fig. 5.2**c). The corollary of this is that if we wish to maintain a given volume flow rate through pipes of successively decreasing diameter, then the applied pressure gradient must be increased by the fourth power.

There is a further instructive experiment that we may perform which will demonstrate how the velocity of flow varies across any cross-section in the fully developed region of the pipe, i.e. it demonstrates the velocity profile of the flow. Suppose that we rapidly inject a narrow streak or filament of dye across the pipe as in **Fig. 5.3**; as the flow proceeds downstream, this filament will be transported with it and will become stretched out as shown. Each small elemental length of the streak will be transported

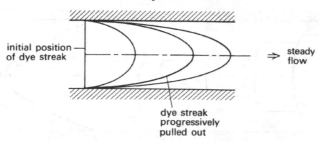

Fig. 5.3. The effect of flow progressively stretching a dye streak initially injected as a straight line across the tube. The successive lines indicate the position reached at different times after injection.

downstream at the rate of the local fluid velocity. Thus, if we record the position of the dye streak at various times after injection, it is possible to compute the velocity profile. It is also possible to introduce small blobs or fluid packets of marker dye at various radial distances from the axis and to watch their movement in a similar manner. Each isolated element will be seen to move with the local fluid velocity as before. It will also be noticed that as it is transported the packet will retain the same radial position; it will not drift towards the axis or wall nor will it corkscrew or spiral down the tube. This indicates that the flow is unidirectional or axial. It is also symmetrical; that is, the flow at any given radial distance from the axis is the same no matter which radial ray we look at. This is called *axisymmetric flow*. As will be confirmed below theoretically, the velocity profile is in practice parabolic in shape with the maximum velocity along the axis, the velocity reducing progressively to zero at the wall.

It was Poiseuille in 1840 who, as a first step towards understanding the mechanics of the circulation, published a quantitative study of the properties of steady flow through long capillary tubes, and flow conditions in the region far from the entrance of the tube are now named after him. He studied the relationship between the driving pressure and the flow rate for different liquids at different temperatures in tubes of different diameters. From careful analysis of his experimental results, he found that

$$Q = KPd^4/L,$$

where K was a constant that depended upon the fluid and the temperature. This 'law' was subsequently derived from the basic equations of fluid mechanics by a number of researchers who related K to the viscosity of the fluid, and Poiseuille's law for steady flow through a rigid, horizontal tube remote from the entrance is generally written in the form

$$\Delta p = 128 \frac{\mu L Q}{\pi d^4}, \tag{5.1}$$

where μ is the fluid viscosity.

By considering the physical processes operating on the fluid flowing in this region of the pipe it is possible to show how *Poiseuille's law* was derived. To do so we may

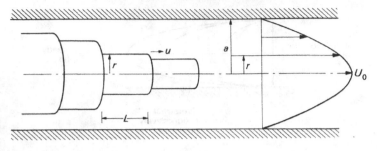

Fig. 5.4. Poiseuille flow in a straight tube showing the parabolic distribution of velocity.

apply a force balance to the fluid in a section of the pipe. In Chapter 4 we wrote down the groups of forces we need to consider in any force balance (Equation (4.5), p. 39). In *Poiseuille flow* we are considering a steady flow in a straight line, so no fluid is subjected to any accelerations, and the left-hand side of Equation (4.5) is zero. In addition, if we consider flow in a horizontal pipe, gravitational forces are not relevant and the body force term is also zero. Thus, we are left with a force balance which says

$$\text{viscous force} = -\text{pressure gradient force.}$$

Consider the flow to be like a set of thin concentric shells of fluid sliding over one another as shown in **Fig. 5.4**, and let us look at the force balance on a shell of inner radius r and thickness dr. If the pressure drop over a short length L of the cylinder is Δp, then the net pressure force acting on the shell is given by the product of the pressure difference and the cross-sectional area of the shell:

$$\text{pressure force} = \Delta p \times \pi[(r+dr)^2 - r^2].$$

Since dr is small,

$$\text{pressure force} = \Delta p \times \pi 2r\,dr.$$

The net viscous force acting on the shell will be given by the difference in shear force on the two surfaces of the shell. The shear stress at any radial position is given by (p. 37)

$$\tau = \mu \frac{du}{dr}.$$

Therefore, the downstream shear *force* on the inner side of the shell is $2\pi L r \mu (du/dr)$, evaluated at radius r. The rate of change of this quantity with radius is given by

$$2\pi L \mu \frac{d}{dr}\left(r \frac{du}{dr}\right).$$

Thus, for small changes dr in radius we can say that the difference in shear force between the two radii r and $(r+dr)$, i.e. the net viscous force on the shell, is given by

$$2\pi L\mu \frac{d}{dr}\left(r\frac{du}{dr}\right)dr.$$

Hence, equating the two forces on the shell,

$$\Delta p \times 2\pi r\,dr = \frac{d}{dr}\left(r\frac{du}{dr}\right)dr \times 2\pi L\mu$$

Thus:

$$\mu d\left(r\frac{du}{dr}\right) = \frac{\Delta p}{L}\,dr.$$

If we now integrate this equation (i.e. sum the effects of all the shells) we obtain Poiseuille's law (Equation (5.1)). We can also show that

$$u = U_0\left(1 - \frac{r^2}{a^2}\right), \tag{5.2a}$$

where u is the velocity at any radial position r, U_0 is the centre-line velocity and $a = \frac{1}{2}d$ is the tube radius. We may define the average velocity $\bar{U}$ of the fluid as

$$\bar{U} = \frac{Q}{\pi a^2}$$

and it can be shown that Equation (5.2a) can be rewritten as

$$u = 2\bar{U}\left(1 - \frac{r^2}{a^2}\right). \tag{5.2b}$$

Thus, we can see that the velocity distribution in the pipe is parabolic, which explains the dye experiments above, and that the centre-line velocity is twice the average velocity of the fluid.

Because the distribution of the velocity across the pipe is parabolic, it follows that the velocity gradient or shear rate is a linear function of radius:

$$\text{shear rate} \equiv \frac{du}{dr} = \frac{-2U_0 r}{a^2}.$$

At the axis ($r = 0$) the velocity gradient is zero and it increases linearly to the wall, where it has a maximum value of $2U_0/a$. Thus, the viscous shear stress at the wall is given by $\mu \times 2U_0/a$ and is the maximum shear stress.

It is also possible to derive Poiseuille's law on the basis of an overall force balance on a section of the pipe of diameter d and length L. If the difference in driving pressure between the two ends of the section is Δp, the net pressure force applied to the fluid is given by

$$\text{pressure force} = \frac{\Delta p\pi d^2}{4}.$$

The force opposing the applied pressure force is equal to the total viscous retarding force operating over the surface of the cylinder. Thus:

$$\text{viscous force} = \text{viscous wall stress} \times \text{tube surface area}$$
$$= \frac{4U_0\mu}{d} \times \pi dL.$$

Since no other forces are acting,

$$\frac{\Delta p \pi d^2}{4} = -4\pi U_0 L \mu.$$

But $U_0 = 2\bar{U}$ and $\bar{U} = 4Q/\pi d^2$; thus, substituting in the above equation and rearranging,

$$\Delta p = -128\frac{\mu L Q}{\pi d^4}.$$

This can be seen to be Poiseuille's law as given in Equation (5.1). In the equation as expressed above there is a minus sign in front of the right-hand side of the equation. Strictly this is necessary because we are dealing with vector quantities, but the sign can be dropped for convenience as we clearly know the direction in which the forces are operating.

The application of this overall force balance to the flow is instructive, for it demonstrates that it is the viscous retarding force operating at the tube wall which effectively opposes the flow.

Flow in the entrance region

It must be emphasized that Poiseuille flow conditions are pertinent only to regions of a straight pipe which are a very long distance from the entrance and remote from any sources of disturbance, such as bends or constrictions. If we now look at the distribution of velocity of the elements of fluid flowing through any cross-section in the pipe nearer to the entrance, we can see that there are changes with distance along the pipe. At the entrance, all the elements of the fluid are moving with the same velocity and, therefore, have a uniform or flat velocity profile. However, the fluid which comes into contact with the wall of the pipe is forced to be motionless because of the 'no-slip condition' (see Chapter 4, p. 38). Immediately, a gradient in velocity is established between the motionless fluid at the wall and the adjacent fluid elements just within the core. As the flow proceeds along the tube, viscosity progressively modifies this initial blunt profile, as shown in **Fig. 5.5**. The original high-velocity gradient at the wall becomes reduced and progressively more of the core fluid becomes sheared. It is essential to realize that, at the same time, the central portion of the flow is accelerated to maintain the constant flow rate through any

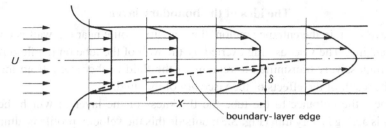

Fig. 5.5. The change in velocity profile with distance along the tube in steady flow indicating the growth of the boundary layer whose thickness is δ at distance X from the entrance. The initially flat profile becomes progressively modified to the established parabolic profile a long distance from the inlet.

cross-section. The gradient in velocity initially to be seen only near the wall becomes evident across progressively more of the tube radius. Ultimately, the velocity can be seen to vary across the whole cross-section of the pipe; it is maximum along the centre line and diminishes progressively towards the walls; the velocity profile is parabolic and Poiseuille flow is set up. Once this fully developed profile is achieved it does not change any further with distance downstream.

The progressive development of the velocity profile in the pipe is related to the changing pattern of pressure gradient with distance down the pipe.

In the entrance region, where the velocity profile is changing with distance, the pressure gradient is initially very high; however, further downstream, where considerable adjustment of the profile has already occurred, the gradient is less. Ultimately the pressure gradient becomes constant, at its lowest value, in the region of the pipe where flow is fully established. If the viscous retarding force at any cross-section before the flow becomes fully developed is calculated, it will be seen to be less than the applied pressure driving force. The excess driving force is taken up in accelerating the flow.

The fundamental difference between entrance and fully developed flow in a pipe is that in the entrance region fluid elements experience acceleration (positive near the tube axis and negative or retarding near the wall). However, since flow is steady, there is no change in velocity with time at any station in the pipe, i.e. there is no *local* acceleration. None the less, individual elements of fluid are accelerated or retarded as they move downstream along the pipe. This is an example of convective acceleration.

In general, flows contain both local and convective accelerations, and the 'forces' (inertial forces) required to overcome the inertia of the fluid and provide these accelerations must be considered. These are the terms which comprise the left-hand side of Equation (4.5) (p. 39).

The idea of the boundary layer

As we have seen, in the entrance region, the velocity profile near the wall is continuously changing. The viscous drag exerted by the wall of the tube on the fluid nearest to it is progressively transmitted to regions of the fluid further away from the wall. This is the result of the effects of the viscosity of the fluid.

Very near the entrance to the tube the thickness of the layer in which the fluid viscosity is acting is very thin (**Fig. 5.5**); outside this the velocity profile is almost flat and viscous effects are small. As the flow proceeds down the tube, the thickness of this layer in which viscosity is acting increases. We call this layer a *boundary layer*.[2] It characterizes the region in which viscosity plays an important role in determining the flow properties.

Boundary layers are established not only within tubes, but whenever a real fluid flows over any solid surface; for example, near the bed of a river or along the hull of a ship. If, instead of forcing a fluid to flow over a body, we pull that body through the fluid, we shall of course still establish a boundary layer over its surface. Thus, if we look at the distribution of velocity around a body falling through water at a fixed speed, the boundary layer will look exactly the same as if we had made the water flow past the body at the same rate. It is only the relative motion of the body and fluid which is important. All real fluids demonstrate these characteristics, but the details of the velocity profile within the boundary layer will depend upon the viscous properties of the fluid and we should not expect the profile to be the same for Newtonian and non-Newtonian materials.

We will now consider the rate at which the boundary layer grows with distance along the pipe; it depends upon the balance between the inertial forces and the viscous drag forces in the region of the wall.

Let us consider a small element of fluid in a region of varying velocity (**Fig. 5.6**). The tangential or shear stress on surface A_1 (area A) is $\mu|\mathrm{d}u/\mathrm{d}y|_1$. Similarly, the shear stress on A_2 (also area A) is $\mu|\mathrm{d}u/\mathrm{d}y|_2$. These two stresses will be slightly different in magnitude because of the change in velocity gradient in the region (see **Fig. 4.7**). The net viscous force on the element will thus be given by the change in stress with distance y from the wall:

$$\mu\frac{\mathrm{d}}{\mathrm{d}y}\left(\frac{\mathrm{d}u}{\mathrm{d}y}\right)A(y_2 - y_1).$$

The viscous forces on the element must balance the inertia forces. The inertia force term is given by

$$\rho \times \text{convective acceleration} \times \text{volume of element},$$

where ρ is the fluid density and the volume of the element is $A(y_2 - y_1)$.

---

[2] Definition of boundary-layer thickness: the boundary layer is that region of fluid in which the velocity is increasing with distance from the wall. Very near the wall the velocity changes very rapidly with distance, but far out in the layer it changes very slowly to the free-stream velocity U_0. For practical purposes we state that the boundary-layer thickness is that distance from a solid surface at which the local velocity u reaches $0.99U_0$.

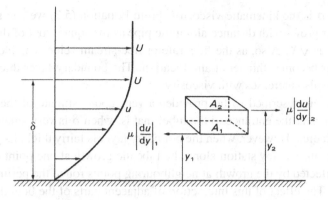

Fig. 5.6. A force balance on an element of fluid within the boundary layer.

If we equate the two forces on the element and attempt to solve the problem mathematically, it will be found that there are not yet mathematical techniques available to solve the equation analytically. It is possible to obtain numerical solutions for particular problems only by using a digital computer. Commonly in fluid mechanics, in order to obtain an idea of the scale or magnitude of the process we are interested in, we give scales to the determining variables corresponding to their magnitude. (We shall see later, in Chapter 6, that this is an example of the technique of dimensional analysis.)

Thus, to obtain an estimate of the magnitude of the viscous force we may relate it to the local thickness of the boundary layer δ and the free-stream velocity U. Then the viscous force is proportional to

$$\frac{\mu U}{\delta^2} A (y_2 - \dot{y}_1).$$

Again, we may give the inertial force a magnitude scale. If the boundary-layer thickness is considered at a distance X from the entrance to the tube, then the time scale for fluid to reach X from the entrance (the convection time) is of the order of X/U. The convective acceleration of the fluid at the location X is then scaled as $U/(X/U)$ or U^2/X.

If we now equate these two scaled forces, we obtain

$$\rho \frac{U^2}{X} = k\mu \left(\frac{U}{\delta^2} \right), \tag{5.3}$$

where k is a numerical constant whose size we must determine experimentally. Thus, we can say

$$\delta \propto \sqrt{\frac{\mu X}{\rho U}} \qquad \text{or} \qquad \delta \propto \sqrt{\frac{\nu X}{U}},$$

where $v = \mu/\rho$ is the kinematic viscosity. From Equation (5.3) we can see that the boundary layer grows with distance along the pipe as the square root of distance (see **Fig. 5.5**), i.e. as $\sqrt{X}$. Also, as the flow rate or free-stream velocity is increased, the boundary layer becomes thinner at any location. The boundary-layer thickness for a given flow rate also increases with viscosity.

The analysis as described so far provides a very good estimate of the boundary-layer thickness near the entrance to the tube; that is, when δ is very small compared with the tube radius. However, when the boundary layer is fairly thick, its growth rate is modified because at any station along the tube the growth at one point on the tube perimeter is affected by the growth at neighbouring points round the perimeter at that cross-section. The effect of this interaction of adjacent parts of the boundary layer is somewhat to retard its development with distance. Nevertheless, from Equation (5.3) we may obtain a scale for the length of tube required for the boundary layer to fill the tube and establish Poiseuille conditions. When the boundary layer fills the tube its thickness δ equals the tube radius $d/2$. Then

$$\frac{X}{d} = k' \left(\frac{Ud}{v} \right).$$

The term Ud/v is known as the *Reynolds number Re* after the British scientist who first performed many important studies of flow in tubes, and it will be considered in greater detail below. The constant k' has been found experimentally to be approximately 0.03. Thus, the entrance length X for steady flow in a straight pipe is given by

$$\frac{X}{d} \approx 0.03(Re). \tag{5.4}$$

Hence, we can see that for a tube of 2.0 cm diameter the entrance length would be approximately 60 cm for a Reynolds number of 1000 – this would be a typical situation for the human aorta if it were a straight pipe and flow were steady at an average velocity of $0.2\,\mathrm{m\,s^{-1}}$.

Equation (5.4) is satisfactory for predicting the entrance length only when the Reynolds number is between 10 and 2500. When Re is very low (less than unity), inertia forces in the motion are negligible and the entrance length is about one diameter of the pipe. When Re exceeds 2500 the flow properties as they have been described so far break down – streamline motion no longer exists and turbulence may be observed (see p. 56).

Thus, we shall see later that flow in the microcirculation can be considered as fully developed, but in large vessels it is of an entrance type – albeit complicated because of the geometry.

Reynolds number

In the force balances we have used so far we have been concerned about the relative magnitude of the inertial and viscous forces operating on an element of fluid. In fact, the ratio between these two forces is an important aspect of all flow problems.

In pipe flow, U and d are the representative velocity and dimension of the flow. Therefore, we can say that the magnitude or scale of the viscous force will be proportional to the product of viscosity and velocity gradient $(= \mu U/d)$. Similarly, the inertial force is proportional to the kinetic energy per unit volume (ρU^2) of the flow. The relative importance of the two quantities can then be expressed as the ratio

$$\text{Reynolds number} = \frac{\text{inertial forces}}{\text{viscous forces}} = \frac{\rho U d}{\mu} = \frac{U d}{\nu}, \tag{5.5}$$

where ν is the kinematic viscosity of the fluid (μ/ρ). The Reynolds number Re is a dimensionless number and, therefore, its magnitude does not depend upon the units in which we express the various parameters *provided* we use a self-consistent set. Throughout this book we will regularly consider flows in terms of the Reynolds number. It is applicable for the characterization not only of conditions in a long straight pipe, but also of any flow situation. It is calculated on the basis of the characteristic velocity and dimension of the system.

Thus, if we are considering flow round a sphere we may use the sphere diameter and approach velocity of the fluid. When the growth rate of the hydrodynamic boundary layer in a pipe was computed in the section above, we obtained Equation (5.3), which may be written as

$$\delta = k\sqrt{\frac{\nu X}{U}}$$

or

$$\delta = kX\sqrt{\frac{\nu}{UX}}.$$

But the term UX/ν can be thought of as the Reynolds number Re_X for the flow based upon the characteristic length or distance X from the entrance of the pipe. We may thus rewrite Equation (5.3) as

$$\frac{\delta}{X} = \frac{k}{\sqrt{Re_\delta}}. \tag{5.6}$$

When Re for a flow is much less than unity we can say that the viscous forces dominate the flow and inertial forces can be ignored. For instance, in the microcirculation, which we will consider to be vessels less than approximately $100\,\mu m$ in diameter, typical Reynolds numbers are less than unity and we can treat the flow as purely viscous. When Re is very much greater than unity then inertial forces dominate the flow and

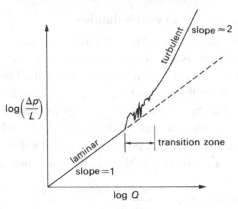

Fig. 5.7. The variation of volume flow rate Q with pressure gradient $\Delta p/L$, plotted on a logarithmic scale, for steady flow in a long tube to indicate the changing relationship as the flow rate is increased.

viscosity has little modifying influence except very close to boundaries. Flow in large arteries and veins provides examples of this.

Turbulence in pipe flow

In 1883 Reynolds published important studies on the flow of fluids in long pipes. He obtained measurements of the pressure gradient along the pipe as the flow rate was increased. At relatively low velocities, the pressure gradient was directly proportional to the flow rate – this is as we would expect from Poiseuille's law (Equation (5.1)). However, at higher flow rates, it became roughly proportional to the square of the flow rate (**Fig. 5.7**). There was a range of intermediate flow rates where the relationship fluctuated with time: at one instant, the slope of the curve in **Fig. 5.7** would be 1, then temporarily it would increase to nearly 2. If great experimental care was taken to ensure that flow entering the pipe was extremely steady and if the pipe was extremely smooth and straight, then the critical flow rate at which the transition takes place from one state to the other could be made to occur at much higher flow rates.

If we consider water flowing in a 1 cm diameter tube, the critical flow rate at which the pressure–flow relationship changes is approximately a litre per minute. This is a small flow rate if we think about typical everyday flow situations like flow in the pipe to a tap or along a hose pipe.

Reynolds performed a second classic set of experiments in which he visualized what was happening within the flow. He studied the flow of water in a glass pipe and introduced a thin filament of coloured water into the main stream at the pipe axis far from the entrance and observed its behaviour. At low flow rates, the injected dye retained its identity and travelled along the pipe axis as a discrete filament. At higher flow rates the dye streak was broken up and became mixed with the bulk of

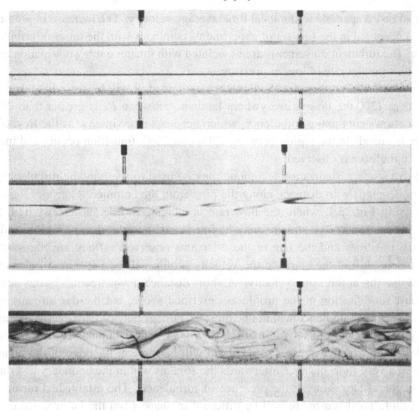

Fig. 5.8. The motion of a filament of dye in a long straight tube. Upper panel: the steady axial motion of a dye streak in a flow at low *Re*. Middle panel: at *Re* just greater than the critical value the steady laminar motion of the dye ceases and short bursts of turbulence can be observed. Bottom panel: at higher *Re* the flow becomes fully turbulent and random motion of the dye streak occurs.

the water (**Fig. 5.8**); the rate of mixing increased as the flow rate was increased. At flow rates just above the critical flow rate, the dye streak could be seen to break up intermittently.

This random motion of the dye indicated that the flow was also flowing with violent random movements. Indeed, if a high-speed cinefilm were taken of the motion of the dye and this were viewed at low speed, it would be seen that individual elements of dye would move randomly in time and all three dimensions of space. However, if we were to compute the *time-average* velocity of packets of dye moving through any one point in a cross-section at some position in the pipe and then repeat the computation at other points in the same cross-section, we would see that there was a well-defined time-average velocity profile. The instantaneous random velocities of the elements would all be comparable to the local time-average velocity. The increase in pressure gradient observed in the first set of experiments coincides with the onset of turbulent motion. The turbulent movements are associated with greater energy dissipation and, therefore, a greater rate of loss of pressure.

If the Reynolds numbers for the flows are computed, it will be seen that when Re is less than 2300 the flow is everywhere laminar, but when Re is greater than 2500 the flow shows continuous turbulence, which becomes more intense as the Reynolds number is raised. In the region between 2300 and 2500, transition occurs and intermittent turbulence is observed.

The dye streak visualization technique may be used to see how the turbulence in the flow develops with distance along the pipe from the entrance. Consider the set-up shown in **Fig. 5.1**, where the flow rate is increased to be sufficiently high for turbulence to occur. Dye streams may be introduced into the flow at various radial and axial positions and the fate of these streams observed as they are transported downstream. Just at the entrance the velocity profile will be flat with all elements of fluid moving at the same velocity. A short distance downstream, viscous action will cause modification of the profile as described above, but the dye streams will still be intact and turbulence will not be present. Further downstream the dye streams will suddenly break up, showing that the flow is no longer laminar but has become turbulent. The first dye streams to break up will be those at the edge of the boundary layer; a very short distance afterwards the streams within the boundary layer and core also break up, showing the presence of turbulence. The established mean velocity profile far downstream will be different in shape from the laminar parabolic profile. The core has a much blunter profile and the velocity gradient at the wall is steeper; the peak velocity, on the axis, is approximately 1.2 times the average velocity.

The entrance length for a turbulent flow is less than that for laminar flow; its size may be estimated on the basis of the following equation:

$$\frac{X}{d} = 0.693 Re^{1/4}. \tag{5.7}$$

Unsteady flow in a very long pipe

Again consider a long straight pipe with laminar flow in it. If now a slowly oscillating pressure gradient is applied to the flow it will slow down, stop and reverse direction, accelerate in that direction, then slow down again. If this takes place gradually enough then the flow will always have a parabolic velocity profile whose magnitude is proportional to the instantaneous flow rate. However, if the pressure gradient is cycled progressively more frequently, then the velocity profile becomes increasingly distorted (**Fig. 5.9**). The inertia of the fluid in the central core prevents the core from following the applied gradient of pressure and the amount by which it lags increases

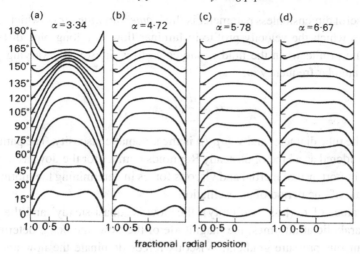

Fig. 5.9. Velocity profiles, at intervals of 15°, of the flow resulting from a sinusoidal pressure gradient in a long straight tube. The value of α in (a) is sufficiently high to indicate that reversal of flow occurs near the wall whilst the core motion is still forwards. The motions in (b–d) indicate the progressive distortion of the profile at values corresponding to the second, third and fourth harmonics of the basic frequency of oscillation in (a). (From McDonald (1974). *Blood Flow in Arteries*, p. 104. Edward Arnold, London.)

as the frequency of oscillation of the pressure gradient is raised still further. At the same time, the amplitude of movement of the core fluid decreases. At very high frequencies we find that there is very little motion of the core, and all flow is constrained to a very thin region near the wall where viscous forces dominate and the flow is less sensitive to inertial forces.

The simplest form of oscillating pressure gradient that we can consider mathematically is the simple sinusoidal pressure gradient. In this case, the instantaneous pressure gradient Δp_t is related to the maximum or peak pressure gradient Δp_0 by the relationship

$$\Delta p_t = \Delta p_0 \sin(\omega t), \tag{5.8}$$

where t is time and ω is the angular frequency ($\omega = 2\pi f$, where f is the frequency) of the oscillation. The instantaneous pressure gradient will vary with time in such a way that at time $t = 0$ it is zero; as time increases so will Δp_t increase to a maximum value of Δp_0 and it will then decrease to zero again; as t increases further, Δp_t will become negative and decrease to a value of $-\Delta p_0$ and it will then rise again through zero and repeat the cycle. The period of time T over which one complete cycle of the pressure oscillation occurs is $2\pi/\omega$. An extended discussion of sinusoidal functions is given in Chapter 8.

A very useful dimensionless parameter is the *Womersley number* α, which indicates the extent to which the velocity profile in laminar flow in a long pipe differs from the Poiseuille profile when the fluid is subjected to a sinusoidally varying pressure gradient of angular frequency ω:

$$\alpha = \frac{d}{2}\sqrt{\frac{\omega}{v}}, \tag{5.9}$$

where d is the tube diameter and $v = \mu/\rho$ is the kinematic viscosity. The parameter α may be considered as a kind of unsteady Reynolds number for the flow, as it indicates the relative importance of inertial and viscous forces in determining the motion within the time scale of one period of an oscillation.

At low values of α (less than unity), the flow is 'quasi-steady' and the velocity profile is parabolic at all times, the magnitude of the flow rate being determined by the instantaneous pressure gradient. Viscous forces dominate the flow and inertial forces can be neglected. At somewhat higher values of α the instantaneous velocity profile looks distorted (**Fig. 5.9**) because inertial forces have begun to influence the flow. Furthermore, the instantaneous flow rate will no longer be the same as predicted on the basis of the instantaneous pressure gradient. The flow will lag behind the applied pressure gradient. We may describe this mathematically as follows. At low values of α, the applied pressure, as given by Equation (5.8), generates a flow rate Q, which is determined by the instantaneous pressure gradient Δp_t and Poiseuille's law. Then

$$Q_t = Q_{ST}\sin(\omega t), \tag{5.10}$$

where Q_{ST} is the flow rate obtained for the pressure gradient Δp_0. At higher values of α (in the range approximately 1–3) the instantaneous flow rate lags behind the instantaneous pressure gradient and then Equation (5.10) becomes

$$Q_t = Q_{ST}\sin(\omega t - \phi),$$

where ϕ is a phase angle (see p. 110); i.e. the instantaneous flow rate is given by the applied pressure gradient at the earlier time $t - (\phi/\omega)$. At even higher values of α (greater than approximately 4) inertial forces begin to dominate the flow to such an extent that not only does the phase lag continue to increase, but also the peak flow rate Q_P becomes progressively less than predicted from the peak pressure gradient Δp_0. The variation of both phase lag ϕ and peak flow rate with increasing α can be seen from **Fig. 5.10**.

In the circulation, the value of α (based on the cardiac frequency) varies over a considerable range; thus, in the aorta α can be larger than 10, whilst in a capillary it is of the order of 10^{-3}. Furthermore, in the circulation, and indeed in many practical flow situations, we are concerned not with Poiseuille flow but with rather more complicated velocity profiles. These may be either of an entrance type or

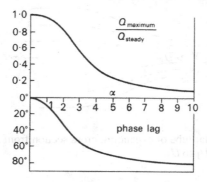

Fig. 5.10. The effect of progressively increasing α on the amplitude (upper panel) and phase (lower panel) of oscillatory flow resulting from a sinusoidally applied pressure gradient. The amplitude is expressed as the ratio of the instantaneous maximum flow rate (Q_{maximum}) to the steady flow-rate (Q_{steady}) at the maximum driving pressure. (From McDonald (1974). *Blood Flow in Arteries*, p. 125. Edward Arnold, London.)

profiles which are disturbed because of geometrical effects. In these cases it is inappropriate to calculate α on the basis of the diameter of the tube or vessel. The parameter should be calculated using, in place of the diameter, the thickness δ of the hydrodynamic boundary layer which would have been present in steady flow, because in reality α is an estimate of the effect of oscillation in distorting the steady flow. Thus

$$\alpha = \delta \sqrt{\frac{\omega}{v}}.$$

Since the boundary-layer thickness is always less than the tube radius ($d/2$) we can see that α will always be lower than anticipated on the basis of established flow conditions. However, in the circulation, it is not possible to determine δ accurately, and so values of α are always quoted on the basis of the vessel diameter.

Effects of constrictions on pipe flow characteristics

So far we have only considered flows within tubes of constant cross-sectional area. In this section we shall look at the effects of flow through constrictions and then in the next section at the properties of flow around bluff bodies and in curved tubes.

Consider a long, straight horizontal tube with a sudden rapid change in cross-sectional area (**Fig. 5.11**) from area A_1 to area A_2. The fluid contained within the pipe is incompressible and is flowing with steady laminar motion, and for the moment we suppose that viscosity can be neglected. Clearly, as the fluid passes into the larger area tube its velocity must fall, since the volume flow rate is the same in both sections:

$$U_1 A_1 = U_2 A_2. \tag{5.11}$$

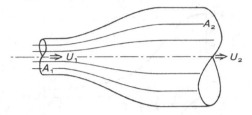

Fig. 5.11. Flow in a circular tube of expanding cross section from area A_1 to area A_2. Note that the velocity falls from U_1 to U_2.

Since there is no viscosity, there is no dissipation of energy, and we may apply Bernoulli's theorem to the flow and make an energy balance between the two sections:

$$\tfrac{1}{2}\rho U_1^2 + p_1 = \tfrac{1}{2}\rho U_2^2 + p_2$$

or

$$p_2 - p_1 = \tfrac{1}{2}\rho(U_1^2 - U_2^2). \tag{5.12}$$

But we know that U_2 is less than U_1, so the static pressure in the larger downstream area is greater than that upstream. The excess kinetic energy associated with the flow has been converted into static pressure. If the flow had been in the reverse direction we would have seen a decrease in static pressure on entry to the region of smaller cross-sectional area. If the fluid was real and of low viscosity then we could apply the same argument, because the pressure loss due to viscous action in a short section of pipe in that region would be very small compared with the pressures involved in the Bernoulli balance (Equation (5.12)).

On the basis of this last assumption, the *Venturi meter* was developed for the measurement of flow. The meter as shown schematically in **Fig. 5.12** consists of an initial contraction from the main tube cross-sectional area A_1 down to a constricted throat of area A_2, followed by a very slow expansion back to the original cross-sectional area. This expansion section is designed to have a divergent half angle of less than $5°$ in order to avoid certain energy losses associated with *flow separation*, which will be discussed later.

If we assume that there are no frictional losses, then from the above discussion we know that p_1 will be greater than p_3, but that pressure will rise again downstream till at the end of the divergent section the pressure will have recovered to p_2, which should equal p_1. However, in reality, friction losses cause p_2 to be slightly lower than p_1, though if the meter is well designed they will be small and may be ignored. The volume flow rate Q may be calculated as follows. It is very important to note here that in this example we are not talking about the mass flow rate but the volume flow rate. If we confused the two we would derive inhomogeneous equations with

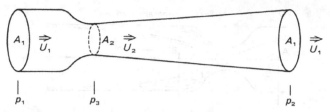

Fig. 5.12. Flow in a Venturi section where initially the tube cross-sectional area is decreased from A_1 to A_2, and then slowly re-expands to area A_1. The pressure measurements of p_1 and p_3 and of p_2 are made upstream of the constriction, at its maximum and far downstream respectively.

inconsistency in the units. Using Equation (5.12) for an energy balance across the convergent section we have

$$p_1 - p_3 = \tfrac{1}{2}\rho(U_2^2 - U_1^2). \qquad (5.13)$$

Also, from Equation (5.11) we know that

$$U_1 A_1 = U_2 A_2,$$

i.e.

$$U_2 = U_1 \frac{A_1}{A_2}.$$

Substituting in Equation (5.13):

$$p_1 - p_3 = \tfrac{1}{2}\rho U_1^2 \left[\left(\frac{A_1}{A_2}\right)^2 - 1\right].$$

But $Q = U_1 A_1$, so

$$p_1 - p_3 = \tfrac{1}{2}\rho \left(\frac{Q}{A_1}\right)^2 \left[\left(\frac{A_1}{A_2}\right)^2 - 1\right]$$

or

$$Q = \sqrt{\frac{2\Delta p}{\rho(1/A_2^2 - 1/A_1^2)}}.$$

Thus, we can see that provided the geometry is known we may compute the flow rate from the pressure drop across the convergent part.

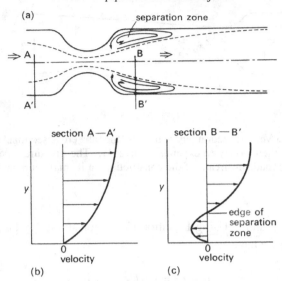

Fig. 5.13. Flow through a constriction followed by a rapid expansion downstream showing the region of flow separation (a). The sections at A–A′ (b) and B–B′ (c) show the velocity profiles at those stations. At the upstream station the flow is everywhere in the forward direction; at the downstream station the fluid near the wall has motion in the reverse direction.

The familiar laboratory water suction pump which fits on to a tap is a device designed rather like a Venturi meter. Water is forced to flow through a constriction, and the fluid acceleration causes a reduction in local static pressure; it is this which acts as the suction pressure for the pump.

If, after the constriction, the tube expands too rapidly, then the phenomenon of *flow separation* occurs. When this happens there is no longer simple streamline motion across the whole of the cross-section. We find that there is forward streamline motion in the core of the tube, but that flow near the walls is sluggish and recirculation of fluid occurs, as shown in **Fig. 5.13**. This separation is a consequence of the increase of pressure along this section (see above, p. 62). In a straight tube, pressure is falling in the direction of the flow and conditions are stable; again, in the convergent section of the meter, pressure is also falling, at an even faster rate. In contrast, in the expansion section of the meter the pressure is rising in the direction of motion and this is adverse to the flow. The effect of this adverse pressure force is to tend to retard the flow. In the core the inertia of the fluid opposes the retardation. However, near the wall the fluid velocity is small, because of the no-slip condition, and its inertia is very low, negligible in comparison with the viscous forces. As a result, when the adverse pressure gradient slows the fluid down, the flow near the wall ultimately reverses. When this happens, there is a drastic change in the nature of the boundary layer,

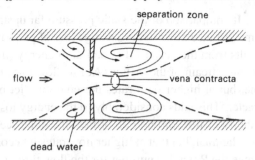

Fig. 5.14. Flow through an orifice showing that the flow continues to be constricted beyond the site of the obstruction, being of minimum cross-sectional area at the 'vena contracta'. Note also that separation occurs upstream of the orifice.

which no longer remains thin and is said to *separate*. The large effect of separation on the flow in the core of the tube can be seen in **Fig. 5.13**. As long as the flow field is laminar, the degree of separation or the size of the zone and intensity of recycling flow will increase as flow rate is increased. The tendency for turbulent flows to separate is less strong than for laminar flows; thus, turbulent flows can withstand more acute geometric changes than can laminar flows.

It is possible to produce constrictions in a tube which are much more severe than the above Venturi type; thus, an orifice inserted into a tube causes severe flow changes. Let us look at the effect of introducing a sharp-edged orifice into a tube as in **Fig. 5.14**, considering first steady laminar flow of a real fluid. As the flow approaches the constriction, the streamlines become bent and the flow accelerates. The fluid velocity in regions near the pipe wall and orifice is extremely low and these regions are effectively 'dead water' zones. As the flow passes through the orifice the jet continues to contract for a short distance further downstream till it reaches the 'vena contracta', where the effective cross-sectional area for flow is a minimum. Thereafter the flow diverges and ultimately fills the tube cross-section again. Immediately downstream of the orifice a separation zone can always be observed. It is characterized by a recirculating vortex-like motion. This results from the extremely rapid increase in cross-section immediately downstream of the orifice which produces a very large adverse pressure gradient for the flow. The flow emerging from the orifice does so effectively as a jet and the rate at which the boundary of the jet expands is relatively slow; as a result, the separation zone extends some distance downstream.

If we measure the static pressure upstream of the orifice where flow is relatively undisturbed and compare this with the static pressure at the level of the vena contracta, the difference is slightly greater than that expected on the basis of an energy balance assuming no viscous effects. Downstream, the agreement is not nearly as good as in the case of the Venturi meter because of the viscous losses associated with the separation of the flow. If comparison of the static pressure far upstream is made with the static pressure some distance downstream, a large loss of pressure (or energy) will be observed. This results from the viscous dissipation of energy (in the form of heat) in the disturbed flow downstream of the orifice. At low flow rates this occurs mainly in the separation zone, but at higher flow rates the emerging jet becomes disturbed and turbulent in character. This adds considerably to the energy losses.

If we study the stability of flow through an orifice, it can be seen that at low flow rates the emerging jet is laminar, but that at higher flow rates it becomes disturbed and turbulent. If we compute the Reynolds number for the flow through the orifice, using the orifice diameter and the mean velocity of fluid flowing through it (from Equation (5.5)), it can be demonstrated that the change in character of the emerging jet occurs for an orifice Reynolds number of 300–400. Thus, it is a very common situation to have an orifice in a pipe where there is nominally laminar flow in the main tube but a turbulent emerging jet.

In the blood circulation, coarctations and stenoses provide examples of constrictions to flow. The coarctation is perhaps most similar to the Venturi constriction and the stenosis to the orifice. The stenosis, the more common condition, is associated with a well-defined audible thrill which results from the noise created by the turbulence in the jet downstream of the orifice (see Chapters 11 and 12).

Flow in curved pipes

We have so far considered the properties of flow in a straight pipe; in a curved pipe, further interesting phenomena can be observed. When fluid flows steadily and without turbulence round a bend, as in **Fig. 5.15**, every element of it must change its direction of motion. That is, it must have a component of acceleration at right angles to its flow direction. It follows that, like a single particle moving in a circle, every element must experience a force in that direction. In this case, the force must be supplied by a sideways pressure gradient, in the plane of the paper of **Fig. 5.15**, acting from the outside of the bend towards the inside. This will act more or less uniformly over the whole cross-section of the tube, and all fluid elements of a given mass will experience approximately the same sideways force to make them turn the corner. Therefore they will all be given the same sideways acceleration. But this means that faster moving elements will change their direction less rapidly than slower moving ones, because they have more inertia.[3] Thus, the faster moving fluid which originally occupied the centre of the tube will tend to be swept out towards the outside of the bend, being replaced by slower moving fluid from near the walls. In consequence, a transverse circulation or *secondary motion* is set up, as shown in **Fig. 5.16**. At the same time, because the faster moving fluid has been swept to the outside of the bend, the axial

---

[3] Alternatively, the magnitude of the lateral acceleration on any particle is u^2/r, where u is the speed of the particle and r is the radius of curvature of its path; thus, if the sideways acceleration is uniform, the paths of the slower moving particles must have smaller radii of curvature.

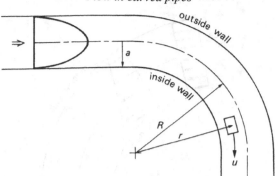

Fig. 5.15. Sketch of the centre plane of a curved tube, radius a, with Poiseuille flow entering. Radius of curvature of centre line is R. A fluid element of speed u, travelling along a path with radius of curvature r, experiences an inward acceleration u^2/r.

velocity profile is distorted from its original symmetric shape, and the greatest velocities occur near the outside wall. This pattern of flow can be seen in a long, continuous bend, far from the entrance, where the flow is fully developed.

An important consequence of the distortion of the Poiseuille velocity profile, together with the presence of secondary motions, is that there are regions in the pipe, in particular near the outside of the bend (**Fig. 5.16**), where the shear rate is considerably higher than in Poiseuille flow. High shear rates mean that the rate of deformation of fluid elements is high, and hence that the rate at which mechanical energy is dissipated by viscosity is also high. Thus, more work has to be done to force a fluid

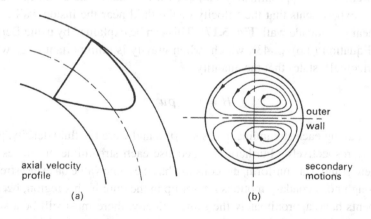

Fig. 5.16. (a) Distortion of the axial velocity profile as a result of tube curvature. (b) Projection of streamlines on to a transverse plane, clearly showing the form of the secondary motion.

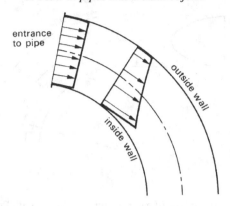

Fig. 5.17. Distortion of a flat velocity profile at the entrance to a curved pipe.

through a curved pipe than through a straight one. In other words, the pressure gradient necessary to maintain a given volume flow rate is greater in a curved pipe than in a straight one. If a straight pipe has a single bend in it, and then continues straight, the flow downstream of the bend will contain secondary motions of the sort just described. These will gradually die away so that, far downstream, Poiseuille flow is once more established; the distance over which the secondary motions and distortions will persist is comparable to the distance for Poiseuille flow to be established at the entrance to a straight pipe. This entrance length is given by Equation (5.4).

When fluid enters a curved tube, a thin boundary layer is present on the wall as in a straight tube. In this case, however, the velocity profile in the core, where viscosity is unimportant, is not approximately flat, even if it is flat at the entrance itself. It is observed in experiments that the velocity of the fluid near the inside wall is greater than that near the outside wall (**Fig. 5.17**). This can be explained by using Bernoulli's theorem (Equation (4.6), p. 43), which, when gravity is unimportant (e.g. when the flow is horizontal), states that the quantity

$$H = p + \tfrac{1}{2}\rho u^2 \tag{5.14}$$

is a constant along each streamline; as usual, ρ, p and u are the fluid density, pressure and velocity respectively. Furthermore, because each streamline originates from a region where p and u are uniform, the constant has the same value on every streamline. Now, although no secondary motions are set up in the core in this region, because all fluid elements have approximately the same velocity, there must still be a sideways pressure gradient for the fluid to be forced round the bend. Hence, p is greater at the outside of the bend than at the inside; so, for H to remain uniform, the fluid velocity u must be smaller at the outside than at the inside.

Secondary motions will not become important until a long way downstream, when the boundary layer is thick and the flow approaches its fully developed form. However, some recent experiments, combined with theoretical work, have shown that the details of the flow in the boundary layer, where weak secondary motions begin immediately, are affected by curvature. This has some interesting consequences, which can best be expressed in terms of the shear rate. If the boundary layer were the same all round the tube cross-section, the skewed velocity profile in the core would result in the wall shear rate being greater at the inside wall than at the outside. However, the incipient secondary motions in the boundary layer cause this distribution to be reversed after a distance along the tube of only about one diameter. From then on the maximum shear rate occurs at the outside wall, as when the flow is fully developed.

Flow past bodies

Our main preoccupation so far has been to investigate the way fluids flow in tubes or systems of tubes, and it has become apparent that the most complex flow patterns arise through the interaction between viscous and inertial forces, in the entrance region of a tube for example. Another very good example of this interaction is provided by the flow of a fluid past a solid body. We examine that example in detail here, both because it permits a deeper understanding of the general properties of a fluid in motion and because many of the phenomena which arise are familiar from everyday experience.

We suppose that the body is totally immersed in the flowing fluid, so that there is no possibility of surface waves being generated; their presence would obscure the important features of the flow. For convenience, we also imagine the body to be a very long cylinder, with the fluid flowing past it at right angles, so that in the middle the influence of the ends is negligible, and there is no component of velocity parallel to the axis of the cylinder. This means that the flow in the middle is the same as if the cylinder were infinitely long; the velocity vector of a fluid element always lies in a plane perpendicular to the cylinder axis and will depend on the position of the element in that plane, and on the time, but will be independent of the distance along the axis of the cylinder. We say that the flow in this case is *two-dimensional*; it is much easier to visualize than a fully three-dimensional flow. Let us further suppose that all other boundaries of the region containing the fluid are so far away that they have no influence on the flow, so that this region is effectively infinite. Finally, we shall first consider only a cylinder of circular cross-section, since that is a convenient shape, which has been the subject of many theoretical and experimental studies. As far as the flow relative to the body is concerned, it is irrelevant whether the fluid is moving past the body (like the wind blowing past a building) or whether the body is moving through the fluid (as an aeroplane moves through the air, or a grain of sand sinks to the bottom of a pool). The flow is normally easier to visualize with the body at rest, although sometimes the other point of view is more helpful. We shall be examining

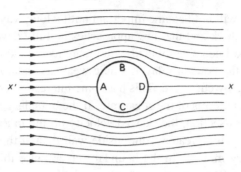

Fig. 5.18. A plane perpendicular to the cylinder axis, showing the streamlines (particle paths) for steady flow of an inviscid fluid past the cylinder. The velocity far from the cylinder is uniform and everywhere parallel to the line $x'x$. The points A and D are stagnation points; B and C are points where, in the absence of viscosity, the fluid speed is greatest.

situations where the body has a constant speed relative to the fluid far away from it. This means that the flow as a whole is steady when the body is taken to be at rest, but is unsteady when the body is moving in a fluid at rest far away.

First of all, we consider a hypothetical fluid with no viscosity and temporarily abandon the no-slip condition. This means that the boundary layers which we would expect to occur on solid boundaries (see p. 52) are absent in this case. Their influence will be considered below. Suppose that the cylinder is at rest and that far away from it the fluid flows forward uniformly with, say, speed U (**Fig. 5.18**). Nearer the body, however, the passage of some fluid is blocked; this fluid is retarded, and forced to flow round the sides of the body.

The paths of fluid elements initially far upstream of the body would approach the cylinder, be deflected above and below, as shown in **Fig. 5.18**, and return to their original paths far downstream. In steady flow, fluid elements move along the streamlines of the motion (lines everywhere parallel to the velocity vector; see p. 42) and therefore **Fig. 5.18** is also a diagram of the streamline pattern. (Alternatively, if the cylinder is thought of as moving, it has to push fluid out in front of it and draw it in behind it as it proceeds, and a flow pattern is set up like that depicted in **Fig. 5.19**.) It is easier – because less energy is required – for the fluid to flow round the sides of the cylinder like this than, for example, for an infinitely long column of fluid to be pushed in front of and pulled behind the cylinder.

There is one streamline which divides the fluid flowing above the cylinder from that flowing below. At the point where this line meets the cylinder the total velocity must be zero although the no-slip condition is not applied: the normal component of velocity is zero because fluid cannot cross a solid boundary, and the tangential component is also zero because, if it were not, the fluid there would be flowing one

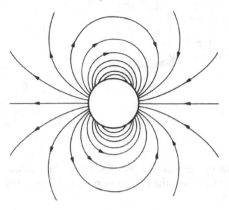

Fig. 5.19. A diagram of the (unsteady) streamlines set up by the motion of a cylinder moving to the left in a fluid otherwise at rest. Fluid is pushed to the sides in front of the cylinder, and moves in behind it again when it has passed.

way or the other round the cylinder. This point, A in **Fig. 5.18**, is called a *stagnation point*. From Bernoulli's theorem (Equation (4.6), p. 43) we know that the pressure at A exceeds the pressure far away, where the flow is uniform (let this pressure be p_0), by an amount equal to $\frac{1}{2}\rho U^2$; this excess pressure can be thought of as providing the force required for the deflection of the other fluid. The velocity of a fluid element which passes close to A, and therefore has near-zero velocity there, must necessarily increase as it is forced sideways by this excess pressure towards the widest part of the cylinder, and experiences an acceleration. The streamlines are closer together in this region than they were far upstream, indicating that the velocity there is greater than U. In fact the velocity of fluid at B or C (**Fig. 5.18**) is equal to $2U$; all fluid particles are accelerated over the forward half of the cylinder, until they pass the widest part. By Bernoulli's theorem, the pressure at B or C is less than p_0 by an amount $\frac{3}{2}\rho U^2$. When there is no viscosity, the flow over the rear half of the cylinder is symmetrical with that over the forward half. The fluid particles are retarded, the pressure along the surface of the cylinder rises (to exceed p_0 by $\frac{1}{2}\rho U^2$ at D again) and far downstream the uniform flow is once more achieved.

The symmetry of the flow leads to the prediction that the net force on the cylinder is zero. In fact, it can be proved mathematically that no body in a uniform stream of inviscid fluid experiences any force in the direction of the stream. This prediction, which was first made by the French philosopher d'Alembert in 1752, is irreconcilable with the everyday observation that all bodies moving through fluids experience some resistance or *drag*, even when the numerical value of the coefficient of viscosity is very small ($\mu = 0.001\,81\,\mathrm{N\,s\,m^{-2}}$ in air at 20°C, $0.1\,\mathrm{N\,s\,m^{-2}}$ in water at 20°C). This difficulty, known as *d'Alembert's paradox*, was a source of great concern in the nineteenth century. Its resolution lies in the fact that all fluids are viscous, and it turns out

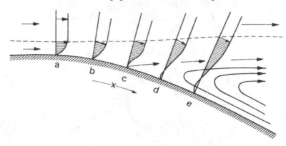

Fig. 5.20. Diagram showing the velocity profiles across the boundary layer, at different positions along the body surface. Note the region of reversed flow downstream of the point c of boundary-layer separation.

that, however small μ is, the presence of viscosity distorts the flow pattern so much that the drag is considerable.

Consider now a real fluid, whose velocity far from the stationary cylinder (of diameter d) is U. The Reynolds number is then equal to Ud/v (where v, equal to μ/ρ, is the kinematic viscosity of the fluid). When this is very large, we expect inertia forces to predominate over viscous forces everywhere in the flow except near the boundaries, where the no-slip condition must be applied. We might, therefore, hope that the flow would be very similar to the flow of an inviscid fluid, except near those boundaries. Fluid elements will be retarded as they approach the stagnation point A (see **Fig. 5.18**) and accelerated again over the forward half of the cylinder. This time, there is a thin region near the walls across which the fluid velocity falls to zero, as required by the no-slip condition, and we can recognize this region as a boundary layer (see p. 52). It will be thinnest at the front of the cylinder (near the point A), but will remain quite thin all the way round to B and C, since the velocity outside it increases with distance along the cylinder surface. (As we have seen, by Equation (5.3), the thickness of a boundary layer increases as the square root of the distance along the boundary, but decreases as the square root of the velocity of fluid outside the layer.) Once the fluid has passed the widest part of the cylinder, however, it begins to be decelerated, and the pressure begins to rise, as described above for an inviscid fluid. This rise in pressure imposes an adverse pressure gradient on the boundary-layer flow, and, as in a rapidly diverging tube, this causes the flow near the wall to come to rest and reverse (**Fig. 5.20**). From the point where reversal first occurs, the region in which the flow does not resemble that in an inviscid fluid, because viscous forces are important, is no longer confined to a thin boundary layer. Separation of the boundary layer has occurred (p. 64). This prevents complete recovery of the pressure from being achieved, so that the pressure behind the cylinder is much lower than that in front. Thus, the body must experience a considerable drag, however large the Reynolds number. The measured distribution of pressure round the

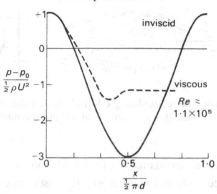

Fig. 5.21. Graphs of the pressure exerted on the cylinder, plotted against distance x from the front stagnation point, measured along the cylinder surface. The broken curve was measured when $Re = 1.1 \times 10^5$, while the solid curve is the theoretical curve for an inviscid fluid. It is the difference between these pressures at the back of the cylinder that is responsible for most of the drag. (From Batchelor (1967). *An Introduction to Fluid Dynamics*, p. 340. Cambridge University Press.)

surface of the cylinder at Reynolds number 1.1×10^5 is shown in **Fig. 5.21**, and is compared with the corresponding distribution for an inviscid fluid. It is the marked difference between the pressure distributions over the rear half of the cylinder which is responsible for a large part of the drag. The other contribution to the drag arises from the direct action on the wall of the tangential viscous stresses associated with the high rates of shear present in the velocity profile across the thin boundary layer (**Fig. 5.20**). For flow round a bluff body like a circular cylinder, however, this direct viscous component of drag contributes about 1% of the total drag for Reynolds numbers above 100.

The character of the flow when it has passed the body depends crucially on the value of the Reynolds number Re. For values below about 10, the idea of a boundary layer, even on the forward half of the cylinder, is not appropriate, and we exclude such values for now. For values of Re between about 10 and about 50, there is a region immediately behind the cylinder in which the flow is confined to a pair of *vortices* (or eddies) where the flow looks like that in a pair of whirlpools (**Fig. 5.22**). Fluid elements are trapped there, circulating round and round while the main flow goes by. The motion in these vortices, where fluid velocities are significantly lower than in the main stream, is driven by the viscous stresses exerted by the main stream at the edges of the vortices. As Re increases (in the range 10–50) the vortices are progressively stretched out downstream.

Beyond the point where the main stream comes together again, there is a long, fairly narrow region where the velocity of the fluid is still less than that outside it in the undisturbed parts of the flow (a typical velocity profile is shown in **Fig. 5.22**).

Fig. 5.22. Diagram showing the recirculating eddies behind a circular cylinder, and a velocity profile across the viscous wake downstream. A and C are stagnation points.

This region is known as the viscous *wake* of the body. Viscous wakes occur whether or not the fluid has a free surface, but are normally observable only at a free surface, where, however, they are often obscured by wave motions. For example, the wake of a ship is visible primarily because of the waves it generates on the sea surface, although this surface is also churned up by vigorous disturbances, partly induced by the propellers, which persist far downstream in a zone analogous to a viscous wake. When the stream of a river flows past the pier of a bridge, waves are of course generated, but, in addition, objects floating in the water can be seen to move more slowly behind the pier than in the main stream. Another familiar example of a wake occurs behind a rapidly moving lorry, where the forward motion in the wake (associated with the low-pressure region immediately behind) tends to pull any small vehicle along with it. The backward motion at the sides of the lorry felt by anyone standing at the roadside (the slipstream) is a reflection of the increased relative velocity at the sides of a body (cf. **Fig. 5.18** or **5.19**). Of course, the Reynolds number in each of these examples is enormous (about 5×10^6 either for a pier of width 2 m in a river flowing slowly at $2.5\,\mathrm{m\,s^{-1}}$ or for a lorry of width 4 m travelling at $15\,\mathrm{m\,s^{-1}}$ (just over 30 miles per hour)), but the gross features of the flows just described are always present.

At values of *Re* above about 40, the symmetrical steady flow depicted in **Fig. 5.22** becomes unstable and the wake is observed to oscillate slowly, with an amplitude which increases with distance downstream (**Fig. 5.23**). As *Re* increases further, the oscillation of the wake moves closer to the cylinder and, when *Re* is near 60, begins to affect the two eddies behind the cylinder. They oscillate together from side to side and appear to shed a small eddy at the end of every half-period, on each side of the cylinder alternatively. The behaviour of the wake at this stage is very striking

Fig. 5.23. Photograph of oscillations starting in the wake of a cylinder (*Re* = 55). (From Batchelor (1967). *An Introduction to Fluid Dynamics*, Pl. 2. Cambridge University Press.)

Fig. 5.24. Photograph of a vortex street ($Re = 102$), with arrows superimposed to show the direction of rotation of the vortices. (From Batchelor (1967). *An Introduction to Fluid Dynamics*, Pl. 2. Cambridge University Press.)

(**Fig. 5.24**). The fluid passing close to the cylinder gathers itself into discrete whirls, or vortices, arranged in two regular staggered rows, one on each side of the centre line. They are shown by the arrows in **Fig. 5.24**. This arrangement, called a *vortex street*, travels downstream at a velocity less than U, and persists for very many cylinder diameters. This periodic pattern of the wake occurs also for a very wide range of Reynolds numbers up to about 2500. For still larger Re the wake is turbulent, and the discrete vortices cannot be easily identified; but even so, the underlying periodic structure persists up to extremely large values of Re (at least 10^5).

Associated with the periodic vortex shedding there is a fluctuating pressure field in the neighbourhood of the body. This exerts a fluctuating force on the body, which may cause it to vibrate at the frequency of the vortex shedding if it is not rigidly fixed. These vibrations may in turn accentuate the intensity of the fluid mechanical oscillations, and so on. If the frequency of these oscillations is one at which the body can naturally vibrate with only a small stimulus, a potentially dangerous *resonance* may result (see Chapter 8, p. 123), in which the body oscillates more and more vigorously. Such resonance has been the cause of a number of engineering disasters, in which structures have collapsed in a wind (a famous example is the Tacoma Narrows bridge, near Seattle, USA, which collapsed soon after being built, in 1940). On a smaller scale, the periodic shedding of vortices is responsible for the humming of telephone wires in the wind, and is the mechanism by which sounds are produced in a number of musical instruments, particularly the human voice. The shedding of vortices in blood by obstacles such as the aortic valve or a stenosis will also generate fluctuating pressures which may be important (see Chapter 12).

The frequency f with which vortices are shed behind a cylinder can be related only to the parameters which determine the flow pattern: that is, the velocity U, diameter d, viscosity μ and density ρ. A simple quantity with the dimensions of frequency which can be derived from these is U/d, although more complicated quantities can be constructed by multiplying this by anything else which depends only on the dimensionless Reynolds number ($Re = \rho U d/\mu$). We may conclude, however, that the dimensionless quantity fd/U can depend only on Re, for a given shape of body. In fact this quantity, called the *Strouhal number*, is approximately constant (0.2 for a

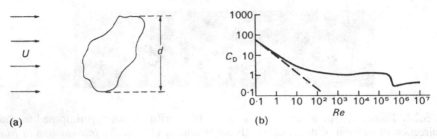

Fig. 5.25. Definition of the diameter d of a cylinder of non-circular cross-section, for use in Equation (5.16), which determines the drag coefficient C_D. (b) Graph of C_D against Re for a circular cylinder (logarithmic scales). The broken curve is the theoretical curve for small values of Re (accurate up to $Re = 0.2$). C_D is approximately constant for values of Re between 200 and 200 000. (After Fox and McDonald (1973). *Introduction to Fluid Mechanics*, p. 408. John Wiley, New York.)

circular cylinder) for a wide range of Reynolds numbers; above approximately 10^3. Over this range, the frequency of vortex shedding from a given body, and hence the pitch of the sound produced, increases in proportion with the flow velocity. The noise of the wind becomes more highly pitched as its speed increases.

The drag force exerted by a fluid flowing past a body, such as a circular cylinder, can be measured experimentally. In order to obtain a formula which can be applied to cylinders of different diameter in flows of different speed, we usually divide the drag by a quantity which also has the dimensions of force, and which can be thought of as a scale for a typical dynamic pressure force acting on the body. This quantity is

$$\tfrac{1}{2}\rho U^2 dL, \tag{5.15}$$

where L is the length of the cylinder being used, which has to be very long for the flow to be approximately two-dimensional. The ratio of the total drag force to this quantity is called the *drag coefficient*, and is given by

$$C_D = \frac{\text{Drag}}{\tfrac{1}{2}\rho U^2 dL}. \tag{5.16}$$

For cylinder cross-sections other than circular, the drag coefficient is similarly defined, only then the 'diameter' d has to be taken to be the maximum width of the cylinder in a direction at right angles to the oncoming flow (**Fig. 5.25**a). For any given shape of cylinder, the drag coefficient, like the Strouhal number, can depend only on Re. The relationship between C_D and Re for a circular cylinder is plotted on logarithmic scales in **Fig. 5.25**b.

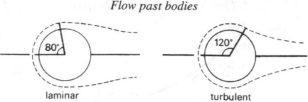

Fig. 5.26. Diagram indicating the extent to which turbulence in the boundary layer inhibits boundary-layer separation, thus reducing the drag on the body.

For very small values of Re, C_D falls rapidly as Re increases. This is because, in that range, inertia forces are negligible compared with viscous forces, so that the 'typical' inertial or dynamic pressure force with which we have made the drag dimensionless in Equation (5.16) is not typical of the motion at all. In this case there is a balance between viscous and pressure forces; no boundary layer can be identified, since the presence of the body distorts the free stream equally in all directions. In fact, the streamlines approximately show fore and aft symmetry, just as in an inviscid fluid. Of course, the low Reynolds number flow is not similarly associated with zero drag, but, as we see from **Fig. 5.25**b, with a very large, viscous drag. The details of the flow at low Reynolds number can be analysed theoretically, on the assumption that the inertia is very small, and results in a prediction of the drag coefficient which is plotted as the broken curve in **Fig. 5.25**b. It can be seen to be accurate for values of Re less than 0.2.

We can also see from **Fig. 5.25**b that for values of Re between about 200 and about 200 000 the value of C_D is remarkably constant, at a value of about 1.2. This shows that the dynamic pressure force (Equation (5.15)) is certainly 'typical' of the flow. It reinforces the above statement that it is the change in the pressure distribution resulting from boundary-layer separation which has the greatest influence on drag; the direct viscous drag is very small. The sudden fall in C_D at a value of Re just over 200 000 is associated with turbulence developing in the boundary layer on the front of the cylinder (the wake will have been turbulent for much smaller values of Re). Rather than increasing the drag, as we might have supposed because of the increased viscous dissipation inevitable in a turbulent flow, it causes it to fall. The reason for this is that turbulence in the boundary layer inhibits boundary-layer separation, by some mechanism as yet unknown. It causes the point on the cylinder surface at which the boundary layer separates to move towards the back (**Fig. 5.26**), which has the effect of reducing the area over which the reduced pressure in the separated region can act, so that the net difference in the pressure force between front and back must fall. Fascinating though this aspect of fluid dynamics is, we shall not discuss it further since it cannot be relevant to the circulation, where the maximum Reynolds number for pipe flow is less than about 10 000.

5. Flow in pipes and around objects

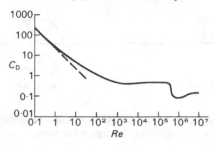

Fig. 5.27. Graph of C_D against Re for a sphere; it is qualitatively similar to that for a cylinder. The broken curve is again the theoretical result for small values of Re. (After Fox and McDonald (1973). *Introduction to Fluid Mechanics*, p. 406. John Wiley, New York.)

We have concentrated so far on two-dimensional bodies because the flow patterns past them are easy to visualize. However, in many everyday examples of flow past bodies, the body, and the flow round it, is three-dimensional. In fact, the main features of the flow past three-dimensional bodies are similar to those already described, but it is important to recognize the few differences. The three-dimensional body most commonly studied is the sphere; this is because the flow past it is both more convenient to produce experimentally and more tractable to analyse theoretically than any other. Furthermore, many everyday examples of approximately spherical bodies come to mind: a large balloon rising through the air (large Re), a small grain of sand settling in water (small Re), a ball-bearing in oil (small Re), a ball in air (quite large Re), etc. In this case, the Reynolds number is again defined in terms of its diameter d and velocity U relative to the fluid far away, together with the kinematic viscosity v (i.e. $Re = Ud/v$). The drag coefficient is again the total drag divided by a typical dynamic pressure force, which for a three-dimensional body is usually taken to be $\frac{1}{2}\rho U^2$ times the area presented to the flow ($\frac{1}{4}\pi d^2$ in the case of a sphere). Thus:

$$C_D = \frac{\text{Drag}}{\frac{1}{2}\rho U^2 \frac{1}{4}\pi d^2}.$$

The graph of C_D against Re is shown in **Fig. 5.27**, and can be seen to have a shape similar to that for a cylinder. Theoretical analysis at low values of Re shows that

$$C_D = 24/Re$$

and the curve corresponding to that is also drawn in **Fig. 5.27**. The fixed eddies which form behind the sphere at higher Reynolds numbers, up to about 130, are of course not like the long, straight vortices formed parallel to a cylinder. They are instead ring shaped, since the overall motion is symmetric about an axis which passes through the centre of the sphere in the direction of the oncoming flow. (A more familiar example of a ring vortex is given by a smoke ring, in which particles can be seen to rotate about a ring-shaped core as it advances.) At still higher Reynolds numbers, the ring-vortices are shed and create a complicated flow pattern in the wake. This is not an ordered motion like a two-dimensional vortex street, because the vortex remains a continuous closed loop, which becomes very distorted because of its tendency to be shed from different parts of the sphere at different times. Despite the apparent confusion in the flow behind the sphere, the perturbations in the force exerted by the fluid on the sphere are dominated by oscillations of a given frequency, leading to possible sound and vibrations at that frequency.

6

Dimensional analysis

On a number of occasions we have used the concept of dimensional homogeneity to test the validity of a mathematical expression of a physical law. In this chapter we will explore this concept more thoroughly and show how useful it is; in addition we will show how the concept may be used in the derivation of physical laws.

In most cases, it is easy to see how we may convert a law into dimensionless form; such a regrouping often tells us something of the relationship between the physical variables involved in the law. For example, let us again look at Poiseuille's law (Chapter 5, p. 47). As a dimensional statement we recognize it as

$$\Delta p = \frac{128\mu LQ}{\pi d^4}.$$

If we define $\bar{U}$ by

$$Q = \frac{1}{4}\pi d^2 \bar{U}$$

and substitute for Q and divide both sides by the product $\rho\bar{U}^2$, then we may write the equation in the dimensionless form

$$\frac{\Delta p}{\rho\bar{U}^2} = 32\left(\frac{\mu}{\rho\bar{U}d}\right)\left(\frac{L}{d}\right)$$

or, substituting for Reynolds number (*Re*),

$$\frac{\Delta p}{\rho\bar{U}^2} = \frac{32(L/d)}{Re}.$$

With Poiseuille's law written in this dimensionless form we are led to the concept of *similarity*. Thus, we can look at the flow conditions in two pipes of differing diameters, with fluids of differing viscosity flowing at different flow rates. It is possible that flow conditions will be 'similar'; and this will be so if respective conditions are such that each of the groups $(\Delta p/\rho\bar{U}^2)$, (L/d) and *Re* are the same. If one or more of the groups are different in magnitude then the flows will be dissimilar.

Similarity and the idea of scale models

The concept of similarity gives us great power in experimental research, for it enables us to design and interpret experiments on scale models of the system we would really like to study.

If the real system is inconveniently large or small, we may scale quantities appropriately so that it is possible to study a similar dynamic condition in a model of reasonable size. For pipe flow studies we can scale down very large tubes to a convenient size, and very corrosive materials which it would be unpleasant to handle in a laboratory may be replaced by safer materials such as air, water or oil of the appropriate viscosity and density.

Prototype aircraft designs do not have to be flown to test their aerodynamic proficiency; it is possible to test small-scale models in wind tunnels. By adjusting the wind speed in the tunnel and the model orientation, it is possible to predict how a full-scale aircraft would react to dynamically similar changes in flight conditions.

Again, the microcirculation is so small in physical size that quantities we would like to measure such as pressure drop are too small for instruments to deal with readily. For this reason large-scale models of the microcirculation have often been constructed. In fact, much of our knowledge of the mechanics of the microcirculation has been derived from studies on enlarged models (see Chapter 13).

However, when we apply such a modelling approach to any problem we must be aware that it will break down if by changing the scale of one of the parameters beyond reasonable limits we need other additional laws to describe conditions. For example, it is easy to envisage experimental situations in fluid flow where the pressure gradient is increased to such a degree that the fluid must be treated as compressible, and the simplifications associated with incompressibility are no longer valid.

We may also use dimensional analysis and dimensionless groups when studying complicated problems, to tell us which groups of variables should be held constant at any time and to what extent other variables should be altered. It helps greatly in the presentation of results, since they can be best presented as plots of data grouped in dimensionless form. Then we can see the response of the system to changes in the appropriate groups of variables without being concerned about the units in which the variables have been expressed. This is particularly valuable if comparisons are being made of results obtained by different workers.

Some examples of scaling in biological systems

The examples of scaling described above relate to mechanical processes; it is, however, quite possible to scale gross physiological behaviour of animals. For instance, we can scale such things as metabolic rate, anatomy and body chemistry as functions of, say, body size. Such correlations can be impressive in their closeness; often there are also deviations whose existence can be instructive.

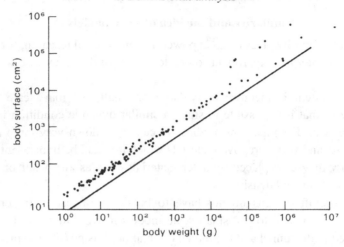

Fig. 6.1. Body surface area of vertebrates in relation to body weight. The solid line represents the surface area of a sphere of density $1\,g\,m^{-3}$. (From Lightfoot (1974). *Transport Phenomena and Living Systems*, p. 348. Wiley–Interscience, London.)

A simple and perhaps surprising relationship is that between body weight and shape. It can be seen for vertebrates (for an enormous range of weight) that the outer surface area is about twice that of a sphere with the same volume (**Fig. 6.1**).

The volume of individual organs in relation to body volume also shows remarkable constancy in mammals. Thus, the volume of the heart is approximately 0.5 to 0.6% of body volume; however, the hearts of greyhounds and race horses are exceptionally large. Indeed, when one considers the cardiovascular requirements of these animals, this might be expected.

The effect of body size on mammalian blood takes two forms. First, the haemoglobin of smaller animals releases oxygen more readily at a given pH, thus helping to maintain the increasing metabolic rate as size decreases. Second, the effect of pH decrease (or increase in $p(CO_2)$) on the oxygen release – the *Bohr effect* – is greater for smaller animals, as can be seen from **Fig. 6.2**. There are two interesting deviations on this figure. It can be seen that the Bohr effect is unusually large in the horse – a creature accustomed to the exertion of running for long periods. The chihuahua is a relatively new breed of dog developed for fashion, not function, and it exhibits a small Bohr effect; this result of rapid selective breeding would appear to be a distinct disadvantage.

A method of obtaining homogeneous relationships between variables

In many problems we can guess which variables are involved, but unless we know the laws governing the process, it is not immediately obvious how they are related. By

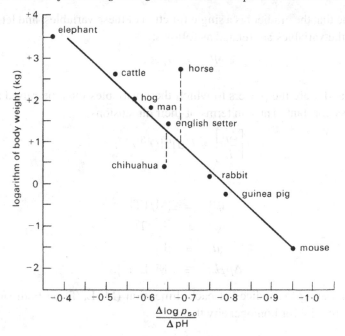

Fig. 6.2. The effect of acid on the unloading of oxygen from the blood (the Bohr effect) is greater in small animals than it is in larger ones. (From Lightfoot (1974). *Transport Phenomena and Living Systems*, p. 352. Wiley–Interscience, London.)

applying the principles of dimensional analysis we can often find out. We know that for any physically realistic description of a process the equation must be dimensionally homogeneous, and thus we must group the variables in such a way that homogeneity is achieved. The technique for doing this is known as the *indicial method* and two simple examples will serve to show how it may be used.

For the first example let us look again at the problem of flow in a tube. We know that the pressure gradient along the tube $(\Delta p/L)$ will depend upon the diameter of the tube d, the volume flow rate Q and the fluid viscosity μ. If we are concerned with flow far from the entrance of a straight tube we know that the fluid elements experience no acceleration. Thus, we are not concerned with fluid inertia, and fluid density is not relevant.

Hence, we can say that the pressure gradient is a function of viscosity, volume flow rate and tube diameter or

$$\frac{\Delta p}{L} = f(\mu, Q, d).$$

Let us assume that the gradient is a single function of these variables, and let us further assume that the variables are related as follows:

$$\frac{\Delta p}{L} = k\mu^{\alpha} Q^{\beta} d^{\gamma},$$

where α, β and γ are the powers to which these variables must be raised and k is a dimensionless constant. Thus, in terms of their dimensions,

$$\left[\frac{\Delta p}{L}\right] \equiv k[\mu]^{\alpha}[Q]^{\beta}[d]^{\gamma},$$

but

$$\begin{aligned}
[\mu] &\equiv [M/LT] \\
[Q] &\equiv [L^3/T] \\
[d] &\equiv [L] \\
[\Delta p/L] &\equiv [M/L^2 T^2].
\end{aligned}$$

Thus, by comparing the indices of each dimension (M, L, T) on both sides of the equation, we require for homogeneity that

$$\begin{aligned}
\text{for mass} \qquad 1 &= \alpha \\
\text{for length} \qquad -2 &= -\alpha + 3\beta + \gamma \\
\text{for time} \qquad -2 &= -\alpha - \beta.
\end{aligned}$$

Solving these three equations simultaneously, we have

$$\alpha = 1, \qquad \beta = 1, \qquad \gamma = -4.$$

or

$$\left[\frac{\Delta p}{L}\right] \equiv k[\mu][Q]/[d]^4.$$

Thus, the only dimensionally homogeneous relationship between the variables is

$$\frac{\Delta p}{L} \propto \frac{\mu Q}{d^4}$$

or

$$\frac{\Delta p}{L} = k\frac{\mu Q}{d^4},$$

and this must be the form the relationship actually takes. This, of course, we recognize as Poiseuille's law, where, by theoretical calculation, the constant k has been shown to be $128/\pi$.

As a second simple example we may consider the case of very slow flow past a sphere (see Chapter 5, p. 77). In this case again we may ignore the effects of inertia and consider the viscous effects only. The force F exerted by the fluid on the sphere will depend upon the size of the sphere (radius a), the velocity u of flow past the sphere and the fluid viscosity μ. Thus

$$F = f(a, u, \mu).$$

Considering the dimensions of the variables,

$$[F] = [a]^{\alpha}[u]^{\beta}[\mu]^{\gamma};$$

but

$$
\begin{aligned}
[F] &= [ML/T^2] \\
[a] &= [L] \\
[u] &= [L/T] \\
[\mu] &= [M/LT].
\end{aligned}
$$

Comparing indices for each of the dimensions M, L and T

$$
\begin{aligned}
1 &= \gamma \\
1 &= \alpha + \beta - \gamma \\
-2 &= -\beta - \gamma.
\end{aligned}
$$

Thus

$$\alpha = 1, \qquad \beta = 1, \qquad \gamma = 1$$

and, therefore,

$$F \propto ua\mu.$$

If we now consider the particular case of a sphere slowly falling under the influence of gravity through a viscous fluid (for example, the slow sedimentation of fine sand particles in water), then the force exerted by the sphere will be proportional to its volume and the difference in density between the sphere and fluid ($\rho_B - \rho_F$). Then

$$F \propto (\rho_B - \rho_F)ga^3.$$

Thus

$$k(\rho_B - \rho_F)ga^3 = ua\mu$$

or

$$u = k(\rho_B - \rho_F)ga^2/\mu. \qquad (6.1)$$

This in fact is *Stokes' law* for slow flow round a sphere. The value of the constant k in this equation can be shown to be $\frac{2}{9}$.

7

Solid mechanics and the properties of blood vessel walls

The walls of blood vessels are elastic and can change their size or shape when different forces are applied to them. These forces include both the pressures and shear stresses exerted by the blood, and the constraints imposed by surrounding tissue. In this chapter, therefore, we both outline the basic principles governing the mechanics of deformable solids and show to what extent they are applicable to blood vessel walls, rather than leaving the application to a later chapter. The essentials of solid mechanics are of course contained in Newton's laws of particle motion; a solid material, like a fluid, can be thought of as split up into a large number of small elements, to each of which the laws can be applied. Again, the forces on the elements consist of long-range body forces and short-range stress forces; it is in the relationship between the stresses and the deformations of the material that solid and fluid mechanics differ.

Definitions of elastic properties

We should begin with a few definitions. An elastic material is one which deforms when a force is applied to it, but returns to its original configuration, without any dissipation of energy, when the force is removed. This means that all the elements return to their original positions. The first understanding of elasticity was obtained by Robert Hooke (the English astronomer and physicist) in 1678, from experiments with metal wires. In his experiments, a uniform cylindrical wire of unstretched length l_0 and cross-sectional area A_0 was suspended from a fixed support, and different weights W were attached to the other end (**Fig. 7.1**a). The extension l' was measured and Hooke found that, over a wide range of applied weights, the extension was proportional to the weight ('*ut tensio sic vis*') (**Fig. 7.1**b). More recent experiments by, for example, Thomas Young (1773–1829, the English physician, physicist and Egyptologist) have made this law (still called Hooke's law) more precise, so that it applies to a series of wires with different cross-sectional areas, but all made of the same material. It is then found that the extension per unit length (l'/l_0, usually called the longitudinal *strain*) is

86

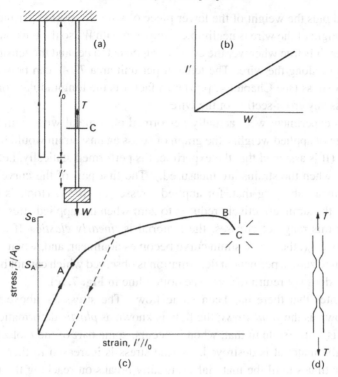

Fig. 7.1. (a) Hooke's wire-stretching experiment. A metal wire is stretched from a fixed support. When no weight is hung from it, its length is l_0; when a weight W is hung from it, it is extended by an amount l'. The point C is an imaginary cut, used to define the tension T in the wire, which is the force exerted by the upper piece of wire on the lower piece (or vice versa). A force balance on the lower piece shows that T must be equal to W. (b) The results of an experiment like that of (a): extension l' is proportional to weight W, as long as W is not too large. (c) The results of a complete wire-stretching experiment, expressed in the form of a graph of stress ($= T/A_0$, where A_0 is the initial cross-sectional area of the wire) against strain ($= l'/l_0$). For stresses less than S_A the two are directly proportional. For greater stresses, plastic flow takes place, which means that the strain does not reduce to zero when the stress is reduced to zero (dotted curve). Once the maximum applied stress S_B has been reached, flow continues even when the stress is reduced below S_B. Fracture occurs at C. (d) Schematic side view of a wire in which necking has begun.

proportional to the tension T in the wire divided by the initial cross-sectional area A_0. In order to define the tension in the wire we suppose it to be divided into two parts by an imaginary cut, marked as C in **Fig. 7.1**a. The total downward force exerted by the piece of wire below C on the piece of wire above C is then called the tension T. By Newton's third law, the upward force exerted *by* the upper piece of wire *on* the lower piece of wire also has magnitude T. Furthermore, if we consider the force balance on the lower piece of wire, we see that T must be equal to the weight W hanging from the lower end plus the weight of the lower piece of wire. In most experiments of this nature, the weight of the wire is negligible compared with W, so the tension T is equal to W. This result is true wherever the cut C is supposed to be, and the tension T in the wire is uniform along the wire. The tension per unit area T/A_0 can be seen to have the form of a stress (see Chapter 4, p. 31); in fact it is the longitudinal normal stress exerted across any cross-section of the wire.

If Hooke's experiment were actually performed on a metal wire, with the widest possible range of applied weights, the graph of stress against strain would be as shown in **Fig. 7.1**c. (It is assumed that the experiment is performed statically, i.e. that there is no motion when the strains are measured.) The first part of the curve, up to the point A, is linear, showing that, for applied stresses up to S_A, Hooke's law holds. Furthermore, the strain effectively reduces to zero when the applied stress is reduced to zero. For this range of stresses, the material is *linearly elastic*. If the stress is increased above S_A, the stress–strain curve becomes nonlinear, and, when the stress is reduced again to zero, a permanent deformation is observed which cannot be reversed; i.e. the strain does not return to zero (see dotted line in **Fig. 7.1**c).

This indicates that there has been some flow. The stress S_A above which this occurs is known as the *yield stress*; the flow is known as *plastic* deformation. Plastic deformation is irreversible in that, when it occurs, some part of the molecular lattice structure of the material is destroyed. As the stress is increased further, plastic deformation continues until the material eventually breaks on reaching the point C on the graph. After the maximum applied stress S_B is reached, the material will continue to flow even if the stress is reduced somewhat. This is because, in this range of strains, the cross-sectional area of some parts of the wire is observed to diminish rapidly ('necking'; see **Fig. 7.1**d) so that the stress, measured as tension divided by undistorted area, is no longer a good representation of the actual stress, i.e. tension divided by actual area. The actual stress at the position where the area is smallest continues to rise from B to C.

The above description is still oversimplified, because if a stress in the elastic range (i.e. less than S_A) were maintained for a very long time (several days in the case of steel wires), it would be observed that the strain very gradually increases. This implies that a very slow flow is taking place. If the stress were then removed, the strain would not reduce exactly to zero, but would still have a small value determined by the amount of flow which has taken place. This phenomenon is known as *creep*. It differs from plastic flow in that it is reversible: if a compressive stress were then applied for a long time, a reverse slow flow would be observed, and eventually the strain measured when no stress is applied could be reduced to zero. Creep does not irreversibly destroy the molecular structure of a material in the way that plastic flow does. The relative motion of neighbouring molecules resembles that in a viscous fluid, and can be reversed by reversing the applied stress. As in the flow of a viscous fluid, mechanical energy is dissipated during creep. An associated phenomenon occurs if, instead of holding the applied stress constant over a long time, we maintain a constant displacement for a long time. It is observed that the stress required to maintain it falls slowly from its initial elastic value. This phenomenon is called *stress relaxation*. Both creep and stress relaxation are examples of what is called *visco-elastic* behaviour. The fact that every material is viscoelastic to some extent means that no material is truly elastic in the sense of our definition. However, for very many materials, like steel in the wire experiment, or stone, or concrete, the amount of creep which occurs in the periods of interest is negligible, and they can be regarded as elastic materials for a wide range of applied stresses. As we shall see, the walls of blood vessels are visco-elastic, and this has a significant effect on the mechanics of blood flow within them.

Once an experiment, like the wire experiment, has been performed on a material, we can say that, over a certain, known range of stresses (and times), the stress–strain curve is linear. In this range we can write

$$\frac{T}{A_0} = E\frac{l'}{l_0}.$$

(7.1)

The constant of proportionality E is called the *Young's modulus of elasticity* of the material under investigation. The quantity E is a characteristic of the material, and in the simplest materials is independent of the manner in which the stresses are applied; it has the units of stress, Nm^{-2}. Note that Equation (7.1) would apply equally well to the deformation of a cylindrical solid under a longitudinal compressive stress (like that applied to the tibia when one is standing upright or to a nail struck by a hammer); in that case both T and l' would be negative.

In the wire-stretching experiment already described, most materials experience a lateral contraction simultaneously with the longitudinal extension. For example, if the wire were circular, with undeformed radius a_0, then, after stretching, its radius may be $a_0 + a'$, where a' is usually negative (**Fig. 7.2**; we keep to the convention of describing a contraction in length as a negative extension). For most simple materials, the lateral strain, a'/a_0, is directly proportional to the longitudinal strain:

$$\frac{a'}{a_0} = -\sigma\frac{l'}{l_0},$$

(7.2).

where σ is another constant characteristic of the material called *Poisson's ratio* (after the nineteenth-century French natural philosopher). Poisson's ratio is a pure number (with no units) and is generally positive. If the volume of the circular cylindrical element of **Fig. 7.2** does not change during a small deformation, then σ must take the value $\frac{1}{2}$. This is because the new volume $\pi(a_0 + a')^2(l_0 + l')$ must equal the old volume $\pi a_0^2 l_0$, and as long as a'/a_0 and l'/l_0 are small, that requirement

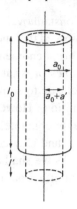

Fig. 7.2. Illustration of the reduction in a wire's radius $(-a')$ which accompanies an extension in its length (l'). They are related by Equation (7.2).

reduces to

$$\frac{a'}{a_0} = -\frac{1}{2}\frac{l'}{l_0},$$

which is Equation (7.2) with $\sigma = \frac{1}{2}$. In general, if a material is incompressible it has Poisson's ratio $\frac{1}{2}$. In no material can it have a value greater than $\frac{1}{2}$, since that would imply an increase in the volume of a cylinder after compressive stresses are applied to its ends. Almost all biological materials, in particular those comprising the walls of blood vessels, are effectively incompressible.

Equations (7.1) and (7.2) describe the response of a solid to a particular type of deforming stress, a longitudinal tension or compression, as long as it lies within the range in which the response is elastic. Like the equations governing the relationship between stress and rate of strain in a fluid, however, they can be simply generalized to cover small three-dimensional deformations.

In addition to longitudinal extension or compression, with the associated lateral strains (**Fig. 7.2**), a piece of material could be subjected to shearing stresses, which tend to cause layers of material to slide over each other (**Fig. 7.3**a). It could also be subjected to bending (**Fig. 7.3**b), or torsion (twisting) (**Fig. 7.3**c), or any combination of these. In each case, the strains would be directly proportional to the

(a)   (b)   (c)

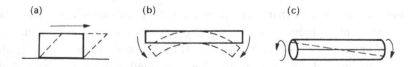

Fig. 7.3. (a) Shear stress. (b) Bending stress. (c) Torsional stress.

applied stresses, and the constants of proportionality could be expressed in terms of the Young's modulus E and Poisson's ratio σ of the material. For an elastic solid of the sort described here (i.e. a metal), E and σ completely determine the response.

The properties of blood vessel walls

We must now examine whether the materials comprising the blood vessel walls obey Hooke's law (stress proportional to strain), and, if not, how they depart from it. When classical elasticity theory is applied to a material, it is assumed that the material is effectively *homogeneous* and *isotropic*. That is, the values measured for E and σ do not vary within the material (homogeneous) and are independent of the direction in which the stress is applied (isotropic). This means that, in whatever way a small specimen of the material is cut, and from whatever region, the same values of E and σ are obtained in an experiment. Now biological tissues, in particular the walls of blood vessels, are neither homogeneous nor isotropic. They normally consist of a number of different materials, arranged in a complicated manner, which is described in detail in Chapters 12–15. We summarize here the properties of the most important materials.

About 70% of the walls of arteries and veins consists of water, which is not elastic except in its ability to withstand compression. However, the rest of the material consists of a mesh of fibres which do have elastic properties. There are three sorts of fibres which determine the elastic properties of vessel walls as a whole: these are elastin, collagen and smooth muscle. Elastin is a rubber-like substance, with a Young's modulus of approximately $3 \times 10^5 \, \mathrm{N\,m^{-2}}$; a graph of stress against strain for pure elastin fibres is given in **Fig. 7.4**a. Collagen is much stiffer than elastin, with a Young's modulus of approximately $10^8 \, \mathrm{N\,m^{-2}}$; a stress–strain graph for collagen is given in **Fig. 7.4**b. Although elastin is so compliant that it can be stretched to twice its unstretched length, the value of the stress at which it eventually breaks (its *tensile strength*) is less than 5% of that of collagen. Smooth muscle has a Young's modulus similar to that of elastin; however, the exact value measured in an experiment depends on the level of physiological activity. It varies from about $1 \times 10^5 \, \mathrm{N\,m^{-2}}$ when the muscle is completely relaxed, to about $2 \times 10^6 \, \mathrm{N\,m^{-2}}$ in the active state. It is therefore difficult to say what its Young's modulus is in vivo, since the degree of activity is not usually known. In large arteries, at least those of dogs, elastin and collagen together constitute about 50% of the dry mass. In the intrathoracic aorta the ratio of elastin to collagen is about 1.5, while in other arteries it is about 0.5; in veins it is as low as 0.3. Similar figures apply in humans, although the transition to an excess of collagen over elastin occurs more peripherally, in the lower abdominal aorta. The amount of smooth muscle in an artery also depends on the position of the artery: the more peripheral it is, the larger its proportion of smooth muscle, which is about 25%

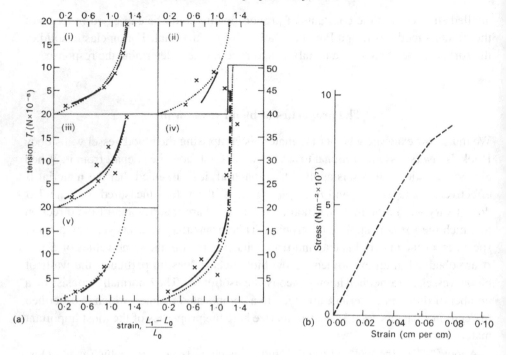

Fig. 7.4. (a) Tension–strain graphs of five fresh elastin fibres in Ringer–Locke solution at 37°C. The measured points (×) are compared for each fibre with the individual regression curve (solid black) and the common regression curve (stippled). Ordinate: tension per fibre. Abscissa: strain (From Carton, Dainauskas, and Clark (1962). Elastic properties of single elastic fibres. *J. Appl. Physiol.* **17**, 549.) (b) Graph of stress against strain for collagen fibres. (From Benedict, Walker and Harris (1968). Stress–strain characteristics and tensile strength of unembalmed human tendon. *J. Biomech.* **1**, 58.)

of the dry mass in the thoracic aorta, and reaches 60% of the dry mass in very small arteries and arterioles. The basic structure of veins is similar to that of arteries, except that they are both thinner and contain less elastin.

The overall elastic behaviour of a blood vessel is determined by its dimensions and by the elastic properties of its constituent materials. The most important of these are the internal diameter d and the wall thickness h, which of course are different for different vessels. Typical values of these quantities in a number of different vessels, measured in vivo in dogs (or, in the case of very small vessels, in other mammals), are given in **Table I** at the back of the book. The table also contains much more information that is pertinent to the various mechanical phenomena observed in the circulation, and this will be discussed in detail in subsequent chapters; we shall constantly refer to **Table I**. The quantities h and d both diminish towards the periphery. An interesting observation is that the ratio of wall thickness to diameter, h/d, does not vary very much throughout the large systemic arteries, having a value around 0.07. It increases in the more muscular arterioles, but is smaller again in the capillaries, and is much smaller in veins and pulmonary arteries than in the systemic arteries of comparable diameter.

An important constraint on the motion of blood vessel walls in response to the variable forces exerted on the inside of them by the blood is provided by the tissue through which they pass and to which they are attached. This has the effect of tethering the walls, and the mechanical properties of the tethering medium are significant (see Chapter 12). Capillaries, in particular, owe almost all their mechanical properties to the surrounding medium, because the wall itself consists of a single layer of endothelial cells.

Since the composition of blood vessel walls is not homogeneous, and since the material is arranged in such a way that it is most unlikely to be isotropic, classical elasticity cannot be applied directly and a single Young's modulus cannot be defined. However, it is still necessary to describe the response of the wall to different applied stresses, and a quantity like the Young's modulus would be very convenient for this purpose. What we do is to measure the deformation of the wall as a whole in response to known applied stresses, and to infer from those measurements the value which the Young's modulus would have if the material were homogeneous and isotropic. We lump together all the different constituent parts of the wall and obtain an *effective* Young's modulus.

Two sorts of experiments have commonly been performed on excised arteries and veins. In one, a given length of vessel is distended by different, known, transmural pressures (the length being held fixed) and the changes in diameter are measured. As we shall see later (p. 102), a distending pressure in a blood vessel requires the presence of a tensile stress in the wall, directed round the circumference, which may be called circumferential stress or hoop stress (see **Fig. 7.10**b). There will also be a longitudinal stress. The hoop stress is the stress from which, together with the change in diameter, E is inferred using Hooke's law (Equation (7.1)). In the other experiments, the piece of vessel is maintained at a given diameter and stretched longitudinally, as in the wire-stretching experiment. Here, the value of E is calculated from the longitudinal stress and strain. The results of such a pair of experiments on pieces of artery are shown in **Fig. 7.5**a and b.

An obvious feature of these graphs is that, unlike those for metals (**Fig. 7.1**b), they are not straight lines. So even when we lump all the components of the wall together, the linear relation (Equation (7.1)) is not appropriate, and we cannot uniquely define a Young's modulus. We can overcome this difficulty if we are interested only in small deformations about some equilibrium state. This need not be a completely unstretched state, because the tube may be already distended radially and stretched longitudinally. Let us suppose that there is an equilibrium configuration, in which the

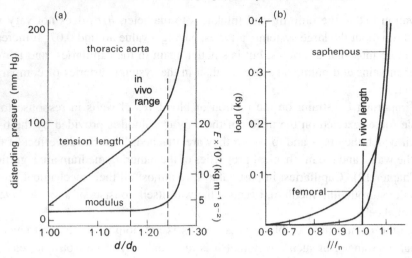

Fig. 7.5. (a) Graph of distending pressure against diameter for a piece of artery of fixed length; d_0 is the unstretched diameter. Note the nonlinearity of the graph, on which the normal in vivo range is marked. Also plotted is the effective (incremental) Young's modulus E relating circumferential stress to circumferential strain. This increases dramatically for distending pressures greater than normal. (From McDonald (1974). *Blood Flow in Arteries*, p. 260. Edward Arnold, London.) (b) Graph of longitudinal tension against length for pieces of two arteries of fixed diameter; l_0 is the in vivo length. Again note the nonlinearity of the graph. (From McDonald (1974). *Blood Flow in Arteries*, p. 261. Edward Arnold, London.)

stress (either longitudinal stress or hoop stress) is S_0 (**Fig. 7.6**) and the strain has a corresponding value e_0. Then, for small departures from this state, the stress–strain curve can be approximately represented by the straight line tangent to it at the point where stress equals S_0, and the small deformations can be analysed in terms of a Young's modulus given by the slope of this tangent. The stress–strain curve is then said to have been *linearized* (see Chapter 8, p. 124). For example, let us consider a longitudinal stretching experiment in which small deformations take place about an initially stressed state with tension, length and cross-sectional area of the specimen equal respectively to T_1, l_1 and A_1. Equation (7.1) can still be used to determine an effective Young's modulus E if in its place we write

$$\frac{T - T_1}{A_1} = \frac{E(l - l_1)}{l_1},$$

where $T - T_1$ is the small increase in the tension and $l - l_1$ is the small increase in the length. The value of E thus obtained is sometimes called the *incremental* Young's modulus. Its value clearly depends on T_1, l_1 and A_1, and it is always important to record the reference values at which E is measured.

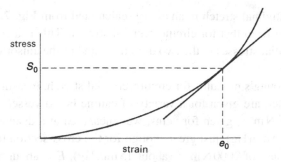

Fig. 7.6. Sketch showing the linearization of a nonlinear stress–strain curve. For small departures from the equilibrium state in which stress is S_0 and strain e_0, the curve can be approximately represented by its tangent at that point; its incremental Young's modulus is then the slope of that tangent.

Now all blood vessels in vivo are tethered to the surrounding tissue, and are subjected to a considerable longitudinal stretch all the time: when a segment of artery is excised and released, its length decreases by 30–40%. Furthermore, the mean blood pressure in all arteries is considerably greater than that just outside: the excess is approximately $1.3 \times 10^4 \, \mathrm{N \, m^{-2}}$ (100 mm Hg) at the level of the heart, increasing hydrostatically below that level and decreasing above. Thus, the arteries are naturally in a state of radial distension and of longitudinal stretch. The points representing the in vivo state are marked on **Fig. 7.5**a and b. It can be seen that the slope of the stress–strain curve, i.e. the incremental Young's modulus, is considerably greater in the in vivo state than in the unstretched state. Thus, in order to analyse small deformations about the in vivo state we must use the correct incremental Young's modulus (this is also plotted on **Fig. 7.5**a). We shall see in Chapter 12 that the variable pressure and shear stress exerted by the blood do cause fairly small deformations of the wall about the stretched, in vivo state; they can be related to the variable stresses by use of the appropriate incremental Young's modulus. From now on, whenever a value of E is quoted for blood vessel walls, it will be the incremental value for the in vivo state. It will also usually be the value relating changes in diameter to changes in pressure (and hence hoop stress) rather than the value for longitudinal stretch. This is because the tethering inhibits longitudinal wall motions more than radial ones, and the latter turn out to be the more important.[1] It should be noted, however, that the Young's

---

[1] E is calculated from the equation

$$E = \frac{\Delta p_i}{\Delta d_e} \cdot \frac{2 d_e d_i^2 (1 - \sigma^2)}{(d_e^2 - d_i^2)} \qquad (7.3a)$$

where Δp_i is the change in internal pressure of the vessel, Δd_e is the corresponding change in the external diameter d_e, d_i is the internal vessel diameter and σ is Poisson's ratio, known to have the value 0.5. If the vessel is thin-walled, its wall thickness $h = \frac{1}{2}(d_e - d_i)$ is small compared with d_i (which is simply denoted by d). In these modulus for longitudinal stretch in an artery (calculated from **Fig. 7.5b** for example) is usually different from that for circumferential stretch. This is a consequence both of the different initial stresses in the two directions and of the anisotropic structure of blood vessel walls.

Values of the Young's modulus for circumferential stretch, measured at the in vivo length and diameter, are given for a variety of canine blood vessels in **Table I**. The value of $0.7 \times 10^5 \, \text{N m}^{-2}$, given for veins, was measured at a distending pressure of $650 \, \text{N m}^{-2}$ (about $5 \, \text{mm Hg}$), just great enough for the cross-section to be circular. At a transmural pressure of $2000 \, \text{N m}^{-2}$ (about $15 \, \text{mm Hg}$), E is about $5.0 \times 10^5 \, \text{N m}^{-2}$, similar to the value for systemic arteries. Veins operate physiologically over a wide range of pressure, and E may vary by a factor of 100. When we wish to analyse large deformations, for example in a collapsing vein, we shall have to recognize this variation of E with cross-sectional area. Note that the value of E for systemic arteries increases peripherally; i.e. peripheral arteries are stiffer than proximal ones. The values of E given in **Table I** were measured on excised segments of blood vessel, so that no effects of smooth muscle tone are included in these measurements. Its presence in vivo would tend to increase the value of E for an artery or vein at a given diameter, although the value of E may be reduced where the contraction of smooth muscle causes a reduction in diameter, because the point of interest would move further down the stress–strain curve (**Fig. 7.5a** or **7.6**), and the slope of the curve would be less there.

The reason for the rapid increase in stiffness of blood vessel walls as their distension increases is thought to be associated with the different arrangement of collagen and elastin fibres within them. At low strains, most of the collagen fibres are slack, and not straight; all the stress is then borne by the elastin fibres. As the strain increases, the collagen fibres straighten out and increasingly take up the stress. Since these are much stiffer than the elastin fibres (**Fig. 7.4a and b**), the wall therefore becomes much stiffer. The situation is analogous to that of a balloon being blown up inside a string bag. It is relatively easy to start with, but extremely difficult once the bag is taut.

A further difficulty in applying elasticity theory to the walls of blood vessels lies in the fact that they exhibit significant visco-elastic behaviour (see p. 89). Steady

circumstances, Equation (7.3a) can be reduced approximately to

$$\frac{2\Delta d}{d\Delta p} = \frac{d(1 - \sigma^2)}{Eh}.$$

But $2\Delta d/d\Delta p$ is equal to $\Delta A/A\Delta p$, where A is the cross-sectional area of the tube, and this quantity is defined as the *distensibility* D of the tube (see Chapter 12, p. 273). In the case of a material for which σ^2 is negligible compared with unity, Equation (7.3a) is approximately equivalent to

$$D = \frac{1}{E(h/d)}. \tag{7.3b}$$

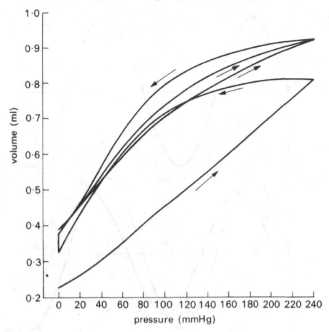

Fig. 7.7. Three consecutive pressure–volume curves of a femoral artery. The sequence of inflation and deflation is shown by the arrows (the last two deflations followed the same course). The vessel appears to become rather less distensible during the series of inflations. (From Bergel (1961). The static elastic properties of the arterial wall. *J. Physiol.* **156**, 451.)

deformations do not occur instantaneously when a steady stress is applied, but the immediate initial deformation is followed by a slower continuing deformation (creep) to an equilibrium value. Similarly, the stress required to maintain a given deformation falls gradually from its initial value to a new equilibrium level (stress relaxation). It is thought to be primarily the smooth muscle which is responsible for the visco-elastic properties of artery walls, although collagen, despite its great stiffness, also exhibits visco-elasticity in experiments in vitro. Elastin, on the other hand, is known to be purely elastic; it is this elastic component in blood vessel walls which prevents creep or stress relaxation from continuing indefinitely. Visco-elasticity is important only when the stress or strain varies with time; for example, if a segment of artery is subjected to a cyclically varying strain of given frequency and amplitude, the stress required for a given strain is greater during extension than during recoil (**Fig. 7.7**). This phenomenon is an example of *hysteresis*.

The properties of visco-elastic materials cannot usually be represented by a small number of constants in the way that linearly elastic materials can be represented by the Young's modulus and Poisson's ratio. Indeed, it is not usually possible to describe

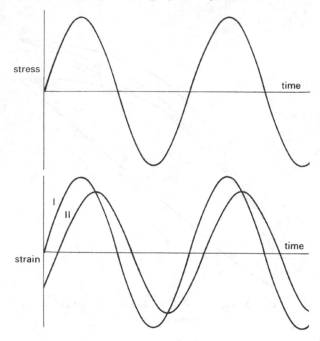

Fig. 7.8. When a cyclically varying stress is applied to a material, the resulting strain stays in phase if the material is purely elastic (curve I). However, if it has visco-elastic properties, the strain lags behind the stress, and varies with a smaller amplitude than if it were purely elastic (curve II).

in exact mathematical terms how a visco-elastic material will respond to unsteady applied stresses of different kinds. And when it is possible, the mathematics is often so complicated as to be of little help. Sometimes, however, a simple approximate description of the behaviour of the material can be made, and is very useful. This is true of blood vessels when the deformations are small and vary cyclically with time under the action of cyclically varying applied stresses (i.e. blood pressure and viscous shear stress, which vary with time during each cardiac cycle). The response of the strain to changes in stress is delayed slightly, so that as the applied stress varies cyclically, the variations in strain lag behind the stress, and have a smaller amplitude than they would have if the material were purely elastic (**Fig. 7.8**). (See Chapter 8, p. 109 for a precise definition of amplitude.) It is possible to represent this behaviour by means of two constants: a 'dynamic Young's modulus' E_{dyn} and an 'effective viscosity' η. As we shall see in Chapter 8, p. 126, the variations with time of any cyclically oscillating quantity (the stress in this case) can be represented as the sum of a number of purely sinusoidal oscillations (**Fig. 8.14**, p. 127). If a single such oscillation in the

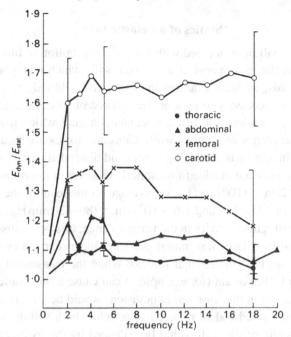

Fig. 7.9. The ratio of the average values of the dynamic Young's modulus and the static Young's modulus E_{dyn}/E_{stat} as a function of frequency in hertz measured in four canine arteries. (From McDonald (1974). *Blood Flow in Arteries*, p. 270. Edward Arnold, London.)

stress has angular frequency ω and amplitude S, then the corresponding oscillation in strain will have amplitude $S/\sqrt{E_{dyn}^2 + \eta^2\omega^2}$, and the phase angle by which it lags behind the stress will be approximately $\eta\omega/E_{dyn}$, measured in radians. Using these formulae, the values of E_{dyn} and η can be inferred when a known, cyclically varying stress is imposed on a piece of artery, and the resulting time-varying strain measured.

Some results of measurements of E_{dyn} are shown in **Fig. 7.9**. It is found that E_{dyn} is usually greater than the value of E measured in static experiments. For frequencies below about 2 Hz (i.e. for $\omega/2\pi$ less than $2\,s^{-1}$), E_{dyn} varies with frequency, approaching the static value as the frequency approaches zero. It is approximately constant for frequencies between 2 Hz and 20 Hz, and exceeds the static value by a ratio which varies from about 1.1 in the thoracic aorta of a dog to about 1.8 in the carotid artery. The viscous component $\eta\omega$ also varies with frequency, but for frequencies above 5 Hz the quantity $\eta\omega$ is approximately constant and takes a value between 0.1 and 0.2 times E_{dyn}. The value of $\eta\omega$ tends to be greater for smaller, more muscular arteries, which are therefore more visco-elastic.

Statics of an elastic tube

In Chapter 12 we shall be concerned with the pulsatile motion of blood in large vessels, together with the movements of the vessel walls which take place at the same time. In other words, we shall examine the *dynamics* of blood and of vessel walls. First, however, it is necessary to look at the balance of forces exerted on the walls of a tube which is completely at rest and contains a fluid which is also stationary; i.e. we look at the *statics* of a vessel wall. Consider, for example, an artery that is stretched longitudinally by the tethering forces and distended by an internal pressure which exceeds the pressure outside (atmospheric pressure in many cases) by approximately $1.3 \times 10^4 \, \mathrm{N\,m^{-2}}$ (100 mm Hg), on average, at the level of the heart; it would vary between about 1.0×10^4 and $1.6 \times 10^4 \, \mathrm{N\,m^{-2}}$ (80–120 mm Hg) during the cardiac cycle. We shall ignore gravity in the force balance; this is because when a blood vessel is distended by a large transmural pressure, the stress forces set up are very large compared with the gravitational forces. When the transmural pressure is not large, gravity can be important (for example, it can cause a thin-walled flexible tube in air to collapse), and in that case our conclusions would be strictly accurate only if the vessel were both filled and surrounded by a fluid which had the same density as itself; then the weight of the wall would be balanced by the hydrostatic pressure in the fluid. In the circulation there is an additional reason why gravity cannot always be neglected, since the pressure inside blood vessels varies hydrostatically (quite apart from the pressure variations associated with blood flow). For example, vessels in the foot will have an internal pressure which, in a human standing upright, can be greater than that at the level of the heart by about $1.3 \times 10^4 \, \mathrm{N\,m^{-2}}$ (100 mm Hg). In the head, the blood pressure is correspondingly smaller. The pressure outside blood vessels is approximately uniform, and close to atmospheric pressure except in the thorax where it can vary from about $3 \times 10^3 \, \mathrm{N\,m^{-2}}$ (23 mm Hg or 30 cm H$_2$O) below atmospheric pressure to about the same amount above it during heavy breathing. Together with the fact that the transmural pressure varies between blood vessels of different types at the same level (being greater in arteries than in veins), this means that there is a wide variation of transmural pressure in the cardiovascular system. As we shall see, this proves to be important in determining whether or not certain blood vessels may be close to collapse.

Let us now examine the static balance of forces on a small element of the wall of a tube, neglecting gravity except in so far as it determines the pressure within the tube. We call the internal and external pressures p_i and p_e respectively, and suppose that the tube has a circular cross-section with internal and external radii r_i and r_e (see **Fig. 7.10**a). The wall thickness h is equal to $r_e - r_i$. We consider the equilibrium of an element of axial length Δl, subtending a small angle θ at the axis (**Fig. 7.10**b). The net forces on this element result from perpendicular (normal) stresses on its six faces; no tangential stresses can be present because of the symmetry of the arrangement.

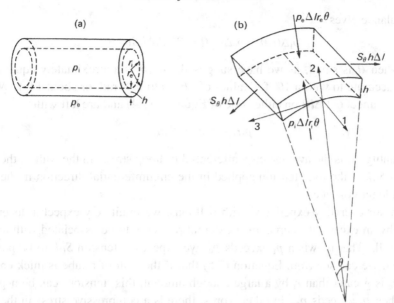

Fig. 7.10. (a) Sketch of a thick-walled tube, with internal and external radii r_i and r_e respectively; wall thickness $h = r_e - r_i$; internal and external pressures p_i and p_e. (b) A segment of the wall of a thick-walled tube, of length Δl, thickness h and subtending an angle θ at the axis of the tube. The directions of the coordinate axes are parallel to the tube axis (1), radially outwards (2), and tangential to a cross-section of the tube (3). The forces on the element are marked with heavy arrows; those on the curved faces come from the internal and external pressures; those on the flat faces parallel to the axis of the tube come from the 'hoop' stress S_θ. The tensile forces on the flat ends balance out and are omitted.

The net components of force acting on the element in three mutually perpendicular directions must be zero; we choose these directions to be parallel to the axis (1), radially out from the axis (2), and tangential to the tube in a plane perpendicular to the axis (3). The force balance in the axial direction (1) shows that the longitudinal tension T on each end must be the same (equal to the area of an end times the average normal stress over that end). Similarly, the force balance in the tangential direction (3) shows that the normal forces on the two flat faces of the element parallel to the axis must also be the same (equal to the area of a face, $h\Delta l$, times the average normal stress S_θ over that face). However, because these faces are inclined to each other, and hence to the radial axis (2), they also have a component radially inwards, equal to $2 \times S_\theta h\Delta l \times \sin\frac{1}{2}\theta$. The other contributions to the radial component of force come from the internal and external pressures; the outwards force from the internal pressure is p_i times the area of the inner curved face ($\Delta l r_i \theta$), and the inwards force from the external pressure is p_e times the area of the outer curved face ($\Delta l r_e \theta$). Thus, the radial force balance gives

$$p_i \Delta l r_i \theta - p_e \Delta l r_e \theta = 2 S_\theta h \Delta l \sin \tfrac{1}{2}\theta. \tag{7.4}$$

Now, when θ is small, as we have supposed, $\sin \tfrac{1}{2}\theta$ is approximately equal to $\tfrac{1}{2}\theta$ (that is accurate to within 1% for values of $\tfrac{1}{2}\theta$ up to 0.24 rad, or about 14°). We can therefore cancel out $\Delta l \theta$ on either side of Equation (7.4) and are left with

$$p_i r_i - p_e r_e = S_\theta h. \tag{7.5}$$

The quantity S_θ is the average circumferential or hoop stress in the wall of the tube, whereas $S_\theta h$ is the total tension applied in the circumferential direction in the wall, per unit length of tube.

From our everyday experience with balloons, we intuitively expect a distension, caused by an excess of internal over external pressure, to be associated with tension in the wall. That is, when p_i exceeds p_e, we expect the tension $S_\theta h$ to be positive. However, we can see from Equation (7.5) that if the wall of a tube is thick enough, i.e. if r_e is greater than r_i by a large enough amount, this 'tension' can be negative, even when p_i exceeds p_e. In other words, there is a compressive stress in the wall. This conclusion is quite independent of the elastic properties of the wall material; they become important in estimating the effect which the compressive stress has on the shape of the tube cross-section, as we shall see on p. 104. First, though, let us calculate whether the hoop stress in blood vessels is a tension or a compression.

Consider, for example, an artery within which the pressure exceeds atmospheric by $1.3 \times 10^4 \, \mathrm{N\,m^{-2}}$ (about 100 mm Hg), and outside which the pressure is atmospheric ($10^5 \, \mathrm{N\,m^{-2}}$). If it is like the dog's femoral artery, for example, whose properties are given in **Table I** at the back of the book, it will have an internal radius of 0.002 m and wall thickness 0.0004 m. Then, from Equation (7.5), we calculate the wall tension $S_\theta h$ to be negative (i.e. a compression); its value is about $-14.0 \, \mathrm{kg\,s^{-2}}$. If the internal pressure were greater, by only 50 mm Hg (as in the foot, perhaps) the wall stress would be a tension; if it were smaller, the compression would be more marked. In fact, if one calculates the wall stress in a great variety of blood vessels by this means, one concludes that the vast majority of them have a compressive hoop stress in their walls. The exceptions are the biggest vessels (aorta and vena cava), and other large vessels below the level of the heart, where the internal pressure is relatively high. The largest compressive stresses occur in small arteries and arterioles, whose walls are very thick, so that r_e exceeds r_i by a large amount.

The fact that the underlying hoop stress may be compressive is not necessarily very important for the mechanical behaviour of blood vessels in vivo. This is because in the circulation we normally wish to analyse small departures from some equilibrium state. As we have explained earlier, it does not matter if there is some initial stress present in that equilibrium state or not, as long as the elastic properties of the material are represented by the *effective* Young's modulus, measured for small departures from the same equilibrium state. For this reason, it is common to quote values of tension and pressure *relative* to a given equilibrium state, and to ignore the absolute values. In physiology, we usually quote pressures relative to atmospheric. The corresponding tensions are quoted relative to their value when the pressures both inside and outside the tube are atmospheric; i.e. when p_e and p_i are both equal to p_{atm}. When that is the case, the hoop stress, say S_0, is given by Equation (7.5) to be

$$S_0 = \frac{p_{atm}(r_i - r_e)}{h} = -p_{atm}.$$

That is to say, all normal stresses are compressive and equal to atmospheric pressure. If, during a departure from this situation, p_e remains atmospheric, if p_i exceeds p_{atm} by an amount p', and if S_0 exceeds $(-p_{atm})$ by an amount S', then Equation (7.5) can be used to give

$$S'h = p'r_i \quad \text{or} \quad p' = \frac{S'h}{r_i}. \tag{7.6}$$

This is the well-known 'law of Laplace'. It shows that the pressure required to distend a tube against a given tension in the wall is inversely proportional to the radius of the tube. A balloon, for example, is more difficult to blow up when its radius is very small than when it is somewhat larger. In the circulation, the law indicates that, relative to atmospheric pressure, the tension required to balance a certain distending pressure decreases as the radius of the tube decreases. This is sometimes used to explain why the walls of capillaries can be so thin: they have only a small tension to support. However, that argument ignores the dominant tethering effect of the surrounding tissue, and may not be relevant.

The law of Laplace, then, can be used to analyse small departures from the equilibrium state in which all pressures are atmospheric. It cannot be used to give an accurate description of the absolute stress level in a vessel wall, unless the distending pressure is so large that the stress S_0 (Equation (7.5)) is large compared with atmospheric pressure. This is usually not the case. If we come to analyse when vessels are likely to collapse, it is important to know the absolute stress level. By collapse of a vessel, we mean a drastic decrease in its cross-sectional area, which occurs spontaneously in response to a small change in applied stresses. It would also usually involve a change in cross-sectional shape, which may go from the circular to the elliptical or to more complicated shapes (**Fig. 7.11**). A tube will clearly collapse if the pressure outside it is very much greater than the pressure inside. The reason for the collapse is associated with the large compressive circumferential stress which, from Equation (7.5), must accompany such pressures. The phenomenon is similar to that of a straight rod which buckles (i.e. bends or crumples) when sufficiently large compressive forces are applied to the ends. The straight configuration is still theoretically possible, but it is *unstable*, in that, if it is very slightly disturbed from the equilibrium position, the departure from equilibrium grows rapidly with time. The buckled condition is *stable*.

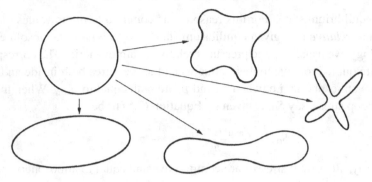

Fig. 7.11. Some of the shapes into which the cross-section of a circular tube might collapse.

In an analogous phenomenon, a pencil can theoretically stand on its point, but it is unstable, since any small departure from that configuration grows rapidly, and it falls over. (When it is lying on its side it is stable.) Similarly, the circular configuration of a tube wall is unstable if there is a large enough compressive stress within it. The wall then buckles, to form a distorted shape (**Fig. 7.11**), which is stable; the tube has collapsed.

The magnitude of the compressive stresses required for collapse and the shape into which a tube collapses depend on all the elastic properties of the wall. In particular, it is the degree to which the wall material can resist severe bending which governs the collapse, and this depends on the Young's modulus, Poisson's ratio and wall thickness. The degree to which the tube is stretched longitudinally and tethered also influences its collapse. A very floppy tube with fairly thin walls under very little longitudinal tension collapses very easily, i.e. with a very small compressive stress in its walls. A blood vessel in vivo, however, is stretched and tethered and has fairly thick walls; collapse will be resisted. It is nevertheless observed that some veins and pulmonary arteries do normally have elliptic cross-sections, which indicate a degree of collapse associated with the compressive stress in their walls. Other situations where blood vessels collapse in vivo are considered in Chapter 14.

8

Oscillations and waves

Simple harmonic motion

When blood is ejected from the heart during systole, the pressure in the aorta and other large arteries rises, and then during diastole it falls again. The pressure rise is associated with outward motions of the walls, and they subsequently return because they are elastic. This process occurs during every cardiac cycle, and it can be seen that elements of the vessel walls oscillate cyclically, with a frequency of oscillation equal to that of the heartbeat. The blood, too, flows in a pulsatile manner, in response to the pulsatile pressure. In fact, as we shall see in Chapter 12, a pressure wave is propagated down the arterial tree. It is therefore appropriate in this chapter to consider the mechanics of pulsatile phenomena in general, and the propagation of waves in particular.

Let us examine first the oscillatory motion of a single particle. Suppose that the particle can be in equilibrium at a certain point, say P, but when it is disturbed from this position, it experiences a restoring force, tending to return it to P. There are many examples of this situation, as when a particle is hanging from a string and is displaced sideways (a simple pendulum) or when the string is elastic and the particle is pulled down below its equilibrium position. In cases like these, the restoring force increases as the distance by which the particle is displaced from P increases. In fact, for sufficiently small displacements, the restoring force is approximately proportional to the distance from P (see p. 124). If the particle is displaced and then released, it will return towards P, but will overshoot because of its inertia. The displacement will then increase in the opposite direction, so that the restoring force increases, slowing the particle down until it stops and then comes back towards P. Again it overshoots, and so on. The particle in fact oscillates about P.

The mathematical analysis of this oscillation is very simple, and we give it here in full because we believe that it helps in gaining a complete understanding of the mechanics of oscillatory motion. Suppose the particle has mass m and oscillates backwards and forwards in a straight line about its equilibrium position P (**Fig. 8.1**). If

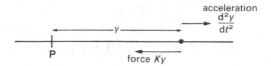

Fig. 8.1. A particle moving along a straight line under the action of a restoring force equal to K times its displacement y from a position of equilibrium P. Its acceleration from P is d^2y/dt^2. The equation governing its motion is (8.1).

y is the distance of the particle from P, then the magnitude of the restoring force, directed towards P, is proportional to y, equal to Ky, say, where K is a positive constant, measured in appropriate units. If we adhere to the convention that y should be considered positive for displacements to the right of P and negative for displacements to the left, then the force on the particle in the *positive y*-direction is $-Ky$. Newton's law of motion states that this force is equal to the mass of the particle times its acceleration in the positive y-direction (this acceleration is equal to d^2y/dt^2; see Equation (2.5), p. 12). That is to say

$$m\frac{d^2y}{dt^2} = -Ky,\tag{8.1}$$

which can be rewritten

$$\frac{d^2y}{dt^2} + \omega^2 y = 0,\tag{8.2a}$$

where $\omega^2 = K/m$. Equation (8.2a) is thus applicable to the oscillations of any particle about a position of equilibrium as long as the oscillations are small enough for the restoring force to be proportional to y. It represents a balance between the restoring force on the particle and its inertia; these two properties are essential for the presence of any oscillatory motion.

Perhaps the most common example is that of the simple pendulum, in which a particle (mass m) hangs almost vertically on a string of length l (**Fig. 8.2**). If we suppose the string to be inextensible, and if we neglect all air resistance and other frictional forces, then the only forces on the particle are its weight, equal to mg and directed vertically downwards, and a force F supplied by the string (the tension in the string), which is directed along the string. Since we do not know the value of F a priori, it is most convenient to consider the balance of forces in a direction at right angles to the string, because F has no component in this direction. If we let θ be the angle which the string makes with the vertical (measured in radians), the only force acting in the given direction is a component of the weight of the particle, equal to $mg\sin\theta$, and directed along the arc towards the equilibrium position (marked P on **Fig. 8.2**). The component of acceleration in the required direction is obtained by

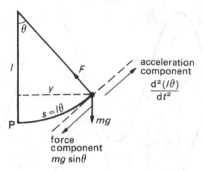

Fig. 8.2. A simple pendulum. The particle hangs on an inextensible string of length l and oscillates in a vertical plane. The forces on the particle are its weight mg, vertically downwards, and the tension in the string F. The most convenient direction for considering the force balance is perpendicular to the string; the components of force and acceleration in this direction are marked on the figure and discussed in the text.

noticing that the particle is actually travelling along the circumference of a vertical circle, of radius l. The component of acceleration tangential to this circle is d^2s/dt^2, where s is the distance travelled along the circumference, measured from P. Thus s is equal to $l\theta$, and the tangential component of acceleration is $d^2(l\theta)/dt^2$ in the direction of increasing θ (see **Fig. 8.2**). The equation 'force equals mass times acceleration' therefore gives

$$mg\sin\theta = -m\frac{d^2(l\theta)}{dt^2},$$

where the minus sign occurs because we want the component of acceleration in the direction of the force, not the opposite direction. This equation may be rewritten

$$\frac{d^2(l\theta)}{dt^2} + g\sin\theta = 0. \tag{8.2b}$$

We notice that if y is the horizontal displacement of the particle, then

$$\sin\theta = y/L.$$

Let us now suppose that θ remains very small all the time; this means that the particle effectively oscillates along a horizontal straight line. It also means that $\sin\theta$ is approximately equal to θ (see Chapter 7, p. 102), so that $l\theta$ is approximately equal to y. In this case Equation (8.2b) reduces to Equation (8.2a), with the quantity ω^2 equal to g/l.

In whatever manner the particle is released, its subsequent motion must be such that y varies with time in a manner consistent with Equation (8.2a) for this particular ω. (In mathematical terms, the function $y(t)$ must be a solution of the differential Equation (8.2a).) However, there are an infinite number of motions consistent with this

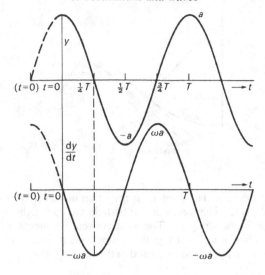

Fig. 8.3. Graphs of y and dy/dt against time when they are given by Equations (8.5) and (8.6). The amplitude of the oscillation in displacement is a and that of the velocity is ωa. The period of the oscillation is T $(= 2\pi/\omega)$. These graphs represent simple harmonic motion. The dotted extensions to the left, with a new origin, represent the graphs of y and dy/dt when they are given by Equations (8.7) and (8.8), with $V = a\omega$.

equation; the particular one appropriate in a given situation is completely determined by the displacement y and velocity dy/dt at the time of release (we could, for example, release the particle from rest at a given displacement, or we could impel it away from P with a given velocity). The most general motion consistent with Equation (8.2a) (its most general solution) is given by

$$y = A \cos \omega t + B \sin \omega t, \tag{8.3}$$

where A and B are any constants. The velocity of the particle, dy/dt, corresponding to this motion is obtained by differentiating Equation (8.3) with respect to time, is

$$\frac{dy}{dt} = -\omega A \sin \omega t + \omega B \cos \omega t. \tag{8.4}$$

The values of the constants A and B may be determined from the initial values of y and dy/dt. As a first example let us consider a simple pendulum, whose bob (the particle) is pulled out a horizontal distance a from its equilibrium position and released from rest there at the initial instant which we will call $t = 0$. That is, at time $t = 0$,

$$y = a, \qquad \frac{dy}{dt} = 0.$$

Now $\sin(0) = 0$ and $\cos(0) = 1$, so the second of these conditions, substituted into Equation (8.4), shows that B must be zero. The first condition, substituted into

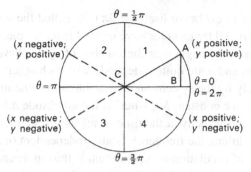

Fig. 8.4. Diagram to illustrate the periodic nature of $\sin\theta$ and $\cos\theta$, as defined trigonometrically. In quadrant 1, both are positive; in 2, $\sin\theta$ is positive and $\cos\theta$ is negative; in 3, both are negative; in 4, $\sin\theta$ is negative and $\cos\theta$ positive. When θ is increased by 2π (360°), $\sin\theta$ and $\cos\theta$ are unchanged.

Equation (8.3), then gives $A = a$. Thus the solution of the problem, the complete motion of the particle for all time, is given by

$$y = a\cos\omega t \tag{8.5}$$

with velocity, again by differentiation,

$$\frac{dy}{dt} = -\omega a\sin\omega t. \tag{8.6}$$

The graphs of these two quantities against time are shown on **Fig. 8.3**. It can be seen that y oscillates cyclically between a and $-a$, and dy/dt oscillates cyclically between $+\omega a$ and $-\omega a$, and passes through zero when y is equal to either $+a$ or $-a$. The quantity a is called the *amplitude* of the oscillation in position; ωa is the velocity amplitude. The *period* of the oscillation, that is the time of one complete cycle, after which the motion reproduces itself exactly (marked T in **Fig. 8.3**), is equal to $2\pi/\omega$.[1] The *frequency* f of the oscillation, i.e. the number of cycles or periods

---

[1] This is a consequence of the trigonometrical definitions of sine and cosine. Let us consider the quantities $\sin\theta$ and $\cos\theta$ as θ takes values which increase from zero, with reference to **Fig 8.4**. In this figure is drawn a circle of unit radius centred on C, the origin of coordinates. The angle θ is represented by the point A, so that the line CA makes an angle θ with the x-axis. The coordinates of A are (x, y). Then $\sin\theta$ is defined in the right-angled triangle ABC as being the length of the opposite side AB divided by that of the hypotenuse CA. That is, $\sin\theta = y/1 = y$. Similarly, $\cos\theta$ is defined as being the length of the adjacent side CB divided by that of the hypotenuse: $\cos\theta = x/1 = x$. As x increases, $\sin\theta$ and $\cos\theta$ change accordingly, until when θ exceeds the value $\pi/2$ radians (90°) the point A moves into the second quadrant, so that x is negative although y is still positive. Thus, $\cos\theta$ is negative and $\sin\theta$ is positive for values of θ between $\pi/2$ and π (180°). In the third quadrant, between π and $3\pi/2$ (270°), x and y are both negative and therefore so are $\cos\theta$ and $\sin\theta$. In the fourth quadrant (θ between $3\pi/2$ and 2π (360°)), x is positive again, so that $\cos\theta$ is positive although $\sin\theta$ is still negative. If θ is increased beyond 2π, the point A moves into the first quadrant once more, and everything starts again: $\cos\theta$ and $\sin\theta$ are both positive. We see that increasing the value of θ by 2π, i.e. going around the origin once to get back to where we started, does not alter the value of $\cos\theta$ or $\sin\theta$. The 'period' of $\sin\theta$ or $\cos\theta$ is 2π. This explains why the distance between successive peaks of the graph of $a\cos\omega t$ (**Fig. 8.3**) is $T = 2\pi/\omega$.

per unit time, is equal to $\omega/2\pi$; ω itself $(= 2\pi f)$ is called the *angular frequency* or *radian frequency*. (NB. All frequencies have units of inverse time, s^{-1}; this is because neither radians, the units of angle, nor the number of cycles have any dimension at all.) The quantities ω and a are quite independent of each other; the frequency ω is determined completely by the system under consideration (the mass of the particles m and the restoring force constant K), whereas the amplitude a is determined solely by the displacement and velocity at the time of release. It is interesting to note that, in the case of the pendulum, the frequency is also independent of m, since ω is equal to $\sqrt{g/l}$. The period of oscillation of a pendulum is thus independent of the mass of the bob; but for this, the operation and adjustment of a pendulum clock would not be as simple as it is.

The solution corresponding to Equations (8.5) and (8.6) in the case when the bob of the pendulum starts off at its equilibrium position, but is given an initial velocity V, can also be written down simply. In that case, at time $t = 0$,

$$y = 0, \qquad \frac{dy}{dt} = V.$$

The first of these conditions, substituted in Equation (8.3), shows that A must be zero; the second, in Equation (8.4), shows that ωB must be equal to V. The complete motion of the particle is then given by

$$y = \frac{V}{\omega} \sin \omega t \qquad (8.7)$$

and

$$\frac{dy}{dt} = V \cos \omega t. \qquad (8.8)$$

In this case, the amplitude of the oscillation in displacement is V/ω; that in velocity is V. The angular frequency is of course still ω. The graphs of these equations can be obtained from **Fig. 8.3** by putting the origin of time a quarter period farther back (see the broken lines on that figure) and by replacing a everywhere by V/ω.

Any motion described by Equation (8.2a), of which those given by Equations (8.5) and (8.7) are particular examples, is called *simple harmonic motion*. Equation (8.3) represents the most general simple harmonic motion with the given angular frequency ω; this can in fact be manipulated into the form

$$y = A' \cos(\omega t - \phi), \qquad (8.9a)$$

where $A' = \sqrt{A^2 + B^2}$ and ϕ is an angle such that $\tan \phi = B/A$. The amplitude of this oscillation is A'; the quantity ϕ is called its *phase*, and the oscillation is said to have a *phase lag* of ϕ behind an oscillation proportional to $\cos \omega t$ alone (Equation

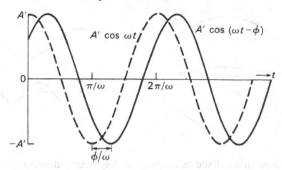

Fig. 8.5. Graphs of Equations (8.9a) (solid curve) and (8.9b) (broken curve). The solid curve has been shifted bodily to the right, by an amount corresponding to the time ϕ/ω; here, ϕ is the phase lead of (8.9a) over (8.9b).

(8.5)). The graph of Equation (8.9a) is given in **Fig. 8.5**, and compared there with that of

$$y = A' \cos \omega t. \qquad (8.9b)$$

The graph has been shifted bodily to the right, as if the quantity ωt had everywhere been diminished by an amount equal to ϕ. In Equation (8.9b) everything happens at a time ϕ/ω earlier than in Equation (8.9a), and the meaning of the phrase 'phase lag' becomes clear. If ϕ is in fact negative, it is said to represent a *phase lead*. The quantity $(-\sin \omega t)$ in Equation (8.6) can be written $\cos(\omega t + \pi/2)$, so that the velocity of this oscillation has a phase lead of $\pi/2$ over the displacement. It can be seen from **Fig. 8.3** that the graph of velocity reaches a given point in the oscillation (e.g. the minimum) at a time $T/2$ before the displacement graph, i.e. at a value of ωt in advance by $\pi/2$.

From **Fig. 8.3** we can retrieve all the details of simple harmonic motion: when the displacement y is a maximum, the velocity, dy/dt, is zero; the restoring force, however, is a maximum, and therefore so is the acceleration, directed back towards P **(Fig. 8.2)**. As the particle returns towards P, y diminishes while the speed increases, until when it passes through P, where the displacement and acceleration are zero, the speed is a maximum. Beyond P, the speed diminishes again while the displacement and acceleration increase to their maximum values on the other side. The process continues indefinitely in the absence of friction.

The simple harmonic motion described above is a motion with one *degree of freedom*; it is completely specified by the determination of how one variable (the coordinate y of the particle) varies with time. A system with two particles on a string, one above the other, both of which can oscillate in the same direction (a double pendulum), or a system of one particle free to oscillate in a plane (like a pendulum whose string is elastic) has two degrees of freedom, since two coordinates are required to

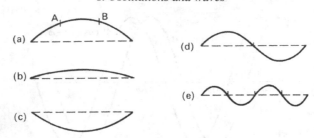

Fig. 8.6. If a stretched string, fixed at its ends, is displaced sideways as in (a), it will subsequently oscillate backwards and forwards, occupying a range of positions of which two are shown in (b) and (c). Such motion is a *standing wave*; the oscillation shown is the *fundamental*, with angular frequency $\omega_0 = (\pi/l)\sqrt{F/M}$. In (d) and (e) are depicted the forms of two of the higher harmonics, with one and three intermediate nodes respectively, and with frequencies $2\omega_0$ and $4\omega_0$ respectively. A general oscillation consists of the superposition of many harmonics.

specify the motion. More general systems of particles have larger, but finite, numbers of degrees of freedom. On the other hand, in any continuous system, like a violin string, the water in a tank or the walls of an artery, the coordinates of an infinite number of elements (e.g. fluid elements) must be specified for a complete description of the motion. Such a system, therefore, has an infinite number of degrees of freedom. The situation is not as complicated as it sounds, however, since all the elements are joined together, and in fact wave motions in continuous systems can be described quite simply.

Simple waves

As a simple example, consider a stretched elastic string, fixed at each end, and suppose that at some initial instant it is plucked or struck or given a small sideways displacement in some other way, and then released (**Fig. 8.6**a). Again, we neglect all friction and suppose that there is no energy dissipation by any means. The tension in the string provides a restoring force which tends to restore every displaced element to its equilibrium position (element AB, for example, clearly has a net downward force on it). Each element has mass, i.e. inertia, and therefore keeps moving until it has passed its equilibrium position; it is then slowed down by the restoring force, and subsequently executes simple harmonic motion as described above. At various times the string may look like **Fig. 8.6**b or c. This motion, in which the string is fixed at the ends and every particle executes simple harmonic motion, is called a *standing wave*. The frequency of the oscillations can be estimated by dimensional analysis. The only quantities it can depend on are: the restoring force, provided by the tension F in the string, with units of force ($\mathrm{kg\,m\,s^{-2}}$); the inertia, supplied by the mass per unit length of the string M (units $\mathrm{kg\,m^{-1}}$); and the geometry, i.e. the length

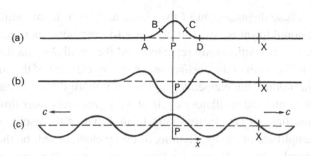

Fig. 8.7. Propagation of a travelling wave on an infinitely long stretched string. The point P (at $x = 0$) is forced to oscillate from side to side in simple harmonic motion. The initial displacement depicted in (a) causes the element BC to exert forces on elements AB and CD, tending to displace them. The element BC is returned towards equilibrium and overshoots. A little while later the string looks like (b). Later still it looks like (c), the disturbance propagating outwards with speed c.

l of the string (units m). The only way to get a frequency (units s^{-1}) from these three quantities is in the form $\sqrt{F/M}/l$, so the frequency must be proportional to this quantity. In fact, there are an infinite number of possible oscillation frequencies of a finite string fixed at its ends, which take the values $\omega = \sqrt{F/M}/l$, $2\sqrt{F/M}/l$, $3\sqrt{F/M}/l$, etc., so although there are an infinite number of frequencies, they are separate from each other. They are called the *natural frequencies* of the string, in that however the string is disturbed initially, its subsequent motion will consist of the superposition of a number of oscillations each having one of the natural frequencies (see p. 126). The *fundamental* oscillation is that with frequency $\sqrt{F/M}/l$; if the string moved with that frequency alone, it would oscillate backwards and forwards in the manner suggested by **Fig. 8.6**a–c. An oscillation with a higher natural frequency is said to be a higher *harmonic*; the first harmonic, with frequency $2\sqrt{F/M}/l$, oscillates in such a way that the midpoint of the string remains fixed (**Fig. 8.6**d), at what is called a *node*. Subsequent harmonics have more fixed points or nodes (**Fig. 8.6**e). An arbitrary oscillation, being a superposition of the fundamental and all its harmonics, will not generally have a node.

Now consider a string which is very long, and one element of it, say the element BC in **Fig. 8.7**a, is continuously made to perform simple harmonic motion with angular frequency ω and some small amplitude a. Immediately after the start of the motion, the string will be deformed as shown in **Fig. 8.7**a. Because of the tension in the string, the element BC exerts a net upwards force on each of the neighbouring elements AB and CD, which begin to move upwards. They will still be moving upwards as the element BC begins to return towards its equilibrium position, and after a little while the situation will look something like **Fig. 8.7**b. The elements AB and

CD then in turn cause displacement of the elements next to them, while themselves tending to be restored towards, and subsequently to overshoot, their equilibrium positions. This process is continuously repeated, and the oscillatory motion propagates along the string in the form of a *travelling wave*. Every element of the string executes simple harmonic motion, but elements progressively more distant from the origin P of the disturbance enter the oscillatory cycle at a progressively later time; that is, the *phase* of their oscillations lags increasingly behind that of the oscillation at P. The frequency and amplitude of the oscillations of every element will be the driving frequency ω and amplitude a. We can use dimensional analysis again to estimate the speed with which the wave is propagated in each direction along the string. The units of the wave speed (which we may call c) are $\mathrm{m\,s^{-1}}$, and the only quantities on which it can depend are F, M and ω. There is only one way in which this can be done, and we see that c must be proportional to $\sqrt{F/M}$ (more precise calculations show that c is actually equal to $\sqrt{F/M}$, which is independent of the frequency ω.

This is not the place to go through the mathematical derivation of wave speed, etc., which is quite complicated, but we shall frequently wish to use the mathematical form of a travelling wave, and therefore set it down here. Consider a point X situated at a distance x to the right of P (**Fig. 8.7**), so that x is positive, and the wave will propagate past the point from left to right. At first there is no disturbance at X; the disturbance reaches X after a time ct, and, after the 'wavefront' has passed, the element of the string at X performs simple harmonic motion, with frequency ω and amplitude a. Then the transverse displacement of the element at X, at time t (we shall call the displacement y), is given by[2]

$$y = a\cos[\omega(t - x/c)]. \tag{8.10}$$

This equation represents a pattern of displacement which is travelling to the right with speed c. To see this, consider the graph of y against x at two successive times, t_1 and t_2 (**Fig. 8.8**). At time t_2, the displacement at position $x = x_1 + c(t_2 - t_1)$ is the same, for any x_1, as that at time t_1 and position x_1; this is because the quantity $(t - x/c)$ has the same value, $(t_1 - x_1/c)$, in each case. That is, the whole pattern has shifted a distance $c(t_2 - t_1)$ to the right during the time interval $t_2 - t_1$, and hence is travelling with speed c.

The *wavelength* λ of the wave represented by Equation (8.8) or **Fig. 8.8** is the spatial distance occupied by one complete oscillation. It is the distance travelled by the wave during one period of its oscillation. The period is $2\pi/\omega$, so the wavelength,

---

[2] Equation (8.10) implies that y is zero, and dy/dt is equal to $-\omega a$, at the origin (P) at time $t = 0$. If this were not the case, either an adjustment would have to be made to the time we called $t = 0$, so that Equation (8.10) would be true, or a constant phase angle ϕ would have to be inserted in the brackets:

$$y = a\cos[\omega(t - x/c) - \phi].$$

Then $y = a\cos\phi$ at $x = 0$ and $t = 0$.

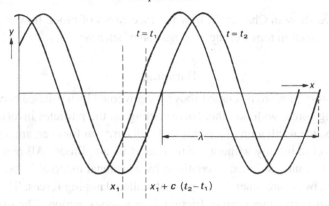

Fig. 8.8. Graphs of y against x, as given by Equation (8.10), and depicted at two successive times t_1 and t_2. The curve at time t_2 has been shifted bodily to the right, by a distance $c(t_2 - t_1)$; to see this, notice that the value given by Equation (8.10) at position x_1 and time t_1 is the same as that given at position $x_1 + c(t_2 - t_1)$ and time t_2. Thus the wave propagates with speed c; the wavelength is λ.

equal to the wave speed times the period, is

$$\lambda = 2\pi c/\omega. \tag{8.11}$$

As with simple harmonic motion of a single particle, the occurrence of waves requires the presence of both restoring force and inertia. In the above example, the restoring force is provided by the tension in the string, and the inertia by its mass per unit length. In the case of waves on the surface of a liquid (except for very short waves, where surface tension is important), the restoring force is gravity, and the inertia is provided by the density of the liquid.[3] Sound waves in a fluid are a balance between the density of the fluid and its compressibility, which causes any small local compression to spring back towards equilibrium. Pressure waves are also propagated along the elastic walls of arteries, where the restoring force comes from the elasticity of the walls and the inertia is that of the blood within (and to some extent that of the

---

[3] The speed of propagation c of surface waves on a deep, infinite ocean when surface tension is unimportant can depend only on the acceleration of gravity g, the density of the water ρ and the wavelength of the waves λ. By dimensional analysis, therefore, c must be proportional to $\sqrt{g\lambda}$. Similarly, if surface tension dominates and if gravity is unimportant, c can depend only on ρ, λ and σ, the surface tension of the air–water interface, which has dimensions $[\mathrm{MT}^{-2}]$. In this case dimensional analysis shows that c must be proportional to $\sqrt{\sigma/\rho\lambda}$. Both gravity and surface tension will be important if these two estimates of the wave speed are of similar magnitude, i.e. if the wavelength is approximately equal to the value λ_c, where

$$\lambda_c = \sqrt{\sigma/\rho g}.$$

For an air–water interface, the value of σ is approximately $0.07\,\mathrm{kg\,s}^{-2}$, so λ_c is approximately 0.002–$0.003\,\mathrm{m}$ (2–3 mm). If the wavelength is much greater than λ_c, surface tension will be unimportant, and this is the case for most waves observed on the sea.

wall itself); we show in Chapter 12 how the mechanics of blood flow in arteries can largely be described in terms of the propagation of such waves.

Damping

All the oscillatory motions discussed above are governed by a balance between inertia and a restoring force, with no other forces acting on the particles involved. In such circumstances, the oscillation of a pendulum will carry on for ever, and the travelling wave will travel to infinity without diminution of its amplitude. All real oscillations die out, however, unless forced to continue by additional external forces, and this is because there always are other forces present, called damping forces. These damping forces all result from some form of frictional, or viscous, action. The oscillations of a pendulum are damped by air resistance and by friction in the bearing; water waves and sound waves are damped by viscous stresses; vibrating strings are damped by air resistance (viscous stresses in the surrounding air) and slight visco-elastic behaviour of the material from which the string is made. Perpetual motion is impossible.

One way of considering damping is in terms of energy. A standing wave, or any other oscillation in which particles execute simple harmonic motion with no damping, has a fixed amount of mechanical energy. This follows from the principle of conservation of energy (Equation (2.10)); for example, the sum of the kinetic and potential energies of the pendulum of **Fig. 8.2**, whose motion is given by Equations (8.5) and (8.6), can be shown to be a constant, equal to

$$E = \tfrac{1}{2}a^2\frac{mg}{l}. \tag{8.12}$$

When a damping force is present, however, it is always directed so as to resist the motion, and the work done by it is negative; i.e. mechanical energy is taken from the system by the friction. This energy is converted into heat, and although it is possible to convert heat back into mechanical energy, you can never get as much mechanical energy out as you put heat in (this, roughly speaking, is the second law of thermodynamics). If the damping is very gradual, so that the system almost executes simple harmonic motion, then the total energy will still be given by Equation (8.12), but the amplitude a of the oscillations (the only variable in that equation) will gradually diminish with time, until it becomes undetectable.

When the oscillations are small enough for the restoring force to be directly proportional to the displacement (the situation considered above), then it is usually an equally good approximation to represent the damping force as directly proportional to the rate of displacement, i.e. to the velocity. In the case of a single particle undergoing simple harmonic motion, like the simple pendulum, the additional force to be inserted on the right-hand side of Equation (8.1) may be written $-C\,dy/dt$, where C is a constant. This is proportional to the velocity dy/dt, and in the opposite direction to it. The equation of motion of the particle then has three terms, instead of the two

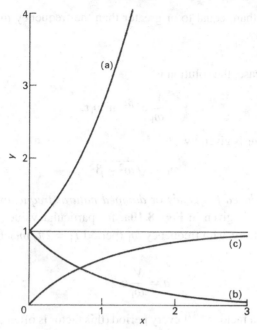

Fig. 8.9. Graphs of (a) $y = e^x$, (b) $y = e^{-x}$, (c) $y = 1 - e^{-x}$ against x.

in Equation (8.1): the mass times the acceleration is equal to the sum of the restoring force and the damping force. The equation corresponding to (8.1) is therefore

$$m\frac{d^2y}{dt^2} = -Ky - C\frac{dy}{dt}.$$

This can be rewritten in the form

$$\frac{d^2y}{d^2t} + 2\beta\frac{dy}{dt} + \omega^2 y = 0, \qquad (8.13)$$

where $2\beta = C/m$ and $\omega^2 = K/m$, which corresponds to Equation (8.2a) in the case without damping.

Rather than consider the most general motion consistent with Equation (8.13), let us examine once more the case where the particle starts off at the point of equilibrium ($y = 0$) with a given velocity V; in symbols this condition is

$$y = 0, \qquad \frac{dy}{dt} = V \qquad \text{at} \qquad t = 0. \qquad (8.14)$$

The complete motion in this case when there is no damping ($\beta = 0$) is given by Equations (8.7) and (8.8). The solution of Equation (8.13), and hence the motion of the particle, when there is damping, is of a different form according to whether the parameter β is less than, equal to or greater than the frequency ω. We consider the three cases separately.

(a) $\beta < \omega$. In this case, the solution is[4]

$$y = \frac{V}{\omega_1} e^{-\beta t} \sin \omega_1 t, \qquad (8.15)$$

where the quantity ω_1 is given by

$$\omega_1 = \sqrt{\omega^2 - \beta^2}$$

and is called the *reduced frequency* or *damped natural frequency*. The graph of y against t in this case is given in **Fig. 8.10a**, for particular values of β and ω. The motion is still oscillatory, with frequency ω_1 (period $T_1 = 2\pi/\omega_1$), but with amplitude a given by

$$a = \frac{V}{\omega_1} e^{-\beta t}, \qquad (8.16)$$

which is reduced by a factor $e^{-\beta T_1}$ every period (this factor is often called the *damping factor*). This case is an example of a damped oscillation.[5]

(b) $\beta > \omega$. In this case the motion is given by

$$y = \frac{V}{2n} e^{-\beta t} (e^{nt} - e^{-nt}), \qquad (8.17)$$

where $n = \sqrt{\beta^2 - \omega^2}$ and is less than β. The graph of y against t is given in **Fig. 8.10b**, for particular values of β and ω. In this case there are no oscillations: the pendulum swings out and swings back, coming gently to rest at its starting point again. Strictly speaking, the motions described by Equations (8.15) and (8.17) do not die away until an infinite time has elapsed, since $e^{-\beta t}$ does not become zero until t becomes infinity. However, in practice, such motions become undetectable after a finite time; for instance, the amplitude given by Equation (8.16) is reduced by a factor

---

[4] The number e is approximately equal to 2.718. The graphs of (a) $y = e^x$, (b) $y = e^{-x}$ and (c) $y = 1 - e^{-x}$ are plotted in **Fig. 8.9**. The quantity e^x (the 'exponential function') becomes infinitely large, and the quantity e^{-x} tends rapidly to zero, as x becomes infinitely large. Indeed, e^{-x} tends to zero more rapidly than any inverse power of x, such as x^{-100} or $x^{-1000000}$. The significance of e^x is that it is its own differential coefficient, i.e. $d/dx(e^x) = e^x$ and $d/dx(e^{-x}) = -e^{-x}$. That is why exponentials occur very frequently in the solution of problems like the present one.

[5] The performance of physiological pressure transducers with fluid-filled needles or catheters attached to them is often analysed in this way. Here, the diaphragm of the transducer acts as a spring and overshoots when it is displaced by a pressure signal, the fluid in the catheter providing most of the inertia; thereafter, the overshoot is damped out by the viscosity of the fluid in the catheter. The magnitude and rate of decay of the overshoot are of interest because they obviously determine the fidelity of the transducer in recording pressure changes. For the liquid-filled systems with stiff diaphragms which are usual in circulatory studies, a decaying oscillatory response occurs, as in **Fig. 8.10a**. If the damping factor and the damped natural frequency can be determined from calibration manoeuvres, the performance of the system can be predicted over a wide frequency range.

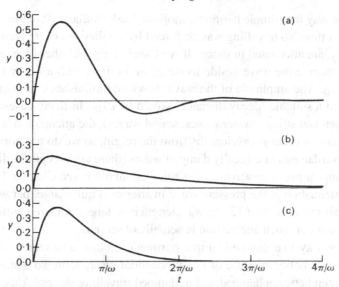

Fig. 8.10. Graphs of y against t when y is given by: (a) Equation (8.15), with $\beta = \frac{1}{2}\omega$; (b) Equation (8.17), with $\beta = 2\omega$; (c) Equation (8.19), with $\beta = \omega$. In each case V/ω was taken to be one unit of length.

$1/100$ by the time $\beta t = 4.6$, so a suitable measure of the decay time is

$$t = 4.6/\beta. \tag{8.18}$$

The completely damped motion described by Equation (8.17) (called an *over-damped oscillation*) is the sum of two exponential decay terms. The one which decays more slowly is $e^{-(\beta-n)t}$, so the time for decay by a factor $1/100$ is $4.6/(\beta-n)$. It is interesting to note that if β is *very* large compared with ω, a case of extremely strong damping, then n is approximately equal to β, and $\beta-n$ is extremely small. In fact, $\beta - n$ is approximately equal to $\omega^2/2\beta$, so the decay time is $8.6\beta/\omega^2$; this is very large compared with $2\pi/\omega$, the period of the undamped oscillations (ω is called the *undamped natural frequency*). So if you want to damp out unwanted oscillations as quickly as possible, the answer is not to make the damping term β as large as possible; in fact β should exceed ω by as little as possible for the fastest decay (without overshoot).

(c) $\beta = \omega$. This is the critical case which separates (a) and (b), and the motion is given by

$$y = Vt\,e^{-\beta t}. \tag{8.19}$$

Its graph is drawn in **Fig. 8.10c**. This is not much different from **Fig. 8.10b**; this case is significant only because ω is the smallest value β can have without oscillations in the motion. The motion is said to be *critically damped*.

In the same way that simple harmonic motions (and, similarly, standing waves) are damped out in time, so travelling waves, forced by oscillations of a given amplitude and frequency, are attenuated in space. If we travel along with the wave, at the wave speed c, we observe the wave beside us decaying in time, with a decay time which we may call T_D. The amplitude of the wave therefore diminishes with distance from the origin, and a suitable 'decay distance' would be cT_D. In many cases of interest (waves on stretched strings, water waves, sound waves), the attenuation is slight and the waves persist for many wavelengths from the origin, so we do not normally need to consider overdamped or critically damped waves (there are exceptions, like surface waves on syrup or pressure waves in the microcirculation: see Chapter 13, p. 375). The rate of attenuation of the pressure wave in arteries is quite large, per wavelength, but, as we shall see in Chapter 12, the wavelength is so large compared with the length of the artery that not much attenuation is actually observed.

In the same way that damped simple harmonic motion was seen to differ from simple harmonic motion because of the exponential decay term, so a similar difference can be seen between damped and undamped travelling waves. A damped wave of frequency ω, corresponding to the pure sinusoidal wave of Equation (8.10), would be represented by an equation relating the displacement y to x and t of the following form:

$$y = ae^{-kx/\lambda}\cos[\omega(t - x/c)]. \tag{8.20}$$

This is of exactly the same form as Equation (8.10), with the propagating oscillations represented by the cos[ ] term, except that the amplitude of the oscillations decays exponentially as x increases. Here, λ is the wavelength and k is a constant which describes the attenuation: the amplitude $ae^{-kx/\lambda}$ is reduced by a factor of $1/100$ in a distance of $4.6/k$ wavelengths. Alternatively, we can say that the wave is attenuated by a factor e^{-k} (the *attenuation factor*) each wavelength.

Wave reflections and resonance

Our discussion of travelling waves, damped or not, has been based on the supposition that the medium in which they are being propagated (the stretched string in the example) is infinitely long, so that there is no impediment to the continued propagation of the waves. In fact, of course, no real system is infinite in extent, and we should consider what happens when a travelling wave meets an obstruction. Suppose that the long stretched string of the above example, oscillated at one end, is attached securely at the other end ($x = l$) so that no displacement is possible there (**Fig. 8.11**). We ask what becomes of the displacement y, given by Equation (8.10), since that would oscillate between $+a$ and $-a$ at every point, including $x = l$. It is more convenient if we let the origin of coordinates be at the point $x = l$, so for simplicity we write $x = l + x'$; x' is the new coordinate, and is actually negative in the string. Then

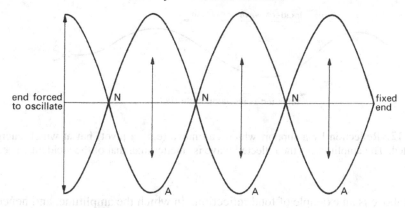

Fig. 8.11. A finite string, of which one end is fixed and the other forced to oscillate with frequency ω. The superposition of the incident wave on to the reflected wave forms a standing wave. No displacement occurs at the nodes N; at the antinodes A the amplitude of the oscillation is a maximum.

Equation (8.10) becomes

$$y = a\cos[\omega(t - x'/c)], \tag{8.21}$$

and the attachment where y must be zero is at $x' = 0$. The presence of the attachment has the effect of superimposing on to the original wave (called the *incident wave*) another wave, with the same amplitude and wave speed, propagating in the opposite direction. This is the *reflected wave*. The complete equation for the displacement is now

$$y = a\{\cos[\omega(t - x'/c)] - \cos[\omega(t + x'/c)]\}. \tag{8.22}$$

At the fixed end ($x' = 0$) and at every half wavelength from it ($x' = -\pi c/\omega, -2\pi c/\omega, -3\pi c/\omega$, etc.) the displacement is zero at all times; halfway between these nodes are points where the displacement oscillates with double the original amplitude (*antinodes*; see **Fig. 8.11**). In fact, the motion is a standing wave; Equation (8.22) can be rewritten as

$$y = 2a\sin\omega t \sin\frac{\omega x'}{c}. \tag{8.23}$$

For every value of x', y oscillates in simple harmonic motion, the amplitude of the oscillation being proportional to $\sin\omega x'/c$, which is of course zero at the nodes and equal to one at the antinodes. The oscillations do not propagate, and clearly represent a standing wave.

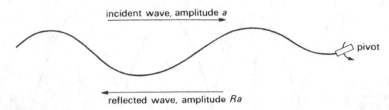

Fig. 8.12. Reflection by a support which can move (e.g. a pivot) but at which energy is dissipated. The amplitude of the reflected wave is smaller than that of the incident wave.

The above is an example of total reflection, in which the amplitude, and hence the energy content, of the reflected wave is equal to that of the incident wave. Usually, however, not all the energy of the incident wave is reflected; some of it is either absorbed or transmitted. Suppose that the stretched string is attached not to a fixed support, but to a support which can move, but whose motion is resisted by some frictional force which dissipates energy (**Fig. 8.12**). Then some of the energy of the incident wave will be used up in feeding this dissipation, and the reflected wave must have less energy, i.e. a smaller amplitude. This is an example of *partial reflection*. If, say, the amplitude of the reflected wave of Equation (8.22) were only 80% of that of the incident wave, the displacement would be given by

$$u = a\{\cos[\omega(t - x'/c)] - R\cos[\omega(t + x'/c)]\},\qquad(8.24)$$

where R is a constant, sometimes called the *reflection coefficient*, equal to 0.8.[6] The incident wave is also partially reflected if it encounters some discontinuity in the propagating medium, like a local change in the density or thickness of the string, or a local constraint on its movement, like a bead attached to the string in the middle (**Fig. 8.13**). In this case a wave continues to be propagated down the string on the other side of the discontinuity, but it too has reduced amplitude. This is the *transmitted wave*; its amplitude would be equal to a times a number T less than one. The quantity T is sometimes called the *transmission coefficient*; more commonly it is T^2 that is so called. If no energy is absorbed, $T^2 = 1 - R^2$.

Any discontinuity in a system can be a source of reflections, and if a system contains many such discontinuities, the wave from any point will be a complicated superposition of the incident wave, the reflected and transmitted waves at each discontinuity, the waves generated by subsequent reflection and transmission of the reflected and transmitted waves, etc. The process of reflection, re-reflection, etc. does not go on for ever, primarily because of the attenuation of the waves, and also in part

[6] Here, R is the amplitude of the reflected wave, relative to that of the incident wave. More commonly the reflection coefficient is defined as the rate at which energy is transmitted in the reflected wave, relative to that in the incident wave, and that quantity is equal to R^2. It is important to know which definition is being used when people refer to the reflection coefficient.

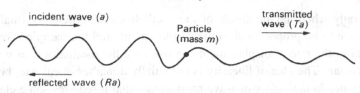

incident wave (*a*)

Particle
(mass *m*)

transmitted
wave (*Ta*)

reflected wave (*Ra*)

Fig. 8.13. Reflection and transmission of a wave at a discontinuity in the string (here, a particle of mass *m* attached to it, which resists displacement because of its inertia).

because of energy absorption at end-points such as the rough pivot in the above example, or a sloping porous beach in the case of water waves, which absorbs almost all the incident energy.

We saw on p. 113 how a finite stretched string, fixed at each end, can support free oscillations at certain natural frequencies. If a point of the string is oscillated at a given frequency ω not equal to one of the natural frequencies, then waves will propagate from it towards each end, will be reflected there, re-reflected at the other end, etc. The final motion will be a complicated standing wave with frequency equal to the forcing frequency ω. If ω is equal to one of the natural frequencies, however, the situation is quite different. In that case, the phases of the incident and reflected waves are such that they strongly reinforce each other on each reflection, and the amplitude of the oscillation would continue to grow until the dissipative effects, which would also be growing, become so large as no longer to be negligible. In addition, the system would cease to be linear (see below). Energy is continuously being put into oscillations of such a frequency that they can maintain themselves at a constant amplitude in the absence of energy input. The effect of all this extra energy is to increase the amplitude dramatically. The phenomenon is known as *resonance*. The amplitude of a resonant oscillation is in practice limited by frictional or dissipative effects (or nonlinear effects), but unless these are very strong, the ultimate amplitude is still much greater than that of the original forcing. Resonance may be desirable or undesirable. Musical instruments are so designed that of the wide range of frequencies present in the forcing (e.g. the oscillations of a reed) only a few resonate and are thereby amplified into audible sound. In the arterial system, however, resonant oscillations are clearly undesirable, since unusually large pressure oscillations would affect blood flow and might damage the tissues.

Linearity

The properties of waves, as outlined above, will be applied in Chapter 12 to the wave motions which are present in the mammalian cardiovascular system. All those properties, and some others, will be seen to have widespread relevance. It is important to realize, however, that the relatively simple considerations discussed so far are applicable only when the amplitude of the oscillations is sufficiently small for the restoring force to be proportional to the displacement, and the damping force to be proportional to the rate of displacement. In this case the oscillations, or waves, are said to be *linear*. The idea of linearity is essentially a mathematical one, but it is of such importance in the theory of wave propagation that it requires some elaboration here.

Suppose that a particular system operates in such a manner that two variable quantities are always related to each other in some way. These might be the velocity and pressure of blood flowing in an artery, measured at a particular station, or the displacement and acceleration of a particle in oscillatory motion, or the mass and oxygen consumption of a living organism; a quite general type of system is envisaged. Let us call the two variables u and p. The relationship between u and p is linear if the graph of u against p is a straight line and u is directly proportional to p; that is:

$$u = ap, \tag{8.25}$$

where a is a constant. A more general definition of a linear system can be derived from this: if u_1 and p_1 are one pair of corresponding values for u and p, and if u_2 and p_2 are another corresponding pair, then the system is linear if $(u_1 + u_2)$ and $(p_1 + p_2)$ are also a corresponding pair. A system represented by Equation (8.25) is clearly linear by this definition, whereas a system represented by, say, $u = p^2$, whose graph is a parabola, is not.

Similarly, the system represented by Equation (8.2a) and the associated conditions at the time of release, which describe simple harmonic motion, is also linear. For instance, we saw that the solution

$$y = a \cos \omega t$$

corresponds to an initial displacement a (Equation (8.5)) and that the solution

$$y = \frac{V}{\omega} \sin \omega t$$

corresponds to an initial velocity V (Equation (8.7)). It is easy to verify that the sum of these solutions, i.e.

$$y = a \cos \omega t + \frac{V}{\omega} \sin \omega t,$$

corresponds to a state in which the initial displacement is a and the initial velocity V. Again, the reflection of sinusoidal waves described above is a linear process, in that a doubling of the amplitude of the incident wave would result in a doubling of the amplitude of the reflected wave, etc. Furthermore, the system in which the waves are propagated, the stretched string, must be linear in order for the total displacement to be represented simply by the sum of the displacements associated with the incident and reflected waves separately.

The equations governing almost all real physical systems are actually nonlinear. For instance, the actual equation describing the oscillations of a simple pendulum is (8.2b) rather than (8.2a), and it reduces to (8.2a) only if θ is small enough for $\sin\theta$ to be approximately equal to θ. The error in this approximation is proportional to θ^3,[7] which is very much less than θ if θ is very much less than unity: when $\theta = 0.1$, θ^3 is equal to 0.001. As long as θ remains small enough, the real nonlinear system, described by Equation (8.2b), will behave almost as if it were linear, and we obtain a good approximation to the behaviour of the real system by analysing the much simpler linear system. The process of converting a small-amplitude nonlinear system to a linear one is called *linearization*. It is a process which is very frequently employed in many fields of application, including the analysis of blood flow in arteries, because linear systems are so much easier to analyse. However, in all practical applications, it is important to verify that the neglected nonlinear parts of the equation describing the real system are indeed small compared with the linear ones. If they are not, the nonlinearities are important, and their effect must be assessed.

One type of system which can usually be treated as linear, but in which nonlinearities become important in certain circumstances, is that of a finite stretched string, forced to oscillate at a frequency close to one of the natural frequencies. A wave is generated at the source of oscillations and is repeatedly reflected in such a way as to reinforce itself (resonance) so that its amplitude becomes very large. In these circumstances the linear approximation under which the wave was originally analysed ceases to be valid. The detailed, nonlinear, analysis of a resonant system is usually very complicated.

The equations of fluid mechanics are nonlinear, because of the term representing convective accelerations: Bernoulli's equation (Equation (4.6), p. 43) is an example. This equation relates the pressure p to the velocity u of a fluid in steady motion when viscosity can be neglected. Suppose that fluid is flowing in a horizontal pipe (so that gravity is unimportant) of non-uniform cross-sectional area (so that u varies along it; see p. 41); if the pipe is also elastic, this can be used as a model of blood flow in large arteries, as we shall see in Chapter 12. Suppose too that the flow is oscillatory, and that Bernoulli's equation can be modified to take account of the unsteadiness by a term proportional to the local acceleration du/dt. It would then become

$$\frac{p}{\rho} + \tfrac{1}{2}u^2 + k\frac{du}{dt} = \text{constant},\qquad(8.26)$$

where k is a constant (with the dimension of length) and ρ is the fluid density, also supposed constant. In fact, the local acceleration term in the real equation describing unsteady flow is more complicated than the one given here, but it is, like this one, linear, so that Equation (8.26) is an adequate model for the present discussion. If the term representing convective accelerations, $\tfrac{1}{2}u^2$, were to be neglected, then Equation

---

[7] Because an approximation to $\sin\theta$ that is better than merely θ is in fact $\sin\theta = \theta - \theta^3/6$.

(8.26) would reduce to

$$\frac{p}{\rho} + k\frac{du}{dt} = \text{constant.} \tag{8.27}$$

From the above definition, Equation (8.27) is linear while (8.26) is not. Suppose that the fluid velocity were oscillating everywhere with angular frequency ω and were proportional, say, to $\cos\omega t$. If (8.27) were the governing equation, this could be associated with a pressure oscillation of the same frequency, proportional to $\sin\omega t$. But if (8.26) were the governing equation, a solution in which u was proportional to $\cos\omega t$ (and hence du/dt to $\sin\omega t$) would require that u^2 be proportional to $\cos^2\omega t$, which is equal to $\frac{1}{2} + \frac{1}{2}\cos 2\omega t$. Thus, a velocity waveform of a single frequency ω must be associated with both the mean pressure and a component of frequency 2ω, as well as the component of frequency ω. This makes the analysis of non-sinusoidal oscillations much more difficult, as we shall see.

Fourier analysis

A general periodic wave motion is not usually purely sinusoidal. The graph of the displacement of a string against position x at any fixed time, or the graph of blood velocity in an artery against time at a fixed position, may have a very complicated shape (see the last panel of **Fig. 8.14**). Such a shape can, however, be expressed as the sum of a number of sinusoidal oscillations of different phase and amplitude, and with frequencies which are multiples of the fundamental frequency $\omega\ (= 2\pi/T)$, where T is the overall period of the oscillations. The process of splitting up a complicated waveform into its sinusoidal components is called Fourier analysis. The different frequency components which make up the complete wave are again called the harmonics of the fundamental (cf. p. 113). The number of harmonics required for an accurate representation of the complete oscillation may be large or small. For example, the waveform shown in **Fig. 8.14**, which is similar to a velocity waveform in an artery, can actually be represented by a mean term, a fundamental and three harmonics. Its equation is

$$u = A_0 + A_1\cos(\omega t + \theta_1) + A_2\cos(2\omega t + \theta_2) + A_3\cos(3\omega t + \theta_3) + A_4\cos(4\omega t + \theta_4) \tag{8.28}$$

where $A_0 = 0$, $A_2/A_1 = 0.97$, $A_3/A_1 = 0.47$, $A_4/A_1 = 0.14$, $\theta_1 = -58° = -1.01$ rad, $\theta_2 = -151° = -2.64$ rad, $\theta_3 = +124° = 2.16$ rad, $\theta_4 = +86° = 1.5$ rad. The way the three components add up to form the complete wave is illustrated in **Fig. 8.14**.

Fourier analysis can be applied to a waveform of any shape. It is a process which considerably simplifies the analysis of oscillations and waves in linear systems, but is of little value in nonlinear systems. Suppose that we wish to calculate the blood pressure p in an artery from a measured oscillatory velocity u, by the use of Equation (8.27). We can express the velocity as the sum of sinusoidal wave components, as in

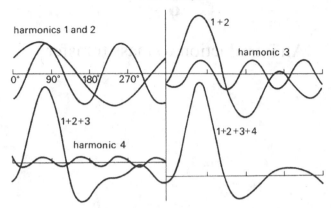

Fig. 8.14. A periodic wave-motion, similar to the graph of blood velocity against time in an artery. Diagrams show how the mean, the fundamental and three harmonics can be superimposed to make up the complete waveform. The mathematical form of each term is given by Equation (8.28). (From McDonald (1974). *Blood Flow in Arteries*, p. 129. Edward Arnold, London.)

Equation (8.28). We can also use Equation (8.27) to calculate the pressure oscillation corresponding to a single component of the velocity oscillation; i.e. corresponding to

$$u = A_n \cos(n\omega t + \theta_n)$$

the equation, with a zero on the right-hand side, gives

$$p = \rho k n \omega A_n \sin(n\omega t + \theta_n).$$

(The constant term on the right-hand side of Equation (8.27) is equal to the mean value of the left-hand side, i.e. p_0/ρ.) We can finally add up all the sinusoidal pressure modes, derived in this way, to obtain the composite pressure waveform, and this addition is permitted since the system is linear. In the nonlinear system represented by Equation (8.26), however, all the frequency components interact with each other, as in the example considered above. It is not permitted to superimpose the solution obtained by solving the equation with a single sinusoidal term on the right-hand side, even if that solution could be obtained simply, which it cannot.

The pressure waves which exist in large arteries are approximately linear, as we shall see in Chapter 12, and their principal features can be derived from linear theory. There are some features, however, which can be explained only with the aid of non-linear ideas. These include the steepening of the pressure waves in some circumstances as they propagate down the arterial tree, and the large-amplitude oscillations which may be heard in arteries which are constricted by a high-pressure cuff (the so-called Korotkoff sounds; see p. 459). Nonlinear behaviour is also seen in veins (see Chapter 14).

9

An introduction to mass transfer

The term 'mass transfer' or 'mass transport' encompasses a vast range of processes involving the movement of matter within a system. It is not possible to provide a simple definition of its scope, except to state that we are concerned with the movement of particular molecular species within a system and with the factors which affect the movement. We can introduce the subject by means of two simple examples, although these in no way describe its full breadth.

A puddle of water on the road surface slowly evaporates, the liquid water progressively being transferred as vapour to the air above it. The rate of evaporation depends upon such factors as humidity, the ambient air temperature relative to the ground and the speed of the wind over the surface of the puddle.

If a crystal of copper sulphate is dropped into a beaker of water it slowly dissolves in the water and produces a concentrated solution around the crystal surface. In time this dissolved material diffuses further and further into the surrounding water. The speed with which the crystal dissolves and the rate of transfer of dissolved copper sulphate to the bulk of the water phase can be modified by a number of factors. For example, the process would occur more quickly in hot water than cold, or if the beaker were stirred.

These examples illustrate the two most fundamental mechanisms of mass transport within a single medium (or *phase*): that is, *diffusion* and bulk motion or *convection*. They also illustrate the important class of phenomena – the transport of *substances* across an interface between phases, here the gas–liquid and liquid–solid interfaces. In biological systems we often encounter the transport between two liquid phases separated by a membrane, for example the cell membrane. Such transport can be by bulk motion and by diffusion, the details depending on the structural and chemical properties of the membrane and factors such as concentrations and pressures within the phases. In this chapter we will show how simple mass transfer processes may be examined quantitatively and estimates obtained of the behaviour of more complicated systems.

It is necessary to begin with some definitions. The first quantity to consider is *concentration*, which may be defined in various ways, in particular as mass concentration, mass fraction, molar concentration and mole fraction (see Chapter 3, p. 27 for a definition of a mole). All are equally admissible, but since we are concerned mainly with transfer between and within the liquid and solid phases we will use molar concentration. The molar concentration c_A of species A in a solution is defined as the number of moles of A per unit volume and is usually expressed in the units $mol\,l^{-1}$. If we now measure the amount of material A moving in a given direction, we may define its *molar flux J* as the number of moles per unit time passing through a plane of unit area at right angles to its path. Molar fluxes are usually expressed in the units $mol\,s^{-1}\,cm^{-2}$. The definition of flux involves specification of the direction in which the movement is measured; a flux is a vector quantity and we must always be aware of this.

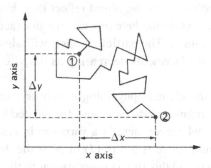

Fig. 9.1. Illustration of Brownian motion of a small particle on the surface of a liquid. Each line represents an individual motion. After some time the particle initially at station 1 has moved to station 2. Δx and Δy represent the projected displacements in the x- and y-directions.

Diffusion

In 1827 the Scottish botanist Robert Brown performed a very simple but important experiment which led to a deeper understanding of the behaviour of molecules in a liquid. Brown discovered that grains of pollen on water moved about ceaselessly in a random manner in all directions (**Fig. 9.1**). This completely random or erratic motion of the pollen grains has subsequently been observed with small suspended particles of all kinds and is now termed *Brownian motion*. The motion of these particles is caused by the motion of the molecules of the liquid in which the particles are immersed. The molecules are in a continual state of random motion and, as a result, they continuously collide with one another and the suspended particles. The haphazard movement of the particles reflects this random molecular movement; whilst the individual molecules and their displacements are too small to observe directly, the consequential motion of

Table 9.1. *Comparison of molecular diffusivities of various substances at room temperature*

Species	Diffusivity ($cm^2 s^{-1}$)
O_2 in air	0.21
H_2 in air	0.41
O_2 in water	1.8×10^{-5}
H_2 in water	5.1×10^{-5}
Acetate in water	0.8×10^{-5}
Low-density lipoprotein in water	10^{-7}–10^{-8}

the particles can be seen. The displacement of the particles results from the impact of a very large number of molecules at any instant of time.

The motions of the molecules of the liquid reflect their kinetic energy, which increases with temperature. As would be expected, the displacements of the particles also increase with temperature. The particle motion will also depend upon the viscosity of the liquid, the displacements decreasing as the viscosity of the medium is raised.

In 1855, Adolf Fick, the German physiologist (whose name is also given to the dilution method for determining cardiac output), formulated his famous law of mass transfer. This states that whenever there is a variation in concentration of any material (A) within a medium, there is a flux (J) of A in the direction of diminishing concentration (x-direction) and this flux is proportional to the concentration gradient (dc_A/dx). Thus

$$J = -D_A \frac{dc_A}{dx}. \tag{9.1}$$

The constant D_A is called the *diffusivity* or *diffusion coefficient* of the material. This equation has been shown experimentally to be applicable to the diffusion of both molecular and particulate material when in low concentration (see p. 132).

The values of diffusivities vary over a very wide range; typical values for diffusion in gases are very much higher than in liquids. Diffusivities also depend upon the size of the molecules or particles compared with the molecules of the medium in which they are diffusing (see **Table 9.1**) and, like the intensity of Brownian motion, upon the temperature. It can be shown[1] that if the absolute temperature of the medium is T, then

$$D = \frac{kT}{\mu},$$

where μ is the fluid viscosity (which also depends on temperature) and k is a constant dependent upon the molecular characteristics of the medium.

---

[1] This assumes that the molecules do not change shape or size as temperature changes; thus, factors such as alteration in the degree of hydration are not included.

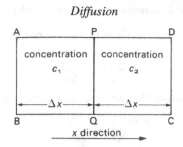

Fig. 9.2. Rectangular area of surface; see text for details.

At first sight it would appear difficult to relate the random motion of the pollen grains in Brown's experiments to the formal law of Fick. However, it is possible to demonstrate a relation between the probable displacement (Δx) of a particle or molecule in a given time to the diffusivity D of that material. Imagine a water surface with pollen grains floating on it and undergoing Brownian motion over the surface. Consider a rectangle ABCD drawn in the surface with a line PQ bisecting it (**Fig. 9.2**) so that the lines AB and CD are each a distance Δx from the line PQ. We wish to compute the number of pollen grains which will cross the line PQ from either side during a small time interval. The concentration of pollen grains in the rectangle ABQP is c_1 (the concentration is defined here as the number of grains per unit area); the concentration in CDPQ is c_2. If the average displacement of grains is Δx in a time interval of t, then the particles arriving at PQ from the left can only be those whose initial position is at AB or closer to PQ. Similarly, particles crossing PQ from the right were originally no further away than the line CD. Moreover, only half of the particles which could cross the line from either side will in fact do so because, on average, half of them in each compartment will be moving away from PQ, not towards it. Thus, the total number of particles moving to the right across the line PQ is given by

$$\text{number} = \tfrac{1}{2} \times \text{area ABQP} \times \text{local concentration}$$
$$= \tfrac{1}{2}c_1 \times \text{area ABQP}.$$

Similarly, the total number moving to the left is given by

$$\text{number} = \tfrac{1}{2}c_2 \times \text{area CDPQ}.$$

Hence, the net number of grains moving from left to right in time t is

$$\text{number} = \tfrac{1}{2}c_1 \times \text{area ABQP} - \tfrac{1}{2}c_2 \times \text{area CDPQ}$$
$$= \tfrac{1}{2}(c_1 - c_2) \times a\,\Delta x,$$

where a is the length of the lines AB and CD. The net flux n (number of particles per unit time and unit length of PQ) is given by

$$n = \frac{\Delta x}{2t}(c_1 - c_2).$$

If we multiply top and bottom of the right-hand side of this equation by Δx,

$$n = \frac{(\Delta x)^2}{2t} \frac{c_1 - c_2}{\Delta x}.$$

But if we consider concentration changes over a small distance,

$$\frac{c_2 - c_1}{\Delta x} = \frac{dc}{dx}.$$

Thus,

$$n = \frac{-(\Delta x)^2}{2t} \frac{dc}{dx}. \tag{9.2}$$

Comparing Equation (9.2) with the mathematical expression of Fick's law (Equation (9.1)) we can see that

$$D = \frac{(\Delta x)^2}{2t}. \tag{9.3}$$

Thus, the diffusion coefficient and the probable distance travelled by a particle or molecule in a given time are directly related.

While Fick appreciated that the driving force for diffusion was derived from the kinetic energy of the molecules, he was unaware of any physical significance of the diffusion coefficient. It was later shown by Nernst that, if the molecules are assumed to be spherical, the diffusion coefficient is related to the frictional forces operating on them thus:

$$D = \frac{RT}{fN}, \tag{9.4}$$

where R is the Universal Gas Constant, T is the absolute temperature, N is Avogadro's number (the number of molecules per mole) and f is the frictional force opposing the motion of a molecule. If the diffusing molecule is large compared with its surrounding molecules, then Stokes' law (Equation (6.1), p. 85) may be used to describe the frictional force:

$$f = 6\pi\mu a, \tag{9.5}$$

where μ is the fluid viscosity and a is the 'effective' radius of the molecule – commonly called the *Stokes–Einstein radius*. This is not the actual radius of the molecule, but the radius of an equivalent spherical molecule, exhibiting the same diffusivity. Equations (9.4) and (9.5) can be combined to give an expression for a:

$$a = \frac{RT}{6\pi\mu DN}. \tag{9.6}$$

Thus, this radius can be determined from a measurement of the diffusivity in free solution.

In the discussion so far we have implicitly assumed that the concentration of the diffusing species is low and the solution is said to be *dilute*. The diffusing molecules are considered to move about and interact only with the molecules of the carrier fluid. However, as the concentration increases so do the molecular interactions of the diffusing material. The effect of these interactions is to decrease the activity of the molecules and hence the diffusivity to some degree. A discussion of the detailed mechanism by which this happens and any indication of the magnitude of the effect are both beyond the scope of this book; the reader should simply be aware of the effect and its possible importance in situations where mass transfer is occurring in concentrated solutions.

The colloidal state

A precise definition of this state is difficult and is not necessary for our purposes. Colloidal solutions, or *sols* as they are generally known, may be solutions of either very large molecules or aggregates of smaller molecules. The distinction between a sol and a particulate suspension (or an emulsion, which is a suspension of liquid droplets) is arbitrarily made according to whether or not the solute molecules or particles are visible under a light microscope (in a sol they are not).

It is often found that the individual large molecules of a sol are not in 'simple' solution but that solvent molecules are adsorbed on to their surfaces, resulting in complexes much bigger than the original molecule. The tendency for large molecules to imbibe water depends upon the electric charge distribution on the surface of the molecules and on the pH of the environment. Progressive removal of the free solvent from a sol ultimately causes it to solidify into a *gel*. When this happens, the absorbed solvent remains trapped within the solid matrix and, if removed, causes destruction of the gel structure.

As can be imagined, sols are common in biological systems, for example in the *interstitial space* between cells (see p. 363). This space contains various large molecules which have a very great affinity for water and, as a result, are considered to be present in colloidal solution.

Mass transfer coefficients

In many practical situations, particularly in biology, it is not convenient and indeed often not possible to consider transport rates simply in terms of the molecular diffusivity. This is particularly true when we have to consider diffusion through cell membranes, as we often know neither the effective area for diffusion nor the thickness of the membrane. To overcome such difficulties we state that the flux (J) of material A across a membrane for a given concentration difference across it (Δc_A) is given by

$$J = K\Delta c_A, \tag{9.7}$$

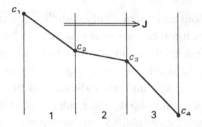

Fig. 9.3. A schematic representation of a three-compartment membrane showing the concentration profile across it under steady-state conditions.

where we define K as the *mass transfer coefficient*. Thus, all diffusion pathways are lumped together and assessed by the one overall coefficient. The resistance of the membrane is then defined as $1/K$.

In many biological situations diffusion occurs through a series of cell membranes and liquid layers. Let us consider an example with three layers, as shown schematically in **Fig. 9.3**. The overall concentration difference across the composite system is $c_1 - c_4$, but the distribution of concentration within the membranes is not a simple linear function of position, depending rather upon the relative mass transfer coefficients of the three sections. At steady state we know that the flux J through all three zones must be the same. Then applying Equation (9.7) to each of the three zones we have

$$\frac{1}{k_1} = \frac{c_1 - c_2}{J}$$

$$\frac{1}{k_2} = \frac{c_2 - c_3}{J}$$

$$\frac{1}{k_3} = \frac{c_3 - c_4}{J},$$

where k_1, k_2, k_3 are the mass transfer coefficients for the three zones. If we add these three equations together,

$$\frac{1}{k_1} + \frac{1}{k_2} + \frac{1}{k_3} = \frac{c_1 - c_4}{J}.$$

We may define an overall mass transfer coefficient K for the composite membrane on the basis of the overall concentration difference:

$$\frac{1}{K} = \frac{c_1 - c_4}{J}.$$

Thus, we can see that

$$\frac{1}{K} = \frac{1}{k_1} + \frac{1}{k_2} + \frac{1}{k_3}.$$

Since $1/k_1$ is the resistance of zone 1, etc., we see that the overall resistance to transfer is the sum of all the component resistances. Furthermore, if one of the component resistances is very large compared with the others, then the overall resistance and the overall mass transfer rate will be determined by the transfer across that section. The difference in concentration across that section will be close to the overall concentration difference and that section of the membrane will be said to be the rate-controlling section.

Diffusion through pores and membranes

In later chapters we shall be concerned with the diffusional transport of molecules across the walls of blood vessels. This is very complicated, and here only a simple introduction is provided to diffusion and filtration through pores and membranes. We shall consider first some very elementary aspects of the process and then *restricted* or *hindered diffusion*.

Consider a container filled with a solution of a substance at a concentration c_1 separated from a second container by a porous filter of area A_m, the second container being filled with a similar solution but of concentration c_2. The concentration gradient existing because of the difference in concentration on the two surfaces of the filter will allow a diffusional flux of solute, from the container of high concentration to the other, with *counter diffusion* of the solvent. Provided the pore diameter of the filter is very large compared with the size of the solute molecules, we may readily compute the flux on the basis of Equation (9.1); the effective area for diffusion is the pore area of the filter. The *porosity* of the filter is the pore area per unit area of filter or membrane A_p, and Equation (9.1) may be written as

$$J = -A_p D \frac{\Delta c_A}{\Delta x}, \tag{9.8}$$

where J is still the flux per unit area of filter and Δc_A is the concentration difference across the filter of thickness Δx.

If the filter is exchanged for others of the same porosity but successively smaller pore diameter, the flux of solute will at first be unaltered. However, beyond a certain critical size the pores will be too small to let the solute molecules pass through, and the flux will reduce to zero; smaller molecules than the solute can still pass through unimpeded, and are said to be sieved off.

This is the mechanism underlying the process of *dialysis*, in which a membrane separates a solution of both large and small molecules from pure solvent. The passage of the large molecules is prevented, but the smaller solute molecules can still pass through the membrane, leaving the original solution relatively rich in large molecules. The capture of solute molecules by the membrane surface and the subsequent transport through it and release on the other side may depend upon a number of complex changes in the orientation of the molecules comprising the membrane. There are a number of molecular 'models' of biological membranes designed to describe different properties of membrane transport phenomena; none are yet fully agreed upon and we shall not discuss this subject further.

Restricted diffusion The arguments presented in the previous section are really only applicable to diffusion through porous filters and membranes in which the pore diameter is many times the diameter of the solute or solvent molecules. It is not true to say that diffusion of molecules occurs unhindered till the molecules are as big as the pore. When the pore size becomes comparable to the random thermal displacements of the molecules, diffusion becomes restricted both because of collisions between the molecules and the pore edge and because of short-range electrical forces between the molecules and the wall. Thus, the effective pore area (A_s) available for transport decreases to zero as the molecular size increases.

Many of the molecules involved in diffusional transport through biological membranes are comparable to the size of the water-filled pores through which they pass. Thus, a pore of 4 nm radius will restrict the diffusion of glucose (radius 0.37 nm) compared with its free diffusion by approximately 35%; it will even restrict the diffusional transport of water (radius 0.15 nm) by about 15%. In order to describe this process quantitatively, let us first rearrange Equation (9.8) in the form

$$\frac{A_p}{\Delta x} = \frac{J}{D \Delta c_A}.$$
(9.9)

The left-hand side of the equation, the pore area per unit path length, is given in terms of readily measurable quantities, provided individual pores are of far greater diameter than the diffusing molecules. As pore dimensions decrease there is a geometric hindrance for molecules entering the pore and also an increase in resistance to diffusion along it. As a result, D should be replaced by D_R (a restricted diffusion coefficient) and the true pore area A_p by an effective pore area A_s. Equation (9.9) may then be rewritten as

$$\frac{A_s}{\Delta x} = \frac{J}{D_R \Delta c_A}.$$
(9.10)

This gives an expression for D_R in terms of the apparent pore area, and enables us to compute the flux of molecules through a membrane with small pores as long as A_s and A_p are known. The problems of doing this in practice are discussed in detail in Chapter 13.

Many biological and synthetic membranes also demonstrate the property of *semi-permeability*. This means that they are capable of selectively preventing the passage of some molecules and ions whilst permitting the passage of others.

The earliest explanation of semi-permeability was that the membrane acted simply as a sieve or porous filter, retaining the large molecules but allowing passage of the

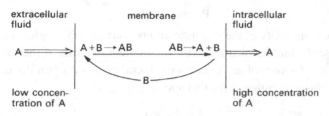

Fig. 9.4. A schematic view of a cell membrane showing the mechanism of active transport of molecules of A from outside to inside the cell. The carrier molecule B is retained within the membrane.

small ones. Whilst this mechanism is the basis of separation in filtration and in passive diffusion through membrane pores, it is now known not to be the sole cause of semi-permeability. It is possible for solvent molecules to pass through the membrane but for dissolved solute molecules to be prevented from passing through even though they may be smaller in diameter than the solvent molecules.

This apparently anomalous behaviour can result from a number of causes. For example, if the solute molecules have solvent molecules hydrated on to them, their effective size may be increased markedly; the solute molecules may also be aggregated. Electrical interactions between solute molecules and the pore walls may also prevent their passage.

Perfect semi-permeable membranes are capable of preventing the passage of the solute molecules completely; many membranes are not perfect and do allow some passage, and they are said to be 'leaky'. The cause of this leakiness can be either inherent in the membrane or the result of physical weakness of the membrane so that it breaks down or partially ruptures because of high pressure differentials set up across it by osmosis (see p. 139).

Active transport Many biological membranes can transport materials faster than expected or against concentration differences by processes which require metabolic energy. This is called *active transport*, and is essential for the maintenance of the correct environment for cells. Many details of the mechanism are poorly understood, but the basic mechanism appears to be the same for all transported substances and to depend upon the existence of carriers within the membrane (**Fig. 9.4**). Molecules of substance A in the extracellular fluid are adsorbed on to the membrane and there combine with the carrier B to form the complex AB. This is then transported across the membrane, and at the intracellular side the complex is split and A is released to the inside of the cell. The carrier B then diffuses back to the outer surface of the membrane to combine with more A. Energy is consumed in the formation and breakdown of the chemical bonds in the membrane.

Permeability

Because of the complexity of the structure and properties of biological membranes, it is usual to describe molecular transport through them in terms of their *permeability*. Thus, if the flux of some solute A across the membrane is J when the concentration difference across the membrane for that solute is Δc_A, then

$$J = P_A \Delta c_A, \tag{9.11}$$

where P_A is the membrane permeability for solute A. The coefficient P_A in this equation is a mass transfer coefficient as defined above in Equation (9.7). The membrane permeability for any substance will depend upon many factors, in particular the solute concentration and temperature, and it is important to report these whenever referring to a permeability. If concentration is not quoted, it is assumed that the solute is in *dilute solution* (see p. 132). Temperature has an important influence on permeability, since it determines the amount of energy available to permit transport across the membrane. This is because energy is required for the solute to be adsorbed on to the membrane surface and to cause conformational changes in the molecular structure of the solute or membrane.

Filtration through membranes

Thus far we have considered only diffusional transport through membranes. Transport can also occur by filtration through the pores if there is a pressure difference across the membrane, and this will be complementary to any diffusional transfer. A feature of this process is the ability of the mobile species to move independently of any concentration gradients if the pressure gradient is large enough. If the pores of the membrane are small enough to prevent passage of some of the constituents of the fluid, then the filtration process can itself cause concentration differences across the membrane. The significance of this will become clear in Chapter 13 when transcapillary exchange is considered (p. 399). If the excluded molecules are macromolecules, such as proteins, then the process is known as *ultrafiltration*; if the excluded molecules are small, such as inorganic salts, it is known as *reverse osmosis*. However, the principle of operation is the same in both cases; it is purely a question of scale.

If the pressure difference across the membrane is Δp and the resulting flux of filtrate is J, then for simple membranes

$$J = F \Delta p, \tag{9.12}$$

where the constant F is called the *filtration coefficient*, analogous to the mass transfer coefficient in Equation (9.7). The filtration coefficient depends upon such factors as the number of pores per unit area, their size, shape, length and tortuosity. For a given geometry, F also decreases as the viscosity of the fluid is increased. Equation (9.12) could be derived theoretically if one assumed that flow within the pores obeyed Poiseuille's law, though the value of F could not be accurately predicted.

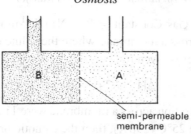

semi-permeable
membrane

Fig. 9.5. Schematic view of two chambers A and B separated by a rigid semi-permeable membrane. Compartment B contains an aqueous solution of protein; compartment A contains no protein.

Osmosis

The phenomenon of osmosis (from the Greek word meaning 'to push') is of fundamental importance in all considerations of mass transfer in biological systems.

Imagine that a rigid porous filter separates two reservoirs as shown in **Fig. 9.5**. If an aqueous protein solution is introduced into chamber B and chamber A is filled with water, then a concentration difference for the protein is established across the barrier. However, if the pore size of the filter is sufficiently small, then protein molecules will be unable to diffuse through the barrier into chamber A. The water in chamber A will, however, diffuse into chamber B, reducing the concentration difference. As a result, there will be an increase in the volume of fluid in chamber B.

If both chambers have tubes let into the top, as shown in **Fig. 9.5**, then the liquid level in B will rise as water diffuses into the chamber and that in A will fall. In time, the levels will cease to change, indicating that there is no further net flow of water. At this time an equilibrium has been achieved between the 'force' driving diffusion and a 'force' due to the excess hydrostatic pressure in the compartment. At equilibrium, this excess pressure is called the *osmotic pressure* (Π) of the protein solution at the equilibrium concentration. The symbol Π is an accepted convention and should not be confused with the mathematical symbol for the number 3.142. If the membrane allows passage of small solute molecules but inhibits the passage of large colloidal molecules, then the osmotic pressure which is established is called the *colloid osmotic pressure* or *oncotic pressure*.

Thus, we can see that osmosis occurs whenever two solutions of differing concentration are separated by a barrier which allows passage of solvent but not the solute; it can result either from separation due to diffusion or from separation during filtration.

It can be shown that the osmotic pressure of a dilute solution whose total molar solute concentration is c (i.e. c is the sum of the molar concentrations of all solute species) at a given absolute temperature T is given by *van't Hoff's law*:

$$\Pi = cRT \tag{9.13}$$

where R is the Universal Gas Constant $= 8.31 \, \mathrm{N \, m \, K^{-1} \, mol^{-1}}$. Thus, the osmotic pressure difference $\Delta\Pi$ across a membrane where the solute concentrations are c_1 on one side and c_2 on the other is given by

$$\Delta\Pi = RT(c_1 - c_2). \tag{9.14}$$

If the difference in concentration across a membrane were $1 \, \mathrm{mol \, cm^{-3}}$ and the temperature of the solutions were $25°\mathrm{C}$ ($298 \, \mathrm{K}$), then the osmotic pressure difference across the membrane would be approximately $1360 \, \mathrm{atm}$! Thus, we can see that, in biological situations, even where solute concentrations are fairly low, the osmotic pressure can be quite considerable; for example, the osmotic pressure of plasma is approximately $3.3 \times 10^3 \, \mathrm{N \, m^{-2}}$ ($25 \, \mathrm{mm \, Hg}$).

Consider again the two-chamber model of **Fig. 9.5**, with a piston inserted into the tube above chamber B. By applying a load to the piston it is possible to increase the hydrostatic pressure in the chamber. At some instant, let the concentration of protein in chamber B be c_1, then the osmotic pressure of the solution Π_1 is given by Equation (9.13); if a pressure p_1 (equal in magnitude to Π_1) is applied to chamber B via the piston, then further passage of water into the chamber can be prevented and the protein solutions will not be further diluted. The osmotic transfer of water is stopped. This does not mean that water molecules do not cross the membrane, for they continue to diffuse across it in both directions; there is simply no *net* flux.

Van't Hoff's law is really only applicable to a perfect semi-permeable membrane separating two solutions; if, for any reason, the membrane is leaky then the osmotic pressure difference is less than predicted by Equation (9.13). For instance, if two water-filled chambers are separated by a membrane with pores of $6 \, \mathrm{nm}$ diameter, then the addition to one compartment of an ideal solute of the same molecular diameter as the pore diameter will generate an osmotic flow of water as given by van't Hoff's law. If, however, the same molar concentration of small molecule (say urea, diameter $0.5 \, \mathrm{nm}$) were added to one compartment instead of the large molecule, the resultant osmotic flow would be less than 5% of that predicted because most of the small molecules would be able to diffuse through the membrane.

This led to the empirical modification of the law as

$$\Pi = \sigma cRT, \tag{9.15}$$

where σ is termed the *osmotic reflection coefficient*. Its value varies from one for a perfect semi-permeable membrane to less than zero when the mobility of the solute is greater than that of the solvent.

The relationship between the osmotic and hydrostatic pressures of a solution is very important in biological exchange processes, since the transfer of molecules almost always involves a change in both pressures. Indeed, both capillary filtration and transcapillary exchange involve consideration of the interaction between these two processes and will be discussed in Chapter 13.

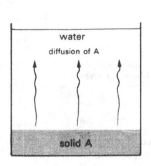
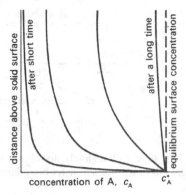

Fig. 9.6. The diffusion of a soluble material A into a stagnant liquid (e.g. copper sulphate into water) showing the progressively changing concentration profile with time in the bulk of the liquid.

A simple mass transfer model

In the circulatory system we are concerned with the transport of materials between the flowing blood and body tissues. Some of the mass transfer occurs by filtration across the vessel walls, but much takes place by diffusion around and through cells. As an introduction to the later discussion on mass transport within the circulatory system, we must first consider theories and models which have been developed to describe some simple dynamic mass transfer processes.

Let us consider in detail what happens when a crystal of copper sulphate (material A, **Fig. 9.6**) is immersed in still water. Initially there will be no material A in the water phase, but after a short while, as A dissolves and diffuses away, the water close to the surface will contain A. In a thin layer very close to the crystal surface, the solution will rapidly become *saturated*, in that no more molecules of A can be contained in it at that temperature. The concentration there is called the *equilibrium surface concentration* c_A^*, and at all times the concentration of A in this layer will be maintained at this level. Further away from the solid surface the concentration (c_A) will be lower but will increase with time. Dissolution and diffusion away will continue to occur until the equilibrium surface concentration c_A^* is attained throughout the aqueous phase. Until that time, there will be a continuously changing concentration gradient (and hence flux) within the water, as shown in **Fig. 9.6**. The decay in concentration with distance from the crystal surface will be roughly exponential and it is possible to determine how fast the concentration profile will change by applying Equation (9.3). For example, the distance (Δx) from the surface at which the concentration takes a particular value (e.g. $0.5c_A^*$) is directly proportional to $\sqrt{Dt}$. Thus, material moves faster as D is increased and the distance travelled increases with the square root of time.

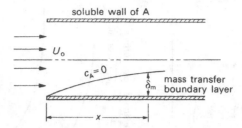

Fig. 9.7. A long tube with a slowly dissolving surface, through which steady laminar flow occurs. The growth of the mass transfer boundary layer is shown.

The interaction of bulk flow and diffusion

There are many situations where mass transfer occurs in the presence of bulk flow of material, and in such cases the transport rate is coupled to the flow conditions. An example is in blood vessels, where we are particularly interested in mass transfer to and from the walls during flow. We must remember that the major resistance to transfer may be located in the wall rather than the blood and, thus, that the latter may not be very important in determining the overall rate of transfer. In order to understand the coupled effect of the flow and the diffusional exchange we will look at a very simplified model. Consider a long straight tube as shown in **Fig. 9.7**; the flow entering the tube has a flat velocity profile which develops with distance along the tube as described in Chapter 5, p. 51. The wall of the tube is made of or contains a material which dissolves very slowly in the fluid, and fluid entering the tube contains none of this soluble material A. As the flow passes over the soluble section, then the wall progressively dissolves. Right at the beginning of the soluble wall ($x = 0$), the concentration of A in the fluid at the wall will be c_A^* (the equilibrium surface concentration), but there will be no A in the core of the fluid. As the flow proceeds downstream, diffusion of A away from the wall will occur and material will penetrate further into the core. However, because the core fluid is flowing downstream, the diffusing molecules will also be convected in that direction. If we map out the radial distance to which A will have diffused at progressive positions downstream (i.e. the distance from the wall to the points where c_A becomes effectively zero as shown in **Fig. 9.7**) the line describes the surface of what is termed a *mass-transfer boundary layer* (cf. the viscous boundary layer in Chapter 5).[2]

Once steady-state conditions have been established we may calculate the approximate rate of growth of the mass-transfer boundary layer with distance in a similar manner to the method used in Chapter 5 for calculating the growth of the viscous

---

[2] In practice, the local thickness (δ_m) of the mass-transfer boundary layer is defined as the distance from the wall at which the local concentration of the species considered differs from its concentration in the core fluid by only 1% of the total difference.

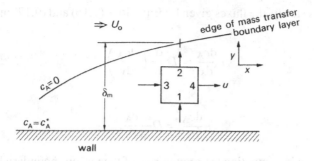

Fig. 9.8. A cubic element of fluid lying within a mass-transfer boundary layer.

boundary layer. Consider a small cubic element of fluid within the boundary layer, as shown in **Fig. 9.8**, and let each face of the cube be of unit area. Then the diffusional flux into the bottom of the cube through face 1 will be given by Equation (9.1), based on the concentration gradient dc_A/dy at the face. The diffusional flux out of the cube through the upper face 2 will also be given by Equation (9.1) using the concentration gradient across the upper face. Thus, the net diffusional flux into the element is given by

net flux = diffusivity × rate of change of concentration gradient with y

or

$$J = D\frac{d}{dy}\left(\frac{dc_A}{dy}\right). \qquad (9.16)$$

This is exactly analogous to the calculation of the net viscous stress on an element in the viscous boundary-layer calculation of Chapter 5.

This diffusional flux into the element is balanced by the convection of the material downstream. The convective flux into the element through face 3 is given by the concentration at the face multiplied by the velocity of flow into the element. We may compute the convective flux out through face 4 in a similar manner. Hence, the net convective flux out of the element is given by

flux = rate of change of (velocity × concentration) with x

or

$$J = \frac{d(uc_A)}{dx}.$$

Provided the rate of mass transfer is low and the molecules diffusing into the element do not contribute to a large change in its volume, then the velocity u is independent of x and

$$J = u\frac{dc_A}{dx}. \qquad (9.17)$$

Since the flow is steady, the fluxes given by Equations (9.16) and (9.17) must be equal. Thus,

$$u\frac{dc_A}{dx} = D\frac{d}{dy}\left(\frac{dc_A}{dy}\right)$$

or

$$u\frac{dc_A}{dx} = D\frac{d^2 c_A}{dy^2}. \tag{9.18}$$

As was the case in computing the growth rate of the viscous boundary layer, we can most easily estimate the growth rate of the mass-transfer boundary layer by ascribing suitable scales or magnitudes to each of the parameters in the equation. Thus, at a distance x from the beginning of the soluble section, the mass-transfer boundary-layer thickness will be δ_m. Across the boundary layer at this location the concentration of A in the fluid falls from c_A^* at the wall to $c_A = 0$ at δ_m from the wall and hence we can scale dc_A as Δc where this equals $(c_A^* - 0)$. The velocity scale within the mass-transfer boundary layer will depend upon whether or not the viscous boundary-layer thickness δ is very much less than δ_m. If the viscous boundary layer is very much thinner, then the appropriate velocity scale will be U_0, i.e. the core velocity of the fluid. On the other hand, if the mass-transfer boundary layer is much thinner than the viscous boundary layer, the velocity *gradient* can be assumed to be constant and a suitable velocity scale is $U_0(\delta_m/\delta)$.

For the case where the viscous boundary layer is very much thinner than the mass-transfer boundary layer we may substitute into Equation (9.18) as follows:

$$U_0\frac{\Delta c}{x} \propto D\frac{\Delta c}{\delta_m^2}.$$

Then

$$\delta_m = k\sqrt{\frac{Dx}{U_0}}, \tag{9.19}$$

where k is a numerical constant whose size must be computed for any given situation. Thus, the thickness of the layer increases with distance from the tube entrance as $\sqrt{x}$. If we compare this with Equation (5.3) (p. 53) they again appear very similar. Indeed, we can see that

$$(\delta/\delta_m)^2 = v/D. \tag{9.20}$$

In the case where the viscous boundary layer is larger, substitution in (9.18) yields the relationship

$$U_0\frac{\delta_m}{\delta}\frac{\Delta c}{x} \propto D\frac{\Delta c}{\delta_m^2}.$$

Hence,

$$\delta_m^3 = k\delta \frac{Dx}{U_0}.$$

If we consider the situation near the entrance of a tube, where δ is given by Equation (5.3), we obtain

$$(\delta/\delta_m)^3 = v/D. \tag{9.21}$$

On the other hand, if we consider the case where the mass transfer is occurring in a section of the tube remote from the entrance in which the velocity profile has become established, then δ is constant and

$$\delta_m \propto x^{1/3}.$$

The boundary layer grows in thickness with distance along the soluble section as $x^{1/3}$.

The Schmidt number The ratio v/D is a dimensionless quantity like the Reynolds number, and it is called the Schmidt number (*Sc*). Schmidt numbers for gases are typically about unity, but in liquids they are much larger. Thus, for oxygen in water at room temperature, the Schmidt number is approximately 600, for urea in water it is 950 and for lactose in water it is 2400. Typically in biological systems, Schmidt numbers for small molecules and ions (e.g. acetate) are of the order of 10^3; as the molecular size increases, so does the Schmidt number. Thus, large protein molecules can have Schmidt numbers of the order of 10^6 in water or plasma.

The enormous size of such Schmidt numbers has very great importance when we consider mass-transfer processes in biological systems. If we calculate the expected thickness of the mass-transfer boundary layer for a large protein molecule in a small vessel of say $100\,\mu m$ diameter (the viscous boundary layer cannot be greater than the vessel radius: $50\,\mu m$), we can see from Equation (9.21) that the layer will be approximately $5 \times 10^{-2}\,\mu m$ or $50\,nm$ thick. That is, it will be only a couple of molecules thick! Clearly, the idea of a continuum breaks down under such circumstances, but the calculation serves to indicate just how thin mass-transfer boundary layers can be in biological situations.

Part II

Mechanics of the circulation

10

Blood

This chapter is concerned with the mechanical properties of the blood and its constituents. We shall examine in Chapters 12 to 15 the flow of blood in blood vessels and its contact with their walls. The mechanics of fluids, discussed in Chapters 4 and 5, provide a background to the material that follows.

Blood is a suspension of the *formed elements* (the various blood cells) and some *liquid particles* (the *chylomicrons*) in the plasma. Plasma itself is an aqueous solution containing numerous low molecular weight organic and inorganic materials in low concentration, and about 7% by weight of protein (**Table 10.1**). The mechanical property of blood which is of principal interest to us is its viscosity. In order to understand what determines the viscosity of whole blood we must first consider what governs the viscosity of simple fluids and suspensions, then the mechanical properties of the plasma (p. 155) and the suspended elements (p. 157), and finally whole blood (p. 169).

Viscosity of fluids and suspensions

It was noted in Chapter 1 that the physical features of liquids, gases and solids are directly related to their molecular structure and that both liquids and gases are classed as fluids, because they flow when a shear stress is applied. The property which relates the rate of shearing to the shear stress is the viscosity (p. 37) and we must now consider the factors that determine the viscosity of a fluid.

It helps in understanding the physics of a liquid if at the same time we consider a gas. Gases are much less dense than liquids; therefore, the molecules of a gas are farther apart than those of a liquid. Thus they change location, by random motion, with greater ease because they collide less frequently. This is also associated with the fact that diffusion coefficients in gases are typically 10 000 times greater than in liquids.

149

Table 10.1. *Outline of composition of plasma*

Material	Concentration (g per 100 ml)	Molecular weight $\times 10^{-3}$	Molecular dimensions (nm)
Water	90–92		
Proteins			
Serum albumin	3.3–4.0	69	15×4
α_1 globulins (including lipoproteins)	0.31–0.32	44–200	
α_2 globulins (including glycoproteins)	0.48–0.52	150–300	
β globulins (including lipoproteins)	0.78–0.81	90–1300	20–50
γ globulins	0.66–0.74	160–320	23×4
Fibrinogen	0.34–0.43	400	$50\text{–}60 \times 3\text{–}8$
Inorganic constituents			
Cations			
Sodium	0.31–0.34		
Potassium	0.016–0.021		
Calcium	0.009–0.011		
Magnesium	0.002–0.003		
Anions			
Chloride	0.36–0.39		
Bicarbonate	0.20–0.24		
Phosphate	0.003–0.004		

Because of this large difference in the mobilities of the molecules in gases and liquids their viscosities arise in differing ways. For the case of a gas, imagine two parallel streams moving at different average speeds. Within each stream individual molecules are moving in random directions and with a wide spectrum of velocities. Some of the molecules from the slower stream will inevitably cross the plane separating the streams and be forced, by successive collisions with molecules in the stream they join, to speed up and hence adapt towards the average speed of the faster stream. At the same time, molecules moving in the opposite direction, across the plane of separation, will be slowed down. Thus the difference in relative velocity of the two streams is reduced. The manner in which its molecules migrate determines the viscosity of a gas; for example, when its temperature is raised, the thermal agitation of its molecules becomes more intense, and so its viscosity rises.

We may think of a liquid as possessing a molecular structure somewhere intermediate between the ordered form of a crystalline solid and the completely random form of a gas. As a solid melts to form a liquid its lattice structure breaks down, but small clusters of molecules with a coherent structure are still present. As the liquid temperature is raised, the increased vibrational energy of molecules within a group causes it to be progressively broken down in size, till at the boiling point the liquid possesses virtually no such groups.

The viscosity of a liquid is very many times higher than that of a gas both because the molecules are packed more closely together and because the relatively large coherent groups provide an increased resistance to deformation. Unlike the situation in gases, the viscosity of liquids decreases with increasing temperature. This is thought to result from the progressive breakdown of the coherent groups despite the effect of increasing mobility of individual molecules within the liquid.

When a suspension of randomly distributed particles (be they rigid, deformable or fluid) is flowing in an apparatus whose dimensions are large compared with those of the particles and the spaces between them, the mixture can be regarded as a homogeneous fluid, as discussed in Chapter 1. By studying the mechanical properties of such a suspension, we can begin to see what determines its viscosity and whether it behaves in a Newtonian fashion. Newtonian behaviour is defined in Chapter 4 (p. 38). We can begin to see, furthermore, what factors determine the viscous behaviour of blood.

Spherical particles If the suspended particles are spherical and non-settling – that is, have the same density as the suspending fluid, which is itself Newtonian – then in any motion the shear stress will be proportional to the strain rate (rate of shear) and the suspension will behave as a Newtonian fluid, unless the concentration of spheres is very high. This is borne out by experiments under steady-state conditions, with suspensions of rigid spheres, as described on p. 385. The effective viscosity (μ_s) of the suspension, which is its viscosity when measured in a particular viscometer under particular conditions (see also p. 174), is independent of the shear rate for volume concentrations (c) of suspended spheres as high as 30%. However, for concentrations greater than about 10% the effective viscosity depends on the method of measurement; that is, on the flow pattern in the viscometer. Therefore, when it is said that a suspension of rigid spheres is Newtonian, what is usually meant is that its viscosity, as measured in a particular way, is independent of shear rate. If the suspended particles are not spherical, or are deformable, or are attracted to one another and tend to aggregate, then the shear stress is not proportional to the shear rate unless the concentration is much less than 30%. A suspension of red cells in plasma, for example, shows a shear dependence of viscosity at concentrations greater than about 12% (**Fig. 10.16**). The effective viscosity of a suspension always exceeds that of the suspending fluid. To see that this must be so, consider a Newtonian fluid in which motion is caused by motion of its boundaries at a steady rate as in **Fig. 10.1**. The fluid will be sheared between the moving boundaries and, as a result of this shearing, energy will be dissipated at a rate proportional to the viscosity of the fluid.

Now suppose that rigid spherical particles are introduced (**Fig. 10.2**). These may rotate in the fluid, but they cannot deform in the same way as did the fluid whose

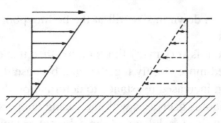

Fig. 10.1. Steady shearing of a Newtonian fluid between parallel plates. There is no slip of the fluid at either boundary (p. 38). The lower plate is stationary with respect to the observer and the length of the arrows indicates fluid velocity with respect to it. Clearly for an observer riding with the upper plate the fluid and lower plate would appear to be travelling in the opposite direction (dotted diagram). There is a linear change in velocity from one boundary to the other and as a result a constant or uniform velocity gradient (see Chapter 4).

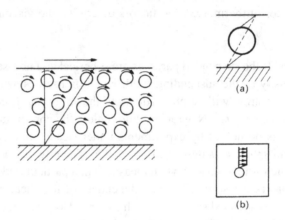

Fig. 10.2. Shearing of a suspension of rigid spherical particles. There is rotation of the particles, but as shown (inset (a)) for a given motion of the boundaries the presence of a rigid particle causes an increased rate of shear in the fluid. It is also seen (inset (b)) that there is a velocity gradient in the fluid adjacent to the particles, because fluid cannot slip at their surface.

volume they occupy. Therefore, deformations in the fluid must be greater for the same motion of the boundaries, and the average shear rate in the fluid increases. Furthermore, since fluid on the surface of the particles cannot slip (Chapter 4, p. 38), there is additional shearing in the fluid close to the particles. Both these effects lead to an increased rate of energy dissipation within the fluid, so that its effective viscosity is increased. As the volume fraction c of suspended particles increases, one would expect the effective viscosity to increase still farther, and this is borne out by experiment. However, unless the concentration of particles is very high the relationship

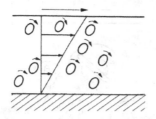

Fig. 10.3. Shearing of a suspension of liquid or deformable particles. The particles rotate as in **Fig. 10.2** but also become deformed and orientated in the direction of motion.

between shear rate and shear stress at any given concentration is constant – that is, the fluid is Newtonian.

A suspension of liquid drops or deformable particles also shows an increase in viscosity with increase in volume fraction, though it is smaller in amount for a given increase in concentration than in the case of rigid particles. This is because the drops can accommodate some of the deformation and the additional rate of strain imposed on the suspending fluid is therefore less. However, as the shear rate increases in such a suspension, the drops not only deform, but also progressively align with the flow (**Fig. 10.3**). This means that the shear stress does not increase linearly with shear rate; thus, the viscosity shows dependence on shear rate and the suspension is therefore non-Newtonian. This topic is dealt with in more detail shortly when we consider the behaviour (also non-Newtonian) of rigid non-spherical particles in suspension.

The behaviour of suspensions of rigid or deformable particles may also be complicated and rendered non-Newtonian by interaction between the particles, due to fluid disturbed by one particle disturbing the motion of another, or to attractive or repulsive forces between the particles themselves. The only exact theory we have relates to dilute suspensions of non-interacting particles. The effective viscosity μ_s of a dilute suspension of equal-sized, rigid, non-interacting, neutrally buoyant, spherical particles in a fluid of viscosity μ_0 was calculated first by Albert Einstein, in 1906. He predicted that, provided the volume concentration c of particles is small compared with 1, the relative viscosity of the suspension μ_r, which is the ratio of its effective viscosity to that of the suspending fluid, is given by

$$\mu_r = \frac{\mu_s}{\mu_0} = 1 + \tfrac{5}{2}c. \tag{10.1}$$

This result has been verified by experiment for values of c less than about 0.1. For greater values the complicated mutual interaction of particles has to be taken into account and this involves terms proportional to c^2. In 1932 Einstein's result was extended by G. I. Taylor to suspensions of liquid drops, which are forced to remain spherical, for example by surface tension. The result corresponding to

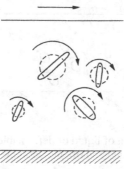

Fig. 10.4. Tumbling motion of asymmetric particles in a sheared suspension with schematic representation of the swept volume for each particle.

Equation (10.1) is

$$\mu_r = \frac{\mu_s}{\mu_0} = 1 + c\left(\frac{\mu_0 + \frac{5}{2}\bar{\mu}}{\mu_0 + \bar{\mu}}\right), \tag{10.2}$$

where $\bar{\mu}$ is the viscosity of the liquid in the drop. This reduces to Equation (10.1) as $\bar{\mu}$ becomes infinite; that is, as the drop becomes effectively rigid.

Asymmetric particles The shearing of a suspension of rigid spherical particles causes the particles to rotate with constant angular velocity. If, however, the suspension consists of asymmetric particles, then, though they rotate when the suspension is sheared, they are found to do so with a time-varying angular velocity. The angular velocity of an asymmetric particle is highest when its long axis is at right angles to the direction of the flow and lowest when it is parallel with the flow. It can be seen, therefore, that the particle spends most of its time with its long axis roughly parallel to the plane of shear. The effect of this 'lining up' of the particles by the shear force is to decrease the viscosity of the suspension, making its behaviour non-Newtonian.

The relative viscosity of a suspension of randomly orientated rigid asymmetric particles exceeds that of a suspension of rigid spherical particles at the same volume concentration c. This is because, during the tumbling motion induced by the shear force, each particle sweeps out a volume of fluid greater than its own volume, but which is intermediate between those predicted from its longest and shortest dimensions (**Fig. 10.4**). The relative viscosity of a suspension of rigid asymmetric particles increases both with increasing concentration and increasing asymmetry of the particles. The relationship is given by

$$\mu_r = \frac{\mu_s}{\mu_0} = 1 + Kc, \tag{10.3}$$

where K (a geometric factor) is greater than $5/2$ (see Equation (10.1)) and increases as the asymmetry of the particles increases.

A further complication occurs if the asymmetric particles are flexible. For reasons which we need not explore here, they will tend to bend during part of their cycle of rotation in a sheared flow (**Fig. 13.24**). As a result, the volume of fluid they sweep out is decreased as the shear rate goes up, and hence so is the effective viscosity of the suspension.

Under suitable conditions of particle size, temperature and viscosity of the suspending fluid, Brownian motion (p. 129) will ensure that asymmetric particles in suspension are randomly orientated. As we have seen, the shearing of such a suspension will reduce its effective viscosity, if the rate of shear is sufficiently high to overcome this effect of Brownian motion. This behaviour is of interest because blood plasma contains an appreciable amount of such proteins as fibrinogen and the globulins, which have markedly asymmetric molecules. These molecules are so large compared with those of water that they may reasonably be regarded as being in suspension and might, therefore, impart non-Newtonian properties to the plasma. In practice, as we shall now see, this does not happen.

Viscosity of plasma Plasma is a pale yellow transparent fluid, which is obtained by removing the cells from blood that has been prevented from coagulating. Typical values for the concentrations and dimensions of some constituents of normal human plasma are given in **Table 10.1**; the materials selected are those particularly likely to influence the mechanics of plasma, or whole blood. *Serum* has the same appearance as plasma, but is obtained by allowing blood to coagulate and removing the clot. Thus, it lacks the plasma constituents involved in clotting, in particular fibrinogen.

Normal plasma behaves like a Newtonian fluid at rates of shear comparable to those in vivo (p. 337), provided that precautions are taken during viscometry to eliminate an air–plasma interface, and thus prevent denaturation of the proteins and mechanical effects due to the interface. This confirms theoretical predictions that there should be an immeasurably small influence of shear rate on plasma viscosity, in the normal range; it seems, therefore, that Brownian motion maintains a random orientation of the asymmetric molecules at these rates of shear. Typical values for the viscosities of normal human plasma and serum, at $37\,^\circ$C, are 1.2 and $1.1\,\mathrm{mN\,s\,m^{-2}}$. The viscosity of plasma is increased in certain diseases, for example in multiple myeloma, a malignant disease of the globulin-forming cells in the bone marrow, when the plasma concentration of globulins may be raised several fold. It is not at present certain whether this abnormal plasma behaves as a Newtonian fluid.

The viscosity of normal plasma varies with temperature in just the same way as does that of its solvent, water; a $5\,^\circ$C increase of temperature, in the physiological range, reduces plasma viscosity by about 10% (**Fig. 10.5**).

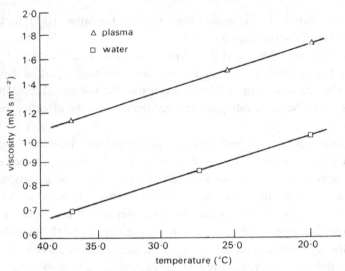

Fig. 10.5. Temperature dependence of plasma viscosity compared with that of water. (After Cokelet (1972). 'The rheology of human blood'. In *Biomechanics, its Foundations and Objectives* (eds Fung, Perrone and Anliker), chapter 4. Prentice-Hall, New Jersey.)

Osmotic pressure of plasma

We set out in Chapter 9 the physical basis of osmotic pressure and now provide some quantitative information on the osmotic properties of plasma. We also begin to examine the influence of plasma osmotic pressure on the mechanics of the circulation.

The osmotic pressure of a solution is determined by the number of particles present and by the temperature. By definition, a one molar solution of a substance contains an Avogadro's number (6.02×10^{23}) of particles per litre (or 1 mol l^{-1}). An ideal molar water solution at $0\,°C$ will have an osmotic pressure of $2.27\,MN\,m^{-2}$ (22.4 atm). The particles can be ions, molecules or aggregates of molecules, and so in an aqueous solution like plasma it is important to distinguish between electrolytes, which will dissociate into ions, and non-electrolytes. For example, a one molar solution of sodium chloride in water will become a two molar solution of ions if it is fully dissociated, and its osmotic pressure at $0\,°C$ will be $4.54\,MN\,m^{-2}$ (44.8 atm). A litre of a one molar solution of a non-electrolyte contains one *osmole* and that of a similar solution of an electrolyte, such as sodium chloride, two osmoles, if dissociation is complete.

The osmotic pressure of the plasma affects the mechanics of the circulation in several ways. An alteration of the osmotic pressure difference across the membrane of a blood cell will cause a shift of water and a change of the cell volume. For example, a reduction of the concentration of sodium chloride external to red cells will cause them to swell. An extreme case is the sphered red cell, which is much stiffer than the normal disc-shaped cell, because of its inability to deform without stretching its membrane

(p. 161). The change both in shape and flexibility will affect the mechanical properties of whole blood, as we shall see later in this chapter (p. 177). Furthermore, a change in plasma osmotic pressure will alter the *haematocrit*, the volume concentration of red cells in whole blood (p. 158) by redistributing water between the intravascular and extravascular spaces. This in turn will alter the mechanics of whole blood.

It is customary to describe a solution which has the same osmotic pressure as normal plasma (approximately $0.7\,\mathrm{MN\,m^{-2}}$ at $0\,^\circ\mathrm{C}$) as being isotonic, and solutions with higher or lower osmotic pressures as being respectively hyper- or hypo-tonic. The colloid osmotic pressure is another commonly used term, applied to the osmotic pressure due to the presence of macromolecules, as opposed to small molecules or ions. There is a small colloid osmotic pressure difference between intravascular and interstitial fluid (i.e. across the walls of the capillaries) because there is less protein in the latter. It is this pressure difference, together with the transcapillary hydrostatic pressure difference, which largely determines the equilibrium distribution of fluid between the intravascular and interstitial compartments (p. 400). It is albumin which mainly determines the colloid osmotic pressure of plasma. The contribution of fibrinogen to the osmotic pressure is very slight because its molar concentration and, hence, its osmotic pressure are only about one-fiftieth those of albumin.

The suspended elements

The elements suspended in plasma are the blood cells and the liquid droplets called chylomicrons. These droplets, which have diameters in the region of 0.2–$0.5\,\mu\mathrm{m}$, are part of the fat transport mechanism of the circulation. They are present, however, in such low concentration as to have no significant effect on the viscosity or osmotic pressure of the plasma. We discuss them, therefore, no further than to mention that they are principally present after fatty meals, enter the circulation from the lymphatics of the intestines and are removed by mechanisms which include transfer of the fat to water-soluble carrier molecules, the lipoproteins (**Table 10.1**).

The blood cells

Several types of cell are present in the circulating blood, but we focus attention particularly on the erythrocytes, or red cells, because they are the only cells which significantly influence the mechanical properties of blood. They do this because they are present in such high concentration (approximately $5 \times 10^6\,\mathrm{mm^{-3}}$), comprising about 45% of its volume (**Table 10.2**).

For comparison there are normally only one or two *white blood cells*, or *leucocytes*, per 1000 red cells. The *platelets* are more numerous than the leucocytes (50–100 per 1000 red cells); but they have a negligible effect on the mechanics of normal blood, compared with the red cells, because they are so small; the volume of a platelet is only about one-tenth that of a red cell, as may be seen from **Table 10.2**.

Table 10.2. *Cells in blood*

Cell	Number per mm^3	Unstressed shape and dimensions (μm)	Volume concentration (%) in blood
Erythrocyte	4–6 $\times$ 10^6	Biconcave disc 8 $\times$ 1–3	45
Leucocytes Total	4–11 $\times$ 10^3		
Granulocytes			
Neutrophils	1.5–7.5 $\times$ 10^3		
Eosinophil	0–4 $\times$ 10^2	Roughly	
Basophil	0–2 $\times$ 10^2	spherical	1
Lymphocytes	1–4.5 $\times$ 10^3	7–22	
Monocytes	0–8 $\times$ 10^2		
Platelets	250–500 $\times$ 10^3	Rounded or oval 2–4	

Red cells We must briefly consider some aspects of the physiology of the red cells, before examining their mechanical properties. Virtually all of the haemoglobin in blood (about 15 g per 100 ml) is within the red cells. It is the haemoglobin (m.w. 67 000), consisting of the protein globin united to the iron-containing organic molecule haem, which gives the cells their red colour, and because of its ability to combine reversibly with oxygen, endows blood with its large capacity for carrying oxygen. The O_2 capacity of whole blood at sea level is about 20 ml per 100 ml, whereas that of plasma is only about 0.3 ml per 100 ml.

The red cells are formed in the bone marrow and when immature they possess nuclei. The nucleus is, however, shed in most mammals as the cell matures, and nucleated red cells are normally absent from the circulating blood. Despite the lack of a nucleus, and of mitochondria,[1] red cells are metabolically active. Studies with isotopically labelled red cells show that their average life span is about 120 days. The maintenance of a steady concentration of the red cells in blood requires, therefore, that about 0.8% of the total population is destroyed and formed daily. Apparently, red cells become more fragile as they age, and it seems that they are destroyed, in vivo, partly by mechanical fragmentation and partly by engulfment by white blood cells and other 'scavenger' cells, a process called *phagocytosis* (p. 165). The mechanical fragmentation of the cells is believed to occur mainly in the microcirculation; it is there that the blood is subjected to the highest levels of shear stress (p. 337). We shall be devoting a lot of attention to the mechanics of the red cell, because of its important contribution to the mechanics of whole blood. However, the study of red

---

[1] These bodies are major intracellular sites of metabolic activity particularly involving oxidative energy release.

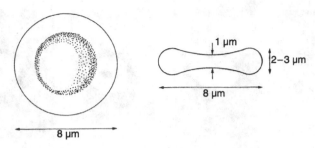

Fig. 10.6. Diagram of an unstressed erythrocyte.

cell mechanics has proved a difficult field and there is still lack of agreement on several subjects.

When suspended unstressed in plasma, or other isotonic fluid, the red cell is a highly flexible bi-concave disc with the shape and dimensions shown in **Fig. 10.6**. Its density is approximately $1.08 \times 10^3 \, \text{kg m}^{-3}$ and it is believed to consist of a very thin membrane and a liquid interior, which is apparently an almost saturated solution (approximately 32% by weight) of haemoglobin; we shall review the evidence for these properties later. Some of the components of the red cells are listed in **Table 10.3**. The viscosity of the fluid interior is thought to be about $6 \, \text{mN s m}^{-2}$, which is about five times greater than that of blood plasma. It is suggested that this may alter in some diseases, such as sickle cell anaemia (p. 168), when the haemoglobin may become crystalline.

Studies of the red cell membrane suggest that it contains a phospholipid bilayer, about 7.5 nm in thickness. The membrane accounts for no more than 3% of the mass of the cell, is highly permeable to the anions chloride and bicarbonate, and to water, but has a very low permeability to the cations potassium and sodium, and to macromolecules, including the plasma proteins. The cell possesses an Na^+–K^+ exchange

Table 10.3. *Composition of the red cell*

	Percentage of mass
Water	65
Membrane components (protein, phospholipid, cholesterol)	3
Haemoglobin	32
Inorganic	
Potassium	0.420 g per 100 ml
Sodium	0.025 g per 100 ml
Magnesium	0.006 g per 100 ml
Calcium	small amount

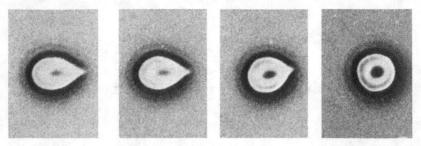

Fig. 10.7. Red cell made to adhere to a glass slide and then subjected to a shearing stress. The direction of the flow is from right to left and the level of the shear stress is greatest on the left. (From Hochmuth, Mohandas and Blackshear (1973). Measurement of the elastic modulus for red cell membrane using a fluid mechanical technique. *Biophys. J.* **13**, 754.)

pump, that pumps Na^+ out and K^+ in, against electrochemical gradients (compare **Tables 10.1** and **10.3**). It metabolizes glucose to produce energy for this pump and, because it lacks mitochondria, this metabolism is primarily *anaerobic*.

Several observations support the notion that the interior of the red cell is liquid. For example, the cell is capable of extreme distortion, as when it travels along a capillary with a diameter smaller than its own. The form that it then adopts, blunt in front and sharpened at the rear (**Fig. 13.33**a), suggests that there has been a flow of its contents into the blunt forward part. A similar conclusion is reached after a study of the behaviour of a red cell that has been made to adhere to a glass microscope slide by coating the slide with serum albumin. If the adherent cell is exposed to a shear stress, then it deforms (**Fig. 10.7**), and with increase of the shear stress it elongates and the cell contents appear to be squeezed out of the stretched tail into the cell body.

These and other observations (including electron microscopy) indicate that intact cells have no internal structure; that is, that the contents of the red cell are liquid. They suggest, furthermore, that the shape of the red cell is determined by the properties of its membrane. Support for that view comes from observations with red cell *ghosts*. If the osmotic pressure of the fluid in which red cells are suspended is gradually lowered, they not only become spherical, but lose their contents, first the low molecular weight materials and then, through rupturing of the membrane, the haemoglobin, a process known as *haemolysis*. If cells ruptured in this way are placed in a suitable isotonic solution the membrane seals, they absorb the solution and return closely to resemble normal red cells in shape and size. The name ghost arises because of the lack of haemoglobin.

There is, however, one piece of experimental evidence which has been used to suggest that the red cell has an internal structure; it can be cut into fine pieces, without losing its contents. To account for this observation it has been suggested that the cell interior contains a network of flexible material. However, neither electron microscopy nor ghost studies provide any support for this; possibly the cell membrane seals about the fragments as they are formed.

Since the thickness of the cell membrane is only some 7.5 nm, and the smallest radius of curvature of the red cell is about 1 μm (**Fig. 10.6**), thin-shell theory can be applied to examine its behaviour. This predicts that the membrane has negligible stiffness to bending, compared with its stiffness on stretching (see also p. 392). As a result, if the interior of the cell is liquid and, hence, able to offer viscous, but not elastic, resistance to deformation, the cell is expected to be very flexible.

The great ability of the red cell to deform (it can, for example, traverse a tube 3 μm in diameter and 12 μm in length without rupturing) is due, however, as much to its shape as to the properties of the membrane and to the liquid interior. Like other non-spherical bodies, the bi-concave red cell can deform into an infinite variety of shapes without changing its volume or surface area; that is, without stretching its membrane. The significance of this is most readily appreciated if one compares the mechanics of a normal and a sphered red cell. It can be shown that a sphere contains the maximum volume for a given surface area. Thus, the area of a sphered cell's membrane must increase if the cell is deformed, so that the tangential stiffness of the membrane determines the cell's flexibility. It is indeed found experimentally that sphered red cells cannot deform to the same extent as normal ones, and cannot, for example, traverse the same narrow channels without rupturing. However, with very severe deformation, as in a highly sheared flow, the membrane will become tense, tending to limit deformation of the cell, and may rupture as described below.

The Young's modulus of the red cell membrane has been determined and is in the region of $10^6 \, \mathrm{N\,m^{-2}}$. These measurements were made with the membrane *stretched* nearly to breaking point, and most biological tissues exhibit an increasing Young's modulus with increasing strain (see Chapter 7). Studies in which the motion of a red cell, as it returns to its *unstressed* shape after deformation, is compared with that of a scaled model cell whose properties are known yield a value for the Young's modulus of the membrane of about $10^5 \, \mathrm{N\,m^{-2}}$; that is, less by a factor of 10.

The red cell membrane has the property that however it is deformed it appears to maintain a constant surface area and thickness. For example, there is no observable change in surface area when a red cell is sphered osmotically. On the other hand, red cells elongate markedly when they are exposed to a high shear stress, as during tethering experiments (**Fig. 10.7**) or in a viscometer (**Fig. 10.9**).

We have not so far attempted to explain why a red cell which has a liquid interior and possesses a disc shape when suspended unstressed in an isotonic solution is capable of swelling and sphering. Theoretically, a cell whose membrane is uniformly extensible cannot behave in this way. As illustrated in **Fig. 10.8**, the change from a disc to a sphere requires a reduction of the circumference of the cell at its equator and hence wrinkling there, as the poles pop out. This is actually seen with a model red

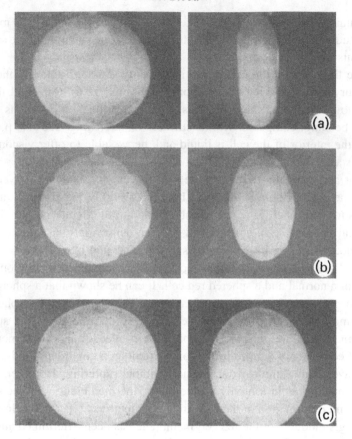

Fig. 10.8. (a) Thin-walled rubber model of red blood cell filled with water and immersed in water. With increase of transmural pressure (b) the poles bulge out and there is buckling at the equator. At large internal pressure (c) the enlarged cell is smooth again and almost spherical. (From Fung (1966). Theoretical considerations of the elasticity of red cells and small blood vessels. *Fedn. Proc. Fedn. Am. Socs. Exp. Biol.* **25**, 1761–72.)

cell which has uniform membrane properties, but not with real red cells. The observed behaviour of the red cell on sphering can be explained if the Young's modulus of the membrane and/or its thickness is greater at the equator than at the poles, but there is as yet no direct evidence on this point.

We now turn to the deformation of red cells which is induced by shear stress and consider observations that have been made when a sheared flow is set up in a dilute suspension of red cells. It is obvious that a cell must deform if it is to enter a very narrow tube and flow along it, and it will be apparent from the discussion of non-rigid suspended particles earlier in the chapter that a blood cell must deform in a sheared flow. Direct observations with a travelling microscope confirm this

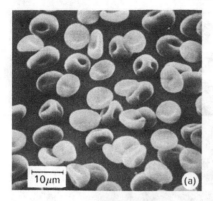 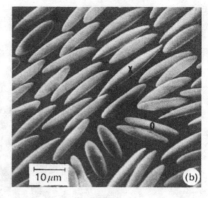

Fig. 10.9. Red cells fixed with glutaraldehyde while being subjected to a shearing stress in a viscometer. The stress is $10\,\mathrm{N\,m^{-2}}$ in (a) and $300\,\mathrm{N\,m^{-2}}$ in (b). (From Sutera, Mehrjardi and Mohandas (1974), personal communication.)

(Chapter 13). When a suspension is sheared the red cells are seen to deform and spin, because of the velocity gradient (shear rate), but the rate of deformation and spin depend on the shear rate and the concentration. At a shear rate of about $1\,\mathrm{s^{-1}}$ the cells bend as they spin, but with increase of the shear rate they cease to spin as a whole and instead travel down the tube with their major axis at an angle to the flow. Close examination suggests that they then have constantly changing contours, consistent with rotation of the cell membrane about the interior, rather like the moving tread of a tank. In that case the cell would be behaving to some extent like a liquid droplet (p. 153).

Experiments with laminar flow have suggested that a shear stress of about $200\,\mathrm{N\,m^{-2}}$ is required to haemolyse red cells, but this level is uncertain and some workers claim that it is contact with foreign surfaces, rather than the level of the shear stress, which primarily causes haemolysis. **Figure 10.9**a and b shows electron micrographs of very dilute (0.2%) suspensions of human red cells, fixed by glutaraldehyde while being subjected to different levels of shear stress in a concentric cylinder viscometer. There is mild deformation of the cells at $10\,\mathrm{N\,m^{-2}}$, but marked deformation at $300\,\mathrm{N\,m^{-2}}$, when the cells are greatly elongated (up to 20–$30\,\mu\mathrm{m}$). When the cells were fixed after the shear stress had been returned to zero, most had a normal appearance, implying that they had suffered no permanent change. A shear stress of $200\,\mathrm{N\,m^{-2}}$ is, however, probably a very high shear stress for the circulation, since a shear stress of $40\,\mathrm{N\,m^{-2}}$ will rupture or tear away the endothelial cells which line the aorta. In intensely turbulent flow, such as may occur downstream of a stenosed aortic valve, haemolysis is also thought to take place, accounting for the anaemia sometimes seen in this condition. However, the magnitude and duration of the shear stresses in this situation remain unknown.

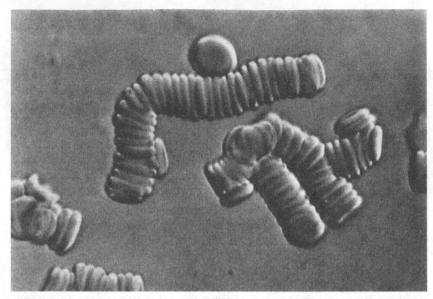

Fig. 10.10. Profile view of erythrocytes forming rouleaux. (From Bessis (1973). *Living Blood Cells and their Ultrastructure* (transl. Weed), p. 141. Springer Verlag, Heidelberg.)

A further property of red cells is their tendency to aggregate. If blood is smeared on to a microscope slide, the red cells can be seen to attach themselves side by side to form what are described as *rouleaux* (**Fig. 10.10**). This process begins after 10–20 s, and the rouleaux continue to increase in length for 1–2 min. The mechanism of rouleaux formation is still incompletely understood. It appears that the red cells attract one another and that the attraction is determined by charged groups on the surface of the cells. The process also depends on the presence of certain large asymmetric macromolecules in the external environment, particularly fibrinogen and globulins. The process will not occur in their absence, for example with red cells suspended in saline or serum albumin at normal concentration. Furthermore, it occurs progressively faster with increasing concentrations of fibrinogen and globulins. Thus, in a steady sheared flow the proportion of the cells aggregated into rouleaux at any instant depends on the concentration of these materials.

The greatest diameter of red cells when studied in rouleaux exceeds the unstressed value and cell thickness is reduced correspondingly; this is consistent with the presence of attracting forces between the cells. Red cells which have been sphered or hardened, for example by glutaraldehyde, will not form rouleaux, possibly because of their inability to undergo the necessary deformation.

Rouleaux can develop in vivo, as well as in vitro, at low rates of shear and their formation shows the same dependence on the concentration of fibrinogen and globulins.

They are important because of their influence on the viscosity of blood at low rates of shear (p. 174) and on its sedimentation velocity (p. 171).

White cells We discuss the white cells, or leucocytes, only briefly, because of their very small effect on the mechanical properties of blood; they are present in blood in very low concentration, constituting about 1% of its volume. They play a crucial role in defence against infection, both by acting as scavengers of micro-organisms and by forming antibodies. Details of these cells can be found in text books of physiology and haematology.

There are three varieties of white cell present in the circulating blood, the granulocytes, lymphocytes and monocytes, though these varieties are subdivided (**Table 10.2**). The different cells can be distinguished by their microscopic appearance and their affinity for various organic dyes.

White cells are normally roughly spherical in shape, with the diameters of the different varieties of cell ranging from about 7 to 22 μm (**Table 10.2**). Electron microscopy shows extensive internal structure, including the presence of mitochondria; and judged on the basis of volume, white cells are considerably more active biochemically than are red cells, exhibiting both aerobic and anaerobic metabolism. They are capable of active amoeboid movement, in which they project pseudopodia. They are also capable of *phagocytosis*, the engulfment of foreign material, including bacteria and ageing or fragmented cells. Smaller particles from macromolecules downwards can also be taken up and extruded, but this occurs by a different mechanism, known as *pinocytosis* (p. 421).

Rather little is known of the mechanical properties of the white cells. It has been argued that they are stiffer than red cells, because in a collision between a red and white cell in flowing blood it is the former which mainly deforms. There is a need, however, to take into account the visco-elastic properties of the cells. Thus, it has been observed that where a white cell is caused to flow from a wide tube into a capillary which is smaller than its unstressed diameter, 10–20 s are required before the cell has deformed sufficiently to flow along the capillary. White cells can be made to swell and rupture if they are placed in a suitable hypotonic solution.

Platelets The platelets are formed in the bone marrow, from very large cells called *megakaryocytes*. When viewed in vivo, for example in a capillary, they are small round or oval bodies, 2–4 μm in diameter and with a volume of 5–10 μm^3. They lack nuclei, but electron microscopy reveals a membrane about 8 nm thick, and inclusions similar to those found in other cells, including mitochondria, microfibrils and microtubules.

If platelets come into contact with ADP (adenosine diphosphate, an organic molecule important in many energy-releasing pathways in the body), they aggregate; ADP is regarded currently as the agent which normally promotes platelet aggregation

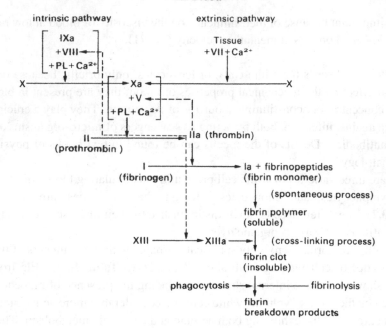

intrinsic pathway   extrinsic pathway

Fig. 10.11. Later stages of clotting process (see text). The factors are numbered according to the international nomenclature of clotting factors. Synonyms for those not identified on the figure are: V – proaccelerin; VII – proconvertin, or serum prothrombin conversion accelerator; VIII – antihaemophilic factor; IX – Christmas factor; X – Stuart–Prower factor; XIII – fibrin stabilizing factor. Activated derivatives are denoted by the suffix a. PL – plasma; Ca^{2+} – calcium ions. (After Hardisty and Weatherall (1974). The haemostatic mechanism. In *Blood and its Disorders*. Blackwell Scientific Publications, Oxford.)

in vivo. It is released from damaged red cells and connective tissue (collagen) at sites of injury, but its chief source for platelet aggregation is probably the platelets themselves, so that platelets adhering to damaged tissue may well release ADP for further platelet aggregation.

In the presence of thrombin (see next section, Blood coagulation, and **Fig. 10.11**), collagen, ADP or adrenaline, platelets rapidly lose their rounded shape, putting out many small protuberances (pseudopodia). This shape change, which is reversible, precedes aggregation. When aggregation occurs, the cells come into close contact and they extrude much larger pseudopodia (**Fig. 10.12**); these changes are irreversible.

Platelet aggregation is one of a number of processes, including clotting and vascular constriction, which operate to stop bleeding from damaged or cut vessels; such processes form important body defences against injury. There is evidence to suggest that platelet aggregation is active also in plugging spontaneously occurring small leaks in the normal circulation.

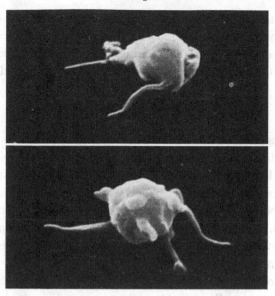

Fig. 10.12. Platelets having developed large pseudopodia (scanning electron micrographs). (From Bessis (1973). *Living Blood Cells and their Ultrastructure* (transl. Weed), p. 381. Springer Verlag, Heidelberg.)

Blood coagulation

Blood coagulates when it is in an abnormal environment, for example a glass test tube, a damaged blood vessel or a wound. The process of blood coagulation is exceedingly complicated and still incompletely understood. We shall therefore set out only the essential stages.

A blood clot consists of a mesh of *fibrils*, which are composed of the protein *fibrin*, and red and white cells and platelets (in the same relative concentrations as in whole blood) entangled in the mesh. There is very little fibrin present in normal blood; it is produced during coagulation, as shown in **Fig. 10.11.** This figure merely illustrates the important stages as they are at present understood; it is based upon a recent review to which the reader is referred. Note that the penultimate major step (conversion of prothrombin to thrombin) can be activated by two pathways: one (extrinsic) depending on the release of clotting factors by tissue damage and the other (intrinsic) dependent on the contact of blood with foreign surfaces. The anticoagulant substance heparin interferes with these pathways and also inhibits the reaction between thrombin and fibrinogen.

If fibrin is formed in blood vessels, in vivo, it may be removed either by phagocytosis or by being broken down, a process called fibrinolysis. Fibrinolysis, like fibrin formation, involves a complicated series of enzymic reactions.

After the formation of a clot it retracts (that is, the fibrin strands shorten), the clot becomes firmer and serum is squeezed out. It seems that the platelets adhering to the fibrin strands are involved in this process. They have been shown to contain a contractile actomyosin-like protein, which is similar to that in muscle.

Thrombosis

When blood coagulates in a blood vessel during life the process is called thrombosis. The process differs at different sites in the circulation, probably being strongly influenced by the local fluid mechanics. Thrombi, like clots, consist of the various elements of the blood trapped in a fibrin meshwork and the mechanism of production of the fibrin is probably similar to that in clotting, involving the chain of processes shown in **Fig. 10.11**. However, the initiating mechanism of thrombosis is usually somewhat different. Because of this and because of the fluid mechanical influence, the structure of thrombi not only differs from that of clots but may also be different in different parts of the circulation. The closest resemblance to clotting occurs in venous thrombosis, especially if the blood flow is very sluggish; this type of thrombosis is particularly likely to occur in older patients who are undergoing surgery or are immobile in bed for long periods. In arteries, on the other hand, the structure of thrombi is different; the first portion of the thrombus to be formed, the 'head', is made up largely of aggregated platelets on which layers of fibrin and blood cells are subsequently deposited.

There are multiple causes of thrombosis and the process can occur anywhere in the circulation, but in arteries at least the crucial initial event is the formation of an aggregate of platelets. Blood platelets do not normally adhere to one another or to vessel walls and, as described earlier, it appears to be the presence of ADP which initiates this. There are also grounds for suggesting that platelet aggregates and damaged tissue may produce material which can initiate blood clotting; these act via the 'extrinsic' pathway (**Fig. 10.11**). The local mechanics of flow may influence the process by affecting both the diffusion of the materials in the blood which are responsible for the platelet aggregation and clotting, and the supply of platelets and other cells to a site of thrombosis. Coagulation of blood within the circulation may be brought about experimentally by the injection of tissue extract or thrombin; thus it is possible that the levels of substances like ADP may, by causing platelet aggregation, influence the likelihood of thrombosis in vivo.

In arteries the commonest predisposing cause is atheroma (Chapter 12), but it seems that the blood must come into contact with collagen or other materials which are deep in the vessel wall, because thrombosis is not seen unless there is extensive damage to the endothelium, with exposure of these materials.

Thrombosis can occur in the microcirculation in such conditions as severe polycythaemia (an abnormally high concentration of red cells in the blood) and sickle cell anaemia. This is a congenital condition in which there is an abnormal haemoglobin in the red cells; the cells become sickle shaped and less flexible if the partial pressure of oxygen in the blood is reduced, as for example on going to a high altitude.

Certain veins in the calf muscles tend to become dilated with increasing age and are then particularly prone to thrombosis. This type of thrombosis especially affects older patients who are undergoing surgery or are immobile in bed for long periods, as noted above. It is believed that sluggish blood flow in the dilated veins is a predisposing factor, particularly because it is found that measures which increase local blood velocity, such as exercising the legs or compressing them rhythmically (p. 462), protect against thrombosis. However, experimental studies suggest that increased coagulability of the blood also plays an important role, though it would not account for the preferential occurrence of the process in certain veins. Thrombosis may also occur in the left atrium of the heart, especially if this is dilated and there is *atrial fibrillation* (small irregular ineffective contractions of its muscle). It is suggested that the mechanism is similar to that in venous thrombosis, in that the blood flow in the atrium is sluggish.

It seems that thrombosis seldom occurs in less than 2 or 3 days in immobile patients, but may occur in an hour in a patient having surgery. However, if a vessel wall is damaged experimentally, platelets can be seen to aggregate within seconds at the site of damage. The rate of growth of the aggregate is then proportional to the local wall shear rate. This is probably because platelets are carried to the site in greater numbers. In contrast, in the patients mentioned above, with venous and atrial thrombosis, wall shear rates are apparently low at the sites of thrombosis. It is important to appreciate that there are also mechanisms operating in vivo to remove thrombi, so that there can be, for example, restoration of the patency of an obstructed vessel.

In normal circumstances the clotting of blood is not the only process, and probably not the most important one, involved in the arrest of bleeding when a vessel is cut. Indeed, patients with congenital absence of fibrinogen in their plasma may lead almost normal lives. It seems that the main factors involved in the immediate arrest of bleeding are actually the formation of a plug of aggregated platelets and constriction of the torn vessel. A clot then forms locally, adhering to the tissues and acting as a seal. As might be expected, patients with a low blood concentration of platelets, or abnormal platelets, are liable to bleed spontaneously from small vessels and to bleed for prolonged periods from wounds.

Mechanical properties of whole blood

We have seen that whole blood may be regarded, from a mechanical point of view, as consisting of flexible red cells suspended in a Newtonian fluid, the plasma. We now consider the sedimentation of red cells in blood, the principles of the measurement of blood viscosity and the viscous properties of blood.

In the discussion that follows it is assumed that the blood is studied in sufficiently large samples for the influence of the walls of the apparatus which confines it to be

ignored, since this assumption simplifies the analysis. In practice, however, blood viscometry involves the use of small samples and the assumption is only partially valid.

Sedimentation of red cells This topic is of interest because the sedimentation of red
cells in blood can influence the mechanics of the circulation and may complicate the
measurement of blood viscosity. Furthermore, observation of the rate of sedimentation of red cells in blood is an important diagnostic test.

In Chapter 5 (pp. 69–79) we discussed the steady flow of a viscous fluid past a rigid
spherical body immersed in the fluid, for the case when inertia is negligible. Such a
flow occurs also when a rigid spherical particle sinks slowly through a viscous fluid
and the walls of the vessel containing the fluid are sufficiently far away to exert no
effect. In this situation the downward gravitational force of the particle is balanced
by the viscous drag force. From this force balance the speed of fall U_0 can be shown
to be given by Stokes' law (Equation (6.1)), i.e.

$$U_0 = \tfrac{2}{9}a^2 g \left(\frac{\rho_p - \rho_f}{\mu} \right),$$

where a is the particle radius, ρ_p and ρ_f (with $\rho_p > \rho_f$) are respectively the particle and
fluid density, μ is fluid viscosity and g is the gravitational acceleration. In particular
we see that the sedimentation velocity of the particle depends on the square of the
radius.

If the particle is released from rest in the fluid, its sedimentation velocity U increases until it attains the above steady value, which is called the terminal velocity
U_0. Its velocity as a function of time t is given by

$$U = U_0 \left(1 - e^{-t/T} \right),$$

so that at large times ($t \gg T$) U becomes equal to U_0. The quantity T is a time scale
for the process and is given by

$$T \propto \frac{a^2 \rho_p}{\mu}.$$

Thus, it increases with particle size. Note that when $t = 3T$, U is approximately 95%
of U_0 (Chapter 8, p. 117).

If we consider a rigid spherical particle the size of a red blood cell (8 µm in diameter) we find that T takes a value about 5×10^{-5} s. Therefore, the particle achieves a
velocity which is 95% of its terminal velocity in the astonishingly short time of about
10^{-4} s.

We cannot apply this theory exactly to a red cell, because of its shape and flexibility; nevertheless, the theory for rigid spheres gives some indication of the behaviour
of a single sedimenting red cell. At the moment no theory is available to describe the
sedimentation of many red cells in suspension, but the rigid sphere theory gives some
feel for this problem also. If whole blood, which has been prevented from clotting,
stands in a vertical tube, the cells aggregate into rouleaux and then settle slowly, leaving clear plasma above. The distance moved by the upper boundary of the blood cells

in the first hour is used as a laboratory screening test, called the *erythrocyte sedimentation rate* or ESR. The normal value for this test is about $3\,\mu m\,s^{-1}$, conventionally reported as $10\,mm\,h^{-1}$. In fact the sedimentation velocity is not constant, because there is initially formation of aggregates as noted above and the velocity falls as the cells begin to pack at the bottom of the tube.

In disease the ESR may be increased as much as tenfold, and this increase is associated with a raised concentration of fibrinogen and/or globulins in the plasma. These plasma proteins increase the tendency for red cells to form rouleaux (p. 164), and it is presumably the rapid sedimentation of these aggregates which accounts for the raised ESR. The change is not due to an alteration in the viscosity of the plasma, which is actually increased, and would thus reduce the ESR if it acted alone.

Under certain conditions, with blood flowing in a horizontal tube, the cells have been seen to sediment away from the uppermost surface and to pack more densely near the lowest surface, thus altering conditions at these interfaces. For example, the viscosity will be reduced near the top of the tube and raised at the bottom. This can presumably happen in vivo, if the blood flow rate is very low or there is a change in the plasma proteins leading to increased rouleaux formation.

The sedimentation of red cells could give rise to errors in the measurement of blood viscosity, particularly with a cone-and-plate viscometer (p. 173), which has practically horizontal surfaces. When operated at the lowest shear rates, red cells will tend to aggregate and sediment away from the upper plate towards the lower. This problem does not appear to have been studied in detail.

Principles of measurement of blood viscosity Before proceeding to a discussion of the mechanics of whole blood in a shear flow we must first consider technical problems associated with the measurement of the viscosity of blood. These arise because it is a particulate suspension and may have non-Newtonian behaviour at physiological rates of shear.

If a material is suspected of showing non-Newtonian behaviour then its viscosity must be measured at a number of different shear rates (or, equivalently, the shear rate must be measured at a number of different viscosities). However, this is particularly difficult in non-Newtonian materials, because a viscometer which will give an essentially constant rate of shear throughout a sample of a Newtonian material may not do so in a sample of a non-Newtonian material. As we have seen, it is the presence of red cells which make normal blood non-Newtonian. The technical difficulties they introduce in viscometry are mainly associated with the fact that blood is not a homogeneous fluid. Thus the viscosity of blood will depend on the size of the sample that is studied; particular problems arise if the dimensions of the sample sheared in the viscometer are comparable with those of a red cell, or, if rouleaux form, with their size. This topic is discussed in detail in the chapter on the microcirculation (Chapter 13). No viscometer is perfect, but some overcome these difficulties better than others.

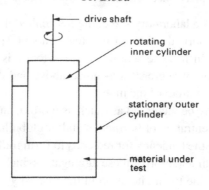

Fig. 10.13. Diagram of a concentric-cylinder viscometer. (From Van Wazer, Lyons, Kim and Colwell (1963). *Viscosity and Flow Measurement*, p. 50. Interscience Publishers, New York.)

Two types of viscometer are commonly used, namely *capillary* and *rotational* viscometers. In the first the fluid is made to flow along a tube of precisely known dimensions, by applying a given pressure difference between the ends. The instrument is constructed so that with a suitable choice of fluid the flow is laminar and effectively fully developed (p. 46). The apparent viscosity of the fluid can then be obtained correctly from Poiseuille's law (p. 47).

A capillary viscometer is well suited to studying Newtonian fluids such as plasma, but is unsuitable for non-Newtonian fluids, including blood, for both the reasons already mentioned. The fluid is subjected to a wide range of shear stresses, varying from a maximum at the wall to zero at the centre line (p. 49). And because a small-bore tube (say 100 μm diameter) has to be used to obtain a reasonable pressure difference with a fully developed laminar flow, the diameter of the tube is only a few times greater than that of the red cells (or of rouleaux). Under these circumstances the apparent viscosity of blood depends on the tube diameter, as discussed later (p. 390).

There are two types of rotational viscometer in common use: the *concentric cylinder* (**Fig. 10.13**) and the *cone and plate* (**Fig. 10.14**). Both consist essentially of two parts, one of which is made to rotate at constant velocity relative to the other, separated by the material under test. Therefore, torque is transmitted to the stationary part by the viscous action of the fluid. This can be measured by the angular displacement of the stationary part, if it is suspended from a wire; it will turn until the torsional force in the wire balances the frictional torque.

Both these instruments have advantages over the capillary viscometer for the study of blood, chiefly because the sample is mostly sheared at something approaching a constant rate if the design of the instrument is correct and the shear stress is sufficiently high. Both instruments also permit the study of a sample over an extended period of time, and of any time-dependent behaviour.

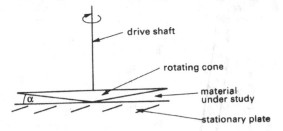

Fig. 10.14. Diagram of a cone-and-plate viscometer. α is the angle between the cone and plate. (After Van Wazer, Lyons, Kim and Colwell (1963). *Viscosity and Flow Measurement*, p. 52. Interscience Publishers, New York.)

In a given concentric-cylinder viscometer, when there is a steady laminar flow of a Newtonian fluid in the gap, the shear rate in the fluid is proportional to the imposed angular velocity and falls across the gap with the square of the radius (**Fig. 10.15**). Therefore, the shear rate can be regarded as constant only if the gap width is much less than the radius. The torque transmitted to the outer cylinder by rotation of the inner (it is conventional to rotate the inner cylinder) is directly proportional to the shear stress and hence to the apparent viscosity of the fluid and the angular velocity.

If the material under study is non-Newtonian, neither the shear stress nor the measured torque is directly proportional to the angular velocity. Equations have been derived for the behaviour of different kinds of non-Newtonian material, but, because these are based on highly idealized models of the materials, it is more appropriate to obtain the apparent viscosity of such materials by calibrating the viscometer with a Newtonian fluid of known viscosity. The test material is then examined at different rates of shear and it will be found that its apparent viscosity is shear-rate dependent.

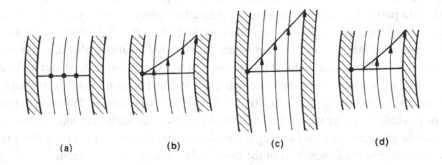

(a)   (b)   (c)   (d)

Fig. 10.15. Fluid velocity in streamlines in the gap of a concentric-cylinder viscometer. The inner cylinder rotates and the outer is stationary, (a) Apparatus at rest; (b) sheared Newtonian fluid – note that fluid velocity falls across the gap (see text); (c) if the gap width is very small compared with the radius of the cylinder, the velocity gradient is almost constant as between parallel plates (**Fig. 10.1**); (d) fluid with a yield stress. (After Van Wazer, Lyons, Kim and Colwell (1963). *Viscosity and Flow Measurement*, p. 75. Interscience Publishers, New York.)

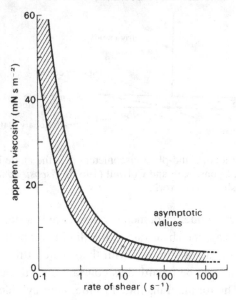

Fig. 10.16. Variation of apparent viscosity of human blood with rate of shear. (After Whitmore (1968). *Rheology of the Circulation*, p. 67. Pergamon Press, Oxford.)

Further difficulty is experienced if the material exhibits a yield stress, since certain parts of the material do not flow when the shear stress is sufficiently low. Normal blood behaves like this at extremely low shear stresses.

In the cone-and-plate viscometer (**Fig. 10.14**) the test material is contained in the space between a cone of large apex angle and a plate with a flat surface normal to the axis. The cone is usually rotated and the torque transmitted by the material to the plate is measured. This instrument has some advantages over the concentric-cylinder device. In particular, in addition to using a smaller volume of test material, it is so designed that for small angles between the cone and plate ($\alpha < 3°$) the shear rate for a Newtonian fluid may be considered constant for all radial positions; this is nearly true even for a non-Newtonian fluid, if its viscosity is not highly shear dependent. As in the case of a concentric-cylinder viscometer, a non-Newtonian material is studied by calibrating the instrument with a Newtonian fluid of known viscosity. However, it may not be possible to ignore the particulate nature of a material such as blood, because of the narrow gap across which shearing takes place. Near the apex of the cone this may be comparable to the dimensions of the blood cells, or aggregates of them.

Viscous properties of blood We now describe experimental results on the viscosity of blood, examining first the dependence on shear rate. The results in **Fig. 10.16**, which are based on several studies with human blood, show the variation of the apparent viscosity with rate of shear. When the shear rate is about $1000\,\text{s}^{-1}$, which is typical

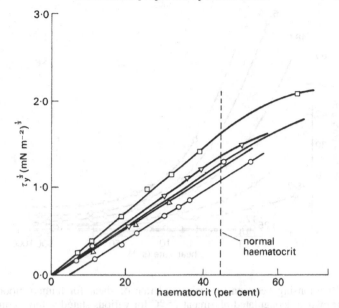

Fig. 10.17. Variation of yield stress τ_y of blood with haematocrit (five different blood samples). When the haematocrit is less than about 5% no yield stress is found. The haematocrit is plotted against the cube root of the shear stress, because of the observed linear relationship. (After Merrill, Gilliland, Cokelet, Shin, Britten and Wells (1963). Rheology of human blood near and at zero flow. *Biophys. J.* **3**, 210.)

for many blood vessels in vivo (p. 337), the non-Newtonian behaviour becomes insignificant and the apparent viscosity approaches an asymptotic value, which is in the range $3-4\,\mathrm{mN\,s\,m^{-2}}$. However, with reduction of the shear rate the apparent viscosity increases slowly, until at a shear rate less than $1\,\mathrm{s^{-1}}$ it rises extremely steeply.

In order to interpret these findings, we first consider the behaviour of blood at very low rates of shear. Rouleaux are known to form in blood at low rates of shear and a tangled network of aggregated red cells can be observed when blood is not sheared. If such blood is subjected to a shear stress below a critical value, the aggregated cell structure is believed to deform, without the blood flowing; that is, it exhibits a *yield stress*. There is certain experimental evidence which supports this idea: careful measurements with normal human blood show a yield stress of about $1.5-5.0\,\mathrm{mN\,m^{-2}}$. This decreases as the haematocrit is decreased, and there is a critical haematocrit below which no yield stress is found, as seen in **Fig. 10.17**. It is presumed that there are then too few red cells per unit volume of blood to permit complete bridging of the sample by a structure of aggregated red cells. The yield stress of blood is increased if the concentration of fibrinogen or gamma globulin in the blood is high, which is expected, because of the tendency for these asymmetric protein molecules to promote rouleaux formation.

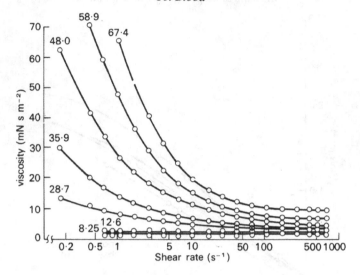

Fig. 10.18. Relationship between viscosity and rate of shear for human blood cells suspended in their own anticoagulated plasma at 25 °C for various stated volume concentrations of the erythrocytes. (From Brooks, Goodwin and Seaman (1970). Interactions among erythrocytes under shear. *J. Appl. Physiol.* **28**, 174.)

The effect of the haematocrit on the mechanics of blood flow has been examined over a wide range of shear rates, and behaviour at shear rates up to $700 \, \text{s}^{-1}$ is illustrated in **Fig. 10.18.** Suspensions of red cells in plasma at haematocrits up to about 12% are found to be Newtonian at all rates of shear, although they have a higher viscosity than plasma. However, as the haematocrit is further raised, not only does the viscosity of the suspension increase, but non-Newtonian behaviour is observed, detectable first at very low rates of shear. We have already seen, in discussing the behaviour of suspensions (pp. 151 *et seq.*), the reasons for this increase of viscosity with haematocrit. The non-Newtonian behaviour of blood at very low rates of shear is partially due to the behaviour of the rouleaux. It has been observed microscopically, with a transparent cone-and-plate viscometer, that the average length of the rouleaux in blood decreases as the rate of shear is increased until the red cells are present individually, and such a change can be expected to be associated with a reduction of viscosity (p. 153). In this context we should mention that rouleaux in a sheared flow have been seen to bend and straighten alternately as they rotate, and that too can be expected to reduce dissipation (p. 154). These are not the complete answer, however, because a suspension of red cells in saline also exhibits non-Newtonian behaviour, and red cells in saline do not form rouleaux.

We must conclude, therefore, that the individual red cells are in part responsible for the non-Newtonian behaviour of blood. We have already noted that liquid droplets in a sheared suspension deform. There is experimental evidence, obtained with dilute

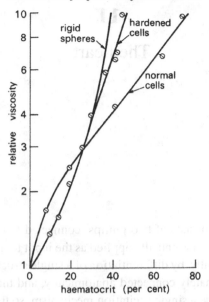

Fig. 10.19. Influence of haematocrit on the relative viscosity of normal and hardened cells suspended in saline (rate of shear 230s^{-1}). (After Whitmore (1968). *Rheology of the Circulation*, p. 81. Pergamon Press, Oxford.)

suspensions of red cells, which indicates that they too deform when in a sheared flow (p. 161). Thus, it seems likely that it is this deformation and rotation of the red cells which contribute to the non-Newtonian behaviour of blood. Strong support for this argument comes from a study with experimentally hardened red cells; a suspension of such cells in saline, even at a haematocrit of 40%, is approximately Newtonian, like a suspension of rigid spheres (p. 151).

These observations suggest that the flexibility of the red cells plays an important part in determining the viscosity of blood and that at a given haematocrit and shear rate the viscosity of a suspension of hardened cells will exceed that of normal cells. In general this expectation is borne out, as shown in **Fig. 10.19**; at a haematocrit exceeding 30% and a shear rate of 230s^{-1} the viscosity of a suspension of experimentally hardened red cells considerably exceeds that of a suspension of normal cells and approaches that of a suspension of rigid spheres. Similar findings are obtained, moreover, with red cells that have been sphered and which, therefore, are less flexible (p. 161), and with the stiffer cells of patients with sickle cell anaemia (p. 168). It is a striking fact that the viscosity of a suspension of rigid spheres tends to infinity as the volume concentration (equivalent to the haematocrit) approaches 60%; that is, it cannot be sheared. A suspension of normal red cells, on the other hand, will flow even at a haematocrit of 98%!

11

The heart

The mammalian heart consists of two pumps, connected to each other in series, so that the output from each is eventually applied as the input to the other. Since they are developed, embryologically, by differentiation of a single structure, it is not surprising that the pumps are intimately connected anatomically, and that they share a number of features. These include a single excitation mechanism, so that they act almost synchronously; a unique type of muscle, cardiac muscle, which has an anatomical structure similar to skeletal muscle, but some important functional differences; and a similar arrangement of chambers and one-way valves. Not surprisingly, the assumption has often been made that the function of the two pumps will also be similar. Thus it has become common practice to examine the properties of one pump, usually the left, and to assume that the results apply to the other also. This may often be unjustified, particularly in studies of cardiac mechanics, with the result that our knowledge of the mechanics of the right heart and the pulmonary circulation remains very incomplete. It must also be remembered that the scope for experiments on the human heart is very limited, and we must rely heavily on experimental information from animal studies. Thus the descriptions which follow apply primarily to the dog heart.

Many factors which affect the performance of the heart are not our concern in this chapter, among the most important being the wide range of reflexes which act on the heart. For example, nerve endings in the aortic wall and carotid sinus are sensitive to stretch, and thus to changes in arterial pressure. A fall in arterial pressure, however caused, will reduce their frequency of initiation of nerve impulses. This change, transmitted through nerves to the brain, alters activity in nerves passing to the heart, causing an increase in its rate and force of contraction. This in turn increases cardiac output and tends to restore the aortic pressure. A number of such reflexes, and a range of hormones reaching the heart via the bloodstream, are continuously modifying its performance in the intact animal, but since they are active physiological mechanisms, with origins outside the heart, we ignore them in this chapter. Thus no attempt is being made to provide a comprehensive description of normal cardiac function; only the muscular and mechanical features of cardiac pumping are explored.

The two pumps which make up the heart must operate over a wide range and be extremely well matched. Cardiac output in humans may increase from a resting level of about $5 \, \mathrm{l min}^{-1}$ to $25 \, \mathrm{l min}^{-1}$ on strenuous exertion, and since we know that the ability of the pulmonary circulation to alter its volume (normally 0.5–1 l) is rather limited, the output of the two pumps must be the same except in the very short term.

The response of the heart in exercise involves an increase in rate of contraction and in output per beat (*stroke volume*). Thus, the muscle fibres in the heart wall appear capable of varying both the time course and the amplitude of their contraction, and it follows that the intrinsic contractile properties of cardiac muscle are of fundamental importance in any study of cardiac mechanics. They have, in the last 20 years, been the subject of intensive study, which was stimulated by the success of previous work on skeletal muscle in quantitatively describing the mechanical features of contraction. It has become clear that profound differences in functional properties exist which complicate the study (and the behaviour) of cardiac muscle, and different experimental methods and terminologies have added to the difficulties. Many interesting problems have emerged in translating the tension-generating and shortening properties of experimentally isolated slips of muscle into the pressure-generating and volume-ejecting properties of the intact ventricle, because the structure and geometry of the ventricle are complicated. Uncertainty about internal energy losses in the muscle and the ordering and synchronicity of activation adds further difficulty here.

At the moment, therefore, although detail of the ultrastructure of the contractile apparatus is becoming available, our understanding of myocardial mechanics remains incomplete. The behaviour of the muscle is still described in terms of crude functional models, and the behaviour of the contracting heart chamber can only be described qualitatively because of the geometrical uncertainties, even though the time course and magnitude of the fluid-dynamic events in the cardiac cycle are being measured with increasing precision in both animals and humans.

Anatomy of the heart

Each of the cardiac pumps consists of a low-pressure chamber (*atrium*), which is filled from a vein system and empties via a non-return valve into a high-pressure chamber (*ventricle*). The ventricle in turn passes blood on through a second non-return valve to an arterial system (**Fig. 11.1**). The right heart receives blood from the body tissues via the *systemic* veins and pumps it into the *pulmonary* (lung) arteries. The left heart receives this blood from the pulmonary veins and pumps it into the systemic (main body) arterial circulation, via the aorta and its branches.

The two atria are comparable in structure; their walls are thin and relatively compliant and they are separated from each other by a common wall, the atrial septum. The veins draining into them communicate with them without valves. The valves separating the atria from the ventricles (*atrioventricular valves*) differ slightly

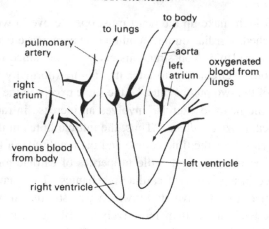

Fig. 11.1. Diagrammatic representation of anatomy and directions of blood flow in the heart.

in structure on the two sides of the heart, the one on the right (*tricuspid*) having three cusps and the one on the left (*mitral*) two. These cusps consist of flaps, attached along one edge to a fibrous ring within the wall of the heart, and with free edges projecting into the heart cavity. They are extremely thin (about 0.1 mm) and are made up largely of a meshwork of collagen and elastin fibres, covered by a layer of cells similar to those lining the walls of the heart chambers and blood vessels (the *endothelium*). The free edges of the cusps of both the mitral and tricuspid valves are 'tethered' to the wall of the respective ventricle by fine fibrous bands (*chordae tendinae*) which connect with slips of muscle (*papillary muscles*) projecting from the wall of the ventricle. Each chorda connects to the free edges of two cusps where they come in contact with each other when the valve closes; thus, tension in the chorda generated by contraction of the papillary muscle at the beginning of systole counteracts the tendency of the increase in intraventricular pressure to turn the valve inside out and allow leakback into the atrium. There appears to be a self-contained mechanism for closure of these valves (see p. 227) in the flow events in late systole.

The exit valves from the ventricles (pulmonary and aortic) are very similar to each other, each consisting of three cusps with free margins reaching to the wall of the valve ring. This arrangement allows opening to the full cross-sectional area of the valve ring without distortion of the cusps. These cusps are not tethered, but can nonetheless support considerable pressure differences (in the case of the aortic valve, approximately $1.3 \times 10^4 \, \mathrm{N\,m^{-2}}$ (100 mm Hg)). They are also extremely efficient, since only trivial backflow occurs through them during closure (which occurs more than 30 million times a year). Not surprisingly, their behaviour has been a source of interest for many years; in the fifteenth century Leonardo da Vinci examined their structure and made speculative drawings of flow patterns through them which have since proved

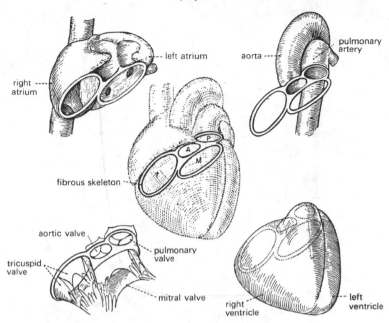

Fig. 11.2. Relationships of various structures in the heart. The 'skeleton' of the heart, consisting mainly of the fibrous valve-rings, also supports both atria and ventricles, and the two great arteries. (From Rushmer (1970). *Cardiovascular Dynamics*. W. B. Saunders Company, Philadelphia.)

remarkably realistic; their mechanism of action has recently been studied in detail (see p. 229).

The four valve orifices in the heart are aligned approximately in a single plane, and the cusps of each are attached at their bases to a stiff ring of fibrous tissue (**Fig. 11.2**). The four rings are in turn connected to each other by fibrous tissue so that the valve apparatuses are set in a stiff framework to which the muscle fibres of each chamber are attached – the atria on one side, the ventricles on the other; and the pulmonary artery and aorta also attach at their origins to this plane of fibrous tissue. The whole heart is contained in a thin fibrous tissue bag, the pericardium, which in turn is attached to other structures within the chest, and through these to the vertebral column. The stress–strain relationship of the pericardium is, as can be inferred from **Fig. 11.3**, highly nonlinear. At normal diastolic heart volumes, the pericardium is only slightly stretched, and probably does not appreciably affect cardiac filling; in certain diseases where fluid accumulates within the pericardium, its constraint can restrict diastolic filling of the ventricles and severely reduce cardiac output (see p. 466).

The left ventricle is the only chamber whose wall structure has been examined systematically. Recent studies of microscopic sections taken serially through the

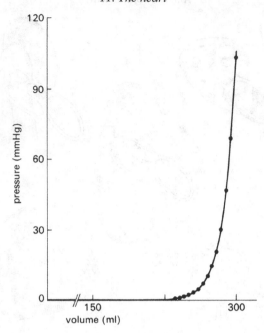

Fig. 11.3. Pressure–volume curve of pericardium of dog. (From Holt, Rhode and Kines (1960). Pericardial and ventricular pressures. *Circulation Res.* **8**, 1171. By permission of the American Heart Association, Inc.)

whole thickness of the wall have shown that older concepts of a ventricular wall made up of clearly defined muscle layers are incorrect, and that there is a well-ordered continuous distribution of fibre orientation (**Fig. 11.4**). The innermost (endocardial) fibres run predominantly longitudinally from the fibrous region around the valves (the base) to the other end of the roughly elliptical chamber (the apex); the fibres slightly further out into the wall lie at a slight angle to the axis of the chamber, so that the fibres spiral slightly as they run towards the apex. This angulation increases in successively deeper fibres so that those approximately halfway through the wall run parallel to the shorter axis of the chamber (i.e. circumferentially); thereafter, the angulation continues, so that the fibres at the outside (epicardial) surface are once again longitudinal. This arrangement gives the ventricle great strength even though individual muscle fibres can only bear tension axially, since a stress applied to the outer wall in any direction can be resisted by at least a proportion of the muscle fibres, and there is no direction or plane of weakness. Note that fibres do not have to terminate at the apex; they can turn and spiral back towards the base like string wound spirally on a stick. As can be inferred from **Fig. 11.4**b, there is a great preponderance of fibres with a circumferential distribution.

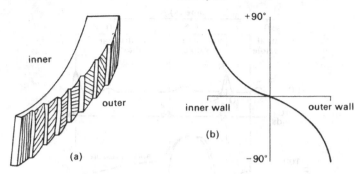

Fig. 11.4. Muscle fibre orientation in wall of left ventricle (LV): (a) the orientation of the long axis of the fibres at successive depths from the outer surface; (b) the way in which the angle made by the fibre with the circumferential plane of the LV changes continuously through the wall. (After Streeter *et al*. (1969). Fibre orientation in the canine left ventricle during diastole and systole. *Circulation Res.* **24**, 339.)

The cardiac cycle

Electrical events The contraction cycle of the heart is initiated in a localized area of nervous tissue in the wall of the right atrium known as the pacemaker or sino-atrial node. This has the inherent property of cyclical depolarization and repolarization (the latter process being dependent upon metabolic energy, derived ultimately from metabolism within the cells). When depolarization occurs in the pacemaker, it spreads at about $1\,\mathrm{m\,s^{-1}}$ into and through the surrounding muscle of the right and left atrial walls (causing atrial contraction) and then into a discrete nervous pathway (the atrio-ventricular bundle, or bundle of His) which passes through the fibrous tissue around the tricuspid valve ring into the muscular septum between the two ventricles; here, it divides and spreads into the muscle mass of each ventricle, terminating in a series of fine fibres amongst the muscle cells (Purkinje fibres). The wave of depolarization spreads through this system rapidly ($5\,\mathrm{m\,s^{-1}}$) and, therefore, depolarization and contraction of all the muscle fibres in both right and left ventricles are relatively synchronous; the ventricular depolarization potential on the electrocardiogram – the 'QRS' component – lasts less than 0.1 s. A number of nervous and hormonal influences which originate outside the heart may act on the pacemaker to cause alterations of frequency, and may modify the conduction velocity of the depolarization wave through the heart; but the orderly sequence of atrial and ventricular contraction which follows pacemaker depolarization is ensured largely by the layout of the conducting pathways.

The cycles of depolarization and repolarization which occur in cardiac muscle generate small electrical potentials, and with suitably located electrodes these can be picked up at the surface of the body and amplified and recorded as the electrocardiogram. A typical tracing is shown at the top of **Fig. 11.5**. Depolarization of

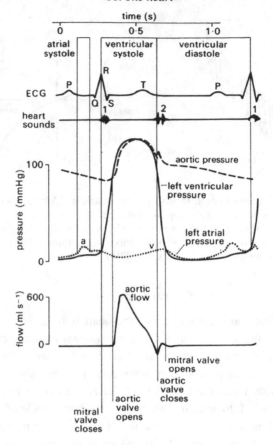

Fig. 11.5. Semi-diagrammatic illustration of the events on the left side of the heart during the cardiac cycle. All pressures related to atmospheric. The origin of the 'a' and 'v' wave in atrial pressure is discussed in the text, p. 186.

the atria produces a small deflection known as the 'P' wave; this is followed after a delay of about 0.2 s by a larger voltage deflection, often multiphasic, known as the 'QRS complex'. This reflects depolarization of the two ventricles, and is followed by a final component, the 'T' wave, which is generated during repolarization of the ventricles. The time relationships between these summed electrical potentials and the mechanical events on the left side of the heart can be deduced from **Fig. 11.5**, which shows the pressure and flow events that take place at a number of locations in the left side of the heart during a complete cardiac cycle.

Mechanical events As has just been mentioned, the onset of ventricular contraction (ventricular muscle depolarization) is signalled electrically by the QRS complex of the electrocardiogram. Both ventricles contract almost synchronously (see **Fig. 11.6**),

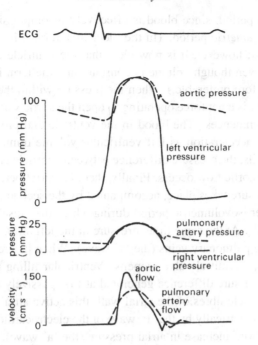

ECG

Fig. 11.6. Semi-diagrammatic illustration of pressure and flow occurring simultaneously on the left and right sides of the heart during the cardiac cycle.

but our description is simplified if we consider events on the left side of the heart only. The sequence of events is illustrated in **Fig. 11.5**. Depolarization is followed after a very short interval by the onset of active tension development (see p. 189) in the muscle fibres of the ventricular wall and an increase in ventricular pressure. At this stage, the aortic valve is still held closed because the pressure in the aorta exceeds that in the left ventricle, and the cusps of the mitral valve are moving together as flow into the ventricle dwindles. Almost immediately, ventricular pressure rises above atrial pressure and a brief period of backward flow from ventricle to atrium occurs, terminated by closure of the mitral valve. This is accompanied by a sound, audible at the chest wall and known clinically as the first heart sound. This marks the onset of systole, the period of ventricular contraction. The second heart sound marks the start of diastole, the period of ventricular dilatation. Note that these periods are defined in relation to the heart sounds and not in terms of muscle mechanics or the electrocardiogram.[1]

In the ventricle, wall tension now starts to rise extremely rapidly until the pressure within the cavity exceeds that in the aorta. There is no change in ventricular

---

[1] Thus, systole does not correspond exactly with the period of ventricular ejection, despite widespread use of the word in that sense.

volume during this period, since blood is effectively incompressible; it is therefore known as the *isovolumetric* period. (In the older literature, it is often referred to as the isometric period; however, it is now clear that the ventricle does change shape during this phase, even though volume is constant, and the term isovolumic, or iso-volumetric, is therefore preferable.) When the pressure within the ventricle exceeds aortic pressure, there is a net force operating to open the aortic valve, and the ejection phase of systole commences. The blood in the ventricle and proximal aorta under-goes rapid forward acceleration as left ventricular volume diminishes. Ventricular wall-tension then falls, the pressure difference between ventricle and aorta is reversed and deceleration of aortic flow occurs. Finally, there is a brief period of backflow be-fore aortic valve closure takes place, accompanied by the second heart sound. There then follows another isovolumetric period during which the muscle relaxes and ven-tricular pressure falls. At the same time, pressure in the left atrium is rising as it fills with blood from the pulmonary veins (the 'v' wave). When its pressure exceeds that in the left ventricle, the mitral valve reopens. Ventricular filling then occurs, under the influence of a pressure difference generated at first passively and then by active shortening of the muscle fibres of the atrial wall; this active atrial contraction (atrial systole) is heralded electrically by the 'P' wave of the electrocardiogram, and marked mechanically by a brief increase in atrial pressure (the 'a' wave). Very shortly after this, activation of the ventricular muscle occurs and the cycle recommences.

In a normal human, the heart rate may range from about $45\,min^{-1}$ (resting athlete) up to slightly above $200\,min^{-1}$ on maximal exercise. Systole is much shorter than diastole at the lower heart rates, occupying about a third of the cycle (**Fig. 11.5**); as the rate increases, there is a much greater shortening of diastole than of systole, until at the highest rates the two may be almost equal. The volume ejected from the ventricle with each beat (stroke volume) is normally in the range $70-100\,cm^3$ at rest; a smaller volume remains in the ventricle – that is, the ventricle ejects only 60–70% of its contents. The variation in stroke volume with exercise is much less than that in the heart rate; thus, increases in cardiac output in severe exercise (fivefold or more) depend much more on rate increase than on stroke volume increase. Blood pressure rises in both the pulmonary artery and the aorta on exercise, but much more modestly than does the flow, because recruitment of additional vessels in the microcirculation, or dilatation of previously constricted ones, lowers the downstream resistance to flow.

Properties of cardiac muscle

Structure Under the light microscope, the myocardium is seen to be made up of elongated muscle cells running in columns and having centrally placed nuclei and abundant mitochondria (**Fig. 11.7**). As in skeletal muscle, the fibres have a cross-striated appearance which is due to the structure of the contractile units, or my-ofibrils, lying within the cells. However, the motor nerve filaments, neuromuscular

Fig. 11.7. Electron micrograph of parts of three cardiac muscle fibres and an adjacent capillary (Cap) in longitudinal section. The two upper cells are joined end to end by a typical steplike intercalated disc (In D). Rows of mitochondria (Mt) appear to divide the contractile substance into myofibril-like units, but, unlike the true myofibrils of skeletal muscle, these branch and rejoin and are quite variable in width. Lipid droplets (Lp) somewhat distorted in specimen preparation are found between the ends of the mitochondria. ×15 000. The structure of a single sarcomere is shown at higher magnification in **Fig. 11.8.** (From Fawcett and McNutt (1969). The ultrastructure of the cat myocardium. *J. Cell Biol.* **42**, 1–45.)

junctions, and the length-monitoring muscle spindles present in skeletal muscle are absent from cardiac muscle; and further points of difference are that the muscle cells branch repeatedly and have abundant collagen fibres between them.

The limit of definition of the light microscope is about 0.25 μm, and at this level further detail is impossible to distinguish. Since there did not appear to be cell membranes running across the fibres, it was assumed for some time that they had a syncytial structure; that is, widespread continuity of cell cytoplasm. However, electron microscopy has revealed that the cell membrane has two layers, with the inner layer

passing across the fibres and dividing them every 50–100 μm into structurally separate cells, about 10–20 μm in diameter. At intervals of about 2 μm all along each
such cell there are extremely fine invaginations of the cell membrane known as T-
tubules, which have been shown to provide for almost simultaneous activation of all
the myofibrils in the cell when the membrane is depolarized.

Numerous other fine details of cell structure have been described, and a great deal
of recent progress has been made in clarifying the biochemical reactions which release
energy for contraction and repolarization. However, we will concentrate only on the
contractile apparatus, since our interest lies in the mechanics of the muscle fibres.

As mentioned previously, the contractile elements of each cell are the myofibrils,
which run parallel to the long axis and show a repeating pattern of cross-striation.
The myofibrils themselves actually consist of bundles of myofilaments, and the cross-
striations repeat themselves because the myofilaments are made up of repeating chains
of *sarcomeres*. The sarcomere is the fundamental contractile apparatus; its structure
was first described in skeletal muscle and has been confirmed in cardiac muscle with
only minor differences. Each sarcomere is limited by two adjacent narrow bands
known as Z bands or lines (**Fig. 11.7**); in between it is subdivided into a wide central
A band, separated from each Z band by narrow, lighter I bands. When the muscle
fibre is stretched, the distance between the Z bands widens, but the A band remains
the same length. All this can be distinguished under the light microscope; sarcomere
length increases with increase of overall muscle length, and in heart muscle at its
diastolic length this is about 2.2 μm; the A band is 1.5 μm and the I bands 0.35 μm
each.

A series of elegant electron microscopy studies has shown these bands to be due to
partially overlapping parallel arrays of filaments, arranged as in **Fig. 11.8**. The thicker
filaments making up the A bands are composed primarily of the protein myosin; the
thinner rods which interdigitate with them are actin. When the muscle is stretched,
the Z bands move apart and the actin rods slide along and partially disengage from
the myosin rods; thus, the I bands widen. When the muscle contracts, the reverse
happens, until at very short muscle lengths the I bands disappear and new dark bands
('contraction bands') appear where the actin rod tips overlap each other at the centre
of the A band. This occurs at sarcomere lengths less than 1.5 μm.

In skeletal muscle, there are fine cross-bridges between the actin and myosin filaments, projecting from each thick (myosin) filament. There is probably one cross-
bridge per molecule, and they are spaced about 6 nm apart, arranged helically at 60°
intervals round the myosin filament. Each myosin filament is surrounded by six actin
filaments and, therefore, makes one cross-bridge with each of these in a length of
approximately 40 nm.

The cross-bridges seem likely to have an important role in the shortening process
of striated muscle, since the filaments themselves are probably too far apart for direct
interaction; one suggestion is that during contraction the cross-bridges may move
back and forth, hooking up to specific sites on the actin filaments and drawing them

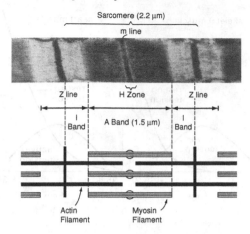

Fig. 11.8. *Top*: sarcomere structure as seen with the electron microscope. The sarcomere is bounded by a pair of Z bands. Within it is the dark central A band (marked at its midpoint by the darker m line) and two paler I bands. *Bottom*: schematic showing the disposition of the rods of the contractile proteins actin and myosin which give rise to this appearance. Note that at this length the actin filaments do not reach past the midline of the sarcomere, so that no 'contraction band' is visible. See text for details. (From Sonnenblick, Spiro and Spotnitz (1964). The ultrastructural basis of Starling's law of the heart. The role of the sarcomere in determining ventricular size and stroke volume. *Am. Heart J.* **68**, 336.)

on before releasing the linkage and moving back to a new linkage site. Thus, during activity they would have a ratchet action; with the cessation of activity the filaments would be free to slide apart passively. To date, this remains a speculative explanation, particularly when applied to cardiac muscle.

Static mechanical properties of cardiac muscle The 'sliding filament' description of sarcomere behaviour outlined above is doubly compelling because it not only fits with the visible ultra-structure of muscle, but also offers an explanation of one of its fundamental mechanical properties: the length–tension relationship. When a muscle is held at a constant length and stimulated electrically it generates tension (active or developed tension) over and above any resting tension present prior to stimulation. If this experiment is repeated with successive small increments in length, the active tension is found to increase successively to a peak and then decline (**Fig. 11.9**). This is true for a wide variety of muscle, and in skeletal muscle (and probably also cardiac muscle) the peak of the length–tension relation comes when the degree of stretch brings sarcomere length to about 2.2 µm. **Figure 11.9** suggests that, in absolute terms (i.e. per unit cross-sectional area), heart muscle is much weaker than skeletal muscle. This may be more apparent than real, since heart muscle contains a greater bulk of non-contractile tissue such as collagen and mitochondria, and the muscle fibres are not all parallel. In muscle which has contracted at sarcomere lengths of less than 2.2 µm, the actin rods reach right in past the midpoint of the myosin rods and overlap

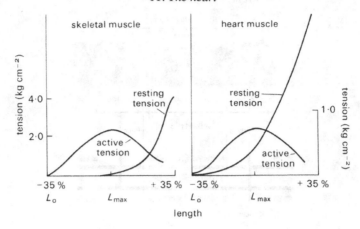

Fig. 11.9. Typical length–tension curves for skeletal and cardiac muscle. In each case resting and active tension was plotted against length as the muscle was held at a series of lengths and stimulated electrically to contract. The ordinates show tension, expressed in kilograms per square centimetre of muscle cross-sectional area. Note the difference in scaling of the two graphs. The abscissae show sarcomere lengths, relative to the lengths L_{max}, at which the maximum active tension was developed; in these experiments L_{max} was 2.2–2.3 μm for both skeletal and cardiac muscle. (After Spiro and Sonnenblick (1964). Comparison of the ultrastructural basis of the contractile process in heart and skeletal muscle. *Circulation Res.* **15** (supplement 2), 14–37. By permission of the American Heart Association, Inc.)

their opposite numbers; it is postulated that this interferes with the formation of cross-linkages so that less than maximum tension can be developed. This overlap lessens as the muscle is stretched, until at a sarcomere length of 2.2 μm the actin and myosin rods cease to have a double overlap, and the maximum number of cross-linkages can be formed. It is at this length that the maximum active tension can be developed. As the sarcomere is stretched beyond this, the actin rods are progressively withdrawn from between the myosin rods, and fewer cross-linkages can form. In this length range, the active tension developed in a contraction declines linearly with length increase, until at a sarcomere length of about 3.5 μm the actin rods are fully withdrawn and the muscle ceases to generate any active tension.

In skeletal muscle this relationship between sarcomere length and tension has been firmly established, and its functional significance is generally agreed. The behaviour of sarcomeres in cardiac muscle has been investigated much less thoroughly; under physiological conditions they appear to operate in the length range from about 1.5 to 2.2 μm – sarcomeres measure approximately 2.2 μm when ventricular diastolic pressure is about 12 mm Hg, which is the upper limit of normal, and thus the sliding filament hypothesis is an attractive explanation of myocardial length–tension relations. However, relatively few observations have been made on the relationship between sarcomere length and developed tension in cardiac muscle, and these are to some extent conflicting. This may be because of technical problems, particularly those of tissue distortion during histological preparation; methods have recently been developed to measure sarcomere length in vivo, and these may resolve the question. At present, it seems well established that no active tension is developed at sarcomere lengths less than about 1.5 μm and that the peak of the length–tension curve normally occurs somewhere between 2.0 and 2.2 μm, as it does in skeletal muscle. In between it is not clear whether increasing tension is the result of successive increments in sarcomere length as in skeletal muscle, or recruitment of increasing numbers of sarcomeres in muscle fibres which were buckled at short muscle lengths and are straightened and then stretched as the muscle lengthens. Furthermore, there is recent evidence that the curve relating tension and sarcomere length (**Fig. 11.9**) may be displaced along the sarcomere length axis in response to changes in the chemical environment of the muscle. We need a more detailed knowledge of the behaviour of actin–myosin cross-linkages to settle these uncertainties.

In skeletal muscle, the linear relationship between muscle length and sarcomere length is maintained until the latter reaches at least 3.5 μm, and the progressive reduction in available cross-linkage sites as the actin and myosin filaments are gradually pulled apart offers an adequate explanation of the linear decline of active tension along the 'descending limb' of the length–tension curve. For heart muscle, the situation at high degrees of stretch is different. Sarcomeres will lengthen to only about 2.6 μm, and even in this range the linear relationship described above does not hold; increments in overall muscle length beyond this point require the imposition of very high resting tensions and probably involve sliding of whole fibres relative to each other. This situation does not arise in the normal heart and seems very unlikely even in the failing or pathological heart; in acute experiments where the relaxed left ventricle was distended with pressures as high as 100 mm Hg (far in excess of the levels reached even in severe heart disease) the sarcomeres in the ventricular wall had an average length of only 2.3 μm.

Dynamic mechanical properties of cardiac muscle The length–tension curve describes an important property of muscle under static conditions – held at a constant length both before and during activity – but it throws no light on the dynamics of muscular contraction, which are of fundamental importance to any understanding of heart muscle performance.

A stimulated muscle goes through a period of mechanical activity (the 'active state') which reflects the release of energy derived from chemical reactions and has measurable properties both of duration and intensity. Enormous progress has been made in elucidating and measuring both the biochemical steps which yield energy, and the mechanical behaviour which is the expression of this energy release. The literature is voluminous, reflecting both the technical difficulties involved in research on the myocardium and its innate complexity, and the subject can only be briefly surveyed here.

The commonest material used for experimental study of the mechanical properties of heart muscle has been papillary muscle, removed from the right ventricles of young animals under anaesthesia. It can be obtained in this way as extremely thin strips a few millimetres in length, and made up of numbers of fairly parallel muscle fibres. When such papillary muscles are mounted in oxygenated, nutrient media of appropriate ionic and osmotic properties, they preserve their contractile properties in response to electrical stimulation for long periods. These contractile properties have been interpreted largely in terms of very simple mechanical models; to demonstrate why such models were chosen it is necessary briefly to describe some early experimental work carried out on skeletal muscle.

Intact skeletal muscles can be removed easily from small animals such as the frog, and a number of workers in the early years of the twentieth century studied these muscles, stimulating them electrically and examining the mechanical properties and heat production during contraction (the latter phenomenon having been demonstrated by Helmholtz over 50 years previously).

As was mentioned earlier, electrical stimulation of muscle leads to tension development. In the resting state, a potential difference of about 90 mV is maintained across the membrane of the muscle cell – the resting potential. An externally applied shock can cause transient reversal of polarity of this potential, followed by slow recovery. This discharge, which is known as the action potential, triggers the release of calcium ions from stores within the muscle cell, and these somehow activate the cross-linkages between the actin and myosin rods in the contractile apparatus of the sarcomeres. This whole process occurs within milliseconds, and the muscle cell is then capable of contracting (i.e. shortening and/or generating tension) for a short time before metabolic processes within the cell remove the calcium ions and the muscle returns to the resting state.

Thus, a single electrical stimulus applied to a muscle fibre causes a short-lived contraction appropriately known as a twitch. A chain of stimuli causes repetitive twitches, and in skeletal muscle if the stimulation frequency is high enough the twitches will fuse together to give a sustained contraction. This is known as a tetanus, and the corresponding train of shocks is a tetanic stimulus.

For a given muscle preparation, the tension generated in each twitch or tetanus will increase with increasing stimulus strength until a maximum is reached which is highly reproducible over long periods of time. If the muscle is held at constant length, the twitch or tetanus is known as isometric; if it is allowed to shorten, the force (if any) opposing shortening is described as the load (or afterload) and if this force is constant, which implies that all accelerations of the load are very small compared with that due to gravity, the contraction is called isotonic. Since maximal isometric contractions were found to be highly reproducible, they were used experimentally as the baseline condition; in this case the muscle generated heat during the course of a stimulation cycle, but since no shortening occurred, no external work was done (work, or energy, is equivalent to force times distance). When muscles were allowed to shorten by a distance x against a load or force P, not only were Px units of work done, but an *extra* amount of heat was released. (This effect of shortening on heat production is known as the 'Fenn effect' after its discoverer; Fenn (an American physiologist who did this work in 1923) also demonstrated the converse to be true – if a muscle was stretched during stimulation, it gave out less heat than when held at constant length. In describing this experiment, Fenn coined a phrase, 'negative work', which has given pain to physical scientists ever since; this is unfortunate, since the implication of the experiment – that the mechanical conditions during contraction control the amount of energy released – is fascinating and appears to have been little explored.)

The explanation of this liberation of excess heat on shortening came some years later when instruments capable of following heat production instant by instant through the contraction and relaxation cycle became available. It was then shown in a famous series of experiments carried out by A. V. Hill that the extra heat associated with shortening is proportional to the distance x shortened; thus, it is equivalent to ax units of work, where the constant a has the dimensions of force. Now the amount of mechanical work done is Px units; so the total extra energy released in shortening is $(P+a)x$, and the rate of release of energy is $(P+a)v$, where $v\,(=\mathrm{d}x/\mathrm{d}t)$ is the velocity of shortening. This rate of energy liberation was found experimentally to increase as load diminished, having its highest value when the load was zero and being zero when the muscle exerted its maximum force in an isometric contraction. The relationships were found experimentally to fit best in the form

$$(P+a)v = b(P_0 - P), \tag{11.1}$$

where b is a constant with the dimensions of velocity. This can be rewritten

$$(P+a)(v+b) = b(P_0 + a), \tag{11.2}$$

where the right-hand side is a constant and a hyperbolic relationship is predicted between the force P and velocity v of contraction. Thus the properties predicted from thermal measurements were open to confirmation by purely mechanical experiments, and were indeed verified when the velocity with which a muscle could shorten isotonically against various loads was examined (**Fig. 11.10**).

The thermal observations were, however, of great importance in another way, since they suggested the first conceptual mechanical model of the muscle fibre. Observations on the course of heat release in the very early stages of stimulation revealed that it was similar for both isometric and isotonic contractions. This suggested that similar mechanical events were occurring in the early stages of both types of contraction, and since the length of the muscle fibre could not change in an isometric contraction, the idea arose of a contractile element in the muscle, which shortened on stimulation but which was linked in series to an elastic element that could lengthen if muscle length was held constant (**Fig. 11.11**a). The force–velocity relationship described above

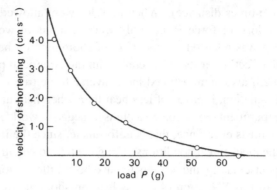

Fig. 11.10. Relationship between load P (grams-weight) and velocity of shortening v cm s^{-1} in isotonic shortening of frog skeletal muscle. The points were obtained experimentally; the line was derived from Equation (11.2) in the text. (From Hill (1938). The heat of shortening and the dynamic constants of muscle. *Proc. R. Soc. B* **126**, 136.)

was assumed to describe the properties of the contractile element, since in steady shortening under isotonic conditions the elastic element would have constant length and would not contribute.

It should be stressed that Hill was examining the properties of skeletal muscle under very particular conditions. First, the muscle was stimulated with trains of high-frequency shocks (tetani), so that a prolonged and maximal response occurred. Thus, each observation was carried out with a constant load and a steady velocity of short-ening. The real physical properties of the series elastic element were not considered, since it was at constant length throughout. Similarly, the time course of development (or decay) of force was ignored. Furthermore, resting tension was very small at the muscle lengths used (approximately 2% of active tension); therefore, a model with two elements was adequate. The addition of a component to account for tension in

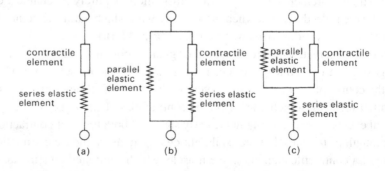

Fig. 11.11. Mechanical models of muscle.

the resting state (parallel elastic component, as in **Fig. 11.11**b and c), which becomes necessary to allow for *resting tension* at longer muscle lengths (**Fig. 11.9**), was not considered in any detail; it was represented as a simple elastic element with no part to play in active shortening, and its nature and mode of coupling to the other two were deliberately left vague. Finally, the exact nature and location of the contractile element and the series elastic element also remained undefined; structures such as tendons might represent a genuine elastic element, or the internal contractile mechanisms might be elastic. Nonetheless, this 'two-element' model of active skeletal muscle achieved widespread acceptance, since it explained a range of mechanical and thermal observations. It was natural, in view of the structural similarity which exists between sarcomeres in cardiac and skeletal muscle, to consider its applicability also to cardiac muscle.

Problems arose immediately. First, cardiac muscle preparations exhibit appreciable tension throughout the range of lengths from which they will contract; thus, the parallel elastic element becomes an essential part of the model, and the force–velocity relation can be examined only incompletely, since forces at and near zero cannot be achieved. Second, and far more important, is the fact that under normal conditions it is not possible to tetanize cardiac muscle like skeletal muscle and get a sustained and highly reproducible isometric contraction. Cardiac muscle repolarizes relatively slowly, and repeated stimuli do not produce a steady, maintained contraction. Instead, they produce twitches which even at high stimulation frequency only partially merge, giving a 'saw tooth' time course of tension or shortening. Thus, an incomplete cycle of relaxation and contraction occurs with each stimulus. At lower frequencies, stimuli evoke twitches which are clearly separated and may be highly reproducible, but neither of these types of response represents a steady state of activity, since the tension-generating and shortening capacity of the muscle (its active state) may be changing continuously during a contraction. A series of twitch responses from a papillary muscle preparation is shown in **Fig. 11.12**.

This greatly complicates the design and interpretation of experiments. The intensity of the active state (however it may be assessed) is obviously an important property of the muscle; but it becomes extremely elusive if it is changing throughout a contraction. The problems are best illustrated by examples. Hill originally defined active state as the tension which the contractile element could bear without changing length; thus, it could only be measured when the contractile element velocity was zero. Even in skeletal muscle this was only easy when tension had reached a sustained maximum value in an isometric tetanus; in this situation the contractile element has moved right along the force–velocity curve (**Fig. 11.10**), shortening progressively more slowly against the growing tension in the stretched series elastic element until it comes to rest; the intensity of the active state is then given by the value of P_0 (Equation (11.1)). If, on the other hand, the contraction is not steady, but takes the form of an isometric twitch in which the active state rises and then falls, it is obviously extremely unlikely

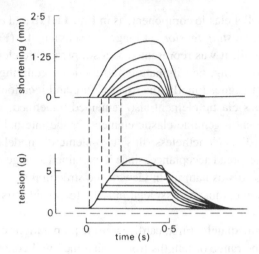

Fig. 11.12. A series of superimposed records showing the length and tension changes that occur in a cat papillary muscle which contracts against a series of different loads. The initial length was held constant; the lower family of curves shows that tension rose to match the load in each case, and then remained constant whilst shortening occurred. The amount of shortening at each load is shown in the upper curve. The final tension curve shows the response when the muscle cannot lift the load; this is the isometric twitch response. (From Sonnenblick (1962). Force–velocity relations in mammalian heart muscle. *Am. J. Physiol.* **202**, 931.)

that the contractile element would be brought to rest by the load just at the time when activity was maximal. But unless this happens, the maximum recorded tension will not correspond to P_0, and the intensity of the active state will be underestimated. The measurement is feasible if some means is devised to hold contractile element length constant, for example by controlled stretching of the muscle during the contraction; but the experimental difficulties are very great.

An alternative approach (which to some extent added confusion, since it introduced another definition of active state) has been to use the unloaded velocity of shortening as an index of active state. This is easier to examine experimentally; the muscle is released from its load at different points during successive twitch cycles and the velocity of shortening is measured immediately after the muscle has sprung back to unloaded length. An example of the results obtained in such an experiment is shown in **Fig. 11.13**; the active state rises to a peak about 150 ms after stimulation (a little before peak tension is reached in an isometric twitch) and then declines steadily.

Recently, however, the techniques based on contractile element velocity have been shown to have a flaw because the duration of active state has been found to depend on length changes in the contractile element; if it shortens at any point in the contraction, the active state wanes earlier. In recent experiments, therefore, the subject has been re-examined by a more sophisticated approach; the stress–strain characteristics of the series elastic element in a muscle are measured first and then computer-controlled

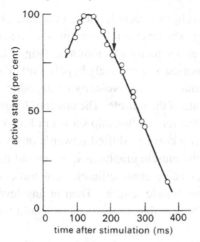

Fig. 11.13. Time course of active state in cardiac muscle as measured by released isotonic contractions. The arrow shows the time at which peak tension was achieved in an isometric twitch. (From Edman and Nilsson (1968). The mechanical parameters of myocardial contraction studied at constant length of the contractile element. *Acta Physiol. Scand.* **72**, 205.)

mechanical feedback is used to pull on the muscle during contraction so that it is continuously lengthened by just the amount necessary to keep the contractile element length constant; the tension on the muscle at each instant through the contraction cycle then defines the active state. This method gives a time course and intensity of active state which differ only a little from the isometric twitch response, the chief difference being a fractionally earlier rise and fall in active state. Even this, however, is unlikely to be the last word, because it has been demonstrated still more recently that the series elastic element has time-dependent properties, and the amount of stretch needed at different times in the cycle to fix contractile element length will therefore not depend solely on instantaneous tension.

These studies, culminating with the paradox that the contractility of a papillary muscle preparation is best defined by an experiment in which it is forced to lengthen, are described in some detail because they highlight the difficulties which arise in the experimental study of cardiac muscle mechanics at this level. It should by now be apparent that at least four variables – time, length, tension and velocity – are important; and a whole range of extraneous factors which affect contractility, such as temperature, oxygenation and ionic environment, need stringent control. Blinks and Jewell[2] have provided a useful recent survey of the subject; they point out that since the standard experiment explores the relationship between a pair of variables, six classes of experiment are needed to explore the interrelationships between time, length, tension and velocity.

[2] Blinks, J. R., and Jewell, B. R. (1972), The meaning and measurement of myocardial contractility. Chapter 7 in *Cardiovascular Fluid Dynamics*, ed. Bergel, D. H., Academic Press.

We do not explore this subject in detail, but certain general conclusions are important enough to need stating. The inverse relationship shown to hold for skeletal muscle
between developed force and velocity of shortening (**Fig. 11.10**) applies also to cardiac muscle (though it is not usually precisely hyperbolic). This must be immediately
qualified by pointing out that the force–velocity relationship is influenced by muscle
length and by the active state of the muscle. The effect of length can be predicted qualitatively from the length–tension relationship shown in **Fig. 11.9**; as the initial length
increases, the force–velocity curve is shifted upwards on its axes. Thus, in order to
present the properties of the muscle graphically, we would need a three-dimensional
plot, with two of the axes being contractile element force and velocity (as in **Fig.
11.10**) and the third being muscle length. Then at any level of active state in the
muscle, the various possible combinations of these would form a three-dimensional
surface within the axes.

In practice, it is difficult to draw such graphs realistically, because they need to
show both positive and negative velocities (i.e. shortening and lengthening), and the
rising and falling limbs of the length–tension relationship, if they are to describe the
whole contraction. In trying to understand how the properties of the muscle interact,
it is easier just to consider two types of contraction – isometric and isotonic.

The simpler example is an isometric twitch, since it avoids muscle length changes.
The contraction starts at some point on the resting length–tension curve (**Fig. 11.9**).
Initially, the contractile element begins to shorten very fast, since it is relatively lightly
loaded; thus velocity rapidly increases. However, since shortening of the contractile element implies lengthening of the series elastic element, force builds up and
the contractile element velocity is then reduced progressively until it becomes zero;
at this point the maximum force (equal to the sum of active plus resting tension in
Fig. 11.9) is being developed, and the muscle has moved along the force–velocity
curve in **Fig. 11.10** to intercept the load (force) axis. Thereafter, relaxation occurs,
and the muscle returns to the original starting point on the resting length–force line.

If the muscle shortens against a load during contraction (isotonic contraction) the
situation is slightly more complicated. At first the contractile element shortens and
develops force, stretching the series element as for the isometric case. Then a point
is reached where the force equals the load; thereafter, force remains constant, and
the whole muscle begins to shorten with the series elastic element at constant length.
The demand for constant force at diminishing muscle length can only be met by a
diminishing contractile element velocity (**Fig. 11.14**); so the muscle shortens more
and more slowly, until it begins to relax, and again returns to the starting point on the
resting length–tension curve.

In the real heart, both isometric and isotonic phases occur during contraction, and,
as will be seen later, the load (i.e. force) is changing continuously during the isotonic
phase; it is easy to imagine that the detailed behaviour during such a contraction
would be extremely complicated. Furthermore, we have ignored a major factor which

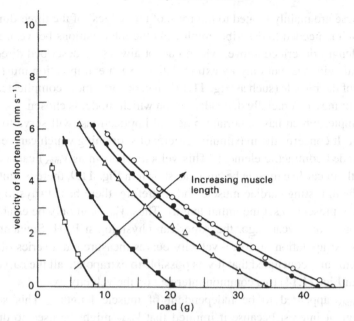

Fig. 11.14. Curves illustrating the length dependence of the force–velocity relationship in cardiac muscle. Each curve is plotted from a series of contractions against different loads such as those shown in **Fig. 11.12**. Successive increases in the initial length of the muscle displaced the curves farther and farther to the right. (After Sonnenblick (1962). Force–velocity relations in mammalian heart muscle. *Am. J. Physiol.* **202**, 931.)

complicates all these relationships – the importance of time. Active state does not remain constant throughout the contraction, but waxes and wanes, so that the force–velocity curve, which has a shape like that in **Fig. 11.10**, represents properties at only one instant; it has grown from the origin following stimulation of the muscle and will collapse back into it as the active state decays. The graphs relating force, velocity and length are all moving with respect to their axes during the contraction and relaxation of the muscle.

This kind of description has some value both in illustrating the complications that exist and in showing that however we examine a contraction (whether in terms of velocity or force) there are two main influences at work in deciding its level. The first is the length of the muscle element and the second is the intensity of the active state. The importance of this latter property is worth stressing; it is easy to see intuitively that a high level of active state will, other things being equal, produce a stronger contraction – however measured. But it is much less easy to see how to express or measure this level, and in many situations we find that there is a property of the muscle – usually termed 'contractility' – which is not precisely measurable or definable. Even if we confine attention to the isolated muscle preparation, there are a number of areas still in dispute. These are mainly related to the form of the 'edges' of the three-dimensional surface which is needed to describe graphically the interrelations between force, velocity and length described above, which cannot always be described directly from experimental evidence, but only by extrapolation. Such extrapolation must be based on a model of the muscle (such as **Fig. 11.11**b or c, or some more complicated variant) and the result may be crucially dependent upon which model is chosen.

One example, which has important practical implications, will serve to illustrate the problem. It concerns the maximum velocity of shortening which can be achieved by the unloaded contractile element. This velocity (known as V_{max}) cannot be measured directly in cardiac muscle because, as shown in **Fig. 11.9**, there is finite tension at all lengths in resting cardiac muscle and, therefore, there has always to be a load (the preload) present to set the initial length. Thus, V_{max} can only be established by extrapolation. Some years ago, the American physiologist E. H. Sonnenblick performed this extrapolation on force–velocity curves measured at a series of different muscle lengths and concluded that it was possible to extrapolate all the curves (which are shown in **Fig. 11.14**) to a common intercept on the velocity axis.

Thus, V_{max} appeared to be independent of muscle length. This suggestion stimulated great interest because it implied that V_{max} might be used to distinguish between the effects which changes in length and in active state intensity (i.e. contractility) have on muscle performance. It will already be clear how useful such a measure would be in isolated muscle experiments; but in examining the performance of the whole heart chamber, where the difficulties of controlling and monitoring these separate influences increase, it would be even more valuable.

However, there are problems in performing the extrapolation which gives V_{max}, because some shape has to be assumed for the force–velocity curve which is being extrapolated, and it is in choosing this that the muscle model becomes important. Now Hill's model of muscle (**Fig. 11.11**a) is inappropriate, because in the heart there is resting tension, so it becomes necessary to correct the measured data, to allow for the effect of a parallel elastic element, before extrapolating. The details of this can be found elsewhere;[3] it suffices here to say that the corrections needed depend upon which form of model is used (**Fig. 11.11**b or c being the simplest possibilities) and affect the values derived for V_{max}. This means that the original hypothesis that V_{max} was independent of initial length, which was based on extrapolation of the curves in **Fig. 11.14** to a common origin on the velocity axis, may be true only as a special case, and the length–velocity relationship at zero load has once more become a subject of dispute.

During the decade or so in which these simple mechanical models have been used in cardiac research, much of the detailed ultrastructure of skeletal muscle has been described, and elegant techniques developed for making measurements at the sarcomere level. Increasingly, the mechanical properties of skeletal muscle are being interpreted

---

[3] Pollack, G. H. (1970). Maximum velocity as an index of contractility in cardiac muscle. *Circulation Res.*, **26**, 111–27.

in relation to the known structure of the contractile apparatus; thus, the sarcomere has become the model. So far, a lack of suitable preparations has prevented this in cardiac muscle research, and the mechanical models continue to be the only useful analogy to real cardiac muscle at present.

Summary Before we go on to consider the behaviour of the intact heart chamber, it is useful to summarize some of the findings on isolated muscle preparations. Such muscles contain considerable numbers of muscle cells, probably all lying reasonably parallel with the long axis. At rest, and over what can be assumed to be their normal operational length range, they resist stretching with a force which rises steeply with increasing length. The structural element responsible for this property has not been identified; in conceptual models it is called the parallel elastic element, and thought of as a spring (though a simple spring is actually a poor representation of the real situation, since the muscle does not oscillate when it is released and exhibits a good deal of creep when it is stretched; both these imply viscous as well as elastic properties). Cardiac muscle cells can be depolarized by an externally applied shock, and this depolarization is followed, after a delay of some milliseconds, by a period of mechanical activity during which the muscle will shorten. If it is loaded to resist shortening then it develops tension, and shortens if and when tension matches the load; the velocity of shortening is then inversely related to the load moved. The ability of the muscle to pull and shorten undoubtedly resides in the structures in the sarcomere, whose length is closely proportional to muscle length over a considerable range. However, the details of the shortening mechanism are unresolved, so again a conceptual model is used, comprising the contractile element, which shortens on stimulation, and is connected to the load (if any) via a series elastic element. The magnitude of force and velocity generated by the muscle increases, in the physiological range, with muscle length; thus, a family of force–velocity curves is needed to describe contractions starting from different lengths. Beyond a certain length, which probably corresponds to some optimal alignment of the protein filaments in the sarcomere, further stretching produces a decline in the force generated by a contraction; but this probably lies outside the normal range of lengths in the intact heart. The individual sarcomere will apparently shorten by a maximum of about 30%, from a resting length of approximately $2.2\,\mu m$ down to $1.5\,\mu m$; but such measurements may be imprecise due to distortion during histological preparation. The time course of contractility of the muscle is limited, as is its intensity. For practical purposes, these limits are probably well described by the isometric twitch response, and both are susceptible to a number of environmental influences. Influences which affect the time course of cardiac contraction (i.e. heart rate) are known as *chronotropic* and those which affect its magnitude as *inotropic*. Inotropic influences include a number of chemical agents, such as hormones, drugs and ions, which may increase or decrease the contractility and are likely to be of great significance in control of the intact heart.

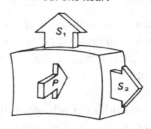

Fig. 11.15. Diagrammatic representation of the static force balance on a small part of the ventricular wall. P denotes the ventricular pressure force, S_1 and S_2 the normal stress forces in wall.

Mechanical behaviour of the intact heart

So far, we have described some properties of the muscle which makes up the heart; but this is not of much practical use in describing the mechanical properties or the pumping action of the heart chambers unless the relevant physical parameters can be redefined in useful form. For example, the load against which the muscle acts is now the force exerted against the wall by the blood in the ventricle. However, this is acting over an area which changes as ventricular volume changes, and to account for this in any balance of forces we should express it as force per unit area – i.e. *pressure*. The tension in the ventricular wall which resists this pressure must also be scaled as force per unit area, and is thus a stress. Similarly, no single length will describe the condition of all muscle fibres, because of the complicated geometry and shape changes the ventricle goes through during the cardiac cycle. Ventricular volume is likely to be the simplest useful index of length; and in deriving this, or any more complicated description of muscle fibre length distribution, the shape of the ventricle will be important. The velocity of shortening of the fibres will be reflected in the rate of change of ventricular volume, which may frequently be measured most easily as the rate of flow of blood into the aorta or pulmonary artery.

The relationship between wall stress and ventricular cavity pressure can be visualized under static conditions by considering the forces acting on a small part of the ventricular wall. The situation is represented in **Fig. 11.15**. The wall is loaded from inside by the ventricular pressure P and from the outside by the external pressure. This is balanced by opposing forces within the wall, the most important components of which are stresses acting parallel to the wall surface (S_1 and S_2).[4] Because they act normally to the applied pressure force P they are called normal stresses; there may also be forces acting parallel to the cut surfaces, which will be tangential stresses.

The relationship between cavity pressure and the wall stresses is not simple for a number of reasons. These include the geometry, but are not exclusively related to it, and a consideration of the simplest possible shape – the sphere – will illustrate this.

---

[4] See also Chapter 7, where a similar force balance is described for blood vessels.

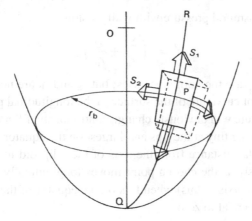

Fig. 11.16. Wall stress in ellipsoidal model of left ventricle. See text for details.

If we wish to use simple elastic theory, we must first define the wall stresses in a way which takes account of possible gradients of stress from the inner to the outer surface of the wall. As in Chapter 7, therefore, we must define stresses as the average values through the wall, since the gradients of stress which must exist are unknown. Thus, the normal stresses (S_1 and S_2 in **Fig. 11.15**) are average values for the whole thickness of the wall. Next, we must assume that the wall is homogeneous and isotropic. Then, for the spherical shape, $S_1 = S_2$.

Now we can apply the law of Laplace, as in Chapter 7 (Equation (7.6)), but bearing in mind that we are dealing with two mutually perpendicular normal stresses rather than the single hoop stress which describes circumferential tension in a tube. Thus, Equation (7.6) becomes

$$p' = \frac{2S'h}{r_i},$$

where p' is the pressure difference across the wall, r_i is internal radius and h is wall thickness; S' is equal to S_1 and S_2. If the material of the wall is not homogeneous and isotropic, a spherical shape is still possible, but S_1 will not equal S_2; then we shall have

$$p' = \frac{(S_1 + S_2)h}{r_i}.$$

If we depart from a spherical shape, the simplest body of revolution to consider is an ellipsoid of revolution (for example, a rugby football). The situation here is potentially much more complicated, because tangential forces have to be taken into account. However, if we choose an element aligned along the principal axes of the ellipsoid, as in **Fig. 11.16**, the tangential forces drop out and we can again write an equation linking transmural pressure with wall tension:

$$p' = h\left(\frac{S_1}{r_a} + \frac{S_2}{r_b}\right);$$

once more, S_1 and S_2 are the normal stresses; but r_a and r_b are not simple radii. The former is the radius of curvature of the surface in the longitudinal plane at the point P. This radius of curvature will obviously change continuously if P moves longitudinally over the ellipsoid, being tiny at the tips and largest on the equator. The second radius r_b is the perpendicular distance from the axis of the ellipsoid to the point P. It will also vary continuously as the chosen point moves longitudinally and will only be a true radius (the minor axis radius) when P is on the equator of the ellipsoid. When it is at the tip it will be equal to zero.

Left ventricular shape and wall stresses It will be apparent from all this that when the body departs from a spherical shape even the simplest elastic theory will predict complicated relationships between transmural pressure and wall stresses. It is thus hardly surprising that in many studies of the left ventricle a spherical shape has been assumed. In the very earliest, thin-walled spherical models were adopted, but these are hopelessly inadequate, not least because even in diastole the left ventricular wall thickness is 20–30% of the external radius, and during systole it is 10–12% thicker than during diastole. The next model to be considered was the thick-walled sphere. This is reasonable in terms of the length changes which the muscle-fibres must undergo; if the contained volume of such a sphere were halved during systole (which is approximately true in the ventricle) then fibres of the innermost wall layer would have to shorten by about 20%. This, is compatible with what we know of the normal sarcomere length range. Of course, the wall will thicken when the ventricle contracts, since muscle is incompressible, and the radius of the outer wall will be reduced by less than 20%. Thus, the shortening which individual fibres undergo will vary continuously through the wall from a maximum at the inside. This is of course a property of all thick-walled structures, as is the stress distribution referred to earlier.

However, although the spherical model is simple, and the length changes reasonable, it is an unrealistic shape to adopt. It has been demonstrated by a number of methods that the shape of the left ventricle is complicated, both in humans and in experimental animals. The clearest demonstrations of this come from ciné films taken during continuous X-ray screening of the heart when radio-opaque water-soluble media are injected through a catheter whose tip lies in the cavity of the ventricle. (This technique, which is called ciné-angiography, is widely used both clinically and in the research laboratory to examine the structure and function of the heart and blood vessels.) The left ventricle is revealed in this way (**Fig. 11.17**) to be shaped rather like a blunted arrowhead, the apex forming the point of the arrow and the two valves forming its base. The inner wall of the ventricle is irregularly corrugated

Fig. 11.17. Outlines of human left ventricular cavity (traced from ciné-angiograms) at equally spaced intervals between end-diastole and end-systole. The outflow tract to the aorta is at the top left and the apex of the ventricle at the bottom right. (From Gibson and Brown (1976), personal communication.)

by longitudinal columns of muscle of varying thickness, and the papillary muscles project into the cavity to complicate its shape still further. In the transverse plane, the ventricle has usually been found to be roughly circular, and at normal diastolic volumes it has a diameter which is about half the length of the long axis.

When the ventricle contracts, the long axis first shortens slightly and the transverse diameter increases correspondingly. This happens before ejection begins, and probably results from contraction of the longitudinally orientated muscle fibres in the innermost part of the heart wall; electrical studies have shown that depolarization of the ventricular muscle starts at the inner wall. When ejection commences, the main change is a shortening of the transverse axes, which reduce by about a third; but this is asymmetric and there is more movement in the antero-posterior semi-axis than in the other; the long axis only shortens slightly. By the end of systole, the ratio of long to short axes is about 2.5:1. The changes which take place during contraction are summarized in **Fig. 11.18**, which is reconstructed from a study in which X-rays were used to follow the movement of radio-opaque markers attached to various points on the ventricular walls.

It is clear from these descriptions that the sphere is an unrealistic shape to adopt, since not only does the ventricle have long and short axes, but also the former hardly varies during the cardiac cycle. Recently, and much more successfully, ellipsoidal models have been used in which the long axis is assumed to be constant in length, volume changes being brought about by shortening of the minor axes; this fits quite well with the observations described above in the real ventricle. The amounts of muscle fibre shortening required in this model are again compatible with known sarcomere properties – halving the contained volume of an ellipsoid of revolution, without shortening of the major axis, means that the minor axis, and therefore the circumferential fibres, must shorten by a maximum of 29% at the inner wall. Finally, this model

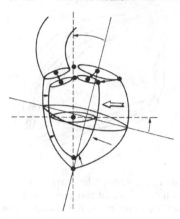

Fig. 11.18. Diagrammatic summary of the external dimensional changes and movement of the left ventricle of the dog from the end of diastole to the end of systole. The arrows indicate the distance and direction of wall movement. (After Hinds *et al.* (1969). 'Instantaneous changes in the left ventricular lengths occurring in dogs during the cardiac cycle'. *Fedn. Proc. Fedn. Am. Socs. Biol.* **28**, 1351.)

predicts wall stresses which are closer to values obtained by direct measurement with force transducers than do spherical models.

However, one limitation still applies: these studies must assume uniform properties through the wall of the model. This ignores both the architecture of the ventricular wall (see **Fig. 11.4**) and the possibility of different degrees of active tension in differently stretched muscle fibres. In a recent and comprehensive study,[5] these factors also have been incorporated.

In this study, it is once more assumed that the ventricle is ellipsoidal, and also that the muscle fibres can only bear tension axially. It is easy to see that the curvature of the muscle fibres in the wall of the left ventricle will be extremely complex, since they have a continuously varying direction through the curved wall (see **Fig. 11.4**). Therefore, fibre orientation and the principal radii of curvature of the ventricular walls were measured in post mortem dog hearts, and the data used to calculate fibre curvatures through the wall. These results were then used to calculate the distribution of normal and radial stresses through the wall, treating the wall as a tethered set of nested ellipsoidal shells, each shell having a single fibre orientation which corresponds to the experimentally determined orientation at that level in the wall. Fibre curvature is found to be greatest in the mid-wall region at the end of systole, and towards the endocardium during diastole. The pressure or radial stress distribution is slightly non-linear through the wall if constant fibre stress is assumed, the pressure decreasing most rapidly in the inner layers of the wall. If constant fibre stress is not assumed in systole

[5] Streeter, D. D. Jr, Vaishnav, R. N., Patel, D. J., Spotnitz, H. M., Ross, J. Jr and Sonnenblick, E. H. (1970). Stress distribution in the canine left ventricle during diastole and systole. *Biophys. J.* **10**, 345–63.

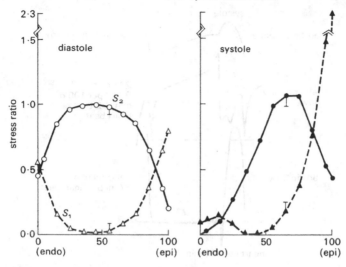

Fig. 11.19. Stress ratios in the left ventricular wall at the end of diastole and the end of systole. Abscissae: position through the ventricular wall from endocardial (0) to epicardial (100) surface. Ordinates: stresses, expressed as fractions of the average stress for the whole wall thickness. S_1 denotes longitudinal stress, S_2 circumferential stress (see **Fig. 11.16**). (After Streeter, Vaishnav, Patel, Spotnitz, Ross Jr and Sonnenblick (1970). Stress distribution in the canine left ventricle during diastole and systole. *Biophys. J.* **10**, 357.)

(and we know – e.g. from **Fig. 11.9** – that active tension is length dependent, and also that systolic sarcomere length varies through the ventricular wall), then pressure distribution across the wall is predicted to be approximately linear. The calculated distributions of normal stresses in the two principal axes during diastole and systole are shown in **Fig. 11.19**; again, the fibre stress is assumed to vary through the wall during systole, but to be constant in diastole. The very great importance of allowing for fibre orientation in studies of this kind is apparent: in models which assume homogeneous wall properties, both longitudinal (S_1) and circumferential (S_2) stresses are found to be highest at the endocardial surface.

As part of the study, predicted values of circumferential forces were compared with values obtained directly in vivo with force transducers; good correspondence was obtained, and even with the assumption of an ellipsoidal shape, such models of the left ventricle are clearly realistic and useful, though we may expect to see them modified as more detailed recent information about fibre geometry in the left ventricle,[6] and more accurate measurements of sarcomere length–tension relationships, are incorporated.

[6] Streeter, D. D. Jr, Powers, W. E., Ross, Alison M. and Torrent-Guasp, F. (1976). Three-dimensional fibre-orientation in the mammalian left ventricular wall. In *Cardiovascular System Dynamics* (eds J. Baan, A. Nordergraaf and J. Raines). MIT Press, Cambridge, MA.

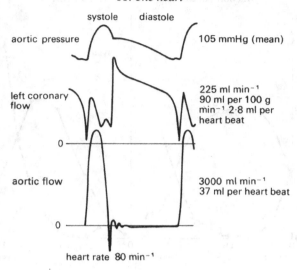

Fig. 11.20. Recordings of pressure and flow in the aorta, and flow in the left coronary artery. Measurements made in a conscious dog, 14 days after surgical implantation of flowmeters. Note that flow in the coronary artery is predominantly diastolic. (From Gregg and Fisher (1962). *Handbook of Physiology.* Section 2, p. 1534.)

It should perhaps be emphasized that interest in a good model for the description of left ventricular wall-stress distribution has a very important practical basis, since the blood supply which nourishes the heart muscle is carried in the coronary vessels which run within the wall and are therefore subjected to the local wall forces. Pressure and flow within the wall are, for technical reasons, difficult both to measure and to interpret, and our knowledge of the distribution of coronary blood flow remains very incomplete. The major part of flow from the aorta into the coronary arterial bed certainly takes place during diastole (**Fig. 11.20**), as might be expected, since intraventricular pressure and, therefore, intramural pressure are low compared with aortic pressure; and this diastolic flow is evenly distributed through the heart muscle. In systole the situation is more complicated, since the pressure within the inner layers of the ventricular wall must be close to that within the cavity, and it is difficult to see how blood flow into these layers can be maintained, since some frictional (viscous) pressure drop must occur in the coronary vessels. In fact, a number of workers have made pressure measurements within this inner (subendocardial) wall region and have found that, during systole, intramural pressures may actually exceed intraventricular pressure. This is quite possible if fibre orientation is not uniform. It could arise, for example, if a group of fibres developed a curvature opposite to that of the main body of the ventricular wall, either naturally or as an experimental artefact, and a similar result can be obtained in mathematical models by allowing this condition.

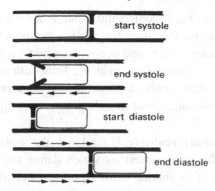

Fig. 11.21. The principle of a check-valve pump, which operates without volume change of the pumping chamber. The pumping action occurs by a change in position of the chamber and appropriate valve action. Emptying of the right ventricular outflow tract in dogs may be accomplished in part by such a mechanism. (From Carlsson (1969). Experimental studies of ventricular mechanics in dogs using the tantalum-labeled heart. *Fedn. Proc. Fedn. Am. Socs. Biol.* **28**, 1324.)

Right ventricular shape The way in which the right ventricle contracts is much less clear because very little study has been devoted to it; but the pumping action is different from that of the left. One wall of the right ventricle – the septum – is functionally part of the left ventricular wall. The other, free wall is much thinner, and has a larger area, so that the cavity of the right ventricle is wrapped around one side of the left ventricle like a pocket, opening at the top into the pulmonary artery and right atrium. The area of the walls is thus large in comparison with the volume contained between them, and a relatively slight movement of the walls towards each other will displace a large volume.

This bellows action certainly takes part in right ventricular ejection, but may be augmented by two other mechanisms. The first has been convincingly demonstrated by X-ray studies of the motion of the free wall of the right ventricle. During systole, the free wall moves downwards, that is to say tangentially to the blood contained in the chamber. During diastole, it moves upwards again. The scale of this movement increases with distance from the apex of the ventricle; thus, the free wall behaves like an elastic membrane, anchored at the apex and stretched longitudinally during diastole; when it contracts, it moves down over the contained blood like a sleeve, so that the blood comes to lie in the outflow tract beyond the pulmonary valve. The principle of this type of pump, which can operate without a change in chamber volume, is illustrated in **Fig. 11.21**.

The other mechanism which may contribute to right ventricular ejection is the contraction of the left ventricle itself, which will exert traction on the free wall of the right ventricle as the radius of curvature of the left ventricular wall decreases.

How much importance this has is conjectural, since it has not been demonstrated radiologically; this is not surprising, since the relative motion which it would cause in any one plane would be small, and would occur very rapidly at a time when the whole heart is moving. There is, however, good evidence from both animal experiments and clinical observations that right ventricular function can be surprisingly well preserved in the presence of severe damage to the muscle fibres of the right ventricular wall.

The mechanics of the entire ventricle If we know the shape of the ventricle, we can begin to calculate for it the quantities which define the behaviour of isolated heart muscle. This way of examining cardiac performance treats the ventricle as a specialized muscle, whose geometry and mechanical properties will be the important determinants of contraction. Since this follows logically on earlier sections, we will deal with it first; but there is a second approach which we shall consider subsequently (although historically its takes precedence). This treats the ventricle as a pump, ignoring detailed mechanics and examining empirically the relationship between input and output and the factors which affect it.

The central feature of this latter approach is an experimentally derived input–output relationship known as Starling's law of the heart, which still remains an extremely valuable basis for understanding many aspects of cardiac function. It is, to a considerable extent, complementary to the muscle mechanics approach, and in recent years the tendency has been to try to unify these two approaches and describe ventricular pumping in terms of the instantaneous relationship between load and muscle properties. As we shall see, the difficulties are great and this sort of synthesis of ventricular function is so far very incomplete.

Performance as a muscle There are strong grounds, as we have seen, for thinking that the shape and movement of the left ventricle are geometrically much simpler than those of the right; and much of what follows relates to the left ventricle (though Starling's law certainly applies to the right ventricle also). If we accept that the shape of the left ventricle can be realistically described by an ellipsoidal figure, whose minor axes shorten during systole but whose long axis is constant, we can begin to calculate tension and velocity of shortening of muscle elements in the wall. Average hoop tension in the plane of the minor axes can be derived from ventricular pressure if wall curvature and thickness are known. The velocity of shortening of a circumferential wall-element in this plane can be calculated from the rate of ejection of blood into the aorta if the shape of the ellipsoid is known. Finally, the instantaneous length of the minor circumference can be calculated if the ejection rate and the initial volume of the ventricle (end-diastolic volume) are known.

Experimentally, all this is an extremely tall order; in fact, some of the required values can only be obtained indirectly. For example, end-diastolic volume cannot at present be measured accurately in the intact heart; it must be calculated from length measurements made with one or other of a variety of techniques such as X-rays, ultrasonics, or strain gauges attached around the ventricle (or, with further assumptions, from the post mortem heart). The situation is further complicated by the fact that the heart contracts neither isometrically nor isotonically: we must assume that shape changes are unimportant in order to treat the isovolumic phase as isometric, and that the afterload is constant throughout ejection if the ejection phase is to be considered isotonic. The former assumption is probably not too bad; the latter is undoubtedly crude, most obviously because aortic pressure is not constant throughout systole – see **Fig. 11.5**. More general assumptions which apply are that the ionic and chemical environment of the heart is held constant during the experiment, that the active state in each muscle fibre begins instantly and remains constant throughout systole and that all muscle fibres are activated synchronously.

In view of these problems, it is not surprising that there have been very few attempts to examine this problem in the normally beating intact heart. More often a special condition has been imposed (end-diastolic volume or aortic pressure held constant, for example, or the contraction rendered isometric by preventing the ventricle from ejecting) so that a particular facet of contractile behaviour, or a particular assumption about it, can be tested. Thus, our knowledge of the intact ventricle, just as that of isolated heart muscle, is pieced together from many different kinds of experiment and is largely qualitative; the research literature is, predictably, just as voluminous and complicated as that which deals with the properties of isolated muscle fibres.

It is difficult for several reasons to decide whether a hyperbolic relationship links muscle force and velocity of contraction. First, calculations of muscle tension become extremely complicated if they are to take into account the complicated geometry and orientation of muscle fibres in the heart wall, as we have seen above. In experimental studies, some average tension is usually calculated to represent muscle force. Second, it will be remembered that the force–velocity relation for muscle strictly applies only to the contractile element in muscle. Where tension can be held constant in a contraction, so that the series elastic component can be assumed to be at constant length, then the relationship will hold for the entire muscle; this condition often applies in isolated muscle experiments and has led to the loose assumption that the hyperbolic relation applies to the whole muscle. In the intact heart, tension changes rapidly during contraction, and there will be times when the series element is lengthening while the contractile element shortens. Allowance for this must be made in calculating contractile element velocity, and to do so involves a whole new series of assumptions, including the use of elastic moduli derived from isolated muscle experiments to describe the series elastic element properties. A further complication in calculating shortening velocity also occurs. Because of the continuous nature of the muscle, shortening velocity has to be thought of as the rate at which a unit length of a given circumference is diminishing. The circumference chosen will be that for which tension is to be calculated; but since the ventricular wall is thick, the shortening per unit length

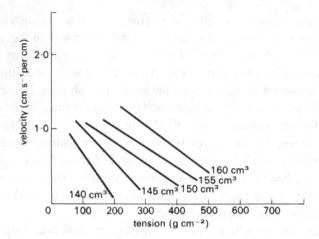

Fig. 11.22. Tension–velocity relationships for the intact ventricle calculated from experimental data for a series of ventricular volumes. Shortening velocity relates to the muscle contractile element and is expressed per centimetre of circumferential length. (After Fry (1962). Discussion. *Fedn. Proc. Fedn. Am. Socs. Biol.* **21**, 991.)

of fibres at different radial positions will vary. This can be overcome by visualizing the wall as a series of infinitely thin shells, all changing volume at an identical rate; a spatial average for the shortening velocity can then be derived. The final limitation of this approach is that the performance of the intact heart can only be examined over a relatively small range of force and velocity.

Despite all these difficulties, it has emerged quite clearly that factors which are important to the performance of isolated cardiac muscle are also very relevant in the intact heart. This can be seen in **Fig. 11.22**, in which the velocity of shortening of the contractile element in the intact heart wall is plotted against the instantaneous tension. In this experiment, the volumes and wall thicknesses were calculated from measurements of surface segment-lengths in known planes, assuming an ellipsoidal shape for the ventricle; shortening velocity was calculated from measurements of the rate of ejection of blood into the aorta, and tension from measurements of left ventricular cavity pressure. Each line is derived from the values obtained at a single instantaneous ventricular volume. The scatter in such experiments is marked (not surprisingly, considering the experimental complexities, let alone the assumptions necessary for the calculations); nonetheless, two conclusions emerge clearly. First, there is an inverse relationship between velocity and tension at each ventricular volume; second, the lines move upwards and to the right with increases in ventricular volume. If we accept the latter as being analogous to increases in fibre length, the similarity between the behaviour of the intact heart in this figure and the isolated muscle preparation in **Fig. 11.14** is striking.

Since results like this have been obtained in a number of different kinds of experiment, we can be confident that tension, initial fibre length and the velocity of shortening of the contractile element are important determinants of function in the intact heart. What remains a problem is that in many experimental situations, and particularly in the clinical situation, these variables will not be known, or calculable, in precise form. It has therefore been necessary to use more easily measured parameters which reflect changes in these fundamental properties. Some of these have already been mentioned; but they assume such experimental importance that it is worth considering them in a little more detail.

We have already seen that ventricular volume is usually used to represent muscle fibre length when we consider the intact heart, and that the relationship between these two is a statistical one – in the sense that volume is representative of some average fibre length – because of the complicated geometry of the ventricle. It has also been mentioned that ventricular volume is difficult to measure in the living heart. However, this is not true of pressure, and it is common practice to use the diastolic ventricular pressure as a yardstick of ventricular volume and, therefore, of resting muscle length; for this reason the end-diastolic pressure is frequently referred to in the literature as the 'preload', by analogy with the resting tension in isolated muscle. Clinically, resort is frequently made to a further approximation. It is assumed that at the end of diastole the atrioventricular valve is still open, but little or no flow is occurring, so that pressure in the atrium will be equal to that in the ventricle. This is particularly useful for the right side of the heart, where the pressure in the veins can be crudely assessed visually by leaning the subject back and using the veins in the neck as a manometer ('jugular venous pressure') or measured accurately by passing a fine catheter from a vein in the arm or neck into the large veins within the chest or the right atrium to measure 'central venous pressure' or 'filling pressure'.

It should be obvious that the use of end-diastolic pressure to define the initial length of muscle in the ventricular wall is a purely qualitative method. For one thing, it assumes that the relationship between ventricular pressure and volume is independent of time – i.e. that it cannot alter in disease states and that there is no 'creep' in the elastic properties of the ventricular wall. This latter assumption does not seem to have been critically tested; it may not be important in short-term experiments, and particularly those involving changes in filling from beat to beat, but creep is certainly observed in isolated heart muscle at rest, and this may be relevant when the heart is distended in diastole for long periods. A further point which is sometimes forgotten is that it is the pressure *difference* across the ventricular wall ('transmural pressure') which dictates how stretched it is, and if the chest is intact then the pressure outside the ventricle varies with breathing. During inspiration, for example, the pressure within the chest becomes sub-atmospheric, and so may that within the ventricle if its transmural pressure is low. This is not too extreme an example; the end-diastolic pressure range in the normal left ventricle is of the same order as the swings in intrathoracic pressure which occur even during quiet breathing.

Nowadays, measurement of end-diastolic pressure in humans with catheters is technically straightforward. Large ventricular volumes are a common feature of human heart disease, particularly heart failure, and therefore all the above considerations are of practical importance. An important advance has been the development of ultrasonic devices that function like depth-sounders on ships and can be applied to the chest wall to examine left ventricular dimensions ('echocardiography'). These can follow ventricular wall movement (and hence the velocity of muscle shortening) throughout the cardiac cycle. Thus, they offer a non-invasive means of assessing chamber size, as well as, in conjunction with catheter pressure measurements, a potential means of examining pressure–volume relationships directly in many experimental and clinical situations.

In summary, the end-diastolic pressure can be extremely useful in describing the initial conditions from which the ventricle contracts in any experiment, although we cannot rely on its absolute level being uniquely related to resting muscle fibre length. In practice, it has been invaluable in studies of the intact heart, particularly as a simple means of monitoring the resting length of the muscle in the heart wall.

Thus, the need for volume measurement, as an index of muscle fibre length, can be at least partially sidestepped in diastole by the use of end-diastolic pressure. In systole, when ventricular volume changes with ejection, the problem crops up again. Here, it is experimentally more manageable, since the rate of ejection of blood from the ventricle can be measured in a variety of ways in experimental animals, for example by the use of flow meters attached to the aorta or pulmonary artery (the method used to obtain the flow records in **Fig. 11.6**); the measured flow rate is obviously related to the shortening velocity of muscle fibres. However, there are still complications. For geometrical reasons the relationship will be nonlinear; for example, if we use the ellipsoidal model with a fixed major axis described earlier, then the minor axis, and therefore segments of the circumference, shorten proportionally to the square root of the volume change, and the absolute shortening is dependent upon the initial length of the axis. Thus, once again, there are geometrical problems in relating the behaviour of the whole ventricle quantitatively to that of the muscle fibres.[7]

In humans, instantaneous aortic flow cannot be measured except in very special circumstances, and the only solution is to measure cardiac output, using a standard method such as indicator dilution. This will give a value for volume flow rate which

[7] If we wish to be really rigorous in examining the applicability of muscle-fibre models to the performance of the intact ventricle, then we must know the velocity of shortening of the contractile element of the muscle, rather than that of the whole fibre; to do this we must either assume that tension is constant through systole (so that the series elastic element is at constant length) or use information about the series element derived from isolated muscle experiments to adjust the velocity of shortening of the whole fibre; even this approach, which was used in deriving **Fig. 11.22**, is obviously not wholly satisfactory, because it assumes that quantitative results obtained in a very artificial experimental situation are applicable generally in the normal heart.

is averaged with respect to time; it can be divided by heart rate to give stroke volume, but it cannot be used to describe instantaneous events during systole.

The load against which this volume of blood is ejected from the ventricle is the third parameter of interest, and is in many ways the most difficult to treat realistically. When the ventricular pressure has risen sufficiently in early systole to just exceed aortic pressure, the aortic valve opens and ejection (muscle shortening) begins. It is easy to see the aortic pressure (which only differs very slightly from ventricular pressure in normal systole – see **Fig. 11.5**) during this phase as analogous to the 'afterload' against which the isolated muscle shortens in **Fig. 11.12**, just as end-diastolic pressure is analogous to the 'preload' which sets the initial length and tension of the muscle.

However, as we saw earlier, we can only describe the load on the ventricular muscle in a way which is analogous to the load applied in isolated muscle-fibre experiments if we replace cavity pressure by wall stress. These are related through the geometry, volume and wall thickness of the ventricle, and the latter two, as well as the pressure, are changing throughout ejection. Ventricular volume goes down fairly steadily throughout systole; but ventricular pressure rises and then falls (**Fig. 11.5**). Thus, if we assume elastic properties for the ventricle, and look at circumferential stress, it may be fairly constant in early systole, but later on the decreasing radius of curvature and increasing wall thickness will tend to reduce it; thus, tension will fall faster than the pressure (see Equation (7.6)). In other words, the pressure–volume relationship for the contracting ventricle will have a different shape from the length–tension relationships of isolated muscle fibres, and the use of systolic pressure as a way of describing the load at muscle fibre level will at best be very crude. Again, we are up against a geometric separation between the quantities which can be measured and those which determine the performance of the muscle fibres.

In summary, the properties which determine the behaviour of isolated samples of ventricular muscle also operate to determine the behaviour of the intact heart muscle; but in the practical study of the heart the properties of the muscle are separated from the properties of the intact chamber by a series of uncertainties – largely geometric – and descriptions of ventricular function cannot usually be presented in terms of the fundamental muscle properties which have been described in this chapter.

Performance as a pump There is, however, an alternative approach, which is to treat the heart as a pump rather than a muscle, describing its performance empirically. This approach is much older; its foundations were laid in the latter part of the nineteenth century by the German physiologist O. Frank, and elaborated some years later in England by E. H. Starling. Thus, it has exclusively experimental origins, and its main ideas predate our knowledge of muscle mechanics.

The main theme of these early experiments was to examine the relationship between the input and output of blood to the heart. Frank worked on the frog's heart, using

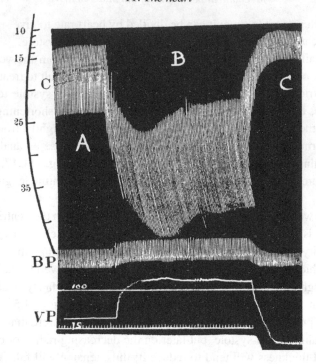

Fig. 11.23. Reproduction of one of Starling's original figures, showing the effect of an increase in venous filling pressure (VP) on diastolic and systolic heart volume in the heart–lung preparation. The records were made with a smoked drum recorder. The top trace (C) is a continuous record of ventricular volume (made by a displacement technique, the ventricles being contained within a rigid chamber); calibration in millilitres. Middle trace (BP) is arterial pressure in mmHg with calibrations below at 100 and 75 mmHg. Lower trace (VP) venous pressure (uncalibrated). Time marker in seconds. (A) Control period. Venous pressure 95 mmH$_2$O. Ventricular output 520 ml min^{-1}. (B) Step increase in venous pressure to 145 mmH$_2$O. Note corresponding increase in both ventricular diastolic volume and stroke volume. Ventricular output 840 ml min^{-1}. (C) Step decrease in venous pressure to 55 mmH$_2$O. Both diastolic and stroke volumes fall below control. Ventricular output 200 ml min^{-1}. (From Patterson, Piper and Starling (1914). The regulation of the heart-beat. *J. Physiol.* **48**, 465.)

rather crude methods; but Starling achieved an enormous experimental success in developing the heart–lung preparation, in which the working heart and lungs of an anaesthetized animal could be disconnected from the circulation to the rest of the body. Thus, the blood ejected from the left ventricle into the aorta was collected in a reservoir and returned to the heart via the right atrium. With this preparation, it was possible to control independently both the filling pressure of the atrium and ventricle and the aortic pressure. Starling found that as the inflow of blood to the atrium was increased, so the atrial pressure gradually rose, and so did the cardiac output (**Fig. 11.23**). The relationship was not linear; successively smaller increments of output followed each equal increment in venous pressure, until in some cases the output actually began to diminish. He went on to demonstrate that this effect occurred on each side of the heart, so that when the right side of the heart increased its output in response to an increase in filling, this extra output was applied to the left side of the heart where the same effect followed. In a further series of experiments, in which changes in ventricular volume were also measured, he showed that when the heart had to do more work because the aortic pressure was raised, the stroke volume at first fell, so that the ventricle was enlarged; thereafter the stroke volume was restored to its previous value.

From our present knowledge of the mechanics of cardiac muscle (which dates fom the late 1950s) we can see that Starling was demonstrating the behaviour of cardiac muscle when it shortens from different initial lengths, and we have at least some clues, from the ultrastructure of the sarcomere, about how this may work. It is a measure of the insight which Starling brought to bear on the behaviour of the heart that he concluded (in 1914): 'The law of the heart is therefore ... that the mechanical energy set free on passage from the resting to the contracted state depends on the area of "chemically active surfaces", i.e. on the length of the muscle-fibres.'

Thus Starling's law of the heart (or the Frank–Starling mechanism, as it is alternatively known) described the important relationship between initial filling of the heart (i.e. muscle length) and the volume of blood ejected. The relationship was expressed in very loose mechanical terms – words like 'energy' and 'work' being poorly defined – but it clearly established the way in which the output of the two 'series' pumps which make up the mammalian heart are matched, and how they can respond to changes in filling pressure (preload) and arterial pressure (afterload) through the same basic mechanism. In brief, Starling showed experimentally that both sides of the intact heart are intrinsically capable, over a wide range of conditions, of ejecting whatever volume of blood is returned to them.

Starling himself was well aware of the limitations of this description of cardiac function. The obvious one is that it is not quantitative; even in a single heart, the absolute level of the relationship between filling pressure and cardiac output may vary. For example, in the second experiment mentioned above, the increase in ventricular volume which follows a step-up in aortic pressure may be followed, after a few beats, by gradual reduction of ventricular end-diastolic volume (i.e. filling pressure) back to its previous level, without any drop in cardiac output.

Starling thought this was due to an increase in coronary blood flow, and thus 'improved nourishment of the muscle'; in fact, the effect occurs even if coronary flow is held constant and its cause is probably related more to improved *distribution* of blood in the myocardium, so that the inner layers are better nourished. For our purposes, it serves merely to demonstrate that the relationships between filling pressure and output of the ventricle may be influenced by the load, and if we wish to express this graphically we need a series of curves showing the relationship under

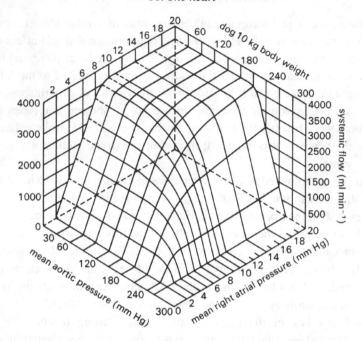

Fig. 11.24. Cardiac output in the dog heart–lung preparation, expressed in terms of venous filling pressure and aortic pressure. Note that over the physiological range of pressures cardiac output is sensitive to changes in filling pressure (preload) but not arterial pressure (afterload). (From Herndon and Sagawa (1969). Combined effects of aortic and right atrial pressures on aortic flow. *Am. J. Physiol.* **217**, 65.)

different loads – i.e. different mean arterial pressures. Again, this can be done using three axes to the graph, as in **Fig. 11.24**.

Since Starling's original experiments, it has been shown that the fall in output which may follow when an already high filling pressure is further increased (and which was regarded in clinical circles for some years as the basic mechanism of heart failure, attributed to elongation of the muscle fibres beyond some functional limit) is actually due to incompetence of the atrioventricular valve, which is overstretched by the great distension of the ventricle so that a portion of the ejected blood passes back into the atrium.

The experiments have also been repeated, in a wide range of conditions, using left ventricular end-diastolic pressure as the index of filling and stroke work as the index of the 'forcefulness' of contraction (**Fig. 11.25**). Stroke work is here defined as stroke volume times mean aortic pressure and thus has the correct dimensions of force times distance to describe external work; it was introduced as an alternative measure of ventricular function to Starling's simple stroke volume for two reasons. First, Starling presented many of his conclusions in terms of energy, so that work (which has the same dimensions) seemed an appropriate index to use. Second, work takes account,

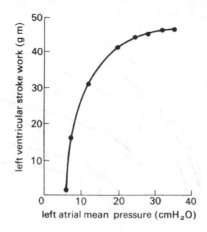

Fig. 11.25. Ventricular function curve. showing the relationship between left atrial mean pressure and left ventricular stroke work. (From Sarnoff (1955). Myocardial contractility as described by ventricular function curves. *Physiol. Rev.* **35**, 107.)

at least crudely, of the load against which ejection occurs, because it includes mean aortic pressure. In the heart–lung preparation where aortic pressure is controlled, and can be held constant, this is valid; but in studies of the intact circulation, stroke work must be used with great prudence, since the aortic pressure is not related in any direct way to end-diastolic pressure. For example, stroke work does not remain constant as aortic pressure is varied; it goes from zero when the ventricle ejects with no opposing pressure, up to a maximum, and then down again to zero when pressure is high enough to prevent ejection completely. Since aortic pressure in the intact circulation reflects the properties of the peripheral circulation as well as the effects of ejection, stroke work may be a highly unreliable index of cardiac performance. There are also other reasons why stroke work is a very imperfect way to describe myocardial performance. For example, it takes no account of either external work in setting the blood in motion or internal energy losses (e.g. due to friction) in the muscle.

When conditions are controlled appropriately, such studies demonstrate (as Starling implied, though never specifically stated) that the curve relating end-diastolic pressure and stroke work (usually known as the 'ventricular function curve') is not a single entity; rather, a family of such curves exists for any particular heart, so that some other influence besides diastolic fibre length must act to determine the performance of the ventricle.

This second determinant is usually described as the contractile state, or contractility, of the ventricle, and factors which affect it, as was briefly mentioned earlier, are known as inotropic factors; they may act either positively as shown in **Fig. 11.26**, displacing the ventricular function curve so that stroke volume or stroke work at a given left ventricular end-diastolic pressure is augmented, or negatively. Such shifts of the curve are usually described as changes in contractility, and may be the end

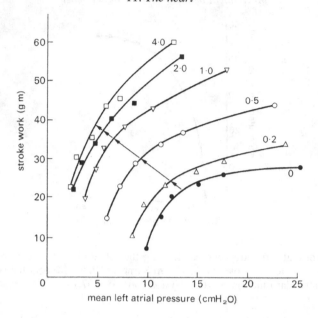

Fig. 11.26. The effect of sympathetic nerve stimulation, which has a strong positive inotropic effect, on ventricular function in the isolated heart–lung preparation. Each curve shows the relationship between mean left atrial pressure and left ventricular stroke work (the 'ventricular function curve') at successively increasing frequencies of stimulation of the sympathetic nerves to the heart (stimulus frequency in shocks per second, marked on each curve). (From Sarnoff and Mitchell (1961). The regulation of the performance of the heart. *Am. J. Med.* **30**, 747.)

result of many different sorts of influence, such as temperature, blood supply, oxygenation and ionic and chemical environment. Among the chemical influences, the most important group of positive inotropic agents (which may well have a final common path in the energy release processes within the muscle cell) are the catecholamines, naturally occurring or synthetic substances similar to the hormone adrenaline. One of these, isoprenaline, has a particularly powerful influence, and is widely used both in the laboratory and clinically for this reason. Stimulation of the sympathetic nerves which supply the heart has a similar effect on ventricular performance (probably acting via adrenaline released at the nerve endings in the heart); a number of drugs, and increased concentrations of calcium and other ions in the circulating blood, all augment the contractility of the heart. Sympathetic nervous activity and the release of adrenaline into the blood from the adrenal gland are normal components of a range of responses (usually categorized as 'fight or flight') in animals, and of both physical and emotional stress in humans. Their importance as a determinant of the behaviour of the normal heart, complementary to the Frank–Starling mechanism, has only comparatively recently been recognized. Starling's original findings, which related venous filling and cardiac output, were immediately accepted, but were only extended in this way after almost 40 years. In fact, the recognition of the effect which adrenaline or sympathetic nerve stimulation has on ventricular function, and the emergence of the concept of 'contractility', roughly coincided with the development of knowledge about the properties of isolated heart muscle.

The concept brought with it great difficulties. The first was one of definition. It is easy to see how, say, sympathetic nerve stimulation may 'set' the ventricular function curve at different levels, as in **Fig. 11.26**; but we cannot define the effect by saying that an increase in contractility causes an increase in stroke volume from a given end-diastolic pressure, because a fall in aortic pressure might do exactly the same thing (see **Fig. 11.24**) and its cause might have nothing to do with the heart. Thus, the definition of contractility in this way first requires a description of the ventricular function curve, and this is obviously very cumbersome. Furthermore, it is extremely difficult experimentally to study contractility in terms of the ventricular function curve, because rigorous control of sympathetic nervous activity, end-diastolic pressure and aortic pressure is necessary before an effect on stroke volume or stroke work can be attributed to a change in contractility. Finally, quantitative comparison between cardiac performance in different experiments is very difficult.

For these reasons, and because of the growing realization that cardiac muscle performance could be described more fundamentally in terms similar to those which applied to skeletal muscle, attempts to describe the heart as a pump lost impetus, and the search began for some means of quantifying the contractility of heart muscle among the force–velocity–length properties we have described earlier in this chapter.

The ultimate objective was, of course, to find some index which could be applied to the intact heart, both experimentally and clinically. We have already seen the difficulties which surround the application of muscle mechanics to the intact ventricle, and it is not surprising that 'contractility' still remains an ill-defined quantity in the intact heart. In isolated muscle preparations, on the other hand, it can be defined: here, the influence of length changes can be eliminated by doing experiments at constant length, and it can be shown that changes in the inotropic state of the muscle are reflected in alterations in the force–velocity curve, as shown in **Fig. 11.27**; again, positive influences give an upward shift to the curve. Thus, an inotropic influence was defined as one which displaced the force–velocity curve when acting on a muscle at constant length. This immediately suggested that one or other of the intercepts of the force–velocity curve might be used as an index of the contractile state of the muscle, because it might be possible to extrapolate to get either the force at zero velocity (isometric force, P_0) or the velocity of contractile element shortening at zero load (V_{max}).

There are insuperable difficulties in trying to use isometric force. We have already seen the difficulties and assumptions which arise in calculating wall force from cavity

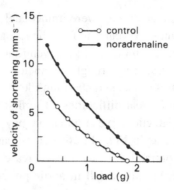

Fig. 11.27. Effect of a positive inotropic intervention (the addition of noradrenaline to the perfusing bath) on the force–velocity relationship of isolated heart muscle. Initial muscle length held constant throughout. (From Sonnenblick (1962). Force–velocity relations in mammalian heart muscle. *Am. J. Physiol.* **202**, 931.)

pressure; and for the present purpose it is also essential to be sure that length changes in the muscle (i.e. changes in end-diastolic volume of the ventricle) do not occur between measurements. (The very powerful effect of initial muscle length on isometric force can be seen in **Fig. 11.14**, where the points of interception on the load axis are isometric contractions.) Nor can we adopt the much simpler, though cruder, approach of equating maximum wall force with maximum developed pressure in the ventricle, because the maximum pressure occurs after the valve has opened (see **Fig. 11.4**), when muscle length is changing.

A much more optimistic view has always been taken of the practicability and validity of using V_{max} as an index of contractility. This was chiefly because V_{max} was thought to be independent of muscle length – implying that all the curves in **Fig. 11.14** can be extrapolated back to the same velocity origin – whereas it can be demonstrated to depend strongly on inotropic state; that is, the curves in **Fig. 11.27** extrapolate back to different velocity origins. The assumption about the length dependence of V_{max} was discussed earlier (p. 200), and has been strongly criticized for the reasons given there; but the idea is still stimulating a great deal of work and controversy because of the practical value which an index of contractility would have and because the alternative approaches have proved fruitless.

Initially, the reasoning was that V_{max}, calculated from the absolute level and rate of rise of ventricular pressure, with assumptions about the properties of the series elastic element, the synchronicity of contraction and the geometry of the ventricle, would be an index of contractility which was independent of length changes in the ventricle (i.e. of the Frank–Starling mechanism). However, doubts have grown steadily in recent years about the constancy of V_{max} with changes in initial fibre length. This, together with the uncertainties and difficulties of the calculations involved, has

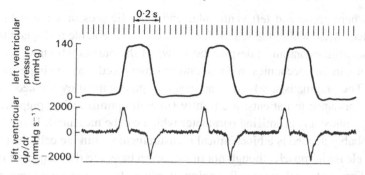

Fig. 11.28. Left ventricular pressure (upper trace) and its rate of change (lower trace). The pressure was recorded with a high-fidelity manometer mounted on a catheter tip and located within the ventricle; the rate of change was derived from this signal by electronic differentiation. (From Gabe (1972). *Cardiovascular Fluid Dynamics*, p. 14. Academic Press, New York.)

directed interest towards simpler, more direct parameters which might change in the same direction as changes in contractility. The one which has received most attention is the rate of rise of ventricular pressure itself, which was one of the measurements necessary for the derivation of V_{max}. The time course of ventricular pressure and its rate of change (dp/dt) are illustrated in **Fig. 11.28**; the rate of rise reaches a peak ('$(dp/dt)_{max}$') late during the isovolumic period, and since this might bear some relation to the maximum velocity of shortening of the muscle contractile elements, and certainly shows large changes in response to inotropic agents, attention has been focused on it. Strictly speaking, dp/dt is not directly measured; it is derived by electronic differentiation from the ventricular pressure signal, which is directly measured. (The more traditional method of drawing a tangent to the upslope of the pressure signal is highly unreliable because the duration of maximum slope is so brief – see **Fig.11.28**.) The pressure measurement must be free of distortion, which is difficult to achieve reliably with long fluid-filled catheter systems; but with modern end-catheter manometers the measurement is practicable in humans and in experimental animals.

Both the interpretation and the usefulness of measurements of $(dp/dt)_{max}$ are currently being debated. The value should, for example, be affected by initial muscle length, since it is an expression of the rate of rise of wall force, and we know that the latter is length dependent. Furthermore, since $(dp/dt)_{max}$ is reached late in the isovolumic period, it might also be reduced if there was early opening of the aortic valve and the onset of shortening – e.g. if there was a low aortic pressure. In other words, it might be susceptible to afterload as well as preload. Most (but not all) investigations have suggested that $(dp/dt)_{max}$ is affected by both these factors, and it is certainly far from being universally acceptable as an index of the contractile state of the ventricle. But it may prove to be of value in certain defined circumstances; for example, where aortic and left ventricular end-diastolic pressures are stable or can be controlled. Meanwhile, the search for a generally acceptable and quantitative index of contractility continues, despite the views of the purists who have studied the problems of cardiac mechanics in detail and are convinced that no single such index can exist. The pragmatists, who are increasingly preoccupied by the need to assess cardiac performance in patients, are rightly (though optimistically) prepared to test empirically almost any promising parameter which can be measured. In the long run, what is probably needed is a biochemical measurement within the cell; the processes of energy release in muscle, though not our concern here, are the subject of intensive and rewarding study and are steadily being clarified. In the meantime, however, the problem has to be tackled as a mechanical one.

Summary The performance of the intact heart has been examined from two main points of view. The first, which takes historical precedence, looks at the heart as a pump and examines the relationship between input (usually measured in terms of venous or atrial pressure) and output (usually measured as stroke volume), without attempting to understand the underlying mechanisms. In this view, blood entering each chamber of the heart stretches it and the contraction which follows ejects a volume of blood which is proportional to this stretch. This is known as the Frank–Starling mechanism (p. 215) and will account not only for control of heart output in response to venous return, but also for matching of output on the left and right sides of the heart. However, it is obviously an incomplete description of cardiac performance, because it takes no account of the load ('afterload') against which the heart must work in order to eject the stroke volume; to allow for this the relationship has been restated in terms of ventricular end-diastolic pressure and stroke work, which is defined as the product of stroke volume and mean aortic pressure, and is actually an incomplete measure of the external work done by ventricular muscle. This relationship is known as the 'ventricular function curve'; it is not unique for any particular heart chamber, since a third major influence, the contractile state of the muscle, also affects performance. Thus, a family of ventricular function curves reflecting the influences of preload, afterload and contractility are needed for a complete description of the heart's behaviour.

Conceptually, this approach has been tremendously useful – it is easy to visualize a curve relating input and output, whose position on a graph is adjusted upwards and downwards in response to variations in overall volume demand by the circulation. Experimentally, however, its usefulness is limited; in addition, since it treats all the parameters as time averages, it cannot be applied at all to describe the course of events within a single beat. Thus, it offers a clear, but very incomplete, description of events.

The second approach is to examine the heart in terms of muscle mechanics and to try to equate the properties described in isolated muscle-fibre experiments with the performance of the whole chamber. A series of extremely difficult and complicated transformations must be made to do this because of the complicated geometry of the ventricles and the fibre orientation – in fact, so far, it has only been seriously attempted for the left ventricle. A considerable body of specialized evidence now exists to support the view that the properties found in isolated muscles *do* determine the behaviour of the intact chamber – but this is not yet a practicable approach to the assessment of cardiac performance at a clinical and experimental level because the necessary measurements are too difficult, and many of the assumptions too insecure. Unfortunately, these limitations tend to be forgotten in the growing clinical demand for useful indices of cardiac performance, both for assessment of disease processes and for the evaluation and guidance of surgical and medical treatment.

One final point needs to be stressed. All the functional descriptions in this chapter relate to the heart isolated from its circulation; we examine input and output pressure and flow exclusively from the point of view of the heart. In the intact animal, the behaviour of the heart is continually modified, through the mechanisms we have discussed, by outside influence. Thus, we may sometimes be in a position to predict the effect on the heart of a particular circulatory adjustment, but this is not true in reverse; the circulation is in a continuous state of reflex adaptation, both to the demands of the body tissues for blood and to 'feedback' mechanisms which respond rapidly to a range of stimuli, like pressure and chemical factors in the blood. In other words, many phenomena occur in the circulatory bed of the body totally independently of the heart's performance.

Fluid mechanical aspects of cardiac function

Right heart The factors governing flow into the right atrium are complicated, but the ultimate driving force is the arterial pressure. The pulsation which is seen in the aortic pressure has been largely smoothed out, and the mean level greatly reduced, during the passage of blood through the systemic microcirculation, but it is still high enough at the venous end of the capillary bed to generate a flow towards the heart. The smoothing which occurs means that this is basically a steady driving force; but a number of factors acting unsteadily superimpose oscillations upon the flow. Some of these, such as the 'muscle pump' (p. 461), are peripheral in origin and cause oscillations of pressure which propagate primarily towards the heart. Others are secondary to pressure swings which occur within the chest due to breathing. Finally, there are pressure changes within the atrium itself, due to contraction of the heart, which add further pressure oscillations; these propagate back through the veins because of the absence of valves at the venous entrances to the heart (see Chapter 14).

Thus, flow enters the right atrium in an unsteady stream and the pressure there, and in the adjacent veins within the thorax, fluctuates throughout the cardiac cycle even in a subject breathing quietly and at rest (**Figs 11.5, 11.29** and **14.20**). The pressure waveform has two main components, the 'a' and 'v' waves. The former is due to

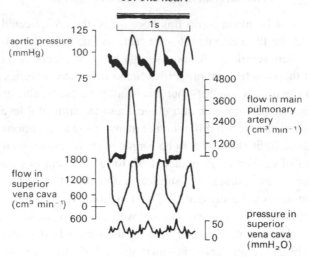

Fig. 11.29. Simultaneous records of aortic pressure, pulmonary artery flow and superior vena cava (SVC) flow and pressure. Note that, in the SVC, maximum flow towards the heart (positive signals) occurs during ventricular systole and maximum pressure (caused by right atrial contraction) coincides with a brief reversal of flow direction. (From Brecher (1956). *Venous Return*. Grune and Stratton Inc., New York.)

atrial systole; the latter shows a slower rise, coinciding with ventricular systole and terminating with a rapid fall (the 'y' descent) when the tricuspid valve opens and flows into the ventricle takes place early in diastole. These fluctuations are visible in the jugular vein in the neck when a subject lies flat or nearly flat; and in disease, their form may give invaluable information about the functional state of the right heart. The mechanisms of venous flow into the thorax are dealt with in more detail in Chapter 14.

Flow within the right side of the heart has not been studied in detail. In view of structural similarities, it seems reasonable to assume that the valves behave like those on the left side, which will be dealt with on pp. 227 *et seq.* On the other hand, the mechanics of right ventricular contraction are certainly different from those on the left, both in the way in which the volume change is achieved and in its time course, which is slower (see **Fig. 11.6**). At the beginning of systole, pressure in the right ventricle rises more slowly than in the left; but the mean pressure in the pulmonary artery is so much lower than that in the aorta (10 mm Hg versus 100 mm Hg) that the pulmonary valve opens before the aortic valve and ejection begins several milliseconds sooner. Both the acceleration and deceleration of blood in the pulmonary artery are much slower than in the aorta, and the peak flow rate is also less. Thus, the whole waveform is smoother and blunter (**Fig. 11.6**), ejection takes longer and the valve closes later than the aortic valve. The second heart sound, therefore, has two components (**Fig. 11.5**).

The peak Reynolds number in the pulmonary artery, which is calculated from Equation (5.5), using peak systolic blood velocity, pulmonary artery diameter and the kinematic viscosity of blood, is somewhat lower than in the aorta. Quoting exact figures is misleading, since there is a very wide range, probably even in normal animals at rest, and certainly as measured under different experimental conditions; but peak Reynolds numbers between 2000 and 4000 for the main pulmonary artery, and 2500–7000 for the aorta, are realistic. The question of the occurrence of turbulence in the circulation is considered later (pp. 321 *et seq.*); the problem has not yet been experimentally studied in the pulmonary artery under normal conditions.

Left heart Pressure and flow events in the pulmonary veins and left atrium differ in some respects from those in the systemic veins and right atrium, and are dealt with in Chapter 14. For our present purpose, which is to examine fluid mechanical events in the left atrium and ventricle, they can be considered as a reservoir, which fills up during ventricular systole and begins to empty into the ventricle at the beginning of diastole when the pressure gradient across the mitral valve becomes negative due to both rising atrial and falling ventricular pressure. The velocity of flow through the mitral valve in diastole rises rapidly, and may exceed $1\,\mathrm{m\,s^{-1}}$, with Reynolds numbers correspondingly high (about 8000); the velocity then falls off until late in diastole, when there is a further brief acceleration due to contraction of the left atrium. Velocity then falls rapidly before the valve closes and ventricular contraction begins.

Mechanics of mitral valve closure It has been well established experimentally that there is very little backflow through the mitral valve when the ventricle contracts at the beginning of systole, and a number of flow visualization and model studies have demonstrated that the mitral valve leaflets are actually moving towards each other (i.e. beginning to close) during the period of forward flow produced by atrial systole. By the time the pressure gradient across the valve orifice is reversed, they are sufficiently near together to close almost immediately. This apparent paradox was originally attributed to early contraction of the papillary muscles, which pulled on the chordae tendinae and thus on the valve cusps (see **Fig. 11.2**); but for a number of reasons this seems unlikely, and recent examination of the problem with ciné-angiography and model experiments suggests that a fluid-mechanical mechanism is responsible, and that the papillary muscles are responsible only for preventing the valve from turning inside out and leaking backwards during ventricular systole.

The controlling mechanism which is suggested to cause the early closure of the mitral valve depends upon the presence of a ring vortex in the ventricle during diastole. This is formed behind the cusps of the valve, where the fluid which has passed through the valve orifice and been deflected from the apex of the ventricle up the ventricular walls is trapped and circulates as shown in **Fig. 11.30**. In model experiments it has been shown that the movement of the valve cusps towards each other is dependent on

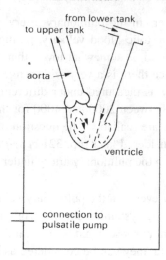

Fig. 11.30. Diagram of model left ventricle showing diastolic flow patterns. (From Bell-house (1972). 'The fluid mechanics of heart valves'. *Cardiovascular Fluid Dynamics*, p. 275. Academic Press, New York.)

the strength of this vortex, and that regurgitation through the valve increases greatly if the vortex strength is reduced either by the delaying of ventricular contraction after the end of forward diastolic flow so that the vortex dies away or by greatly increasing the size of the model ventricle so that only a very weak vortex is formed.

The mechanical reasons why the valve begins to close while there is still forward flow through it can be understood quite simply. We refer to **Fig. 11.31**, which is a schematic representation of a ventricle in which the region occupied by the vortices is significantly larger than that enclosed between the valve cusps. During the

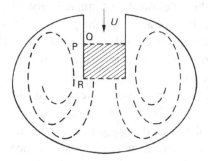

Fig. 11.31. Schematic representation of mitral valve orifice and left ventricle. Flow through the valve from the left atrium during diastole occurs in the direction shown by the arrow, with velocity U. Vortices are set up behind the valve cusps.

approximately steady state of forward flow which exists before deceleration, the pressures on either side of the cusp, at the points marked P and Q for instance, are equal and the valve cusps are therefore stationary. When the pressure difference across the valve orifice causes the flow through it to decelerate, the larger size of the vortex regions, together with the inertia of the swirling fluid, means that the corresponding changes in the velocity of flow in these regions are smaller than those through the valve. In consequence, the pressure on the outside of the valve cusp, at P say, changes less rapidly than that on the inside, at Q. Roughly, we can say that the pressures between P and R on the outside of the cusps are unchanged, while there must be a pressure drop from R to Q on the inside to provide the observed deceleration. Therefore, the pressure at P exceeds that at Q, and the valve will tend to close. It should be emphasized that this analysis is very rough and ignores many details of the motion which would influence the magnitude of the pressure differences across the valve cusps. However, it probably does describe the essential mechanism involved in valve closure.

Mechanics of aortic valve closure The structure of the aortic valve has already been briefly described. It has three cusps, attached to a circular ring of fibrous tissue which forms the 'outflow tract' at the base of the left ventricle; each cusp has free margins right to the wall, and the valve can open without distortion to expose the full cross-sectional area of the outflow tract. Immediately behind each cusp of the valve is a rounded pouch in the aortic wall, known as a sinus of Valsalva; and off two of these open the left and right coronary arteries, which supply blood to the muscle of the heart wall itself. Each valve cusp is extremely thin (0.1 mm) and is composed largely of elastic tissue and collagen, covered with a layer of endothelium similar to that lining the arterial walls.

The anatomy of the pulmonary valve is very similar, the only significant difference being that there are no coronary arteries opening off the sinuses of Valsalva in the pulmonary artery. As we have seen, the absolute pressure within the pulmonary artery and its time course both differ somewhat from those in the aorta, and the same is true of the amplitude and time course of flow. In the mechanics of valve closure, these differences may or may not be important; to date the experimental work on models, and the very few visualization studies which have been carried out in vivo, all relate to the aortic valve, so that the description given here applies primarily to it.

Direct information about the behaviour of even the aortic valve is very scanty. Ciné-angiography in normal animals and humans certainly confirms that the aortic valve does not obstruct forward flow during systole, but does not reveal any detail of valve-cusp behaviour. In one or two high-speed ciné studies of radio-opaque particles, circular motion in the sinuses of Valsalva has been reported, but the existence of a vortex mechanism controlling the movement of the aortic valve, although a very old idea, has had to be established largely by model studies. It has of course been clear for a long time that some mechanism controls the movements of the valve, since

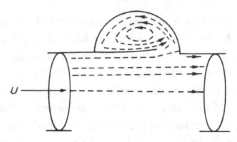

Fig. 11.32. Streamlines in aortic root at peak systole. (From Bellhouse (1972). The fluid mechanics of heart valves. *Cardiovascular Fluid Dynamics*, p. 264. Academic Press, New York.)

measurement of instantaneous flow in the proximal aorta (e.g. **Fig. 11.5**, bottom waveform) has consistently shown that less than 5% of the stroke volume normally flows back into the ventricle during valve closure. Furthermore, some flow occurs into the coronary arteries during systole (**Fig. 11.20**), implying that the orifices of the coronary arteries are not occluded by the cusps of the aortic valve.

Studies of realistic models of the aortic valve region, both by direct visualization and by measurements of pressure and velocity distribution in unsteady flow, have established that a vortex is generated behind each cusp of the valve during systole. As the valve is swept open, the ridge at the downstream end of the sinus acts as a stagnation point, and flow is diverted from the main stream into the sinus, where a vortex is set up (**Fig. 11.32**). This controls the position of the valve cusp so that it holds a stable position during acceleration and peak systolic flow, without either blocking the coronary arteries or obstructing forward flow. During flow deceleration the valve cusps move evenly towards closure, and finally shut with very little leakage backwards at the end of the systole, as in the case of the mitral valve. When the experiments were repeated with a model in which the sinuses were not present and the tube downstream of the valve was a simple cylinder, the valve became much less efficient, closing irregularly and late, with a good deal of backward leakage.

An approximate theoretical analysis of the operation of the aortic valve has been made by Bellhouse.[8] It involves two important assumptions. The first is that the effects of viscosity can be neglected, so that pressures can be related simply to local velocities and accelerations (as in Bernoulli's theorem, Equation (4.6)). This is justified because the whole cardiac cycle takes less than 1 s, whereas the time for viscosity to cause the flow in the vortex to decay significantly is about 20 s. The second assumption is that velocities within the vortex (i.e. its strength) are directly proportional to the aortic velocity just downstream of the sinuses of Valsalva. This assumption is

---

[8] Bellhouse, B. J. (1972). The fluid mechanics of heart valves. Chapter 8 in *Cardiovascular Fluid Dynamics* (ed. D. H. Bergel). Academic Press.

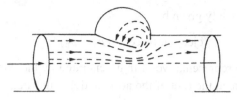

Fig. 11.33. Streamlines during valve closure. (From Bellhouse and Talbot (1969). The fluid mechanics of the aortic valve. *J. Fluid Mech.* **35**, 721.)

not at all easy to justify, especially in the unsteady state during valve closure, when the flow is observed to spread out rapidly behind the closing cusps (**Fig. 11.33**). Some of it goes to filling the expanding cavity behind the valve, and maintaining the vortex there, and the rest generates a very smooth flow downstream in the aorta. If, however, this assumption is accepted, then the mechanism of valve closure can be shown to be very similar to that already described for the mitral valve. Certainly, a theoretical analysis of the closure, which uses the observed time-dependent flow rate through the valve and the observed rate of closure of the valve to predict local pressures in the aorta and the sinuses, gives results which correlate very well with those measured in the model. The great stability of cusp position during the accelerating part of systole can also be explained quite simply if a further assumption is made. This is that the strength of the vortex varies with the size of the region behind the cusp in the sinus. If the cusp is displaced so that it protrudes farther into the sinus than normally (**Fig. 11.32**), the strength of the vortex goes down. This means that the pressure there goes up, from Bernoulli's theorem, and hence tends to push the cusp back into its equilibrium position. The same argument works in reverse if the cusp is displaced outwards. Although the theoretical analysis of the motion of a valve cusp is based on important assumptions which cannot readily be justified a priori, the function of the vortex in the control of cusp position seems clearly established.

Ejection from the left ventricle So far, we have said nothing about the pressures and flows which are generated in the aorta by left ventricular ejection. These will be dealt with at length in Chapter 12, where it will become clear that aortic pressure and flow are strongly influenced by downstream conditions as well as upstream ones. Here we outline very briefly the link between ventricular ejection and aortic events in systole by considering the balance of forces acting on the blood in the ventricle.

The aortic valve opens as soon as the pressure in the ventricle exceeds that in the root of the aorta. Before that time the ventricle contracts isovolumically, but afterwards the volume of the left ventricle (and perhaps its shape) as well as the pressure in the blood in contact with its wall change with time. The force which causes the blood to be accelerated out into the aorta comes from the difference between the pressure generated by the contracting ventricular muscle and that in the aorta. Its magnitude is approximately given by

$$\text{ejection force} = (p_v - p_a)A, \tag{11.3}$$

where p_v is a pressure representative of that generated by the contraction of the ventricle, p_a is the pressure at the root of the aorta and A is the area of the orifice into the aorta.

The time course and magnitude of the pressure difference between the left ventricle and the root of the aorta can be obtained from **Fig. 12.13**, which shows pressure waveforms measured simultaneously at the two sites with high-precision pressure manometers. Early in systole, left ventricular pressure rises rapidly to exceed aortic pressure, and it is at this point that flow into the aorta starts. The blood continues to accelerate until about halfway through systole, when the pressure records cross again. Thereafter, aortic pressure exceeds ventricular pressure, and the force acting across the aortic orifice (Equation (11.3)) changes sign, and acts to decelerate the flow. The flow does not actually come to rest or reverse its direction until considerably later, because it takes a finite time for the inertia it acquired during the accelerative period of early systole to be overcome by the adverse pressure gradient. Thus, pressure gradient and acceleration are 'in phase', but pressure gradient and flow velocity are not: the velocity lags behind the pressure gradient. When the flow does reverse direction, the aortic valve closes, and this is marked by the notch on the aortic pressure record in **Fig. 12.13** (the dicrotic notch). Thereafter, of course, there is no continuity between ventricle and aorta, and the force balance breaks down.

If we consider the period when the valve is open, however, and assume that the effect of viscosity is negligible (which is justified because there is insufficient time during systole for viscous boundary layers to develop and resist the motion significantly), then the force acting to cause ejection (Equation (11.3)) must be equal, at all times, to the sum of two terms:

(1) the rate at which momentum is carried out of the ventricle;
(2) the rate of change of momentum of the blood in the ventricle.

(This is simply Newton's second law of motion, expressed in terms of momentum; this form of the law is explained in Chapter 2, Equation (2.9).)

The first of these terms can be estimated by considering the blood ejected from the ventricle in a short time δt (**Fig. 11.34**). Its mass is $\rho A u_a \, \delta t$, where ρ is the density of the blood and u_a is the instantaneous velocity of flow at the root of the aorta. The momentum (mass times velocity) of this blood is therefore $(\rho A u_a \, \delta t)u_a$, which is the amount of momentum carried out of the ventricle in the time δt. Hence, the rate at which momentum is carried out of the ventricle (this quantity divided by δt) is $\rho A u_a^2$. This will vary continuously throughout systole.

The second term arises because the momentum of the blood remaining in the ventricle is also changing continuously. This momentum is equal to the sum of the

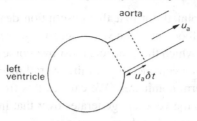

Fig. 11.34. Sketch of the left ventricle and aorta during systole. In a short time δt, blood in the aorta moving with velocity u_a will travel a distance $u_a \delta t$.

momenta of all the elements of blood in the ventricle moving towards the aorta. Although it can only be calculated exactly if we know the shape of the ventricle and its rate of change of volume throughout ejection, we can predict that it will be equal to the mass of blood in the ventricle ρV (where V is the instantaneous ventricular volume) times the velocity of ejection u_a times a scaling factor which we shall call I. Thus, the rate of change of momentum in the ventricle is $d(I\rho V u_a)/dt$. If the shape of the ventricle did not change during contraction, I would be a constant. Since ventricular shape does change, it probably varies, but in either case it can in principle be calculated.

Thus, we can derive the following relationship, which will hold at all times when the aortic valve is open:

$$\frac{p_v - p_a}{\rho} = u_a^2 + \frac{1}{A}\frac{d(IVu_a)}{dt}. \tag{11.4}$$

There is a further, more obvious relationship between ventricular volume V and aortic velocity u_a; since blood is incompressible, the rate of outflow into the aorta must be equal to the rate at which ventricular volume changes; that is,

$$Au_a = -\frac{dV}{dt}. \tag{11.5}$$

It is these equations which link pressure and volume in the ventricle to pressure and blood velocity in the aorta, and in principle they provide a means of examining ventricular performance from measurements in the aorta. If, for example, we know I and A, then systolic pressures and volume changes in the ventricle could be inferred from velocity and pressure measurements in the aorta, which are becoming practicable even in humans by catheter techniques. In practice, there are great difficulties. For example, the value of I has not been determined for any shape of ventricle, nor its variability in vivo explored; furthermore, total ventricular volume V cannot be obtained from Equation (11.5) without knowledge of end-diastolic volume, which is notoriously difficult to measure accurately. So far, this approach has not been explored either theoretically or experimentally.

Finally, it should be pointed out that this description deals only with the forces relating to the motion of blood out of the ventricle, and ignores the very high pressure (aortic diastolic pressure) which the ventricle must overcome to open the aortic valve and initiate the motion. In a consideration of the overall mechanical performance of the ventricle, this latter term dominates. We can see this from **Fig. 12.13** (p. 260); the excess pressure which the ventricle generates over that in the aorta is very small compared with the absolute level of the pressures.

Sounds and murmurs in the heart

Sounds There has been agreement for many years that the normal heart sounds are caused by closure of the valves. However, there has been sufficient confusion and misunderstanding about the basic mechanism to justify a detailed description.

We will take the mitral valve, and therefore the first heart sound as an example. We have seen that at the end of diastole the valve cusps are already close together; the last stage of closure is caused by retrograde flow of blood from the ventricle to the atrium. As the valve closes, the moving blood close to it continues in motion towards the atrium because of its inertia and stretches the valve. This is elastic, and therefore recoils, setting the blood in motion in the opposite direction. Again, therefore, the valve is carried past its equilibrium point and the process repeats itself. This oscillation continues until it is damped out by viscous energy losses in the blood and in the valve cusps themselves.

Of course, the equilibrium position of the valve will be changing throughout this period, because the pressure difference across it is increasing as ventricular pressure rises. This will modify the process because the elastic properties of the valve cusps are nonlinear (stiffness increases with strain, as it does in arteries – and probably for the same reason; see Chapter 12). Nonetheless, this is the basic mechanism which generates high-frequency pressure oscillations in the atrium and ventricle, and it is these which are the primary event in the generation of the first heart sound.

The misunderstandings about this mechanism which have arisen in the past appear to stem from two sources. First, the physical events have been misunderstood (for example, reversal of the pressure difference between atrium and ventricle has been assumed to cause instantaneous reversal of flow, which leads to mistiming because reversal is not instantaneous in a flow like this where inertia is important). Second, limitations in the performance of experimental equipment have caused events to be distorted or mistimed.[9] The best proof that vibration of the valve is the origin of

---

[9] One example, which is particularly apposite here, is a form of distortion known as 'ringing' which may occur in pressure measurements with conventional catheter–manometer systems. It consists of high-frequency oscillations (which may be mistaken for events within the ventricle) and is caused by exactly the same mechanism as the heart sounds themselves. There is an elastic membrane (the manometer diaphragm plus the catheter walls) coupled to a mass (the fluid in the catheter), and when this mass is set in motion by a sudden change in pressure at the catheter tip the diaphragm is displaced, overshoots and the system oscillates briefly. Damping is again supplied by the viscous properties of the fluid.

the heart sound comes from simultaneous recordings with paired microphones in the left atrium and left ventricle; these show that the vibrations which constitute the first sound are of opposite phase in the two chambers. An oscillatory motion of the valve seems the only way to interpret this fact.

All this does not mean that other events do not make any contribution to the heart sounds. The pressure fluctuations in the blood will set the heart walls in motion, and elastic waves will propagate over them (and along the arteries in the case of the aortic and pulmonary valves) in the manner described in Chapter 12. The sum of all these vibrations will contain components covering a wide frequency range. The higher frequencies (above about 50Hz) will be picked up by the stethoscope if they reach the surface of the chest, and by high-fidelity devices such as catheter-tip pressure transducers and catheter-mounted microphones. The low-frequency components (up to perhaps 20–30Hz) can be picked up by conventional catheters and pressure transducers. If we take closure of the aortic valve as an example, the low-frequency vibrations register on aortic pressure recordings and are called the dicrotic notch (**Fig. 11.5**); the high-frequency vibrations are audible at the chest wall with a stethoscope and form the aortic component of the second heart sound. A pulmonary component occurs in the same way, but slightly later because right ventricular ejection lasts longer than left (**Figs 11.5 and 11.6**).

Considering their clinical importance, very little is known about the mode of propagation of the heart sounds. It should be made clear at once that, in a true acoustic sense, they are not sounds at all. Acoustic sound waves are compression waves which travel within a material itself, accompanied by density changes; in liquids they have a very high velocity (about $1500 \, \mathrm{m\,s^{-1}}$ in blood). Vibrations picked up at the chest wall by the diaphragm or bell of a stethoscope induce this type of wave in the air within the stethoscope tubing, and thus reach the ear as true sound; but the vibrations do not travel from their site of origin to the chest wall in this form. It has been shown that the heart sounds propagate along blood vessels like the aorta and through the tissues to the chest wall at velocities less than a hundredth of the acoustic sound speed for blood; in the aorta, for example, the second sound travels at the velocity of the pulse wave (about 5–$10 \, \mathrm{m\,s^{-1}}$), and is rapidly damped out (i.e. 'stopped') if the vessel wall is locally bound so that it cannot move radially. Thus, the second heart sound travels as a pressure wave in an elastic vessel exactly like the pulse wave, and is in fact part of it, though its constituent frequencies lie outside those normally registered by pressure transducers. It is attenuated very rapidly as it passes along the vessel; virtually nothing is detectable 20cm or more from the valve.

The first heart sound (i.e. pressure fluctuations set up by closure of the mitral valve) has been shown to spread over the ventricular wall in exactly the same fashion, moving towards the apex and then reflecting back at similar velocities, and in the same form, as the pressure waves in arteries described in Chapter 12. Finally, it has been shown that both heart sounds can travel either upstream or downstream from their point of origin. Thus, the first heart sound from the mitral valve travels through the left atrium as well as the left ventricle, and the aortic second sound travels along the left ventricular wall as well as the aorta. How quickly the waves die out depends both on their frequency and on the stiffness of the wall along which they are travelling; they are damped out in a shorter distance when they travel through a relatively 'floppy' wall, or when their frequency is high.

These effects of frequency and wall stiffness on attenuation are linked through the wave speed. The frequency dependence is relatively simple to understand, because this type of wave attenuates by a given amount per wavelength (see Chapter 12); thus, the higher frequency (shorter wavelength) waves will die out in travelling a shorter distance (and this will also take a shorter time). If, for example, a wave was travelling along the aorta, or back and forth over the ventricular wall, it might have a wave speed of $5\,\mathrm{m\,s^{-1}}$, and attenuate by 50% per wavelength. Then its amplitude would fall by about 97% after travelling five wavelengths; for a wave at 100 Hz, this would be 25 cm and would take 50 ms. For a wave at 50 Hz, both distance and time would be doubled. Thus, if the vibrations are set up by a transient event like a valve closure, the higher frequency components will fall off rapidly whether the waves are observed locally, decaying as they reflect back and forth, or followed along some particular pathway such as the aorta.

The effect of wall stiffness on attenuation arises because the wave speed is dependent on wall stiffness (see Chapter 12) and falls as the wall becomes floppier; thus, in a floppy tube the wavelength of our 100 Hz wave shortens, and it attenuates in a shorter distance (but not in a shorter time) than before.

When we leave the vascular structures themselves and come to consider the mode of transmission of these waves to the chest wall, the situation is much less well understood. It is presumed that they induce vibrations in the chest wall either by direct contact or via other interposed structures. Careful 'mapping' over the chest wall with multiple microphones has shown that each heart sound is detectable first at a particular site (which corresponds well in each case with the classical clinical site for listening with a stethoscope) and then radiates out over the chest wall, attenuating rapidly. The velocity of spread is again much too slow for a true sound wave, and high frequencies within the wave are attenuated much more rapidly than low; but beyond these facts, the physical nature of the wave has not been elucidated.

Murmurs Most of the foregoing considerations apply also to the vibrations which correspond to audible murmurs, in the sense that they travel relatively slowly through the structures of the chest, primarily appear at particular sites on the chest wall and show rapid attenuation of high-frequency components as they spread. Whereas, however, the origin of the heart sounds in vibrations of the valves is well established, there has been considerable uncertainty about the physical source of murmurs. The classical view has always been that they originated in the pressure fluctuations accompanying turbulence, and that the latter could occur in three main circumstances: when flow took place either through an abnormal valve, or through a normal valve when the chamber beyond was enlarged (jet formation), or finally where the anatomy was normal but there was a very high flow rate across the valve. The extreme difficulties of visualizing flows in the heart satisfactorily, or of making direct measurements which would demonstrate turbulence, meant that most speculation about murmurs remained theoretical; and there are certainly other types of flow disturbance which could act as a source of sound. The question of turbulence in the circulation will be dealt with in more detail when flow in large arteries is discussed (Chapter 12); for the moment it is sufficient to say that direct study of the frequency structure of murmurs, and of velocity fluctuations occurring in the blood in conjunction with them, makes it reasonably certain that murmurs do have their origin in turbulent disturbances. There remains, however, a great deal of uncertainty about the origins of other sounds coming from the heart in disease states; it is not at present possible to be really sure of the physical origin of third or fourth heart sounds, ejection clicks or the opening snap.

Further reading

Bergel, D. H. (ed.) (1972). *Cardiovascular Fluid Dynamics*. Chapter 7: The meaning and measurement of myocardial contractility, by J. R. Blinks and B. R. Jewell; Chapter 8: The fluid mechanics of heart valves, by B. J. Bellhouse. Academic Press, London (2 vols).

Challice, C. E. and Viragh, S. (eds) (1973). *Ultrastructure in Biological Systems*, Vol. 6: *Ultrastructure of the mammalian heart*. Academic Press, London.

CIBA Foundation Symposium (1974). New Series No. 24. *The Physiological Basis of Starling's Law of the Heart*. Associated Scientific Publishers, Amsterdam.

Fung, Y. C., Perrone, N. and Anliker, M. (eds) (1972). *Biomechanics: Its Foundations and Objectives*. Chapter 12: Mechanics of contraction in the intact heart, by J. W. Covell; Chapter 13: Determinants of cardiac performance, by C. Urschel and E. H. Sonnenblick. Prentice-Hall, New Jersey.

Hamilton, W. F. (ed.) (1962). *Handbook of Physiology*. Section 2: Circulation, Vol. 1: several chapters on different aspects of cardiac physiology. American Physiological Society, Washington, D. C.

Hill, A. V. (1970). *First and Last Experiments in Muscle Mechanics*. Cambridge University Press.

Taylor, M. G. (1973). Haemodynamics. *Annu. Rev. Physiol.* **35**, 87–116.

12

The systemic arteries

This chapter deals with the mechanisms of flow in the larger systemic arteries. The pulmonary arteries are specifically excluded, because they have special properties and are dealt with separately; thus, we are concerned here with the aorta and its branches, which supply oxygenated blood to the organs of the body. As in other parts of the book, we take the vascular system of the dog as our primary example because it has been so widely studied experimentally; but we will refer to the situation in the human wherever specific differences of function or structure appear important. Again, we do not deal with active physiological processes, such as reflexes or mechanisms of vasoconstriction which may alter the flow or distribution of blood, but concentrate upon the physical properties of the system which are changed when such processes act.

This book deals with the arterial part of the systemic circulation in two parts: the arteries in this chapter and the microcirculation in Chapter 13. First, therefore, we must describe how and why this subdivision is made, and then we shall provide a brief description of the anatomy and structure of systemic arteries, and of pressures and flows which occur within them. Thereafter we shall introduce the fundamental mechanics which govern events and then successively add the complicating or modifying features which bring us nearer to a complete description of the pressure and flow in the arteries; in doing this, we shall repeatedly refer to the mechanics described earlier in the book.

Traditionally, physiologists have dealt with the systemic circulation in terms of large and small vessels, with a rather arbitrary separation at or about the level of small arteries or arterioles. As will be discussed in Chapter 13, this separation has been made for various reasons, both physiological and mechanical; but it is most convenient and sensible in this chapter to subdivide the circulation on the basis of a fundamental difference in fluid mechanical properties which exists between small and large vessels. This is the difference between flows dominated by inertia and viscous flow, in which fluid inertia is negligible, which have been discussed in Chapter 5. The relative importance of these properties in a flow is expressed through the Reynolds

238

number Re, where

$$Re = \frac{Ud}{v}$$

and is thus dependent upon the velocity U of the flow, the kinematic viscosity v of the fluid and the diameter d of the vessel. We may recall from Chapter 5 that flows with Re much greater than unity will show the influence of inertia, whereas those with Re much less than unity will be dominated by viscous forces. In this chapter we shall deal with high Reynolds number flows in which the dominant forces are inertial. In practice, this means flow in the larger vessels (down to small arteries, approximately $100\,\mu m$ in diameter, where Re falls below unity), because repeated branching causes a great increase in the cross-sectional area of the vascular 'bed' and thus the velocity of flow falls as the blood moves out into the smaller vessels.[1] Chapter 13 covers the mechanics of small diameter vessels where viscosity dominates. It is important to recognize that the boundary between large and small Reynolds number flows may not be fixed in the circulation, because the Reynolds number depends on parameters which may vary widely; for example, increased flow through a particular vascular 'bed' (e.g. through a muscle during exercise) will occur through dilation of small vessels and an increase in blood flow velocity; thus, the Reynolds number at any particular level will increase, and the boundary point will move outwards towards the capillaries. With vasoconstriction the opposite will happen.

Anatomy and structure

The anatomy of large blood vessels The anatomy of the canine aorta and its main branches is illustrated diagrammatically in **Fig. 12.1**, and many of the relevant dimensions are listed in **Table I**. The aorta takes origin from the left ventricle at the aortic valve and almost immediately curves, in a complicated three-dimensional way, through about $180°$, giving off branches to the heart, head and upper limbs (**Fig. 12.2**); it then pursues a fairly straight course down through the diaphragm to the abdomen, where it distributes branches to abdominal organs. Low down in the abdomen it terminates by dividing into the iliac vessels, which supply the hind limbs and (in the dog) the tail.

The aorta tapers along its length; the rate of tapering appears to be quite variable from animal to animal and probably from species to species; but in the dog, the area change fits quite well to an exponential equation of the form

$$A = A_0 e^{-Bx/R_0},$$

where A is aortic area, A_0 and R_0 are the area and radius at the upstream site, x is the distance from that upstream site and B is a 'taper factor', which has been

---

[1] Note that it is a reasonable assumption to regard blood viscosity as constant in these larger vessels, both because vessel diameters are large compared with individual cell diameters, so that blood can be considered a continuous fluid, and because shear rates are high enough for viscosity to be independent of them (Chapter 10).

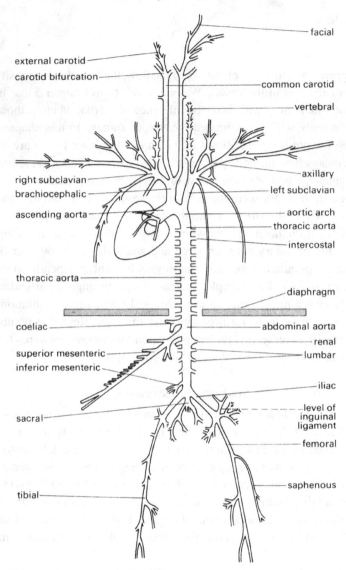

Fig. 12.1. A diagrammatic representation of the major branches of the canine arterial tree. (After McDonald (1974). *Blood Flow in Arteries*. Edward Arnold, London.)

found to lie between 0.02 and 0.05. This is quite marked tapering, as is shown in **Fig. 12.3**a, which is constructed from measurements of post mortem canine aortic casts; the value of *B* corresponding to the taper in this figure is about 0.033. By contrast, the available evidence in humans suggests that the tapering is not as smooth as this equation implies.

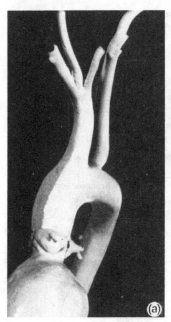
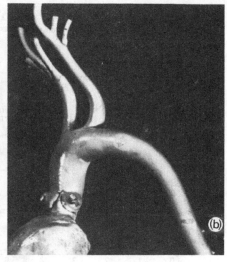

Fig. 12.2. Views from (a) the front and (b) the left side of a cast of the dog aorta, made at physiological pressure. Note the three sinuses of Valsalva and coronary arteries at the origin of the aorta. The figure also illustrates the complicated curvature of the aorta in two planes, and the fact that the angles made by the brachiocephalic and left subclavian branches may be difficult to define, since the vessels go on curving after they leave the aorta.

Branching ratios and angles Although the aorta and other vessels taper individually, the total cross-sectional area of the arterial bed increases with distance from the heart. The data usually reproduced to support this statement are probably misleading for various reasons, including the fact that they were not always measured at physiological pressures, and there may have been some shrinkage resulting from the use of chemical fixatives. Measurements in living animals are, of course, difficult to obtain and our knowledge is still incomplete, even for a single species. However, there is good information about the branches of the aorta in the dog, based on both direct measurements in living arteries and post mortem casts made at physiological perfusion pressures. An 'average' aortic tree (based on such measurements in 10 fairly large dogs) is depicted in **Fig. 12.3**b.

In general, the size of individual branches corresponds well to the amount of flow they conduct (the exception being the renal arteries, which are disproportionately narrow considering that they take 20–25% of the cardiac output). However, if we calculate branching ratios (expressed as the sum of the areas of the daughter vessels at a branch divided by the area of the parent vessel, so that values greater than 1.0 imply an expansion of the vascular bed and values less than 1.0 imply contraction), we find considerable variation, with values ranging between 0.79 and 1.29.

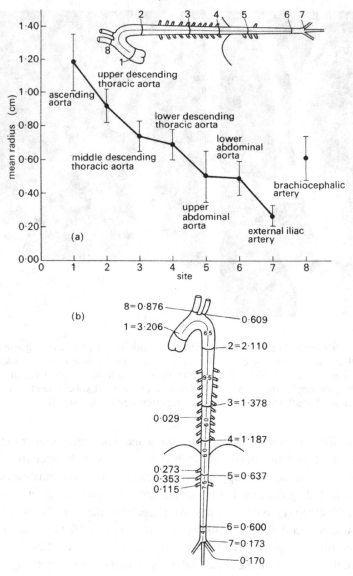

(a)

(b)

8 = 0·876

1 = 3·206

0·609

2 = 2·110

3 = 1·378

0·029

4 = 1·187

0·273
0·353
0·115

5 = 0·637

6 = 0·600

7 = 0·173

0·170

Fig. 12.3. (a) Taper in the aorta. The mean external radius of the aorta at various sites measured at physiological pressure in 10 large dogs. (From Fry, Griggs Jr and Greenfield Jr (1963). In vivo studies of pulsatile blood flow. In *Pulsatile Blood Flow* (ed. Attinger), p. 110. Used with permission of McGraw-Hill Book Company, New York.) (b) Lumen cross-sectional area at various sites in the aorta and its main branches (details of data in (a)). (From Patel, de Freitas, Greenfield Jr and Fry (1963). Relationship of radius to pressure along the aorta in living dogs. *J. Appl. Physiol.* **18**, 1111).

Table 12.1. *Dimensions of blood vessels in the wing of the living bat*

Vessel	Average length (mm)	Average diameter (μm)	Average number of branches	Number of vessels	Total cross-sec area (μm²)	Capacity (mm³ × 10⁻³)	Per cent of capacity
Artery	17.0	52.6	12.3	1	2 263[a]	38.4	10.1
Small artery	3.5	19.0	9.7	12.3	4 144	14.4	3.8
Arteriole	0.95	7.0	4.6	119.3	5 101	4.7	1.2
Capillary	0.23	3.7	3.1[b]	548.7	6 548	1.5	0.39
Post-capillary venule	0.21	7.3		1727.0	78 233	16.4	4.3
Venule	1.0	21.0	5.0	345.4	12 7995	127.9	33.7
Small vein	3.4	37.0	14.1	24.5	27 885	94.7	25.0
Vein	16.6	76.2	24.5	1	4 882	81.0	21.4

(From Wiedeman (1963). Dimensions of blood vessels from distributing artery to collecting vein. *Circulation Res.* **12**, 376.)
[a] Average of individual cross-sectional areas.
[b] Calculated.

There is a tendency for the branches within the chest to have ratios close to 1.0, and for those in the upper abdomen to have slightly higher values showing expansion; but this is largely counteracted by contraction (ratio 0.85–0.90) at the termination of the abdominal aorta, and this is also found in humans. Overall, there is little change; the summed areas of the branches of the aorta in **Fig. 12.3**b are only about 7% greater than that of the ascending aorta. Other experiments put this higher (up to 60% expansion), the scatter reflecting the difficulties of accurately preserving physiological pressures throughout the vascular bed as the casting material sets.

However, there is no doubt that, beyond these early branches, the cross-sectional area of the arterial tree begins to expand dramatically. The most detailed study in the living state has been made in the bat's wing, which can be transilluminated and examined directly with a microscope. The results are summarized in **Table 12.1**. Everything except the capillary dimensions is, of course, scaled down, because even the aorta has a diameter of only about 1 mm in the bat, but the increase in total vascular cross-sectional area with branching is striking.

Comparable studies have not been carried out in other species, and in the dog information has to be pieced together from many sources, largely post mortem preparations such as the casts mentioned above. The pattern which emerges is less precise, but very similar, and is illustrated in **Fig. 12.4**. The same is true in humans, where branching behaviour can also be studied in life, both by observation during surgery and by X-ray visualization of radio-opaque dye injected into vessels as a diagnostic procedure (arteriography). **Fig. 12.5** shows the course and distribution of the carotid artery within the skull demonstrated in this way.

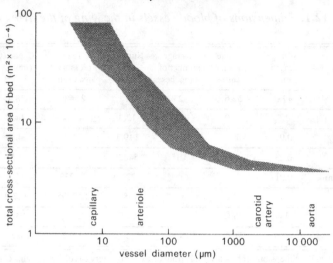

Fig. 12.4. Diagram illustrating how the cross-sectional area of the vascular bed expands peripherally. The scatter is not a consequence of experimental error; it reflects the fact that vessels at each level have varying diameters and degrees of taper. (Data from many sources, quoted by Iberall (1967). Anatomy and steady flow characteristics of the arterial system. *Math. Biosci.* **1**, 375.)

The angles at which branches come off the aorta vary just as branching ratios do, and it is clear that there is little symmetry in the structure of the central arterial tree, either in branching angles, ratios or lengths; this must be borne in mind when we consider how to treat arterial behaviour quantitatively. Among the larger central arteries, symmetrical bifurcation is exceptional; in humans the only example is the aortic bifurcation; in the dog, as **Fig. 12.1** shows, there is none. Patterns of flow through arterial bifurcations are discussed on p. 317.

Further downstream, dichotomous branching with fairly narrow angles is commoner, though there are many exceptions and individual organs may have highly specialized arrangements not found elsewhere. In most organs, anastomoses (i.e. connecting links between different arterial branches) occur, or can develop to bypass a blocked or damaged vessel. One of the most important of these lies at the base of the brain, where several vessels feed into a single ring, known as the circle of Willis (named after the English physician who, in 1664, produced an authoritative account – with illustrations by Christopher Wren – of the anatomy of the nervous system); this in turn supplies branches to many parts of the brain.

The structure of the arterial wall All arteries (and indeed veins – see p. 427) show a common pattern of organization and are made up of similar materials, though the proportions vary in different parts of the circulation. We will first review the

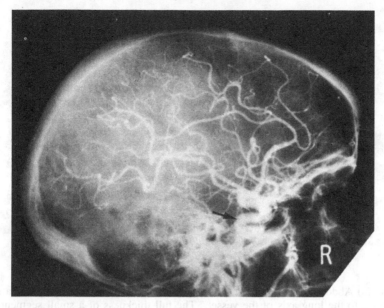

Fig. 12.5. A lateral X-ray of the skull immediately after injection of radio-opaque material into the carotid artery in the neck in a man. The arrow shows the internal carotid artery, which follows a short, winding course within the skull and then divides into numerous branches which supply blood to the brain.

microscopic structure of the artery wall and then discuss the functional importance of the various wall layers and components and some of the effects of ageing.

The arterial wall is made up of three layers: an internal tunica intima, surrounded by a tunica media and then an external tunica adventitia. Each of these layers has a predominant structure and cell type; the important features are illustrated in the examples in **Fig. 12.6**, which shows micrographs of transverse sections of a typical small artery.

The tunica intima has two components. The innermost, which is in contact with the blood, is the endothelium; this is a single layer of cells, which extends as a continuous lining right through the circulation, covering all the surfaces which come in contact with the blood: arteries, capillaries, veins, heart valves and endocardial surfaces. It is mechanically rather fragile (being damaged by high shear stresses, for example) but has considerable powers of regeneration and growth; it regenerates itself continuously and can grow to form a lining over living or synthetic graft materials inserted surgically. The detailed structure of endothelial cells is dealt with in Chapter 13. Surrounding the endothelium is a thin subendothelial layer, containing a few collagen-generating cells (fibroblasts) and collagen fibres. Then comes a layer of branching elastic fibres (the internal elastic lamina) which is particularly well defined in smaller

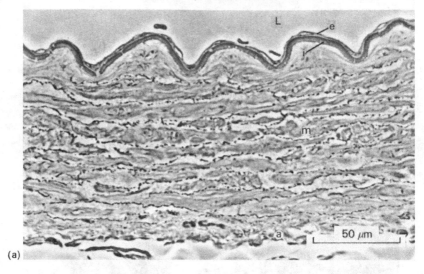

(a)

Fig. 12.6. (a) Abdominal aorta of the guinea-pig. Phase contrast micrograph of a section cut transverse to the long axis of the vessel. The full thickness of a small segment of the wall is shown. L: lumen; e: endothelium; i: internal elastic lamina; m: media; a: adventitia. This is a fairly typical 'muscular' artery. The lamellar structure of the media is well shown, with layers of smooth-muscle cells separated by darker, thinner bands of collagen and elastic fibres. Micrographs by courtesy of Dr G. Gabella.

arteries (see **Fig. 12.6**a) and forms the inner boundary of the next wall layer, the tunica media. The tunica media is usually the thickest layer in the wall, and the one which shows the greatest variation in structure and properties in different regions of the circulation. Because of the differences occurring in this layer, arteries are generally divided into elastic and muscular vessels. This is an oversimplification, because there is no sharp subdivision, and there are also specialized variations in particular organs and species, but it does draw attention to the change in structure which occurs with distance away from the heart. The tunica media of large central arteries (the aorta and its large branches within the chest) is made up of multiple concentric layers of elastic tissue separated by thin layers of connective tissue, collagen fibres and sparse smooth muscle cells which obliquely cross-link successive elastic layers. Smooth muscle cells are long (25–50 μm) and thin (5 μm), often with tentacle-like extensions, and contain contractile filaments of muscle which have the same basic form as those described in the muscle cells of the heart (p. 186). Under the microscope, these myofibrils are less prominent than they are in heart muscle cells because they are less systematically arranged and fewer in number, but their function is identical – they contract in response to depolarization of the cell membrane and generate tension. In large arteries, they tend to be orientated obliquely or longitudinally; in smaller, more peripheral arteries they lie increasingly more circumferentially, forming flat spirals.

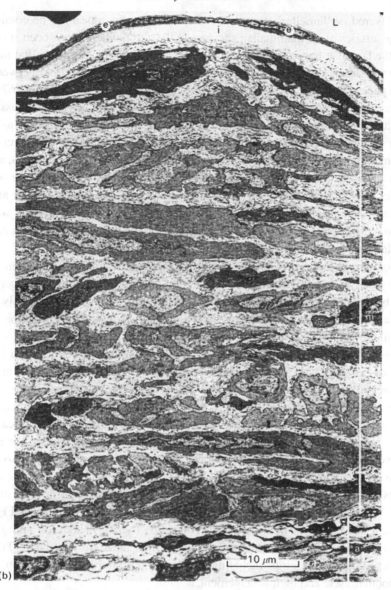

Fig. 12.6. (b) Transmission electron micrograph of the same specimen as in (a) at higher magnification. Note the single layer of endothelial cells, and the pale connective tissue which surrounds the muscle cells of the media. At the bottom of the section, in the adventitia, a few dark-staining fibroblasts and bundles of collagen fibrils are visible. Micrographs by courtesy of Dr G. Gabella.

The layered or 'lamellar' structure of the tunica media appears to be common to the large arteries of all mammalian species. In the aorta, where it has been studied in detail, the lamellae have a fairly constant thickness of about 15 μm, so that the total number present is closely proportional to the radius of the vessel both between and within species; thus, the mouse aorta has five lamellae and the pig aorta over 70.

Further away from the heart, the structure of the media changes and the vessels are known as the muscular arteries, with the media consisting largely of spirally arranged smooth muscle cells (**Fig. 12.6**). Again, these are disposed in multiple layers, with small amounts of connective tissue, collagen and elastic tissue between them, and the number of layers diminishes as vessel radius diminishes; in very small arteries and arterioles there may be only one or two such layers. Throughout the smaller arteries, however, the tunica media remains the predominant element in the vessel wall; its outer boundary is marked by a very thin layer of elastic tissue known as the external elastic lamina.

The outermost layer of the arterial wall, the tunica adventitia, may in some places be as thick as or even thicker than the media, but it is less prominent microscopically. This is because it is composed of loose connective tissue, containing relatively sparse elastin and collagen fibres running in a predominantly longitudinal direction, and it has an ill-defined edge as it merges into the surrounding tissues.

The walls of all arteries larger than about 1 mm in diameter have their own nutrient blood vessels, the vasa vasorum. These take origin either from the parent artery or from a neighbouring one and break up into a capillary network which supplies the adventitia and, in larger vessels, reaches in as far as the inner layers of the tunica media. The nourishment of the tunica intima and the innermost layers of the tunica media depends predominantly on transport of materials from the arterial lumen. Small lymphatic vessels, and fine nerve fibres running to the smooth muscle cells, also ramify through the tunica adventitia.

It has been known for many years that the elastic properties of arteries are nonlinear. They become markedly stiffer as they are stretched (as shown in **Fig. 7.5**, p. 94), and an explanation of this was given on p. 96. There is in addition a progressive increase in Young's modulus which takes place with distance from the heart. This may be due to a number of factors, including the changing proportions of collagen, elastin and smooth muscle and their arrangement.

Elastin and collagen, both of which are made up of long chains of protein molecules, are present in fibrous lattices and sheets in the various layers of the artery wall. The physical properties of both these materials have already been described (p. 91), so will not be repeated here. It is nonetheless worth repeating that in large arteries there is a fairly clear relationship between their properties and the elastic behaviour of the whole artery wall. On the other hand, the precise influence of smooth muscle on the properties of the larger arteries has long been a matter of controversy. This is principally because most studies have been carried out on excised blood vessels, where the state of contraction of the smooth muscle ('vasomotor tone') is likely to differ from that in the animal. However, there remain considerable difficulties even if studies can be performed on intact arteries within the animal. This is because contraction of smooth muscle in the intact circulation so decreases the total cross-sectional area of the vascular bed by constricting small arteries that the blood pressure in the large arteries rises; thus, any diameter reduction due to an active increase in wall tension is masked. Recent work, in which these effects are separated, has demonstrated that the contraction of smooth muscle can, even in the largest arteries where it is relatively sparse, produce considerable change in wall tension and actually constrict the vessel.

In the aorta, these calibre changes are small. For example, if the nerves supplying the smooth muscle of the aortic wall are stimulated electrically (mean blood pressure being held constant), aortic diameter decreases by only about 5%. Most stimuli in the normal animal would be less intense than this, and it seems unlikely that physiological variations in smooth muscle activity would ever change aortic calibre by more than this. In more peripheral (muscular) arteries, the effect of smooth muscle is greater; a similar experiment in the femoral artery, for example, produces a 20% diameter reduction, and cutting the nerves which supply the smooth muscle causes the diameter to increase above its resting level by a similar proportion.

The way in which smooth muscle contraction affects the properties of an artery will depend upon the conditions of the experiment and the way in which the measurements are made. If, for example, the diameter of a vessel is held constant and the smooth muscle stimulated, the wall becomes stiffer and the incremental elastic modulus increases. If, on the other hand, the diameter is not controlled, then smooth muscle activity will constrict the vessel and the elastic modulus may go down, as explained earlier (p. 96). In the normal living animal, smooth muscle contraction stiffens the walls of large arteries (unless the vessel is so stretched that the collagen in the wall is taking much of the load; since the elastic modulus of maximally active smooth muscle is less than that of collagen, smooth muscle contraction would have no effect in this situation).

As was mentioned earlier, smooth muscle cells in large arteries tend to be orientated obliquely or longitudinally, and the cells appear to connect to the elastic tissue, particularly the internal and external elastic laminae of the tunica media. Thus, they alter the wall tension partly by pulling on the elastic tissue and partly by acting in parallel with it and collagen. Their oblique disposition and the fact that they have a limited ability to shorten (by a maximum of about 25% of relaxed length) probably account for their relatively slight effect on the diameter of large arteries even when they are maximally activated. The ability of some vessels, including quite large muscular arteries, to go into 'spasm' when handled experimentally or injured is presumably also a property of smooth muscle. However, it has not yet been adequately explained, because the lumen of an artery in spasm may be virtually obliterated, rather than reduced by the 20–25% which sympathetic stimulation can produce and which the shortening properties of smooth muscle would predict.

In summary, therefore, the effects of smooth muscle on the size and stiffness of large arteries are probably relatively slight under normal circumstances, particularly in comparison with its effects in smaller arteries, where it is the main constituent of the wall, and the near-circumferential arrangement of the fibres means that they can greatly alter the vessel lumen. Its activity there controls both the local distribution of flow and the overall resistance to blood flow, aspects that are dealt with in Chapter 13.

Smooth muscle does, however, have appreciable effects on the dynamic properties of the large vessels, as we have seen in Chapter 7. The viscous properties of the artery (stress relaxation, creep, hysteresis) are primarily attributed to the presence of smooth muscle and greatly complicate wall mechanics (p. 96). One further dynamic property of smooth muscle, the physiological significance of which is undetermined, is the 'myogenic response'. It is a property of smooth muscle cells that they depolarize (and therefore contract) if they are stretched. The time course of this response is slow (too slow for it to modify wall properties during a single heartbeat, for example), but there is no doubt that it can significantly modify the stress–strain properties of the vessel wall. Its physiological role is a matter of continuing debate.

Arterial wall thickness The thickness of the artery wall varies considerably through the circulation, and representative values are given in **Table I**. The measurements show a good deal of scatter; in the case of the small vessels, this is not surprising, because there are considerable problems in holding the vessels at their normal luminal pressures and preventing distortion as they are prepared for study. A further difficulty, which affects large vessels as well as small ones, arises because the tunica adventitia may merge gradually into surrounding tissues so that it is difficult to decide where the artery wall ends. The microcirculation does not appear to have been studied in the dog, and the values for very small vessels in **Table I** have to be drawn from other species – small mammals such as cats, rats and mice. This is probably unimportant, as surveys of small-vessel dimensions in mammals of widely varying body size have shown them to be remarkably similar.

This is far from true, however, in large vessels, where wall thickness varies dramatically with body size among the smaller mammals. **Fig. 12.7** shows the results of a study of this, in which great care was taken to maintain the vessels at their in vivo size and to minimize and correct for distortion during histological preparation. Of course, diameter also varies greatly with body size in these large vessels; this is shown in **Fig. 12.8**, which is taken from the same source. As can be inferred from these two figures, the relationship between vessel diameter and wall thickness, both within and between species, is approximately constant in large arteries; thus, the h/d ratios in **Table I** are independent of body size.

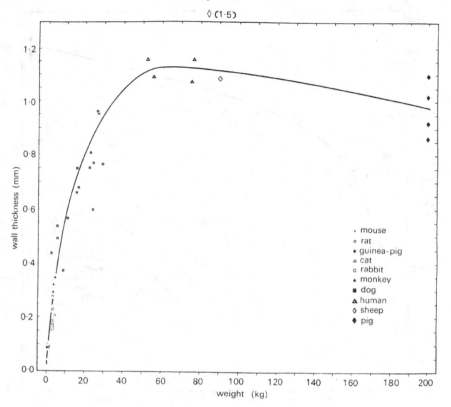

Fig. 12.7. Thickness of the media of the thoracic aorta in 10 mammalian species. Medial thickness increases rapidly with adult body weight in the relatively small mammals; there is little or no increase with body weight in the large mammals. (From Wolinsky and Glagov (1967). A lamellar unit of aortic medial structure and function in mammals. *Circulation Res.* **20**, 99.)

In summary then, in large central arteries the internal diameter d and wall thickness h depend on body size, though the ratio between them (h/d) does not, and is thus independent of species and of site. As we move peripherally into small vessels, the walls become thinner, but h/d actually increases until in the very small arteries the external diameter may be almost twice that of the lumen even when the smooth muscle is relaxed. Finally, in the very smallest vessels in the microcirculation, and particularly the capillaries, the relationship with body size disappears completely, with the diameter and wall thickness being very similar in all mammalian species.

Changes in the arterial wall with age The dimensions and properties of the larger arteries vary not only with body size and with position in the circulation, but also with the age of the animal. Systematic studies of the changes with growth and ageing

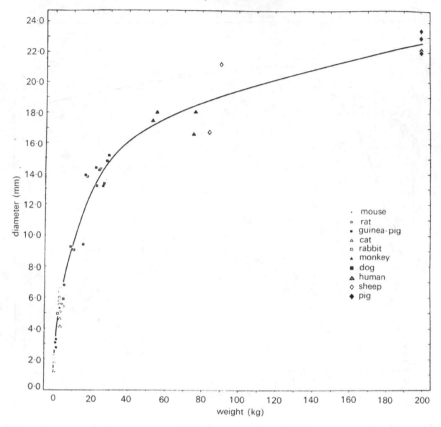

Fig. 12.8. Diameter of the thoracic aorta in 10 mammalian species. Diameter increases more rapidly with adult body weight for small mammals than for large. (From Wolinsky and Glaeov (1967). A lamellar unit of aortic medial structure and function in mammals. *Circulation Res.* **20**, 99.)

have mainly been made in humans, and many of the measurements have been made at unstretched length in post mortem tissues, so that we know more about the histological changes with age than about precise dimensions. However, it is clear that the thickness and the structure of arterial walls are changing slowly throughout life, and these changes are reflected in the elastic properties. It is also clear that comparable changes occur in other species of mammals, though the evidence is less complete.

In the growing animal or child, both the diameter and wall thickness of the large arteries increase steadily. Measurements at physiological degrees of stretch are lacking, and the values in **Table 12.2** are for aortas at zero transmural pressure; h/d in these circumstances is virtually constant at all ages, an observation which has been confirmed repeatedly.

The increase in wall thickness in these central, elastic arteries is largely due to thickening and proliferation of the elastic lamellae in the tunica media, a process

Table 12.2. *Dimensions of unstretched arteries in man at different ages during growth*

Age (years)	Internal diameter d(mm)	Wall thickness h(mm)	h/d
1	11	0.83	0.08
3–6	13	1.14–1.23	0.09
15–18	17	1.41–1.58	0.09
20	19	1.58	0.08

(Calculated from data from Dittmer and Grebe (1959), *Handbook of Circulation*, p. 20. W. B. Saunders Company, Philadelphia.)

which is complete by early adult life and thereafter merges imperceptibly into a different kind of histological change which is usually considered to be degenerative. Here, the elastic elements in the wall begin to fray and fragment, and may become calcified; the collagen fibres increase in number, both replacing muscle cells and proliferating in other parts of the wall. The overall effect is that the diameter of the vessel increases, and its wall becomes thicker and much less distensible. This is 'hardening of the arteries', a process which affects large and medium-sized arteries.[2] It is difficult to know whether to regard it as a physiological process or a disease state, since it occurs universally but shows a wide spectrum of severity.

A more definitely pathological process, known as atheroma, may also occur. This is an increasingly common disease, which is now responsible for about half the deaths occurring in developed countries. Because of this, and because there has been a certain amount of investigation of the possible role of mechanics in its development, this is discussed in some detail on p. 333.

A number of studies have been made of the changes in dimensions and elastic properties which take place in human arteries with increasing age. **Fig. 12.9**, which is taken from one such study, summarizes what takes place. Healthy arteries taken at post mortem from a group of 'young' (<20 years) and 'old' (>35 years) patients were studied; they were held at their living length and inflated to physiological pressures. The aorta increases in diameter and thickens with age, the wall thickness increasing slightly more than the diameter so that h/d is increased. The smaller, 'muscular' branches of the aorta do not change appreciably in diameter, though their walls thicken considerably. Thus, old vessels dilate and thicken, but to different degrees at different sites; the big vessels tend to dilate and thicken, while the smaller ones show only wall thickening.

---

[2] The term 'arteriosclerosis' is often applied to this process. However, confusion arises because this word is also sometimes used to describe the later stages of atheroma.

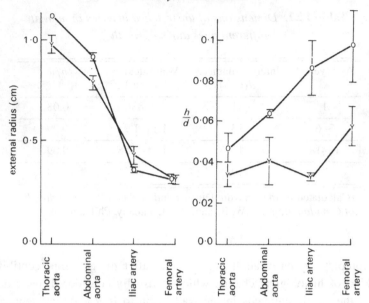

Fig. 12.9. The effect of ageing on radius and wall thickness of human arteries at physiological pressure. Average results from groups of young (Y) and old (O) arteries, plotted against position in the arterial tree. (From Learoyd and Taylor (1966). Alterations with age in the viscoelastic properties of human arterial walls. *Circulation Res.* **18**, 278.)

Accompanying these changes in dimensions are interesting changes in elastic properties. In the thoracic aorta, at physiological pressure ($1.33 \times 10^4 \, \mathrm{N\,m^{-2}}$, i.e. 100 mm Hg) the incremental Young's modulus E (p. 94) increases steadily with age (**Fig. 12.10**a); more peripherally, however, there is no change or a fall (**Fig. 12.10**b). The explanation of this appears to be that the diameter of the aorta increases with age; thus, it is further along its (nonlinear) length–tension curve (**Fig. 7.5**a, p. 94) and E is correspondingly greater.

We can summarize a good deal of this section by saying that the dimensions and elastic properties of arteries vary greatly with age and body size, as well as with their site in the circulation. Much of this is intuitively obvious – if we think of a mouse and an elephant we can all predict a correlation between aortic size and body weight, for example. Nonetheless, a certain amount of detail is essential, because many of the properties are less obviously related, and in the rest of this chapter we shall be examining mechanical phenomena (such as pressure and flow wave-forms) which can only be clearly understood if the behaviour of the arterial wall and the anatomy of the vascular bed are known in some detail. Finally, this section has a cautionary value, since it concerns variability in biological systems. Physical scientists who study the circulation need quantitative information, and are often frustrated by the inability of physiologists to provide it. It may be salutary to illustrate that a simple parameter like

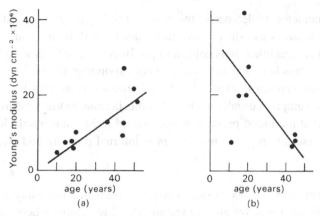

Fig. 12.10. Effect of ageing on elastic properties of arteries. (a) Incremental Young's modulus plotted against age for normal human aorta at a pressure of 100 mm Hg. (b) Similar data for iliac artery. (From Learoyd and Taylor (1966). Alterations with age in the viscoelastic properties of human arterial walls. *Circulation Res.* **18**, 278.)

aortic diameter not only is extremely difficult to measure, but will show variation with species, site, age, body size, blood pressure and smooth-muscle activity. If we wish to calculate the Reynolds number for the aorta, we need not only the vessel diameter, but also two other measurements – blood velocity and viscosity; to calculate pulse-wave velocity using the Moens–Korteweg equation (Equation (12.11), p. 276) we must have diameter, wall thickness, density and Young's modulus for the wall. Now all these measurements will, like aortic diameter, be subject to a number of biological influences, and will therefore show variation. It follows that derived parameters like *Re* may show wide variation even when they are calculated from accurate and compatible measurements. If they are calculated from badly chosen data, this variability will become grotesque.

Blood pressure and flow in systemic arteries

The pressure in all systemic arteries fluctuates, the source of the fluctuations being the pumping action of the left ventricle. A typical aortic pressure record is shown in **Fig. 12.15**, and there are many other examples in this book. Certain conventions are followed in describing arterial blood pressures, and we need to examine these with some care in order to understand exactly what is meant and how relevant the measurements are to the mechanics of flow.

The first convention is that of expressing the pressure in mm Hg. This unit does not appear in any international system, old or new, but arose because the blood pressure is actually measured in mm Hg in the vast majority of cases, using an inflated cuff connected to a mercury manometer to occlude the artery in a limb; this is known as a sphygmomanometer. In defence of the unit, it should perhaps be pointed out that it has been in continuous use since it was introduced (by Poiseuille) in 1828 and has, therefore, seen several other units come and go. In deference to its wide familiarity it is used widely in this book together with the corresponding SI unit.[3]

The second convention in expressing blood pressure is to refer it to atmospheric pressure. The assumption usually made is that this is equal to the pressure just outside the artery, so that the blood pressure is considered to be transmural blood pressure (transmural pressure is $p_i - p_o$, and here p_i is internal pressure and p_o atmospheric pressure).

Transmural pressures We need to know what the transmural pressure is for a variety of reasons. We need it if we are to examine the force balance between blood and wall, and if we are to choose realistic wall properties for the living state when their measurement has to be made post mortem. It also tells us something about the load against which the heart must work, and about the state of the peripheral vascular bed, blood volume, reflex activity and many other physiological events. It does not, however, tell us anything about the force which is being applied to cause motion of the blood – to know that we need the pressure difference between two points in the circulation, just as we need the pressure difference between inside and outside of the arterial wall in order to examine the forces acting across it. The point to be made here is that the pressures measured by a physiologist are often not those needed by an engineer for a physical description of the system. For example, the pressure difference across the wall of a large artery in a limb – the conventional 'blood pressure' – may be on average $1.33 \times 10^4 \, \mathrm{N\,m^{-2}}$ (100 mm Hg); but the pressure difference between the ascending aorta and such an artery, which is the force which actually drives the flow between them, averages only about $0.03 \times 10^4 \, \mathrm{N\,m^{-2}}$ (2–3 mm Hg) and is rarely measured because it is technically extremely difficult to do so.

An important factor influencing transmural pressure is gravity. This is easiest to understand if we consider the situation in a dead animal. All the vascular smooth muscle relaxes and the blood redistributes itself. The increase in volume of the peripheral vessels and small veins is so great that many large vessels are virtually emptied and may collapse.[4] In the absence of gravity, the pressure would be the same, and practically equal to atmospheric, in all vessels.

We can understand the effect of gravity if we consider the dead body propped upright. The blood will now redistribute itself, and above a certain level in the body where the pressure is equal to atmospheric (the 'equal pressure point') all the vessels will be collapsed. From that level upwards, if the vessels are empty, the pressure in them will be equal to the surrounding tissue pressure, and hence approximately equal to atmospheric pressure p_A. Below that level (and above it if the vessels there are not

---

[3] $1 \, \mathrm{mm\,Hg} = 133.32 \, \mathrm{N\,m^{-2}}$.

[4] The term 'collapse' is defined on p. 104; it does not necessarily imply that the vessel is completely empty.

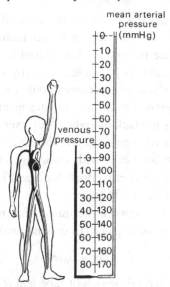

mean arterial
pressure
(mmHg)

Fig. 12.11. Mean arterial and venous pressures in the human circulation, relative to atmospheric pressure. (After Rushmer (1970). *Cardiovascular Dynamics*, p. 196. W. B. Saunders Company, Philadelphia.)

completely empty) the pressure in the vessels p_i is given by

$$p_i = p_A + h\rho g, \tag{12.1}$$

where h is the vertical distance below the equal pressure point (h is negative above that point), ρ is the density of blood and g the acceleration due to gravity. This equation is, of course, identical to Equation (4.2), p. 34, and the pressure we are considering is the hydrostatic pressure.

Hydrostatic pressure is an important force for the redistribution of blood in the living vascular system, as well as in the corpse. The zero level ($h = 0$) in a man standing upright is approximately at the level of the heart (**Fig. 12.11**); thus, veins in the neck are normally collapsed (but not completely so; blood still flows through them) and the 'filling pressure' of the heart (Chapter 11) is close to atmospheric pressure. This filling pressure can be altered either by changes in blood volume or by changes in the volume of the vascular bed in the lower limbs.

The most familiar expression of gravitational effects is fainting, where one of the important features is that when the subject stands up, the volume of blood in the leg veins increases abnormally and the venous filling pressure of the heart falls. This leads to a fall in cardiac output and thus of blood supply to the brain. Falling or lying down relieves the faint by eliminating the force which is holding blood in the leg vessels. One of the reasons why everyone does not faint on standing is that reflex constriction of the veins in the legs normally occurs so that their ability to act as a reservoir is greatly reduced. Another reason is that there is arteriolar constriction, which increases the resistance to flow so that arterial blood pressure does not fall significantly. If vasoconstriction is prevented (e.g. by certain drugs or diseases), then fainting or dizziness on standing becomes common; and it also occurs where the vascular bed is greatly dilated as in hot climates or after a hot bath. The same mechanism may cause pilots to black out when they execute 'high-g' turns in high-performance aircraft if the tension generated in the vessel walls is not great enough to balance the hydrostatic pressure even with vasoconstriction. The vessel walls will then be stretched (becoming progressively stiffer, see **Fig. 14.2**, p. 434) until the tension within them does balance the pressure.

The quantitative relationship between the transmural pressure ($p_i - p_o$) and wall tension is expressed by the law of Laplace (Equation (7.6)), which tells us that

$$T = (p_i - p_o)R,$$

where T is the wall tension per unit length of tube and R is the tube radius. As the transmural pressure and radius increase, a balance can only be struck if T increases more rapidly. Thus, the stability of vessel walls depends upon their nonlinear elastic properties. Note that in distending a vessel it is just as effective to lower p_o as to raise p_i; a normal person can be made to faint if the lower half of his body is subjected to a subatmospheric pressure. Physiologically, the systemic vessels within the chest are subjected to a p_o which swings above and below atmospheric with each breath, and this has important consequences, particularly for the stability of veins (see Chapter 14).

From what has been said so far, it will be clear that the hydrostatic forces are an important component of the forces acting on vessel walls, particularly the veins. The situation in life is illustrated in **Fig. 12.11**, in which the mean arterial blood pressure ($1.2 \times 10^4 \, \mathrm{N\,m^{-2}}$, i.e. 90 mm Hg) is added to the hydrostatic pressure acting in arteries, but not in veins. The transmural pressure in a leg artery may be four times that in the raised arm, while the arm veins will empty towards the heart and collapse when the arm is raised above the level of the chest.

Thus, the transmural pressures encountered in the normal circulation can vary very considerably, a point which is often forgotten in calculations of the force balance on the arterial wall. Nor is the 'safety margin' in arterial pressure very great; as has been explained in Chapter 7 (p. 102), the artery may collapse whilst there is still a slightly positive (i.e. distending) transmural pressure. Here again is a source of biological variation; the standing giraffe has to have a much higher mean arterial blood pressure than a human in order to prevent the collapse of the arteries supplying blood to the brain.

The effects of hydrostatic pressure on the distension, and therefore elastic properties, of arteries and veins tend to be forgotten. This is because clinicians measure blood pressure in the upper arm, which is approximately at the upper chest level, where the hydrostatic component of the transmural pressure in a lying or standing patient is zero. In studies of the circulation where fluid-filled catheter and manometer systems are used, the convention is to align the manometer diaphragm at 'mid-chest' level, which also eliminates the hydrostatic pressure component. Thus, the hydrostatic term in Equation (12.1) ($h\rho g$) is zero, and the 'excess pressure' generated by the heart is the only one described. It is worth remembering, when considering the elastic behaviour of arteries, that a transmural pressure of 200 mm Hg may be entirely normal in dependent limbs.

Unsteady pressure in large arteries The rather complicated patterns of transmural pressure which may arise in the circulation are of interest from the point of view of vessel-wall mechanics, but they should not be allowed to obscure the real nature of the pressure force which drives the flow. This has nothing to do with transmural pressure – flow can occur through an almost completely collapsed tube, as it does for example when blood returns through the neck veins to the heart in an upright subject. Nor is the hydrostatic pressure important; it requires the same amount of energy to force fluid through a tube whether it is horizontal or vertical (provided the size is unchanged). Thus, when we consider the forces governing flow, we are exclusively interested in the pressure generated by the heart, which we shall call the excess pressure p_e. This quantity is the difference between the actual pressure and its hydrostatic component, and is commonly referred to as 'blood pressure'. It is the gradient of excess pressure which drives the flow. The distribution of excess pressure through the circulation is illustrated in **Fig. 12.12**; it can be seen that it is largely dissipated in forcing the blood through the microcirculation. That will be dealt with in Chapter 13; here we are interested in the much smaller pressure gradients necessary to force flow through the large arteries, and in the rest of this chapter we shall be concentrating on them, and on the flows which result. In many parts of this book we shall follow common practice and refer to the excess pressure as blood pressure, or simply pressure, since the meaning will be clear in context.

We will examine the pressures first. The pressures in the left ventricle and in the ascending aorta immediately downstream from the aortic valve are shown in **Fig. 12.13**. These were recorded simultaneously, using a pair of extremely carefully matched pressure transducers, and are therefore a faithful record of events *instantaneously* during the cardiac cycle.[5]

As was described in Chapter 11, pressure rises rapidly in the ventricle at the beginning of systole and soon exceeds that in the aorta, so that the aortic valve opens,

---

[5] We shall see that it is the unsteadiness of pressure and flow which dominates the fluid mechanics of large arteries, and experimental apparatus must be capable of responding fast enough to register such changes faithfully. Thus, in addition to the usual requirements of stability and adequate sensitivity, is added that for a good frequency response. This is often a much more rigorous demand and has not always been adequately met in the past (see p. 267).

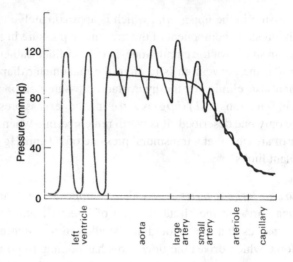

Fig. 12.12. Waveforms of excess pressure, and its mean level, in the arterial circulation.

blood is ejected and aortic pressure rises. During the early part of the ejection phase, ventricular pressure exceeds aortic pressure. About halfway through ejection, the two pressure traces cross and there is now an adverse pressure gradient across the aortic valve, which is maintained as both pressures start to fall. At this point there is a notch on the aortic pressure record (the dicrotic notch) which marks the closure of the aortic valve, and thereafter the ventricular pressure falls very rapidly as the heart muscle relaxes. The pressure in the aorta falls much more slowly. This fall is easy to understand if we remember that the large central arteries, and particularly the aorta, are elastic, and thus act as a reservoir during systole, storing some of the ejected blood which is then forced out into the peripheral vessels during diastole.

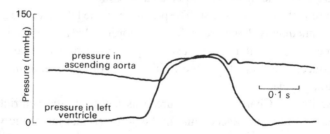

Fig. 12.13. Simultaneous recordings of pressure in the left ventricle and the ascending aorta of the dog. (From Noble (1968). The contribution of blood momentum to left ventricular ejection in the dog. *Circulation Res.* **23**, 663.)

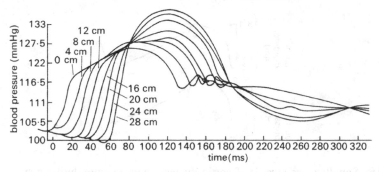

Fig. 12.14. Simultaneous blood pressure records made at a series of sites along the aorta in the dog: 0cm is at the start of the descending aorta. (From Olson (1968). Aortic blood pressure and velocity as a function of time and position. *J. Appl. Physiol.* **24**, 563.)

If we now move slightly downstream and examine the pressures at two points on the axis of the aorta (e.g. points 0 and 4 cm in **Fig. 12.14**) we can see that the pressure records are almost identical in shape at the two sites, but the one at the downstream point is slightly delayed. In other words, the pressure pulse generated by ventricular contraction is travelling along the aorta as a *wave*. We can roughly calculate its velocity from the delay if we know the distance between measuring sites, and we can also work out the pressure difference driving the flow along that segment if we subtract the downstream pressure from the upstream one.

If we make simultaneous measurements at points all along the aorta (or, as is technically much easier, make successive measurements at different sites with a catheter and align them by reference to the signals from a second, fixed, reference catheter) then we see that the pressure wave changes shape as it travels down the aorta (**Fig. 12.14**). It is progressively delayed, of course, but also steepens and increases in amplitude, whilst at the same time losing the sharp dicrotic notch. Thus, we have the paradox that the systolic pressure actually increases with distance from the heart. (This is not a peculiarity of the species; it is also systematically observed in humans and other animals.) In fact, of course, the *mean* level of the arterial pressure falls with distance from the heart; it is difficult to visualize this from the figure, because the fall is only about $0.05 \times 10^4\,\mathrm{N m^{-2}}$ (4 mm Hg) along the length of the aorta, whilst the amplitude of the pressure oscillation between systole and diastole nearly doubles.

This process of amplification of the pressure pulse continues in the branches of the aorta out to the level (in the dog) of about the third generation of branches – for example, the saphenous artery in the lower leg, with an internal diameter of 1–2 mm (this is shown in **Fig. 12.19**). Thereafter, both the oscillation and the mean pressure decrease rapidly to the levels found in the microcirculation.

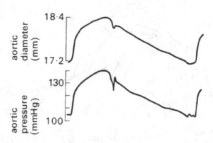

Fig. 12.15. Simultaneous records of aortic diameter and pressure during a single cardiac cycle in the dog. (From Barnett, Mallos and Shapiro (1961). Relationships of aortic pressure and diameter in the dog. *J. Appl. Physiol.* **16**, 545.)

The radial expansion and contraction of the arterial wall which takes place with the passage of the pressure wave has been measured by a number of methods. One such measurement is shown in **Fig. 12.15**, together with the pressure within the vessel, recorded simultaneously. It can be seen that the motion of the wall is virtually in phase with the pressure; in other words, the wall behaves like a purely elastic material, with little evidence of hysteresis due to visco-elastic properties (see p. 97). Note also that the change in radius is small; expressed as a strain (p. 86) it is about 7%. These observations justify the use of linear elastic theory to describe the behaviour of the arterial wall, expressing its properties in terms of a constant elastic modulus E (p. 94). As we shall see in the next section of this chapter, linear elastic theory works very well as a first approximation in describing the wave, and for predicting the wave speed.

Flow In order to complete our description of events in large arteries, before going on to consider the detailed mechanics, we must examine the flow along the aorta and its branches. Predictably, since the pressures and pressure gradients all vary with time, so do the flows; we are everywhere dealing with unsteady motion of the blood. As we shall see, this greatly complicates the physical theory we have to use. Furthermore, it means that we need accurate measurements of instantaneous flow rate or velocity, and very often simultaneous records of pressure, in order to understand events. Such records still represent a very considerable technical achievement, even in experimental animals where surgical access to large vessels is possible. In a normal human, access to the systemic circulation is very limited and we have inadequate information. It is only the increasingly thorough investigation of suspected cardiovascular disease (particularly coronary artery disease) by catheter techniques that is slowly yielding a store of information about normal function.

Figure 12.16 shows the pressure and flow velocity waveforms in the ascending aorta close to the aortic valve. At resting heart rates, the period of forward flow occupies about a quarter to a third of the cardiac cycle. When the heart rate increases,

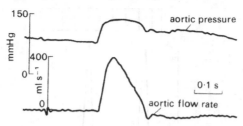

Fig. 12.16. Simultaneously records of pressure and flow rate in the ascending aorta of a resting, conscious dog. (After Noble (1968). The contribution of blood momentum to left ventricular ejection in the dog. *Circulation Res.* **23**, 663.)

for example in exercise, both systole and diastole shorten, but at different rates, with diastole shortening to a much greater extent. Thus, forward flow comes to occupy a greater proportion (up to 50%) of the cycle at high heart rates.

Forward motion of blood in the ascending aorta begins, of course, when the aortic valve opens and blood is ejected from the ventricle. The velocity rises rapidly to a peak and promptly, but more slowly, falls again; there is a brief period of backward flow towards the aortic valve as it closes and then the blood comes almost to rest for the remainder of the cardiac cycle. The maximum velocity observed depends on a large number of factors (not least the conditions of the experiment, such as the type and depth of anaesthesia), but in the normal conscious dog is probably in excess of $1\,\mathrm{m\,s^{-1}}$. (The values quoted in **Table I** are drawn from many sources in the literature and their wide range reflects the many influences which can act.) Acceleration at the beginning of ejection is abrupt and short-lived – the analogy has been drawn of the column of blood being struck by a hammer, and this is not entirely inappropriate, as can be seen from **Fig. 12.17**, where the accelerations and decelerations which the blood undergoes are shown synchronously with the velocity waveform.[6] The peak acceleration is in the region of $5g$ to $10g$ in the normal dog, and may increase up to 50% when the heart is stimulated in exercise or by drugs. Values of velocity and acceleration are not yet well established in humans, but are probably somewhat lower (**Fig. 12.17**).

The time relationships between pressure and flow at the entrance to the aorta are illustrated in **Fig. 12.18**, where the waveforms have been scaled to be of comparable size and superimposed. In early systole, pressure and flow rate rise in phase with each other, a fact which serves to emphasize that it is the pressure *gradient* and not

---

[6] Acceleration is the rate of change of velocity, and is thus given by the slope of the velocity waveform. In mathematical notation it is the differential with respect to time, $\mathrm{d}u/\mathrm{d}t$, of the velocity u. In experimental work, acceleration is frequently derived from velocity signals by means of electronic differentiating circuits which give a voltage output proportional to the rate of change of their input signal. The same means can be used to derive rate of change of pressure from a pressure signal.

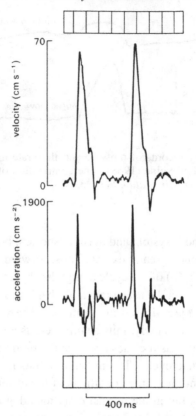

Fig. 12.17. Records of velocity (upper trace) and acceleration of blood flow in the ascending aorta in a normal human. The acceleration record is obtained by electronic differentiation of the velocity signal. (From Bennett, Else, Miller, Sutton, Miller and Noble (1974). Maximum acceleration of blood from the left ventricle of patients with ischaemic heart disease. *Clin. Sci. Mol. Med.* **46**, 49.)

the pressure which drives the flow. Later in systole, the synchrony between pressure and flow rate breaks down, chiefly due to the arrival of reflected components of the pressure wave; this subject will be examined later.

If we examine the velocity waveform at points progressively further from the heart (**Fig. 12.19**, lower records) we see that the amplitude of the velocity waveform decreases progressively; this is in contrast with the peaking and steepening of the pressure wave which we noticed earlier (**Fig. 12.19**, upper records). Later in this chapter (p. 269 *et seq.*) we attempt to explain how these effects, and the others we have described, are brought about.

We find that if the detailed properties of the vascular bed are given, the pressure and flow pulses everywhere can in principle be predicted from either the pressure or the

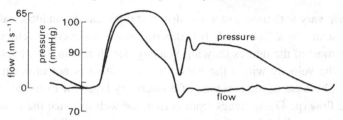

Fig. 12.18. Tracings of experimental pressure and flow records in the ascending aorta of the dog. The flow curve is scaled so that the peak height is comparable to the peak systolic pressure. (After Kouchoukos, Sheppard and McDonald (1970). Estimation of stroke volume in the dog by a pulse contour method. *Circulation Res.* **26**, 611.)

flow pulse at the root of the aorta. In particular, the pressure at the root of the aorta determines the flow there, and vice versa. Furthermore, although we do not have a complete description of the geometry and mechanical properties of blood vessels, there is enough information to permit the relationship between pressure and flow rate throughout the aorta to be predicted with some accuracy.

Terminology The unsteadiness of the flow and the presence of the velocity profile across an artery demand that we carefully define what we mean when we quote a velocity. If at any instant there is a volume flow rate Q throughout an artery of cross-sectional area A, then at that instant the *cross-sectional average* velocity is given by

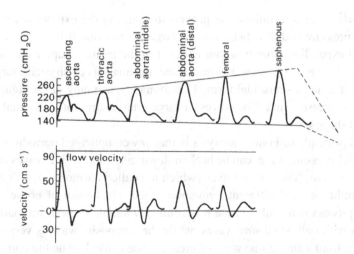

Fig. 12.19. Matched records of pressure and flow velocity at different sites in dog arteries. (After McDonald (1974). *Blood Flow in Arteries.* Edward Arnold, London.)

Q/A. This will vary with time, just as the flow rate does, having an identical waveform to it, merely scaled by a factor $1/A$. It is this *instantaneous* average velocity which is illustrated in most of the figures showing velocity waveforms in this chapter.

However, the velocity within the vessel at that instant is not uniform across the cross-section, because of the presence of a boundary layer for example. In the case of Poiseuille flow (p. 47), it ranges from zero at the wall to twice the cross-sectional average at the centre-line. In large arteries, for reasons we shall be discussing later in this chapter, velocity profiles are very blunt (see, for example, **Fig. 12.39**) with narrow boundary layers close to the wall and fluid outside the boundary layer (the 'core flow') travelling at a uniform velocity which is only slightly greater than the cross-sectional mean. Thus, in large arteries it will be a good approximation (probably within 10%) to equate velocity in the core (e.g. centre-line velocity) to cross-sectional average velocity.

Having seen that we can average the flow velocity spatially to give a cross-sectional average, we must now point out that we can also average the (unsteady) flow velocity with respect to *time*. Thus, we can describe a mean or time-mean flow velocity, either for a particular point in the flow or for the whole flow. In **Fig. 12.38**, for example, the time-mean velocities at a whole series of points across each cross-section of the vessel are plotted, to give the time-mean or mean velocity profile.

Finally, we may in certain circumstances have to use a special form of averaging, known as *ensemble* averaging, to separate random velocity fluctuations such as turbulence which are superimposed upon the instantaneous velocity waveform. This is described, in its context, on p. 327.

Fourier analysis This technique has proved so useful in the exploration and understanding of pressure and flow behaviour in large arteries that it is worth very briefly revising and expanding what was said about it in the earlier chapter on oscillations and wave motion (p. 126). As was shown there, the analysis of an oscillatory wave is very simple if it has a sinusoidal form. Unfortunately, as many of the figures in this chapter show, pressure and flow waves in large arteries are not sinusoidal but have complicated shapes.

The basic principle of Fourier analysis is that any complicated periodic waveform, like an arterial pressure wave, can be broken down into a set of sine waves and resynthesized from them. These sine waves (which are called harmonics) will each have different amplitudes and different frequencies, and will be out of phase with each other. The process is useful because it is relatively simple to perform mathematical calculations with individual sine waves, while the composite wave is very unwieldy. For example, from a single sine wave of pressure one can calculate the corresponding sine wave of flow rate in an artery, on the basis of the simple theory outlined below (p. 277). Then the composite flow-rate waveform can be constructed by adding together all the different flow-rate sine waves, with their different frequencies, amplitudes and phases. This can then be compared with observation. Equally, the pressure or flow-rate wave-form at one position in the artery can in principle be calculated from a knowledge of the pressure waveform at a different position, as long as we also know the geometry and elastic properties of the arteries in between so that the way each component propagates can be calculated. (The gap between principle and practice is largely due to our incomplete knowledge of these properties.) The main limitation on such applications of Fourier analysis is that they are valid only if the system under study is linear, because otherwise the different harmonics do affect each other, and the behaviour of the composite wave cannot be predicted by superimposing separate sine waves. This is a very important restriction, discussed in detail on p. 274. As we shall see, linearity is a good approximation for arteries under physiological conditions.

The actual process of Fourier analysis is extremely laborious if performed by hand, and nowadays is normally done with the aid of a computer. The frequencies of the harmonics into which the wave is to be analysed are dictated by the repetition frequency of the wave itself (the *fundamental* frequency). The first and subsequent harmonics have frequencies which are multiples of this frequency; thus, if the fundamental frequency is 2 Hz, the first harmonic will be at 4 Hz, the second at 6 Hz and so on. In general, the amplitude of successively higher harmonics falls steadily, until ultimately they can be ignored.

The practical test of the adequacy of a Fourier series in this respect is to put all the harmonics together again and compare this result with the original waveform. This is done repeatedly, adding successively higher harmonics until a satisfactory match is obtained. If the original waveform (e.g. blood pressure) oscillates about some non-zero mean level, this level will of course have to be included in the Fourier series as a constant term.

Aortic pressure and flow-rate waveforms can be very accurately described by a constant term plus about 10 harmonics.[7] The amplitude of the harmonics up to the tenth is shown in **Fig. 12.20**. Predictably, in the case of the pressure, the constant term is the largest; very small higher frequency terms are chiefly necessary to describe the sharp fluctuations of the dicrotic notch. In the case of flow rate, the constant term is relatively much smaller and the fundamental much larger. The higher frequencies here are necessary to describe the very rapid rise in flow in early systole and the sharp fluctuations when flow reversal occurs at the end of systole. The accuracy with which a constant term and 10 harmonics describes the original waveform is demonstrated by resynthesis in **Fig. 12.21**. Further from the heart, where the waveforms are smoother, they may be adequately described by fewer harmonics. On p. 127, for example, we showed how the flow-rate waveform may be approximately reconstructed from four harmonics.

---

[7] This means frequencies up to about 20 Hz in the dog and 10 Hz in humans. Measurement and recording systems must therefore have known properties up to these frequencies.

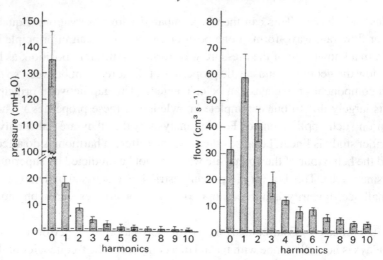

Fig. 12.20. Harmonic content of the pressure and flow curves measured at the root of the aorta. Averaged results from eight dogs. The bars about the mean values represent the standard error of the mean. The horizontal lines near zero are the noise levels of the recording equipment. (After Patel, de Freitas and Fry (1963). Hydraulic input impedance to aorta and pulmonary artery in dogs. *J. Appl. Physiol.* **18**, 134.)

Finally, a note of warning should be sounded; one cannot get more out of a Fourier series than is put in. In other words, if the original waveform is faulty for some reason (e.g. instrumental inadequacy) then so will be the Fourier series describing it. Suppose, for example, that we record arterial pressure with a transducer which has a limited frequency response and can only 'follow' up to the frequency of the fifth harmonic. Then, if we perform a Fourier analysis on the recorded signal, we shall get the impression that no harmonics above the fifth are of significant amplitude. The inverse of this occurs if we know, from testing, that the pressure transducer can faithfully pick up frequencies going higher than that of the fifth harmonic. Then, if our Fourier analysis shows only five harmonics of significant amplitude, we can be sure that we are not misled about the shape of the recorded wave. In fact, this actually gives us a method of checking the validity of recorded waveforms, and has been one of the most useful applications of Fourier analysis in the past. Another related use has been in the correction of the recorded waveform; if we have a pressure transducer which distorts certain frequencies in a known way – a common problem – we can do a Fourier analysis on the recorded waveform, correct the amplitude and phase of harmonics whose frequencies fall in the distorted region and then resynthesize a corrected waveform. Thus, Fourier analysis is an extremely useful practical tool as well as a framework for the understanding of pressure and flow behaviour in arteries.

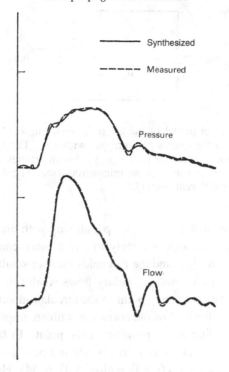

Fig. 12.21. Pressure and flow waveforms in the ascending aorta, as measured and as resynthesized from the constant term and the first 10 harmonics of a Fourier analysis. These obviously describe the waves very adequately. (From McDonald (1974). *Blood Flow in Arteries.* Edward Arnold, London.)

Wave propagation in arteries

In discussing the mechanics of blood flow in large arteries, it is both convenient and physically sensible to regard the given input of the system as the pressure pulse at the root of the aorta (**Fig. 12.18**). The excess pressure p_{ea} here fluctuates about an average value of approximately $1.3 \times 10^4 \, \text{N m}^{-2}$ (100 mm Hg). This is much greater than the corresponding excess pressure p_{ev} in the vena cava (which is close to zero), and it is this arterio-venous pressure difference ($p_{ea} - p_{ev}$) which causes the blood to flow. If the vascular system consisted of a single long, straight, rigid tube, with Poiseuille flow in it, the volume flow rate Q would at all times be uniform (i.e. the same at all stations along the tube) and directly proportional to the pressure difference between the two ends. That is,

$$p_{ea} - p_{ev} = RQ, \tag{12.2}$$

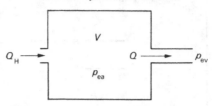

Fig. 12.22. The Windkessel model of the arteries, consisting of a compliant chamber of volume V, proportional to the excess pressure p_{ea} within it. The flow rate Q out of it is proportional to the pressure difference $(p_{ea} - p_{ev})$ between arteries and veins, the constant of proportionality being the resistance of the microcirculation. The flow rate into the system from the heart Q_H may be different from Q.

where R is a constant resistance and Q, p_{ea}, p_{ev} all vary with time (cf. Equation (5.1), p. 47). The same would be approximately true in a more complicated network of tubes, as long as they were rigid and the Reynolds number small enough for entrance effects to be small (cf. p. 53) and secondary flows weak (cf. p. 67). In any rigid system the pressure waveform at any point in the circulation would be similar to and synchronous with that at the root of the aorta, but with an amplitude scaled down by a fixed proportion according to the position of the point. In fact, for the values of Reynolds number occurring in arteries, this would not be a good approximation because of the long entrance region (see Equation (5.4), p. 54). However, if the system were rigid, the gross pressure changes would be almost synchronous everywhere, and the volume flow rate into the system would always be equal to the volume flow rate out.

The Windkessel model Now, in vivo the time course of the pressure pulse is not similar or synchronous everywhere (**Fig. 12.14**), the time course of the volume flow rate (or average velocity) is not the same in the vena cava as in the aorta, and at neither location is it proportional to the applied pressure difference. The reason for these observations lies in the elasticity of blood vessels. When blood is pumped from the heart, and the pressure rises in the large arteries, they expand. When the pressure subsequently falls, they contract again, so that the flow rate through the small peripheral vessels does not immediately fall to zero. This action of the arteries in storing blood during systole and ejecting it in diastole was first described by Stephen Hales (a country parson, who was also the first person to measure blood pressure) in 1733, but it was not until 1899 that the German physiologist Otto Frank proposed a theory based on this idea.

In Frank's model (**Fig. 12.22**) the arteries are represented by a compliant chamber (usually called a 'Windkessel', or air chamber) whose volume V is proportional to the excess pressure within it. The microcirculation is represented in this model by a single constant resistance, with the flow rate Q directly proportional to the pressure difference across it. Equation (12.2) therefore represents the relation between the flow rate through the microcirculation and the arterio-venous pressure difference; and if the excess pressure in the veins is taken to be much smaller than that in the arteries, it may be set to zero without greatly altering the predictions of the model:

$$p_{ea} = RQ. \tag{12.3}$$

The volume of the arterial chamber is related to the pressure within it by

$$p_{ea} = V/C, \tag{12.4}$$

where C is a constant compliance or capacitance; this is a gross oversimplification, even for a single artery, as the discussion in Chapter 7 well shows. Finally, the rate of increase of volume of the chamber dV/dt must be equal to the difference between the rate of flow into it from the heart Q_H and that out of it into the microcirculation Q:

$$\frac{dV}{dt} = Q_H - Q. \tag{12.5}$$

Equations (12.3), (12.4) and (12.5) can be combined to form a single equation giving p_{ea}, V or Q, and this can be solved to relate the pressure in the arteries to the flow rate into them from the heart at any time.

During diastole, when Q_H is effectively zero, this model predicts that the pressure in the arteries falls off exponentially with time; that is,

$$p_{ea} \propto e^{-t/RC}, \tag{12.6}$$

and such a decay is closely realized in practice. However, in early systole, when the inflow Q_H is large, the theory predicts that the rate of pressure rise is proportional to the inflow, whereas experiments show that it is the pressure itself (relative to end-diastolic pressure p_{ed}) which is proportional to the flow rate; that is,

$$p_{ea} - p_{ed} \sim Q_H.$$

This defect demonstrates that the Windkessel model is inadequate to describe dynamic events accurately, even in the central circulation. Its conceptual simplicity is attractive, but it masks the crucial fact that the pressure pulse travels through the circulation as a wave.

The propagation of the pressure wave The principal shortcoming of the Windkessel model lies in the assumption that all arteries are distended simultaneously. As we have seen (**Fig. 12.14**), the pressure peak occurs later in more peripheral arteries than in the aorta, and the pulse is propagated along the blood vessels in the form of a wave. As blood is pumped into the entrance of the aorta, the pressure there rises, the vessel wall is stretched and the tension in it also increases. As the rate of cardiac ejection slows down, the pressure begins to fall and the distended wall returns towards its equilibrium position. The inertia of the fluid keeps it moving forwards after the driving pressure difference has fallen; this causes the first piece of artery wall to overshoot its equilibrium position, and an óscillatory motion is set up. At the same time the next section of wall becomes distended. As this too recoils, the fluid driven out distends a further section of the wall, which also recoils, and so on. Thus, the disturbance is propagated along the arterial system in the form of a pressure wave; it is analogous to a wave propagating along a stretched string, described in Chapter 8 (p. 113). There is a balance between the restoring force, supplied by the elasticity of the artery walls, and inertia, supplied principally by the blood, although the inertia of the vessel wall itself makes a small contribution. As with the waves of Chapter 8, the pressure wave in an artery can propagate in either direction; however, in systemic arteries it primarily originates at the heart and travels distally, although, as we shall see, it is modified by components reflected from the periphery.

The radial wall motions associated with the pressure wave also cause some longitudinal motions of the vessel wall as it is stretched or compressed, but these are secondary and have negligible influence on the propagation of the wave. One can also imagine a different kind of wave altogether, in which the oscillatory wall motions are primarily longitudinal and drive longitudinal fluid motions in the boundary layer by viscous action, thereby providing inertia. There is little physiological evidence for the existence of these waves, and they are believed to have no significant influence on the measured pressures and velocities. They are discussed more fully on pp. 304–306.

Determination of the wave speed We saw in Chapter 8 that the speed of propagation c of waves in a stretched string is governed by the balance between the restoring force (the tension in the string F) and the inertia (the mass per unit length M), and that the wave speed is actually given by

$$c = \sqrt{\frac{F}{M}}. \tag{12.7}$$

Thus, a heavier string has a smaller wave speed than a lighter one at the same tension, and the wave speed increases if the restoring force is increased by stretching it more tightly. A similar balance governs the speed at which pressure waves are propagated along blood vessels. The inertia is supplied by the mass of the blood (and of the vessel walls), and it can be characterized by the blood density ρ. The restoring force comes from the elasticity of the walls; if they are very distensible, the wave speed will be low, whereas if they are stiff it will be high. We saw in Chapter 7 that the most direct measure of the stiffness of a vessel wall is the quantity defined there as the *distensibility D*. This measures the fractional change in cross-sectional area $\Delta A / A$ of a segment of artery of fixed length when a small change in pressure Δp_e is applied, so that D is given by

$$D = \frac{1}{A}\frac{\Delta A}{\Delta p_e}. \tag{12.8}$$

The arguments based on dimensional analysis, which led to Equation (12.7) (Chapter 8, p. 114) can be used here to show that the speed of propagation of pressure waves in blood vessels is proportional to $1/\sqrt{\rho D}$. Indeed, it can be shown by a detailed mathematical analysis of a simplified model of an artery that, as in Equation (12.7), the constant of proportionality is equal to unity, so that

$$c = \frac{1}{\sqrt{\rho D}}. \tag{12.9}$$

As D is diminished, so c increases, and vice versa; that is, stiffer vessels transmit waves more rapidly, as we would expect.

The mathematical analysis which leads to Equation (12.9) is based on the two fundamental principles of fluid flow: conservation of mass and Newton's second law (force = mass × acceleration). These are applied to the motion of a fluid in an infinitely long distensible tube whose diameter does not change unless disturbances like the pressure pulse are imposed. This is already an oversimplified model of an artery, but two further approximations have to be made to obtain Equation (12.9). These are (i) that blood viscosity has no effect on the motion and (ii) that the disturbance or pressure pulse is sufficiently small for the elastic wall properties and the fluid mechanics to be linear. The validity of these approximations must be examined.

Neglect of viscosity By definition, viscosity is important only in regions where the fluid is being sheared, and the greatest shear rate always occurs in the boundary layer close to the vessel wall, since the fluid velocity is zero at the wall itself. For oscillatory flow in a rigid tube, the thickness of the boundary layer is the distance over which viscosity can affect the flow in a time equal to one period of the oscillation, and is therefore proportional to $\sqrt{v/\omega}$, where v is the kinematic viscosity of the fluid and ω the angular velocity of the oscillation. The parameter which governs how much of the tube is occupied by the boundary layer is therefore the ratio of the radius $d/2$ of the tube to $\sqrt{v/\omega}$, i.e. it is the Womersley parameter

$$\alpha = \tfrac{1}{2}d\sqrt{\frac{\omega}{v}},$$

which has already been defined in Chapter 5. Another way of looking at α is to notice that α^2 is the ratio between the time required for viscous forces to diffuse across the whole width of the tube ($d^2/4v$) and a time characterizing the period of the oscillation ($1/\omega$). The same parameter is relevant in an elastic tube.

When α is large, the boundary layer is thin and the volume of fluid retarded because of the no-slip condition on the tube wall is small compared with the volume set in motion by the passage of the wave. In the large arteries of all but the smallest mammals the value of α calculated from the frequency of the fundamental heart rate is considerably larger than unity; for example, a typical value in the aorta of humans is 20, in a dog it is 14, in a cat 8, in a rat 3. Representative values at other sites in a normal dog are shown in **Table I**; the corresponding values in humans are about 50% greater. Thus, the neglect of viscosity is likely to lead to only a small error in the calculation of the propagation speed of pressure waves in large arteries. How big the error actually is and what other effects viscosity might have are investigated on pp. 299–304.

The assumption of linearity Linearity implies that a single-frequency oscillation in pressure is associated with only a single-frequency oscillation in fluid velocity, as discussed in Chapter 8 (p. 123). There are two potential sources of nonlinearity which might be important in the propagation of the pulse wave. They both arise from the fact that pressure fluctuations are associated with changes in cross-sectional area of the vessel. Such area changes mean that the fluid velocity varies along the tube; therefore, there are convective accelerations (p. 40) as well as local accelerations which have to balance the pressure gradient. It is these convective accelerations which are nonlinear, because they involve the square of the velocity rather than the velocity itself, as discussed in relation to Bernoulli's equation (4.6) on p. 125. This is one source of nonlinearity, and it is unimportant if the convective acceleration is small compared with the local acceleration. This requires that the time scale for significant local changes in the fluid velocity (i.e. the wave period $2\pi/\omega$) should be small compared with the time taken for a fluid element to be convected to a position where the fluid velocity is significantly different. This second time scale is equal to the wavelength, $2\pi c/\omega$, divided by a representative fluid velocity $\bar{u}$. Thus, local accelerations always dominate as long as $1/\omega$ is always small compared with $c/\omega\bar{u}$; that is, if

$$\frac{\bar{u}}{c} \ll 1. \tag{12.10}$$

This is approximately satisfied in large arteries, as we show later.

The second source of nonlinearity lies in the fact that the vessel becomes stiffer as it is stretched (see **Fig. 7.5**a, p. 94). The theory leading to Equation (12.9) is linear, and hence applicable, only if the distensibility D is approximately constant throughout the motion; however, D decreases as the distending pressure is increased; therefore, the pulse pressure should be small compared with the mean pressure for linearity to hold.

In fact, this ratio is normally about 0.2 in a systemic artery, so the question arises as to how much smaller than unity it must be for the system to be approximately linear. Similarly, the maximum value of $\bar{u}/c$ is normally about 0.25, and the same question arises. We shall see later (p. 297 *et seq.*) that the nonlinearities of the system have a measurable effect; they are not important for predictions of wave speed, although they do influence the shape of the pressure and flow waveforms.

Comparison of theory with experiment The distensibility D is relatively easy to measure in a segment of artery, especially under static conditions in which increments of distending pressure are applied and the resulting area changes measured step by step. Then, since the density of blood is known, it is possible to predict the wave speed from Equation (12.9).

At first sight it seems a straightforward matter also to measure the speed of propagation of the pressure wave. One would record the pressure pulse at two locations in a blood vessel, a known distance apart, and measure the time taken for a given part of the wave (for example, the pressure maximum or the 'foot' of the wave where the pressure first starts to rise) to travel between them. If the waveform did not change shape as it propagated, that would indeed give an unambiguous measure of wave speed. However, the waveform does change shape. As we shall see in subsequent sections, the change in shape is associated with wave reflections from terminations and branches of the vessels, and with the continuous tapering of individual vessels. Another influence is the fact that different components of the wave have different frequencies[8] and so may travel at different speeds and be attenuated by different amounts in a given length of tube. The presence of reflections, in particular, means that the apparent wave speed as measured in an artery may not correspond closely to the intrinsic wave speed with which waves would travel in an infinite tube with the same elastic properties. In that case it should not be possible to compare the prediction of Equation (12.9) with experiment. In fact, it turns out that if the initial rising part of the wave is used for measurements of wave speed, then everywhere except very near a site of reflection (see p. 278 *et seq.*) there is not enough time for reflected waves to return and distort the wave. We can therefore use this 'foot-to-foot' wave speed as a measure of the intrinsic wave speed in an artery. However, we must remember that there may be significant variations in the elastic properties of an artery along the segment between the two measuring sites, if there are branches for example, so that the wave speed may vary; thus, only an average wave speed for that segment can be measured.

Many workers do not use Equation (12.9) as it stands to predict the wave speed, but instead rewrite it by means of the effective incremental Young's modulus E of the artery wall (see p. 91 *et seq.*). This means interpreting the distensibility in terms of a simple model in which the wall is assumed to behave like a thin, homogeneous membrane. Then D is given by Equation (7.3b), p. 96, to be

$$D = \frac{1}{E(h/d)},$$

[8] Examined by Fourier analysis – see p. 266 and p. 126.

where h/d is the wall thickness to diameter ratio. When this is substituted into Equation (12.9), the predicted value of the wave speed is seen to be

$$c_0 = \sqrt{\frac{Eh}{\rho d}}, \qquad (12.11)$$

where c_0 is merely the label we choose to give to this calculated value of wave speed. This equation is very well known and is called the Moens–Korteweg equation, after two Dutch scientists who published the result in 1878. Actually, it was first derived by Thomas Young in 1809. Equation (12.11) is likely to give less accurate results in practice than Equation (12.9) used directly, because of the interpolation of the simple model of wall elasticity. This can lead to error if used with values of E derived in different ways, largely because of the uncertainty in measurements of h/d. As long as these quantities are derived from distensibility measurements, however, use of Equation (12.11) is justified.

The values of c_0 predicted from Equation (12.11) are given in **Table I** for comparison with the measured values of c. The most striking aspect of the comparison, considering the assumptions about viscosity, linearity, elasticity, etc. that have gone into the theory, is that in arteries the measured value of c differs from c_0 by no more than 15%, which is well within normal experimental error and physiological variability. This suggests that the simplifications introduced into the theory are not sources of great inaccuracy in the calculation of wave speed.

Accuracy of the linearity approximation Using the measured values of the wave speed c and the peak blood velocity $\bar{u}_{\max}$, given in **Table I**, we can check how well the assumption that $\bar{u}/c$ is always small compared with unity is satisfied in practice. In the aorta and main pulmonary artery the maximum value of $\bar{u}/c$ is normally between 0.2 and 0.25, which is considerably less than unity but not completely negligible. In more peripheral arteries, however, two factors combine to make $\bar{u}/c$ smaller: (a) the blood velocity falls and (b) the wave speed increases, because vessels become stiffer towards the periphery (see **Table I**). Thus, in normal dogs (and humans) the assumption of linearity in analysing the propagation of the pulse wave is a good one in all but the largest arteries. This is confirmed by a simple calculation of the ratio of the pressure amplitude Δp_e and the average excess pressure p_e. In large arteries this ratio has a value ranging from about 0.2 in the ascending aorta to about 0.4 in the femoral artery (the 'peaking' of the pressure pulse), but it falls in small vessels.

There are pathological conditions in which the ratio $\bar{u}/c$ can take values very close to unity in large arteries. These include rare connective-tissue disorders, a feature of which is a very floppy arterial wall, so that c is low, and conditions in which the stroke volume is increased, like heart-block, aortic valve incompetence and large arterial shunts such as patent ductus arteriosus. In all of these conditions the upstroke of the pulse wave becomes very steep, and blood velocities greatly in excess of normal are recorded.

Further limitations of the simple elastic model According to the theory outlined above, the pressure wave will be propagated along the artery without change of shape. However, we can see from **Figs. 12.14** and **12.19** that in the circulation the pressure wave does change shape as it propagates through the system. Furthermore, the shape of the flow pulse (the graph of $\bar{u}$ against t) is predicted by the simple theory to be the same as that of the pressure pulse, with an amplitude equal to that of the pressure pulse divided by ρc.[9] In fact, the shape of the flow pulse is completely different (**Fig. 12.18**), and its amplitude falls with distance from the heart, while that of the pressure pulse initially rises (**Fig. 12.19**). This means that although the above theory yields a good prediction of wave speed, and although linearity can reasonably be assumed, some of the other approximations in the simple model are quite inadequate. We must consider four factors which in real life modify the predictions of the simple model.

(1) The tube has been assumed to be of infinite length, whereas blood vessels are in fact relatively short, so that wave reflections will occur at the ends. Indeed, the wavelength of the fundamental wave in the aorta (wave speed $\simeq 5\,\mathrm{m\,s^{-1}}$; frequency $\simeq 1.2\,\mathrm{s^{-1}}$) is about $4\,\mathrm{m}$, while the length of the aorta itself is about $0.5\,\mathrm{m}$ in humans. Hence, only a small fraction of the total wave can exist in the aorta at any given time; it was this fact which enabled Otto Frank, who was well aware of the facts of wave propagation, to justify his Windkessel model. The significance of wave reflections will be examined in the next section.

(2) No artery is uniform along its length. The aorta in particular is both narrower and stiffer at its distal end than it is near the heart (see **Table I**), and it is this tapering which proves to be partly responsible for the peaking of the pressure pulse, seen, for example, in **Fig. 12.19**. The effect of the gradual tapering and

---

[9] This is because, according to the equation of motion, the density times the local acceleration, $\rho\,\mathrm{d}\bar{u}/\mathrm{d}t$, must be equal to the gradient of excess pressure, $-\mathrm{d}p_\mathrm{e}/\mathrm{d}x$. Both $\bar{u}$ and p_e can be represented as the sum of a constant and oscillatory terms like

$$B\cos[\omega(t-x/c)-\theta], \tag{12.12}$$

where B is the amplitude, equal to a velocity scale $\bar{u}_0$ or a pressure scale p_0 according to the quantity concerned. To derive acceleration and pressure gradient from velocity and pressure, we differentiate each such term and obtain

$$\rho\frac{\mathrm{d}\bar{u}}{\mathrm{d}t} = -\omega\bar{u}_0\rho\sin[\omega(t-x/c)-\theta]$$

and

$$-\frac{\mathrm{d}p_\mathrm{e}}{\mathrm{d}x} = -\frac{\omega p_0}{c}\sin[\omega(t-x/c)-\theta].$$

For these two to be equal, $\bar{u}_0$ must equal $p_0/\rho c$. If the wave were being propagated in the opposite direction, the sign of x/c in Equation (12.12) would be reversed and $\bar{u}_0$ would equal $-p_0/\rho c$.

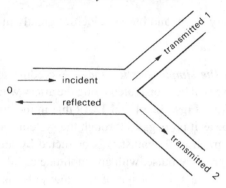

Fig. 12.23. Sketch of an arterial bifurcation. The incident wave is partially reflected in the parent tube 0 and partially transmitted in the daughter tubes 1 and 2.

stiffening of arteries with distance from the heart will be discussed in the next section.

(3) Nonlinear behaviour has already been mentioned. The situations in which it assumes importance are discussed on p. 297 *et seq.*

(4) Finally, no mention has yet been made of the viscous dissipation of energy which accompanies any motion of a viscous fluid or a visco-elastic solid. This will lead to damping or attenuation of the wave, as explained on p. 116; it is discussed with reference to the pulse wave on p. 301.

Reflection and transmission of the wave at junctions

Every point at which the properties of an artery change will be a site of partial reflection of the pressure wave, in the same way that any discontinuity in a stretched string is a source of reflections (see p. 120). Throughout any artery there are many unevennesses and slight bends which will cause some reflection, but the most obvious and potent discontinuities are the junctions between arteries (**Fig. 12.1**). It is reflections which cause the greatest modification of the shape of the pulse wave as it travels peripherally. Therefore, we must examine them in detail, first looking at a single junction in isolation, and then studying wave reflection at a sequence of discontinuities.

Reflection at a single junction We begin by examining a single parent artery which branches into two daughters (**Fig. 12.23**). The daughter tubes may be of different sizes (in the cardiovascular system they usually are) and may come off at different angles. A wave travelling down the parent artery will be partially reflected at the junction, so that a reflected wave travels back up the parent tube and is partially transmitted down the daughters. The properties of the reflected and transmitted waves can be predicted quite simply.

There are two conditions which have to be satisfied by the reflected and transmitted waves at the junction. First, the pressure in each tube must be the same. If this were not the case, a finite pressure difference would be acting on a small volume of blood, which would therefore experience an instantaneously large acceleration causing the pressure difference to disappear. Not only must the mean pressure be the same, but also the oscillatory part of the waveforms must be the same in each tube. If the oscillatory pressure at the junction associated with the incident wave is $\tilde{p}_I$ (we shall always denote the oscillatory part of the pressure waveform by the symbol $\tilde{p}$), if that associated with the reflected wave is $\tilde{p}_R$ and if those associated with the transmitted wave are $\tilde{p}_{T_1}$ and $\tilde{p}_{T_2}$, then at all times it is necessary that

$$\tilde{p}_I + \tilde{p}_R = \tilde{p}_{T_1} = \tilde{p}_{T_2}. \tag{12.13}$$

Second, mass must be conserved. That is, what goes into the junction must come out; that is, the volume flow-rate into the junction from the parent tube must at all times be equal to the volume flow rate out of it in the two daughters. (This assumes that the volume contained within the junction is so small that any change in it can be ignored.) Thus, if we represent the volume flow rate in the direction of propagation of the wave by the symbol Q, with the appropriate subscript, this condition requires that at the junction

$$Q_I - Q_R = Q_{T_1} + Q_{T_2}, \tag{12.14}$$

where Q_R has a minus sign because the reflected wave is travelling to the left in **Fig. 12.23**. Now we have seen already (p. 277) that the cross-sectional average velocity $\bar{u}$ in the direction of propagation of the wave varies with time in the same way as the excess pressure and has an amplitude equal to that of the pressure pulse divided by ρc. Hence, the volume flow rate Q, equal to $A\bar{u}$, where A is the undisturbed area of the tube, is given by

$$Q = \frac{A}{\rho c}\tilde{p}. \tag{12.15}$$

The quantities A, ρ and c are properties of the artery in which the wave is propagating, not of the wave itself. This is also true, therefore, of the quantity $\rho c/A$, which is often represented by a separate symbol, Z:

$$Z = \rho c/A. \tag{12.16}$$

Z is called the *characteristic impedance* of the tube, defined as the ratio of oscillatory pressure to flow rate in a tube when a single wave is travelling along it, and takes the same constant value at every point of the tube.[10] The dimensions of Z are

---

[10] The word 'impedance' comes from electrical network theory and is much used in cardiovascular physiology. It is analogous to 'resistance', in that it measures the amount by which flow through a system, in response to a known oscillatory pressure difference, is impeded by the system. Various different 'impedances' can be defined, and these are discussed on p. 338 in a more mathematical context.

$[ML^{-4}T^{-1}]$, and it can be measured in the units $kg\,m^{-4}\,s^{-1}$. The volume flow-rate condition (Equation (12.14)) can now be represented in terms of the oscillatory pressures and the impedances of the three tubes as follows:

$$\frac{\tilde{p}_I - \tilde{p}_R}{Z_0} = \frac{\tilde{p}_{T_1}}{Z_1} + \frac{\tilde{p}_{T_2}}{Z_2}. \tag{12.17}$$

Equations (12.13) and (12.17) must be satisfied at the junction all the time. They constitute three simultaneous equations which can be solved to give the three unknown quantities, $\tilde{p}_R$, $\tilde{p}_{T_1}$ and $\tilde{p}_{T_2}$, in terms of the pressure at the junction in the known incident wave, $\tilde{p}_I$. The solution of these equations yields

$$\frac{\tilde{p}_R}{\tilde{p}_I} = \frac{Z_0^{-1} - (Z_1^{-1} + Z_2^{-1})}{Z_0^{-1} + (Z_1^{-1} + Z_2^{-1})} \equiv R, \tag{12.18}$$

and

$$\frac{\tilde{p}_T}{\tilde{p}_I} = \frac{2Z_0^{-1}}{Z_0^{-1} + (Z_1^{-1} + Z_2^{-1})} \equiv T, \tag{12.19}$$

where the subscript has been dropped from $\tilde{p}_T$ because, at the junction, $\tilde{p}_{T_1} = \tilde{p}_{T_2}$. The reciprocal of the impedance Z^{-1} is often called the *admittance*, and we shall use this term when it is appropriate.

Equations (12.18) and (12.19) can provide a great deal of information. Let us first consider a particular form of incident wave, a sinusoidal one in which

$$\tilde{p}_I = p_0 \cos[\omega(t - x/c_0)] \tag{12.20}$$

and we choose the junction itself to be the origin of the x-coordinate in each tube. Thus, x is negative in the parent tube and positive in each daughter tube (see **Fig. 12.23**). At this stage it is not necessary to choose a particular wave shape, but the cosine in Equation (12.20) is particularly simple and convenient.

At the junction, $x = 0$, the pressure in the incident wave is, from Equation (12.20), equal to

$$\tilde{p}_I = p_0 \cos \omega t.$$

From the form of Equations (12.18) and (12.19), this means that $\tilde{p}_R$ and $\tilde{p}_T$ are also proportional to $\cos \omega t$ at the junction. Furthermore, since we know that the reflected and transmitted waves are propagated in a similar manner to the incident wave, their dependence on x, as well as on t, follows:

$$\begin{aligned} \tilde{p}_R &= R p_0 \cos[\omega(t + x/c_0)] \\ \tilde{p}_{T_{1,2}} &= T p_0 \cos[\omega(t - x/c_{1,2})]. \end{aligned} \tag{12.21}$$

In the first of these equations, the sign of x/c_0 is different from that in Equation (12.20) because the reflected wave travels in the negative x-direction. The second equation is applicable to either of the daughter tubes, with subscript 1 for tube 1 and subscript 2 for tube 2.

These equations tell us everything about the reflected and transmitted waves:

(1) They are the same shape as the incident wave (sinusoidal).
(2) The amplitude of the reflected pressure wave is R times that of the incident wave, while that of the transmitted wave in each daughter tube is T times that of the incident wave.
(3) The wave speeds of the reflected and transmitted waves are those appropriate to single waves in the tubes concerned (c_0, c_1 and c_2 respectively).
(4) Finally, the total oscillatory pressure in the parent tube is equal to the sum of that in the incident wave and in the reflected wave:

$$\tilde{p} = \tilde{p}_{\mathrm{I}} + \tilde{p}_{\mathrm{R}}.$$

The theory of wave reflections at the junction between a parent or proximal tube and a pair of daughter or distal tubes can readily be extended to a situation in which more than two daughter tubes originate at the junction. Such a situation occurs, for example, where the two renal arteries branch off the aorta, or at the trifurcation of the canine aorta. A similar set of equations shows that just as it is the sum of the admittances (Z^{-1}) of the two daughter tubes which governs the reflected and transmitted waves at a bifurcation, so it is the sum of the admittances of the three or more daughter tubes which governs them at a more complex junction. The quantity $Z_1^{-1} + Z_2^{-1}$ in Equations (12.18) and (12.19) is replaced by $Z_1^{-1} + Z_2^{-1} + Z_3^{-1}$, etc.

The ratio of the amplitudes of the transmitted wave to that of the incident wave (T, Equation (12.19)) is positive. This means that the transmitted pressure wave is in phase with the incident wave at the junction, although its amplitude and wave speed (c_1 or c_2) may change. The amplitude ratio of the reflected wave (R) may, however, be negative as well as positive, in which case the pressure in the reflected wave is at a minimum when that in the incident wave is at a maximum, and vice versa; that is, the waves are 180° out of phase. Thus, the sign of R is important and must be investigated further.

The matching of impedances If the sum of the admittances (reciprocals of impedances) of the daughter tubes is equal to that of the parent tube, then R is zero and there is no reflected wave. In this case the junction is said to be one at which the impedances are matched. We can examine the conditions under which this occurs by using the definition of impedance together with the relation of wave speed to distensibility (Equation (12.9)), which give

$$Z = \frac{\rho c}{A} = \frac{1}{A}\sqrt{\frac{\rho}{D}}. \tag{12.22}$$

If the distensibility is given in terms of Young's modulus and wall thickness-to-diameter ratio by Equation (7.3b), this gives

$$Z = \frac{1}{A}\sqrt{\frac{\rho E h}{d}}. \tag{12.23}$$

So, if the daughter tubes have the same Young's modulus and wall thickness-to-diameter ratio as the parent (i.e. the same distensibility), the impedances will be matched if the sum of their areas is equal to that of the parent. However, if the daughters are slightly less distensible than the parent, as indicated by **Table I** in which E is seen to increase peripherally while h/d remains more or less constant, then their combined area must be somewhat greater than that of the parent for perfect matching. There is some evidence, as we shall see shortly, that many junctions in the cardiovascular system are quite well matched.

It is unlikely that any real arterial junction is perfectly matched, but if a junction is only fairly well matched, so that the magnitude of R is quite small but still measurable, then we can show that the amount of energy in the reflected wave is totally negligible, because it is proportional to the square of the amplitude. The rate W at which energy is carried along in a sine wave can be calculated quite simply, because most energy is transferred through the work done on the blood by pressure forces. This turns out at all times to be equal to the oscillatory pressure $\tilde{p}$ multiplied by the oscillatory flow rate Q, and since $Q = \tilde{p}/Z$, the rate of energy transmission is given by

$$W = \tilde{p}Q = \tilde{p}^2/Z. \tag{12.24}$$

Although $\tilde{p}$ is a pressure which oscillates periodically and has a zero mean value, as in Equation (12.20), its square does not have a zero mean. For example, the mean value of $\cos^2[\omega(t - x/c_0)]$ averaged over a whole cycle, is $\frac{1}{2}$ (see **Fig. 12.24**). Thus, the mean rate of energy transmission by the pressure wave at any cross-section $\bar{W}$ is given by

$$\bar{W} = \tfrac{1}{2}p_0^2/Z$$

when $\tilde{p}$ has amplitude p_0 as given by Equation (12.20).

We can therefore see that the mean rate of energy transfer in a reflected wave, relative to that in the incident wave, is equal to

$$\frac{\tfrac{1}{2}R^2 p_0^2/Z}{\tfrac{1}{2}p_0^2/Z}, \text{ i.e. } R^2,$$

where R^2 is called the *reflection coefficient* (see p. 122). Similarly, the rate of energy transfer in the two transmitted waves, compared with that in the incident wave, is

$$\frac{Z_1^{-1} + Z_2^{-1}}{Z_0^{-1}}T^2.$$

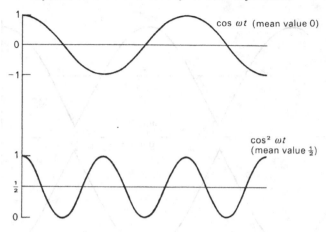

Fig. 12.24. A comparison between the graphs of $\cos \omega t$ and $\cos^2 \omega t$, showing the latter to have a mean value of $\frac{1}{2}$ while the former has zero mean.

This quantity is called the *transmission coefficient*. From the definitions of R and T (Equations (12.18) and (12.19)) we can see that the sum of the reflection and transmission coefficients is unity; that is, no energy is lost at a junction of this type.

Now suppose that a junction is slightly mismatched, so that $Z_1^{-1} + Z_2^{-1}$ differs from Z_0^{-1}, by 5% say. In that case the fraction of energy reflected works out at 0.0006; that is, most of it is transmitted. This means that even when perfect matching is not achieved, the mismatch has to be quite considerable before it has a significant effect on the energy transmission.

These conclusions are relevant to two types of 'junction' which occur in the cardiovascular system: (i) junctions at which very small arteries branch off an otherwise unchanged large one (e.g. intercostal arteries off the aorta); (ii) sites where the area or distensibility of a single tube changes by a small amount – the continuous tapering of most larger arteries can be treated as a succession of such sites, as discussed below (p. 294).

Positive and negative reflection When matching is not perfect, R is either positive or negative. In the extreme case, when the sum of the admittances of the daughter tubes is much smaller than that of the parent (which would occur, for example, if their combined area were very small, and their distensibility were not much different), then R would be approximately 1. In these circumstances, we would expect the reflection to be similar to that at a closed end, across which no flow is possible. At a closed end, the reflected flow wave is 180° out of phase with the incident flow wave, and the amplitude of the pressure oscillation is doubled, the phase of the reflected pressure

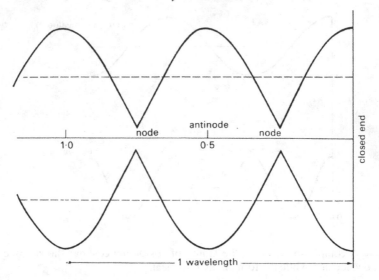

Fig. 12.25. A diagram to show how the amplitude of the oscillations varies with distance from a site of partial (80%) closed-end reflection. Dotted lines represent the amplitude of the undisturbed incoming wave.

wave being unchanged; that is,

$$\left.\begin{array}{rcl} Q_R &=& -Q_I \\ \tilde{p}_R &=& +\tilde{p}_I \end{array}\right\} \quad \text{and so} \quad R = 1.$$

If an observer measured the pressure at a fixed point in an infinitely long elastic tube, in which the single sine wave described by Equation (12.20) was propagating, the observer would measure the sinusoidal oscillation with amplitude p_0. If a closed end were inserted, the observer would continue to measure a sinusoidal oscillation, but the amplitude would depend on the distance from it. At the closed end itself it would be $2p_0$, but would diminish as the observer moved away, until it became zero at a quarter of a wavelength from the end (see **Fig. 8.11**, p. 121; this point is called a node). Farther away, the amplitude would increase again, to a maximum of $2p_0$ at a half wavelength from the end (an antinode) and so on. If instead of a closed end there was a junction or other discontinuity for which R was positive but not equal to unity (0.8, say), the amplitude would not reduce to zero but to $0.2p_0$ at the nodes, and would be only $1.8p_0$ at the antinodes (see **Fig. 12.25**). This would be a partial reflection, of closed-end type.

It is interesting to note that when R is close to 1, T, given by Equation (12.19), is close to 2. This means that the amplitude of the pressure wave transmitted through an almost closed junction is *twice* that in the incident wave. This seems surprising until we recall that the flow-rate amplitude is equal to Z^{-1} times the pressure amplitude (Equation (12.15)), so that even when this is doubled the transmitted flow-rate oscillation in this case is still very small. The doubled pressure amplitude in the daughter tube is merely the response to the doubled amplitude at the closed end of the parent tube.

Another extreme type of reflection occurs when the sum of the admittances of the daughter tubes is much greater than that of the parent; for example, when the combined area of the daughter tubes is very large. In this case R is close to -1, while T is very small. The situation is similar to that at a completely free, open end, at which the oscillation in pressure has to be zero, since the pressure must be the same as that in the chamber into which the tube opens, and the oscillation in flow rate is doubled. Here, the pressure nodes are at the open end and a whole number of half wavelengths from it; the antinodes are at a quarter wavelength from the end, and thereafter at half wavelength intervals. The phase of the pressure wave is now reversed on reflection, while that of the flow rate is unchanged. The phases in all cases of negative reflection, for which R is negative, are of *open-end* type.

In summary, for positive reflection, the phase of the pressure wave is unchanged, that of the flow wave is reversed and the amplitude of the transmitted pressure wave is increased. For negative reflection, the phase of the flow wave is unchanged, that of the pressure wave is reversed and the amplitude of the transmitted pressure wave is reduced.

Physiological evidence of wave reflections There have been a number of studies in which the pressure and flow pulses have been measured at various sites in the large arteries of humans. Waveforms from the ascending aorta, the descending thoracic and abdominal aorta, the common iliac, subclavian, innominate, and renal arteries (in humans) are shown in **Fig.12.26**. On the basis of the above discussion, we can predict certain features of the waveforms which should be observed as a result of wave reflections.[11]

We first calculate the time after the beginning of systole at which (a) the pulse wave first reaches a certain site and (b) reflected waves reach it. The site in the ascending aorta can be regarded as located at the exit from the ventricle. The pulse wave will reach the branches at the arch of the aorta, about 8 cm away, after a time calculated from the wave speed there, about $5\,\mathrm{m\,s^{-1}}$ (**Table I**: measured wave speeds in humans and dogs do not differ greatly). The time is therefore approximately 0.016 s, and the time which will have elapsed before a wave reflected from these junctions will return to the root of the aorta is only 0.032 s. But the initial part of systole during which the blood is accelerated and the pressure and velocity rise rapidly is normally about 0.1 s. Thus, the reflected wave, not likely to be very large in amplitude because the

---

[11] The pattern of reflected waves observed at any site is likely to be species specific. The present discussion is applicable to humans; the corresponding observations for the dog are given in chapter 12 of D. A. McDonald's book, *Blood Flow in Arteries* (Edward Arnold, 1974).

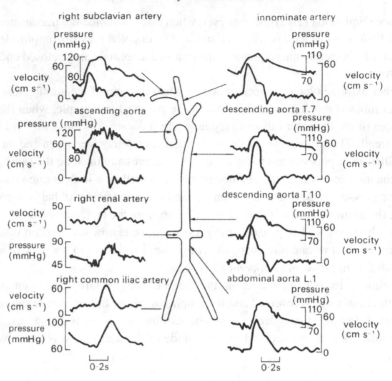

Fig. 12.26. Simultaneous pressure and blood velocity waveforms at numerous points in the human arterial tree. (All were taken from one patient with the exception of the right renal artery and the right common iliac artery.) (After Mills *et al.* (1970). Pressure–flow relationships and vascular impedance in man. *Cardiovasc. Res.* **4**, 405.)

total cross-sectional area downstream of the junctions is not very much greater than that upstream, will first arrive during this period of rapid pressure rise and is unlikely to be perceptible.

Another major site of reflection which the wave encounters as it travels down the aorta, and the one with the largest reflection coefficient, is the aortic bifurcation. Measurements have shown that the total cross-sectional area of the two iliac arteries is commonly less by about 20% than that of the abdominal aorta. Thus, a significant positive reflection is expected. (This would not be true in a dog, where the aortic trifurcation may be quite well matched.) Now the time of travel of the pulse from the heart to this point is the distance, about 60 cm in humans, divided by the average wave speed in the aorta, about $6 \, \mathrm{m \, s^{-1}}$. Thus, the time of travel is about 0.1 s, and the reflected wave will return to the root of the aorta (and, equally, to the measurement sites in the innominate and subclavian arteries) in about 0.2 s. Measurements of the velocity in the core of the tube (representative of the flow rate because the velocity profile is fairly flat; see p. 306 *et seq.*) reveal that the descending part of the waveform has a step in it, representing a momentary enhancement, followed by a slowing down of the deceleration (see **Fig. 12.26**). This step occurs at a time after the start of systole which increases in proportion to the distance *up* the aorta from the aortic bifurcation, and reaches a value of about 0.2 s in the innominate artery. We may therefore interpret it as a mark of the reflected wave. Now the iliac bifurcation is expected to be a site of positive, or closed-end, reflection. This means that the flow rate in the reflected wave is out of phase with that in the incident wave, and the positive peak of the incident wave will correspond to a negative trough of the reflected wave. It is easy to see that such a trough would cause an enhanced deceleration in the aortic waveform, as observed.

The pressure waveforms of **Fig. 12.26** contain so many fluctuations that it is difficult to interpret any one as the reflected wave from the aortic bifurcation. The notch at the end of systole marks the closure of the aortic valve, and is propagated away from the heart with the basic pulse, although it is rapidly attenuated (see p. 301 *et seq.*). Note also how the amplitude of the pressure pulse increases with distance from the heart (by about 60% along the length of the aorta, **Fig. 12.19**), while that of the flow-rate pulse falls. This is consistent with a closed-end reflection at the aortic bifurcation, although it may in part be due to the gradual tapering and stiffening of the aorta, which is discussed in the next section.

Another phenomenon which may occur in a system in which waves are travelling is *resonance*, which was described on p. 123. At a quarter wavelength from a site of closed-end reflection there should be a point (a pressure node) at which the pressure amplitude is a minimum and the flow amplitude a maximum (see **Fig. 12.25**). A similar point occurs every further half wavelength from the reflection site. In between there are the *antinodes* at which the pressure amplitude is a maximum. If either an antinode or a node falls at a second reflection site, for instance at the other end of a finite tube, then another antinode will occur at the original site. The pressure amplitude there, already high because of the positive reflection, is further augmented by each successive reflection. This is the condition called resonance, which should be observed if the length of the system is equal to a whole number of half wavelengths. The wavelength of the basic waveform in the aorta is much larger than the length of the vessel, so we do not expect to see resonance of the whole waveform. However, the fourth harmonic of the basic wave has a wavelength only about twice the length of the aorta in humans, while the sixth harmonic has this property in dogs. We must, therefore, split the waveform into its sinusoidal components by Fourier analysis to see whether the amplitude of the fourth harmonic (sixth in dogs) is disproportionately high. Even for these components, however, no evidence of resonance can be seen in practice (**Fig. 12.20**). This is probably because they are attenuated too rapidly to remain significant after several reflections (see pp. 301–304).

So far we have concentrated on major arterial branches, but there are also sites where small side branches come off a large artery (**Fig. 12.27**); for example, the intercostal and coronary arteries branching off the aorta. Here, the impedance of one

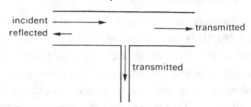

Fig. 12.27. A junction in which the parent tube continues unaltered after a small tube has branched off (cf. the junction between an intercostal artery and the aorta).

daughter tube is the same as that of the parent, while that of the other (or others) is very small. Thus, the sum of the admittances of the daughters is only slightly greater than that in the parent, and the junction is almost matched. This means that a negligible amount of energy is reflected (p. 282), so that the basic waveform in the aorta is effectively undisturbed, while the pressure pulse in the small branch is as large as that in the aorta (Equation (12.19)). This is observed in practice.

Multiple reflections The relationship between the pressure and flow-rate waveforms proximal and distal to a single junction can be explained without reference to the rest of the arterial system. However, for a detailed interpretation of the measured waveforms at any one position, in the ascending aorta for example, it is necessary to recognize that there are many sites of reflection in the system, situated at widely different distances from the measurement point. Therefore, it is necessary to examine the influence of repeated reflections on the wave.

Consider first the double junction illustrated in **Fig. 12.28**. A wave reflected once at junction B is reflected a second time at junction A, and so on. The amplitudes of the

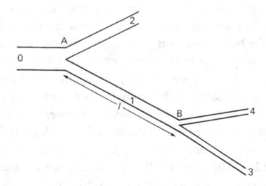

Fig. 12.28. Sketch of a double junction. Wave incident in tube 0 is reflected and transmitted at A. The transmitted wave in 1 is reflected and transmitted at B. The reflected wave in 1 is again reflected at A, etc.

reflected and transmitted waves on each occasion are determined by the characteristics of the junction. If junction B is fairly well matched, the wave reflected from it will have small amplitude. When reflected again at A it will have even smaller amplitude, though not very much smaller, for if A is well matched for waves incident in the parent tube, it will not be so for waves incident in a daughter. In any case, the amplitude of that part of the wave trapped in the tube AB will be very small after only a few reflections, and that will be true even in the absence of attenuation.

The only exception to this conclusion would occur if the length of the tube AB were equal to a whole number of half wavelengths of the wave. In that case resonance would occur. From the point of view of a wave incident from the parent tube of junction A, this means that no energy is required to drive the oscillation in AB, and the waves transmitted at B have the same characteristics as they would have if AB were not there. The tube AB can pass waves on without modification.

If, on the other hand, the length of the tube AB were close to an odd number of quarter wavelengths, the reflected wave would on each occasion be exactly out of phase with its predecessor if the reflection at B were positive, and in phase if negative; the amplitude of the transmitted wave would be diminished for a positive reflection and enhanced for a negative one.

In general, these extremes would not be met with, but the details of the reflected and transmitted waves will still depend on the relative magnitude of the length l of the intermediate tube and the wavelength of the wave in that tube. It is this which governs the time delay, and hence the phase difference, between a wave leaving A and returning there again after reflection at B. This means that the behaviour of a wave as it passes two or more junctions depends on its frequency. Thus, the behaviour of a composite periodic wave, made up of a number of sinusoidal components of different frequencies (cf. **Fig. 8.14**, p. 127), is likely to be quite complicated, and the shape of the wave transmitted through the double junction will be different from that of the incident wave. The only components which will be unaffected, apart from those for which l is approximately a whole number of half wavelengths, are those whose wavelength is much greater than l, for then the two junctions and the tube between will all seem like part of a single reflection site as far as the wave is concerned. In the cardiovascular system, the length of the longest artery, the aorta (60 cm in humans, 40 cm in dogs), is quite small compared with the fundamental wavelength (about 5 m and 3 m respectively), so it is only the higher harmonics which will be affected by this process, and it is their contribution to the total waveform which we expect to change significantly.

In the last section, we were able to calculate the details of wave reflection and transmission at a single junction. It is possible to use the same conditions, continuity of pressure and flow rate at every junction, to calculate similar details, for waves of a single frequency, at multiple junctions. For example, if we know what waves are present in tubes 3 and 4 of **Fig. 12.28**, we can calculate what is happening in tube 1. If we have also calculated the waves in tube 2 in the same way, we can combine the two results, evaluating pressure and flow rate at A, to calculate the details of the wave in tube 0. Furthermore, if that tube itself began at another junction, we can repeat the process to analyse the details in the parent tube of that junction, and so on. In this way we can calculate the pressure and flow rate everywhere in a complete network of branched tubes, as long as we know the relationship between pressure and flow rate in the smallest tubes, and the lengths and characteristic impedances of every tube in the system. In order to do this calculation for the arterial system, we could, for example, assume that the flow rate is proportional to the pressure gradient in the vessels of the microcirculation (i.e. constant peripheral resistance). At present, however, such a calculation is not practicable, because the necessary anatomical data on the intermediate arteries are not available. A calculation has been performed for a model of the arterial system, consisting of several generations of branched tubes. The branches of any one generation were taken to have the same diameter and visco-elastic properties, but their lengths were distributed randomly and the wave speed and cross-sectional area were taken to increase towards the periphery. The results of this calculation will be given below.

Even if we do not know the waves present in the terminal branches of the system, such calculations enable us to predict the relationship between pressure and flow rate everywhere in the system. That is, given the pressure waveform at a point, we can calculate the flow-rate waveform, and vice versa. Also, if we measure the pressure and the flow rate at a point we can infer some of the properties of the system. For waves of a single frequency, the relationship can be described by two constants: one is the ratio between the amplitudes of the pressure and flow-rate oscillations of that frequency, often given the symbol M, and the other is the phase lag θ of the flow-rate oscillation behind the pressure oscillation. Thus, if the component of the pressure oscillation at the entrance to the system with angular frequency ω is $p\cos\omega t$, the flow-rate oscillation will be $(p/M)\cos(\omega t - \theta)$. The complete results for a system can be given by plotting graphs of M and θ against ω, so that the behaviour of any wave made up of sinusoidal components can be described. For example, if the pressure and flow-rate waveforms are measured in the ascending aorta, a Fourier analysis can be performed on each, and the amplitude and phase of every flow-rate harmonic can be compared with those of the corresponding pressure harmonic. Thus, the values of M and θ for each frequency present in the waveforms can be inferred. Many experimental results are expressed in this way.

The quantity M, being the ratio of a pressure and a flow rate, has the dimensions of an impedance. However, since it depends on the position of the point of measurement as well as on the frequency of the wave, it is not the same as the characteristic impedance of the tube in which the measurements are made (i.e. the aorta). Together with θ, it represents the *effective impedance*, at the point of measurement, of the whole system for waves of frequency ω. In fact, M is usually called the *modulus* of the effective (or input) impedance of the arterial system and θ is called its phase. These terms arise from a description of waves in terms of complex numbers, which we have hitherto tried to avoid. They are explained further in the Appendix to this chapter, along with various other quantities which are sometimes given the name 'impedance'. The proliferation of such quantities can be very confusing, and we hope that the Appendix will serve both as a clarification of the subject and as a warning not to use the word 'impedance' without a qualifying adjective like 'characteristic' or 'effective'.

Interpretation of observed pressure waveforms in large arteries The idea of the effective impedance of the cardiovascular system for pressure waves of different frequencies makes it possible to interpret the measured pressure pulse in the aorta and other large arteries much more fully than is possible by mere inspection of the waveforms. Only the gross effects, like the presence of a single powerful reflection site, can be observed in that way. For more detailed information, it is necessary to split the pressure and flow-rate waveforms at the observation site into their different frequency components. The amplitudes and phases of the flow-rate harmonics are then compared with those of the corresponding pressure harmonics and the results expressed in terms of the modulus M and phase θ of the effective impedance. These quantities are then plotted against frequency.

Certain information can be inferred from such graphs. For example, if the system consists simply of a single uniform tube with a termination at one reflection point, then Equation (12.32) (in the Appendix) indicates that, when the measurement site is a quarter wavelength from the reflection site, the intervening piece of tube has no effect on the effective impedance. In that case the magnitude M of the effective impedance is a minimum. Furthermore, if the frequency is increased beyond that for which l is a quarter wavelength, then the phase θ of the effective impedance changes from negative to positive. These considerations have led experimentalists to plot M and θ against ω, and to infer that, when M is a minimum (and θ is zero), there is a site of reflection at a distance of $\pi c/2\omega$ from the measurement site, where c is the independently measured wave speed in the vessel. In a complicated branched network, such interpretations are inevitably much more tentative. The reflection site inferred in this way almost certainly does not correspond exactly to an actual bifurcation because of the influence of all the other junctions. The reflected wave will be a complicated resultant of reflections from many sites. However, if there is, anatomically, one dominant junction (e.g. the aortic bifurcation), and the inferred reflection site is close to it, it can legitimately be regarded as the principal site of reflection.

The sorts of graph which are normally obtained from measurements in the ascending aorta of humans are shown in **Fig. 12.29**. These were taken from the same set of measurements as the waveforms shown in **Fig. 12.26**. There is a minimum of M at a frequency of 3 Hz, and calculation shows that this implies the presence of a reflection site roughly at the level of the aortic bifurcation. Measurements at different sites in the aorta lead to the same conclusion.

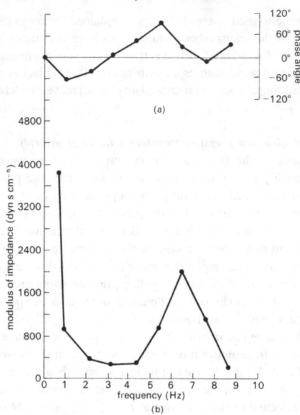

Fig. 12.29. Graphs of (a) phase and (b) modulus of the input impedance of the human ascending aorta, plotted against wave frequency f. The single minimum of the modulus suggests that there is a single effective reflection site at a peripheral location. (From Mills *et al.* (1970). Pressure–flow relationships and vascular impedance in man. *Cardiovas. Res.* **4**, 405)

Another universal feature to be seen from **Fig. 12.29** is the very large value of M (with $\theta = 0$) for very small frequencies. The zero-frequency component of the wave represents the mean flow, and this real, large value of the effective impedance represents the resistance experienced by that mean flow. We therefore see that the 'opposition' to the first few harmonics of the pulse wave is much less than to the mean flow. But these harmonics contain most of the pulse energy. This implies that the amount of work required to drive the oscillatory motions in the aorta is not very much in excess (about 10%) of that required to drive the same mean flow steadily.

Similar measurements in the ascending aorta of a dog are shown in **Fig. 12.30**. They have a slightly different appearance, in that there are two minima of the impedance modulus. These have been interpreted in terms of two distinct reflection sites at different distances from the heart. Since the second minimum is absent in

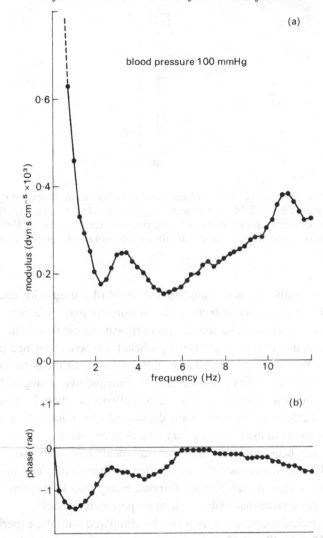

Fig. 12.30. Graphs of (a) modulus and (b) phase of the input impedance of a dog's ascending aorta plotted against wave frequency f. The appearance of two minima of the modulus suggests the presence of two effective reflection sites at different distances from the heart. (After O'Rourke and Taylor (1967). Input impedance of the systemic circulation. *Circulation Res.* **20**, 365.)

the descending aorta, it has been inferred that the second major reflection site is in the upper part of the body; that is, in that part of the circulation supplying the head and forelimbs. This has led to a model of the systemic circulation consisting of an 'asymmetric T' (**Fig. 12.31**), in which two such reflection sites are incorporated. Certainly, the observations in a dog's aorta can be interpreted in terms of such a model,

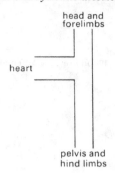

Fig. 12.31. The 'asymmetric T' model of the systemic circulation, proposed by O'Rourke to explain the results of **Fig. 12.30**. The nearer reflection site is associated with the head and forelimbs, the further site with the pelvis and hind limbs. (From O'Rourke (1967). Pressure and flow waves in systemic arteries and the anatomical design of the arterial system. *J Appl. Physiol.* **23**, 148.)

but the system is really so complicated that it should be used with caution. The apparent absence of the upper reflection site in humans may be a consequence of a number of factors, such as: (a) a smaller proportionate blood flow in the upper part of the body, so that the reflected wave is insignificant; (b) better matched junctions in the upper part of the body; (c) a difference in the anatomy of the aortic branches of the two species (three branches from the arch in humans, two in dogs); (d) the fact that the measurements in humans were made on patients in whom heart disease had been diagnosed. Perhaps in healthy humans the second minimum of M would appear.

From measurements in the femoral artery of a dog, an effective reflection site just below the knee has been inferred. Such measurements have also revealed that the ratio between the impedance modulus of the first few harmonics and the resistance offered to the mean flow is larger in the femoral artery than in the aorta. This is a consequence of the smaller distensibility of more peripheral vessels.

The only theoretical prediction which can be compared with the experimental results shown in **Figs 12.29** and **12.30** is that obtained from the model described on p. 288, in which the arterial system is replaced by an assembly of randomly branching tubes. The calculated results are given in **Fig. 12.32**. It can be seen that the main features of the experimental results at lower frequencies are reproduced, but that there is little detailed agreement.

The effect of taper We can use the relatively simple theory for reflections at isolated sites to describe what happens to waves in a tube whose cross-sectional area and distensibility both vary continuously with distance along the tube. The description can then be applied to waves in large arteries like the aorta, which both taper and become stiffer towards the periphery. The effect of the taper, together with the changes in

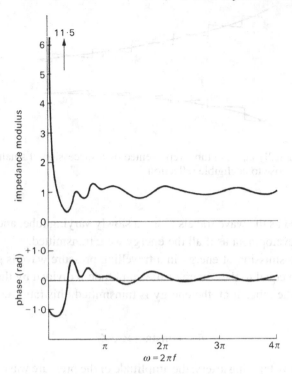

Fig. 12.32. Predictions of the modulus and phase of the input impedance of a systemic arterial tree. They should be compared with the observations recorded in **Figs 12.29** and **12.30**. The units for the modulus are chosen so that the characteristic impedance of the aorta has modulus equal to unity. (After Taylor (1966). The input impedance of an assembly of randomly branching elastic tubes. *Biophys. J.* **6**, 29.)

wave amplitude caused by reflection from the iliac junction, explains why the pressure pulse increases in amplitude (**Fig. 12.19**), although the peaking is rather less than this simple theory predicts.

In order to use the theory, we treat the taper as a series of separate reflection sites distributed throughout the length of the artery (**Fig. 12.33**). All components of the wave will be repeatedly reflected at each site; the combined effect is equivalent to a continuous, gradual modification of the original wave, so that its amplitude, shape and propagation speed vary with distance down the tube. If the variation in wall properties is gradual enough, the analysis of the change in shape of the wave becomes very simple, because each of the reflection sites which make up the continuous variation corresponds to a very small change in the characteristic impedance of the tube. Therefore, the amplitude of the reflected wave at each site is also small, and the amount of energy reflected is smaller still (see p. 282). This means that almost none of the

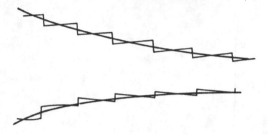

Fig. 12.33. A gradually tapering tube, represented by a succession of small discontinuities, each of which gives rise to negligible reflection.

energy is reflected as the wave travels along a slowly varying tube, and we may analyse the wave's development as if all the energy were transmitted.

The rate of transmission of energy in a travelling pressure wave is given by Equation (12.24) to be equal to the square of the amplitude $\tilde{p}$ divided by the characteristic impedance Z of the tube. If all the energy is transmitted, this ratio remains constant; that is,

$$\tilde{p} = \text{constant} \times \sqrt{Z}. \tag{12.25}$$

Thus, in a gradually tapering artery, the amplitude of the pressure wave is proportional to the square root of impedance.

Finally, we see from Equation (12.22) that the impedance of a tube increases if either its cross-sectional area A or its distensibility D decreases. Thus, in the aorta, where both A and D decrease towards the periphery, the amplitude of the pressure wave is predicted to increase. The amplitude of the aortic *flow* pulse, equal to $\tilde{p}/Z$ (Equation (12.15)), will correspondingly decrease, being proportional to $1/\sqrt{Z}$.

The cross-sectional area of the abdominal aorta is about 40% of the area of the thoracic aorta, as can be seen from the values of the diameters given in **Table I**. Furthermore, the Young's modulus is about twice that of the thoracic aorta. Hence, the impedance of the abdominal aorta is about 3.5 times that of the thoracic aorta (Equation (12.23)), and the amplitude of the pressure pulse is predicted to increase by about 85%. A further increase in amplitude is to be expected as the pressure pulse approaches the isolated reflection site at the iliac junction (p. 286), so the eventual prediction is that the amplitude of the pulse should be at least doubled as it travels the length of the aorta. That the actual peaking is not so great may be attributed to two factors: (i) inaccuracy of the present analysis; and (ii) attenuation of the wave by viscous and visco-elastic energy dissipation. The second of these factors will be examined on pp. 299–304; here, we briefly consider the first.

The present theory accurately predicts the increase in pressure wave amplitude along a tapered tube as long as a negligible amount of energy is reflected. If the tube tapers too sharply, significant energy will be reflected, and the predicted peaking will be diminished. The criterion for this theory to be accurate, then, is that the length of tube over which the taper takes place should be great enough to encompass several wavelengths of the wave. This is clearly not true for the first few harmonics which make up the pressure pulse, for their wavelengths exceed the length of the aorta, and therefore it is not surprising that the actual peaking is less than predicted.

The influence of nonlinearities

We have hitherto assumed that the arterial system behaves in a linear way, so that the propagation and reflection of single components of the composite pressure wave can be treated independently of the other components. This assumption is based on the approximations: (i) that the average blood velocity $\bar{u}$ is always small compared with the wave speed c; and (ii) that the pulse pressure is small compared with the mean excess pressure in the arteries, so that the distensibility does not vary during passage of the pulse. It was shown on p. 276 that in the largest arteries, the aorta and pulmonary artery, these may not be good approximations, and the actual passage of the pulse might be perceptibly modified by nonlinearities.

The nonlinearities take two forms. One, elastic nonlinearity, is a consequence of the curvature of the stress–strain curve (cf. p. 94), which shows that an artery becomes stiffer, i.e. its distensibility D decreases, as the distending pressure is raised. This means that the wave speed c, predicted from Equation (12.9), increases. This prediction has been confirmed both in vitro, using segments of artery, and in vivo, using short trains of high-frequency pressure oscillations, generated in a dog's aorta during diastole, with different values of diastolic pressure. The wavelength of such oscillations is short, so that several cycles can be recorded at a downstream observation site before the reflected wave from the iliac junction returns to that site and distorts the recording. The results of such an experiment are shown in **Fig. 12.34**, and show that there is a significant difference between the wave speed associated with the maximum pressure achieved during a normal pulse in systole and that associated with normal diastolic pressure. Thus, the basic pulse wave itself will move more rapidly in systole than in diastole.

There is also a fluid dynamic source of nonlinearity, which will be important if $\bar{u}/c$ is not small. The wave can be shown to travel with speed c *relative to the local fluid speed*, and not relative to the tube wall. Thus, if the blood has a certain velocity $\bar{u}$, the forward-moving part of the wave will have a speed $c + \bar{u}$, while the backward-moving part, returning from some more distal reflection site, will have a speed $c - \bar{u}$. This has also been confirmed, again by observing small, high-frequency waves superimposed on the basic pulse at different points of the cycle, and measuring their speed of propagation both up- and down-stream. The effect of this phenomenon on the basic pulse itself is to enhance the forward wave speed during systole, when the

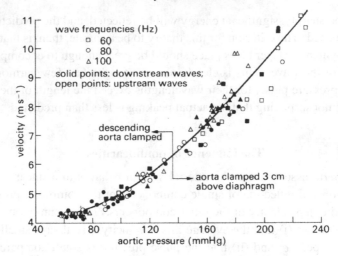

Fig. 12.34. Measured values of the wave speed in a dog's aorta, plotted against pressure in the aorta. High-frequency waves were introduced during diastole; diastolic pressure was controlled by occluding the aorta at different locations. (From Histand and Anliker (1973). Influence of flow and pressure on wave propagation in the canine aorta. *Circulation Res.* **32**, 524.)

blood speed is greatest, relative to its value during diastole. Thus, this too has the effect of speeding up the early part of the wave relative to the later parts.

From both sources of nonlinearity, therefore, we deduce that the forward wave speed during early systole is greater than that during diastole. Thus, as the wave is propagated downstream, the peaks of pressure and velocity tend to overtake the troughs, causing the front of the wave to become steeper. This steepening effect can be seen from simultaneous measurements of the pressure pulses at different positions along the aorta and in other large arteries (**Fig. 12.14**). In normal subjects it is not thought to be of any great importance. However, in patients whose arteries are abnormally distensible (so that c is low) or in whom the amplitude of the pulse is very large (so that $\bar{u}$ and $\Delta p_e/p_e$ are large), it can be quite important. In such patients, the steepening occurs more rapidly, and a situation may arise in which the pressure peak catches up with the trough, and the wavefront becomes very steep.

Phenomena such as this are familiar in other types of wave and cannot be described by linear theory. For example, when the front of a water wave becomes vertical, it breaks, usually in a frothy, turbulent manner, with considerable dissipation of energy. When a sound wave becomes so strong that the pressure peaks catch up with the troughs, a shock wave is formed, in which much energy is dissipated thermally. In each case, a balance is struck between the nonlinear steepening effect and dissipation, so that a steady state is set up. (When the amplitude of a water or sound wave is quite small, the tendency to steepen is not very strong, and the small dissipation associated with the basic wave motion is enough to counteract it.) So, in the arterial system, we can predict that if the amplitude of the pulse is great enough, some sort of 'shock' wave may develop. Some computer models of nonlinear wave propagation in the arteries have predicted the development of 'shocks' in patients suffering from aortic valve incompetence, in whom the output of the left ventricle is very large, and so $\bar{u}$ is large. No-one, either theoretically or experimentally, has yet studied the detailed dynamics of such shocks, nor discovered a mechanism of vigorous energy dissipation analogous to turbulence in a water wave. However, 'pistol-shot' sounds are sometimes observed in the arteries of patients with aortic valve incompetence, which may well be the outward sign of the passage of a 'shock'.

The vessels in the venous and pulmonary arterial systems are rather more distensible than the systemic arteries and experience pressure fluctuations which are a larger fraction of their mean pressure. Therefore, nonlinearities may be of greater significance in those vessels, but there have been very few detailed studies, so little evidence is available. The topics are dealt with in separate chapters devoted to these vessels (Chapters 14 and 15).

Viscous effects

We now have a satisfactory understanding of the mechanisms of propagation of the pressure pulse and of the changes in the pressure waveform which occur as it travels along the arteries. Apart from the steepening of the wavefront, all such changes are described by the linear theory. However, none of the factors considered above can explain the shape of the flow-rate waveform (**Fig. 12.19**), which is predicted by the linear theory to be the same as that of the pressure waveform and in phase with it (see footnote on p. 277). Only when blood viscosity is taken into account can the observed shape be explained.

Viscosity is also important in the attenuation, or damping, of the pressure wave. We have seen already (Chapter 8, p. 116) that waves in any real system must be attenuated, because the inevitable presence of friction or viscosity causes their mechanical energy to be dissipated into heat; in fact, visco-elastic dissipation in the vessel wall proves to be more important than viscous dissipation in the blood.

Effect of blood viscosity on flow-rate waveform The pressure and flow-rate waveforms, measured simultaneously in the femoral artery of a dog, are shown in **Fig. 12.35**. They are approximately the same shape during the acceleration phase of systole, but thereafter they diverge considerably. Furthermore, the peak flow rate has an appreciable phase lead over the peak pressure, while the simple non-viscous theory predicts that the two waveforms should be in phase. The effect of viscosity is therefore considerable.

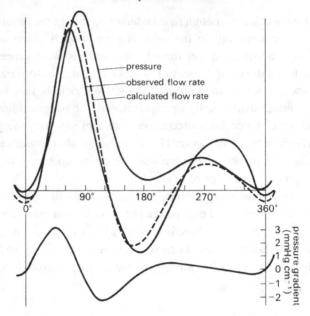

Fig. 12.35. The pressure and flow-rate waveforms measured simultaneously in a dog's femoral artery. Also shown are the waveform of pressure gradient (below) and the flow-rate waveform calculated from it using the theory for a rigid tube (broken curve). (After McDonald (1974). *Blood Flow in Arteries*. Edward Arnold, London.)

To understand how viscosity produces its effect, it is once more necessary to use Fourier analysis, and to consider the relationship between a sinusoidal pressure wave-form and the corresponding sinusoidal flow-rate waveform. In fact, as discussed in Chapter 4, it is the local pressure gradient, not the pressure itself, which determines the flow. However, if a sinusoidal pressure wave is described by the equation

$$\tilde{p} = p_0 \cos[\omega(t - x/c)]$$

(cf. Equation (12.20)) then the pressure gradient acting in the x-direction, $-\partial \tilde{p}/\partial x$, is equal to

$$-\frac{\omega p_0}{c} \sin[\omega(t - x/c)].$$

This is also sinusoidal, with the same angular frequency ω, and has a phase lead of $\pi/2$ or 90° over the pressure (see p. 111 for a definition of phase lead and phase lag; what the phrase means here is that, at a given value of x, the pressure gradient reaches its peak at a time $\pi/2\omega$ in advance of the pressure). We have already discussed, in Chapter 5, how the flow in a straight rigid pipe responds to the action of a sinusoidal pressure gradient. The same analysis proves to be applicable in arteries, because the walls remain approximately parallel even though they are not rigid (see p. 315).

The shape of the flow-rate waveform, in response to a sinusoidal variation of pressure gradient, is also sinusoidal, with the same frequency (because the relationship is linear), but its amplitude and phase depend on the Womersley parameter α (see p. 273). If α is large, the boundary layer is very thin compared with the tube radius, and the flow rate responds to the pressure gradient approximately as in a non-viscous fluid. However, if α is small, viscosity is important throughout the tube, and the flow rate is at all times related to the pressure gradient as in Poiseuille flow. The detailed relationship has been worked out for all values of α. As reported on p. 273, the value of α is large (over 10) in the largest arteries of large mammals, but becomes very much less than unity in the microcirculation.

In order to test the relevance of rigid-tube viscous flow theory in an artery, it is necessary: (i) to split the measured waveform of pressure gradient into its harmonic components; (ii) to calculate the flow rate corresponding to each component of the pressure gradient; and (iii) to combine the sinusoidal flow-rate components into the predicted flow-rate waveform. This can then be compared with the measured one. Such a comparison for the waveform in a dog's femoral artery is shown in **Fig. 12.35**. (The value of α associated with the fundamental heart rate is 1.4, which is not very much in excess of unity.) Only four harmonics were used in making the prediction, but nevertheless the agreement is very good, apart from somewhat smaller oscillations during diastole than are actually observed. It is thus clear that rigid-tube theory is very good for predicting the flow rate corresponding to a given oscillatory pressure gradient, although vessel elasticity has a strong influence on that pressure-gradient waveform, and viscosity does not (see pp. 271–276).

Effect of viscosity on wave propagation Viscosity has the effect both of reducing the speed of propagation of the pressure wave and of attenuating it. Both effects come about because the relative longitudinal motion (slip) between the blood and the wall, which is predicted by non-viscous theory, is prevented in a viscous fluid. This means, first, that a slightly smaller oscillatory flow rate is associated with the same oscillatory pressure gradient; thus, there is an increase in the effective inertia of the fluid, which reduces the wave speed. Second, high shear rates are generated in the boundary layer, causing energy dissipation and hence wave attenuation. Because the thickness of the boundary layer and the shear rates within it depend on the value of α, the magnitude of the wave speed c and the amount of attenuation per wavelength are also expected to depend on α.

The way in which wave attenuation can be described quantitatively was explained on p. 120. A sinusoidally fluctuating component $\tilde{p}$ of the pressure wave will vary with time t and distance x along the vessel as follows:

$$\tilde{p} = p_0 \, e^{-kx/\lambda} \cos\left[\omega\left(t - \frac{x}{c}\right) - \phi\right], \tag{12.26}$$

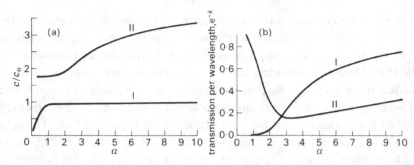

Fig. 12.36. (a) Predicted wave speed c divided by the Moens–Korteweg wave speed c_0 plotted against α. The lower curve corresponds to transverse pressure waves and the upper curve to longitudinal shear waves (see p. 304). (b) Predicted variation of the transmission per wavelength (the attenuation factor e^{-k}) with α. Curve I corresponds to pressure waves and curve II to shear waves. (After Atabek (1968). Wave propagation through a viscous fluid contained in a tethered, initially stressed, orthotropic elastic tube. *Biophys. J.* **8**, 626.)

where p_0 is the pressure amplitude, ϕ is a phase angle, c is wave speed, ω is the angular frequency and λ $(= 2\pi c/\omega)$ is the wavelength. The rate of attenuation is then represented by the constant k, since e^{-k} is the factor by which the amplitude is reduced as the wave travels over a distance equal to λ. A detailed theory of the effects of viscosity indicates that c and e^{-k} vary with α in the manner shown in **Fig. 12.36**.

We see from these curves that the wave speed remains approximately equal to the Moens–Korteweg wave speed c_0 (p. 276) for all values of α greater than about 1.5. Thus, in large mammals, where the value of α corresponding to the fundamental heart-beat frequency is greater than 10, all components of the wave will travel with the same speed in the aorta. In much smaller arteries, however, where the fundamental value of α is less than 1.5, **Fig. 12.36a** shows that the higher frequency components will move with greater speeds. This phenomenon, called *dispersion*, is another way in which a composite wave can change shape as it propagates. As we have seen, however, it is expected to be insignificant in large arteries.

Even in the largest arteries, attenuation is predicted to be measurable. **Figure 12.36b** shows that the predicted attenuation per wavelength decreases with increasing α (i.e. with increasing frequency).[12] However, the attenuation per unit distance increases with increasing frequency, because the wavelength λ decreases. The quantity

---

[12] For large values of α (greater than about 6), the value of k shown in **Fig. 12.36b** is approximately given by

$$k \approx \pi\sqrt{2}/\alpha. \tag{12.27}$$

The corresponding value of the wave speed is

$$c = c_0(1 - 1/\alpha\sqrt{2}),$$

where c_0 is the Moens–Korteweg wave speed.

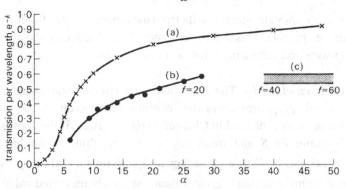

Fig. 12.37. The attenuation factor e^{-k} plotted against α for pressure waves in the dog's aorta: (a) values predicted on the assumption that all attenuation is caused by blood viscosity; (b) and (c) measured values in two series of experiments. f is frequency in hertz. (From McDonald (1974). *Blood Flow in Arteries*. Edward Arnold, London.)

k/λ (see Equation (12.26)) is proportional to $\sqrt{\omega}$, so the higher frequency components die out more rapidly in a given length of tube.

The amount of attenuation in a given length of artery can readily be predicted from **Fig. 12.36**b (or Equation (12.27)). In a dog's aorta, the fundamental wave has a frequency of about 2 Hz (i.e. $\omega = 4\pi$) with an associated value of α of about 11. The corresponding value of k is seen to be about 0.4, and the wave is thus predicted to be attenuated by 33% per wavelength. Since the wavelength is about 3 m, this means attenuation by only 5% in the length of the aorta. The value of k for a wave of frequency 40 Hz is about 0.1 and that for a wave of frequency 100 Hz is 0.06. Waves of these frequencies would thus be attenuated by 24% and 33% respectively in the length of the (dog's) aorta. Now in the experiments already mentioned, in which artificial high-frequency oscillations of between 40 and 150 Hz were superimposed on the pulse wave in a dog's aorta, the attenuation was very carefully measured and was found to be much greater than predicted. The results of the experiments were that k showed no significant variation with frequency, and that its value almost always lies in the range 0.7 to 1.0.

These results are consistent with other measurements made at much lower frequencies, and shown in **Fig. 12.37**. In its prediction of attenuation, therefore, the viscous theory is inadequate, especially for high-frequency waves. The additional attenuation must be associated with dissipative mechanisms in the vessel wall, through the visco-elasticity either of the materials comprising the wall or of the tissue to which the wall is tethered (see p. 93). Before turning to this topic, we should remark that the experiments on high-frequency waves in vivo showed that attenuation was significantly greater for waves travelling upstream against the mean flow ($k = 1.3$–1.5) than for those travelling with the mean flow ($k = 0.7$–1.0). The probable reason for that is the variation in wave amplitude which results from the tapering and stiffening of the aorta with distance (see p. 294). This would have the effect of reducing the attenuation for downstream waves and enhancing it for upstream waves.

Effect of wall visco-elasticity The introduction of visco-elastic, rather than purely elastic, vessel wall properties alters the relationship between excess pressure and cross-sectional area, as explained in Chapter 7 (pp. 96–102). If the applied stress oscillates with amplitude S and frequency ω, the resulting strain has amplitude $S/\sqrt{E_{dyn}^2 + \eta^2\omega^2}$ and exhibits a phase lag of approximately $\eta\omega/E_{dyn}$, where E_{dyn} is the dynamic Young's modulus (greater than a statically measured value of E), and η is the viscous term.[13]

Correspondingly, the distensibility D (Equation (7.3b)) decreases by a factor $E/\sqrt{E_{dyn}^2 + \eta^2\omega^2}$ and the area changes lag the pressure changes by the same phase lag $\eta\omega/E_{dyn}$. The result is a slight *increase* in wave speed (contrasting with the decrease from blood viscosity) consequent upon the decrease in distensibility, and, again, an attenuation of the wave as it propagates. The amount of attenuation per wavelength can again be represented by a constant k (cf. Equation (12.26)), which is approximately equal to 2π times the phase lag $\eta\omega/E_{dyn}$ in radians, as long as that is small. But, as described in Chapter 7, $\eta\omega/E_{dyn}$ is approximately constant for frequencies above 5 Hz and takes a value between 0.1 and 0.2. Thus, k should take a value between about 0.6 and 1.2, and indeed that is what is observed in the experiments already described (**Fig. 12.37**). It seems, therefore, that the attenuation of high-frequency waves is well described by this simple model of wall visco-elasticity. For the lower frequency waves typical of the fundamental and first harmonic (frequency less than 5 Hz), the value of the phase lag, and hence that of k, is smaller. In addition, the contribution to attenuation from blood viscosity is larger at low frequencies, so both effects are of comparable importance for the basic pulse wave. Since the wavelength of that wave is so long, however, the predicted attenuation ($k < 1$) is unlikely to be easy to measure, because of the distortion of the waveform by reflections, etc.

Other types of wave

There are other types of wave which can propagate along arteries in addition to the pressure wave. Suppose that the artery is modelled as a long, thin-walled, untethered elastic tube containing a viscous fluid, and that the mathematical analysis is carried out on the reasonable assumptions that (a) only axial and radial fluid motions are present and (b) that the wavelengths of all waves are very long compared with the tube radius. Then the calculations result in two possible values for the wave speed c,

---

[13] Using complex numbers, this can be expressed by replacing the Young's modulus E (Equation (7.1)) by $E_{dyn} + i\eta\omega$. In the same way, D (Equation (7.3b)) is replaced (approximately) by $DE(E_{dyn} - i\eta\omega)/(E_{dyn}^2 + \eta^2\omega^2)$.

with corresponding values of the attenuation factor e^{-k} (based on the inviscid wave speed c_0). One pair of values, c_1 and e^{-k_1}, corresponds to the pressure wave, and the dependence of these quantities on α is shown in **Fig. 12.36**. The other pair of values, c_2 and e^{-k_2}, are also plotted in **Fig. 12.36**.

These correspond to a different type of wave, often called the fast wave because c_2 is greater than c_1 for all values of α, and becomes infinitely large for infinitely large values of α. Whereas the pressure wave is associated primarily with radial motions of the vessel wall, this second type of wave is associated chiefly with longitudinal motions. Elements of the tube wall oscillate longitudinally backwards and forwards, under the action of a restoring force depending on the longitudinal stretch of the wall. The inertia required to counteract the restoring force is provided by the fluid which is dragged backwards and forwards with the wall through the shearing action of viscosity. For large values of α, the amount of fluid set in motion in this way is confined to a thin boundary layer, and the bulk of the fluid remains at rest. As α increases, the boundary layer becomes even thinner and the inertia of the fluid diminishes. Therefore, the wave speed increases, which explains the large α result in **Fig. 12.36**a. In a real tube, of course, the inertia of the tube wall itself would provide a limiting value above which c_2 could not rise.

At first glance it is surprising that when α is very small this type of wave has a non-zero wave speed and negligible attenuation. It can be understood when we realize that, in this limit, the longitudinal oscillations of a segment of wall cause all the fluid contained in that segment to be carried backwards and forwards as a solid body. The fluid motion is everywhere in phase with the wall motion, and there is no shearing at all. Therefore, no energy is dissipated in the fluid and the wave is not damped.

Yet a third type of wave is possible if there are torsional or twisting oscillations of the tube wall which, through viscous action, drive circumferential shearing motions of the fluid. These provide the inertia to balance the elastic restoring force. Since torsional waves have not been observed, we consider them no further.

In fact, the fast shear waves are not observed in vivo either. The reason for this almost certainly lies in the tethering of artery walls to the surrounding tissue. This tethering inhibits longitudinal (and torsional) wall motions almost entirely, while it still permits significant radial motions to take place. It can be analysed by considering the artery wall to be fastened to a rigid support by springs with certain elastic and viscous properties. Even when these properties are chosen to be much weaker than is actually the case, the calculations show that shear waves are almost totally suppressed.

Tethering of the wall is only one of many different modifications of the simple artery model which research workers have analysed in order to assess their importance. They have, for example, considered the effect of the inertia of the wall itself, which puts a limit on the maximum wave speed of shear waves, and in addition increases the inertia component of the fundamental pressure waves and, hence, reduces the wave speed slightly. They have also examined many different models of wall elasticity other than that of the thin-walled tube of uniform Young's modulus E, including the effects of a thick wall, with anisotropic elastic properties, and subject to different initial stresses, as well as the effect of visco-elasticity. All these factors, however, influence only the relationship between tube distensibility and the elastic properties of the wall. They do not affect the relationship of wave speed to distensibility (Equation (12.9)), except in so far as dissipative forces introduce phase changes and damping and, hence, they do not alter the basic mechanics of wave propagation.[14]

Finally, mention must be made of the one other type of wave which can exist in arteries: a true sound wave. In such a wave, the restoring force comes from the compressibility of the fluid (or solid) in which it is being propagated, and not from the distensibility of vessel walls. Such waves certainly are propagated in blood vessels and other tissues of the body, but since their speed of propagation is very large (about $1500 \ \mathrm{m\,s^{-1}}$), their passage through the cardiovascular system is virtually instantaneous, and their overall dynamic effect is negligible. The neglect of sound waves is equivalent to the assumption, made throughout this book, that blood is an incompressible fluid. The physical nature of the heart sounds is dealt with in Chapter 11.

Flow patterns in arteries

We turn now to an examination of the patterns of blood flow within arteries, with the twofold aim of interpreting measurements of blood velocity and understanding the influence of local blood flow on physiological processes in the artery wall. We shall first consider experimental information about flow patterns within large arteries and then go on to discuss physical mechanisms which lead to these patterns.

Velocity profiles in large arteries In the last few years a number of techniques have been developed for the measurement of blood velocity in arteries in the living animal. Some of these (in particular the electromagnetic method, utilizing cuff or catheter-tip probes) are sensitive to the flow in all, or a considerable part, of the lumen of the vessel, and thus measure a spatially averaged velocity. Others, notably the hot-film and pulsed Doppler techniques, can be used to plot blood velocity from point to point within a vessel. The ultrasonic Doppler technique has tremendous promise, since it can produce an almost instantaneous plot of the velocity distribution across a vessel, without the necessity of inserting a probe into it (and potentially even in the intact animal); but it is still in the process of development, and has so far yielded only a little information about arterial flows; we shall therefore concentrate on the studies with hot-film probes.

The technique of hot-film anemometry employs tiny, velocity-sensitive probes mounted near the tips of hypodermic needles or catheters, which are inserted into

---

[14] These various modifications of the simple elastic artery model are reviewed in detail by R. H. Cox (1969). Comparison of linearized wave propagation models for arterial blood flow analysis. *J. Biomech.*, **2**, 251–65.

the artery. Such probes are sensitive only to flow within a millimetre or less of their surface; thus, they can be used to plot blood velocity from point to point within a vessel. In addition, since they are capable of registering velocity fluctuations up to frequencies in excess of 500 Hz, they can be used to study disturbed and turbulent flows in blood vessels.

The technique has revealed a good deal about the velocity distribution in the aorta of the dog, and a few observations have been made in humans; but even with the smallest probes, it has not been possible to explore the boundary layer itself in either of these species. This is because the boundary layer is very thin. Even the smallest hot-film probes become unreliable near the artery wall, partly because the signal is influenced by the proximity of the wall, and partly because the vessel wall is moving radially. This means that the relative position of the (fixed) probe and the wall changes during the cardiac cycle, and that appreciable radial velocities occur in the blood which distort the recording of the longitudinal velocity. This unreliability of hot-film probes very near vessel walls means that they can only be used to study the velocity profiles in large arteries, and in the dog this means the aorta. Even there, only certain sites are accessible, because of problems in surgical exposure, and because care must be taken to avoid distorting the vessel during measurement; it has been found repeatedly that the velocity profile changes dramatically if the vessel is distorted. A final problem encountered in reporting and interpreting these studies is that different workers have not always precisely defined the sites and the planes at which they have traversed vessels. As will be seen, local geometry has important effects on the flow, and many inconsistencies between different studies or different animals are probably attributable to this. In what follows, we concentrate on the features which have been consistently observed.

If we examine the mean velocity profile (i.e. the velocity profile constructed from time-mean measurements) at some of the sites which have been studied in the dog (**Fig. 12.38**), we can see that it remains blunt throughout the length of the aorta, with only the measurements closest to the wall showing retardation of the flow compared with that in the centre of the vessel. Close to the heart, virtually no retardation is detectable anywhere; the blood moves as a coherent 'slug' and the boundary layer is confined to a wall region too narrow to be observed; in practice, this means that it is less than 2 mm thick.

This is true throughout the flow cycle in the thoracic aorta. In **Fig. 12.39**, instantaneous velocity profiles in the ascending aorta are shown at six times during the cardiac cycle. It can be seen that a marked skew occurs during forward flow; but there is no evidence at any stage of velocities being preferentially lower close to the wall. Some workers have approached the problem of studying these very thin boundary layers by using the pig and the horse, which have larger aortas than the dog. In the pig, radial wall movement was eliminated by splinting the vessel with a rigid external cuff; but even in these circumstances it was not possible to examine the boundary

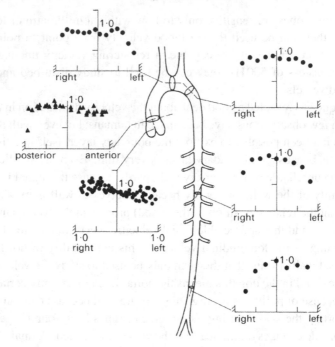

Fig. 12.38. Velocity profiles in the aorta of the dog. For each graph, the mean velocity at each site on the vessel diameter is expressed relative to the centre-line mean velocity. (From Schultz (1972). Pressure and flow in large arteries. *Cardiovascular Fluid Dynamics* (ed. Bergel), vol. I. Academic Press, New York.)

layer in any detail, because it was less than 1.5 mm thick. On theoretical grounds it is expected to be thin, as we shall see later.

These studies have, however, revealed considerable detail about the behaviour of the flow in the core of the vessel. At the entrance of the aorta, a centimetre or two downstream from the aortic valve, the velocity profile is usually symmetric during systole, at least in the plane of the main curvature of the aortic arch (this is the only one accessible to study, and is virtually antero-posterior at that level – **Fig. 12.40**, lower site). This is the form which is to be expected for an entrance flow (see p. 50), but there are several less familiar features.

First, velocity waveforms at this site frequently show low-frequency disturbances during systole, which may be quite variable from beat to beat, or may be so reproducible that they skew the velocity profile on each beat, usually in late systole. Such disturbances may be residues of the eddying known to occur within the ventricle (p. 227), or may arise during passage of the blood through the aortic valve. In studies with catheter-tip probes where records of velocity have been made within the ventricle, disturbances have been seen during systole as well as diastole. This suggests

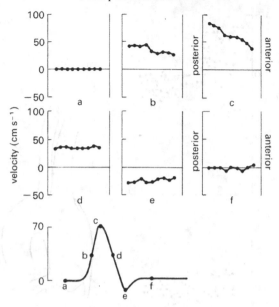

Fig. 12.39. Instantaneous velocity profiles at six different times during the cardiac cycle in the ascending aorta of the dog. The lower figure shows schematically the velocity waveform, with the six times marked on it.

that the disturbances set up in the ventricle during diastole do persist into the ejection phase and may, therefore, be the source of the aortic flow disturbances. However, this sort of evidence is not conclusive because catheters in the ventricle may be jerked about during systole, causing distortion of the velocity signal, and to date the source of these low-frequency flow disturbances in the ascending aorta has not been firmly established.

A second interesting feature of the flow is that this symmetry of the systolic velocity profile is lost as the flow travels up the ascending aorta and into the arch; the flow close to the inner (posterior) wall of the arch is accelerated, producing a marked skew (**Fig. 12.40** upper profiles).

The skewing of the flow as it enters the aortic arch is a constant finding. A number of explanations have been put forward; for example, the velocity profile may be influenced by the proximity of the major branches coming off the aortic arch, which are expected to cause considerable local distortion of the flow (see, for example, **Fig. 12.47**). Another explanation is related to the tight curvature of the aortic arch. In flow through such a bend, a transverse pressure gradient is set up, so that along a diameter in the plane of the curve (i.e. aligned like the spoke of a wheel) the pressure will rise towards the outer wall. In an inviscid flow (such as that in the core of the aorta) this causes relative retardation of the streamlines near the outer wall; the

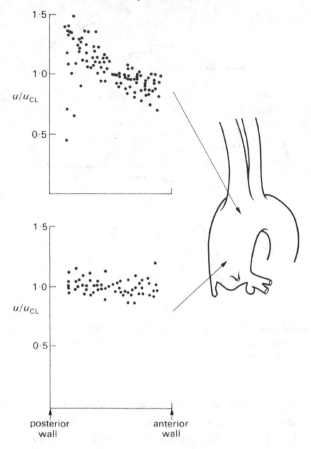

Fig. 12.40. Velocity profiles at peak systolic velocity across two diameters in the ascending aorta of the dog. All velocities u are expressed as a ratio of the centre-line velocity u_{CL} and all radial positions as a ratio of the vessel radius. The profiles at both sites are from several dogs; in every case they were measured in the plane of the main curvature of the aortic arch.

effect is explained in detail in Chapter 5 (p. 68). This explanation would also fit the observed fact that the skew disappears as the velocity falls and recurs during flow reversal in early diastole. It would also explain why the magnitude of the skew varies from animal to animal, since this depends on the local geometry (radius of curvature of the bend and diameter of the vessel). This of course would also be true if the skew was due to vessel branching, and it is quite likely that both mechanisms play a role.

Profiles have also been examined at right angles to the plane of the aortic arch. This plane is not accessible close to the aortic valve, because the pulmonary artery overlies the aorta, but can be examined more distally, at the entrance to the arch (roughly level with the upper site in **Fig. 12.40**). Here again, skewing occurs, with higher velocities towards the left wall during systole. This is unlikely to be a curvature effect, because the vessel is relatively straight in this plane, and we have as yet no clear explanation for it.

During diastole, the cross-sectional average velocity, as measured, for example, by a cuff electromagnetic flowmeter, is very close to zero in the ascending aorta. However, hot-film measurements show that the blood does not actually come to rest during diastole. Local backflows, affecting only part of the cross-section of the vessel, and compatible with drainage into the coronary arteries, can be observed. Also, at the entrance to the aortic arch, flow may simultaneously be moving forwards on one side of the vessel and backwards on the other. This suggests a circulating flow, presumably associated with drainage into the arch branches, where forward flow can be maintained during diastole.

Measurements in the proximal descending aorta can only be made in a plane orthogonal to the main curvature of the arch. Here, too, circulating flow may occur during diastole, but in systole there is strikingly little evidence of any influence from the major branches just upstream. The systolic profiles remain blunt, with a slight skew towards the right wall; the mean profiles are usually symmetrical (**Fig. 12.38**). More distally, the first hint of 'rounding' of the mean profiles, implying viscous influences and the growth of the boundary layer, can be seen.

In the abdominal aorta and the major branches, hot-film probes are usually too big for detailed studies in the dog. Recently, however, measurements have been made in the horse, where access is possible much more peripherally (in a large horse, the aorta may have a diameter of 5 cm at its origin, and major branches like the brachiocephalic or iliac arteries are 1.3 cm across). At first sight, this might seem an obvious solution to the problem of examining the boundary layers also; but in fact the boundary-layer thickness in unsteady flow is independent of vessel diameter and is proportional to $\sqrt{\nu/\omega}$ (see p. 273). The kinematic viscosity of blood at high rates of shear is much the same in the majority of mammals, so that boundary-layer thickness depends on $1/\sqrt{\omega}$, and thus varies inversely with the square root of heart rate. Between different species, therefore, the thickness of the boundary layer will only increase if the heart rate is lower.

In fact, heart rate does tend to fall with increasing body size; therefore, there are good reasons for studying larger animals. The horse, for example, has a heart rate of $30–50\,\text{min}^{-1}$, compared with $80–120\,\text{min}^{-1}$ in the dog; thus, the boundary layer is expected to be 60% thicker in the horse and has proved to be accessible to study, though not yet in detail. **Figure 12.41** shows velocity profiles in the descending thoracic aorta at three instants during systole. They confirm that the boundary layer is thin (3–4 mm) during forward flow; but it must be remembered, in examining the boundary layer itself, that measurements extremely close to the wall may not be very accurate because of wall movement and radial flow. High-frequency velocity fluctuations,

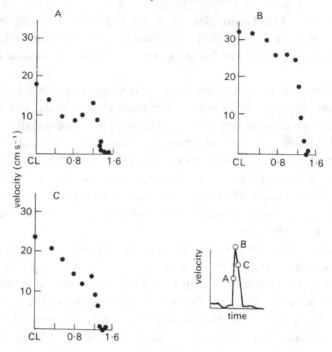

Fig. 12.41. Half-width velocity profiles (from centre-line to right wall) in the descending thoracic aorta (10 cm distal to the arch) of the horse. The inset velocity waveform shows the points in the cycle at which the profiles were constructed. Ordinate: velocity of blood (cm s^{-1}). Abscissa: distance from centre-line (CL) (cm). Vessel diameter 2.69 cm; $\alpha = 19.4$; peak $Re = 2536$. (After Nerem, Rumberger Jr, Gross, Hamlin and Geiger (1974). Hot-film anemometer velocity measurement of arterial blood flow in horses. *Circulation Res.* **34**, 193.)

indicating significant flow disturbance or turbulence, were frequently encountered in the thoracic aorta, but not further downstream.

In the abdominal aorta, several influences affect the velocity distribution. Immediately below the point where the mesenteric, coeliac and renal arteries branch off the aorta (the anatomy in this region is similar to that in the dog, **Fig. 12.1**), velocities fall markedly, reflecting the large diversion of blood into these branches. The velocity profile becomes more rounded, suggesting the influence of viscous forces, and also skewed, presumably in relation to local upstream drainage into a branch. The peak Reynolds numbers in this region lie between 200 and 800, and flow disturbances were not seen. Further downstream, very close to the termination of the aorta, the most striking feature was that the highest velocities were observed off the centre-line of the vessel, so that the systolic profile was M-shaped. It is not clear how such a pattern could be set up during forward flow as a result of the proximity of the branching.

It seems likely to be an oscillatory flow phenomenon, as can be seen occurring in a straight, unbranched pipe in **Fig. 5.9**, p. 59.

A few measurements have also been made in the iliac arteries, distal to the termination of the aorta. There, the profiles during systole are relatively symmetrical and markedly rounded. More distally than this, measurements have not been made with hot-film probes in any species.

Physical mechanisms underlying the velocity profiles Oscillatory flow in a long, straight, rigid tube when an oscillatory pressure difference is applied between the ends has been described both in Chapter 5 and on p. 273 and p. 301 of this chapter. Experimental observations on such a flow, which show very close agreement with theoretical predictions, are shown in **Fig. 12.42**. These observations are not entirely applicable to flow in the aorta because, as will become clear below, the steady component of the aortic flow is not fully developed. The oscillatory components, on the other hand, probably are appropriate.

In this section, we consider steady and unsteady flow near the entrance of a vessel, the effect of vessel curvature and the flow pattern near sites of branching. We shall neglect the distensibility of the artery wall, for although that is of central importance in determining the local pressure gradient as the pulse wave propagates, it does not affect the flow rates and velocity profiles driven by the pressure gradient. The reason is that the wavelength of the wave is very much longer than the distance travelled during any one pulse by an individual fluid element, because the wave speed is much greater than the blood velocity (see p. 276). For instance, the wavelength of the pulse in the dog's aorta is 3–4 m, while the distance travelled by a representative fluid element, which is equal to the mean velocity multiplied by the pulse period, is 10–20 cm. This means that the segment of artery traversed by a fluid element during one heartbeat is approximately uniform in cross-sectional area; the fact that it varies over a distance of one wavelength is irrelevant, particularly since that variation is only about 10%. Elasticity would be important only in circumstances in which passage of the wave caused significant variation of cross-sectional area within a length scale of about 10 cm; this may happen when the wavefront becomes very steep (see p. 298), but does not normally occur.

The entrance region The entrance length for steady flow in a straight tube can be calculated by assessing where the boundary layer comes to occupy the whole radius of the tube (Equation (5.4)). In unsteady flow the mean profile grows with distance from the entrance as it does in steady flow, though it has superimposed upon it everywhere the time-varying oscillatory profile, whose form is dominated by the local value of α. The observer making measurements of the velocity profile at any instant in the cycle will see a velocity profile which is a combination of the steady and unsteady components (as in **Fig. 12.42**). We can calculate the entrance length for the

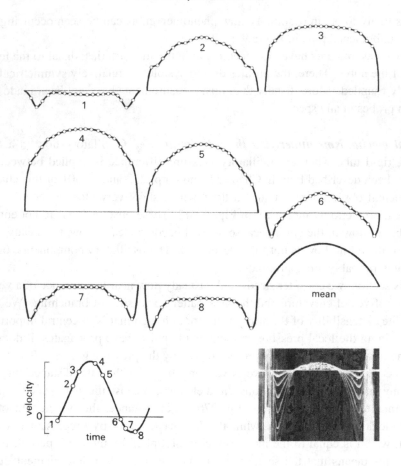

Fig. 12.42. Instantaneous velocity profiles at eight different times during a cycle of oscillating flow in a long, straight, rigid pipe. The times are marked on the flow-rate waveform, which is shown at bottom left and which consists of a sinusoidal oscillation superimposed on a steady flow, so that there is only a short period of backflow. The measurements were made far from the entrance of the pipe, so that the mean flow had a fully developed, Poiseuille profile (also shown). The flow conditions were chosen to be appropriate to those in the aorta: mean Reynolds number $Re = 1785$, Womersley parameter $\alpha = 20.5$, peak velocity 2.74 times mean velocity. The measurements were made by 'marking' the flow with lines of tiny hydrogen bubbles, generated by pulsing electrical current through a fine wire stretched across the flow. These bubbles are carried with the flow and are photographed after a known time interval (inset); the velocity at each point on the diameter during this interval can then be calculated. (Data reproduced by courtesy of Mr P. Minton and Dr M. Clamen.)

oscillatory component of the flow by assessing where the thickness of the boundary layer growing from the entrance (δ_1) becomes equal to the thickness it would have in fully developed oscillatory flow (δ_2). The boundary-layer thickness in steady flow was defined (p. 52) as the distance from the wall at which the fluid velocity reaches

99% of its value outside the layer. This distance is equal to $3.5\sqrt{vx/u}$, where v is the kinematic viscosity of the fluid and u is the core velocity; x is the distance from the entrance (see **Fig. 5.5**, p. 51). The same formula holds at any time for δ_1, if u is the instantaneous core velocity. The oscillatory boundary-layer thickness δ_2, similarly defined, is equal to $6.5\sqrt{v/\omega}$ for an oscillation of angular frequency ω. This leads to the following equation for the unsteady entrance length l_1:

$$l_1 \approx 3.4u/\omega. \tag{12.28}$$

In the aorta of a dog, for example, the heart rate is about 2 Hz (so $\omega = 4\pi$) and the amplitude of the largest unsteady velocity component is about $40\,\mathrm{cm\,s^{-1}}$, so that the largest possible instantaneous value of l_1 is about 10 cm. For most of the cycle, because the instantaneous value of u is small, and for the components of higher frequency, the unsteady entrance length will be considerably smaller (and will be zero during flow reversal).

The mean flow will continue to develop beyond this point. This development is not affected by the oscillatory components, because, far from the entrance, the oscillatory boundary layer is much thinner than the developing mean boundary layer, and thus only influences a narrow region very close to the wall where the mean velocity is in any case very small. We can therefore use the simple relationship of Equation (5.4) ($l = 0.03\,d\,Re$, where d is the tube diameter and Re is mean Reynolds number) to estimate the entrance length for the mean flow. In a dog's aorta, for example, with diameter 1.5 cm and mean velocity $20\,\mathrm{cm\,s^{-1}}$, the mean Reynolds number is approximately 800, and the corresponding entrance length is approximately 36 cm, which is almost the whole length of the aorta. In a human aorta, with $d = 2.5$ cm, $\bar{u} = 20\,\mathrm{cm\,s^{-1}}$, the entrance length is approximately 94 cm, which is greater than the length of the aorta. In these calculations we have neglected all effects of bends and branches, so that the estimated lengths are only approximate. However, the main conclusion remains, that as far as the mean flow is concerned almost the whole aorta is an entrance region. If we consider small arteries on the other hand, both d and the mean Reynolds number are smaller and the entrance length occupies a much smaller proportion of the vessel; for example, in a dog's femoral artery, with $d = 0.4$ cm, $\bar{u} = 10\,\mathrm{cm\,s^{-1}}$ (**Table I**), the value of l is approximately 1.2 cm, a small fraction of the total length.

We have seen that the velocity profiles in the core of the aorta will be more or less flat and the boundary layers thin. Factors which might modify these predictions are the curvature of arteries, especially at the arch of the aorta, and the presence of branches. We now examine these effects.

Curvature of the aorta The steady motion of fluid in a curved tube was described in Chapter 5 (p. 66 *et seq.*). Far from the entrance, the faster moving fluid in the centre of the tube is swept to the outside wall of the bend, because its inertia makes it

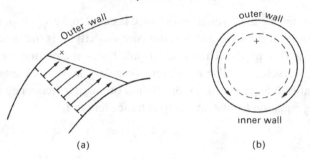

Fig. 12.43. (a) Velocity profile in a curved tube near the entrance (ignoring the boundary layer). + means high pressure, – means low pressure. (b) The pressure difference from + to – drives secondary motions in the boundary layers, as indicated by the arrows.

respond less readily to the transverse pressure gradient which is forcing it round the bend. It is replaced by the slower moving fluid near the walls, which moves round the walls towards the inside of the bend. In this way secondary motions are set up, and the axial velocity profile is distorted in the manner shown in **Fig. 5.16** (p. 67), with higher velocities near the outside wall.

Near the entrance of the tube, on the other hand, the initially flat velocity profile is skewed so that higher velocities occur near the inside wall (**Fig. 5.17**, p. 68); this occurs in the aorta (**Fig. 12.40**). There are thin boundary layers on the walls near the entrance, as usual, but because of the higher core velocity towards the inner wall, the boundary layer is thinner there, and hence the velocity gradients (and thus the wall shear rate) are greater. However, the higher pressures which occur at the outside of the tube (**Fig. 12.43**) to force the rapidly moving core fluid round the bend also act on the slower moving fluid in the boundary layer. This is therefore forced round the walls towards the inside (as in fully developed flow) so that the boundary-layer thickness on the inside increases relative to that on the outside and the shear rate decreases on the inside. Hence, the initial distribution of shear rate is reversed; this is predicted to happen after about one diameter. Such a flow pattern is not discernible in experiments to measure velocity profiles in the aorta, as described above (**Fig. 12.40**), because the boundary layer is undetectable. However, the effect on wall shear rate should be seen in experiments to investigate the transport of substances across vessel walls, when the rate of transport depends on wall shear rate or shear stress (see p. 333 *et seq.*).

This description strictly applies only to steady flow. In oscillatory flow, regardless of the value of α, it will apply very near the entrance. Further downstream, when α is large it will apply only to the mean flow; for smaller values of α (less than about unity), it is expected to apply everywhere. This means that our predictions of the velocity profiles in the proximal aorta are largely unchanged by the presence of oscillations, but the flow further downstream will be a complicated superposition of a

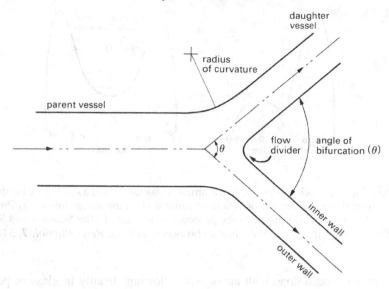

Fig. 12.44. Diagram of a symmetrical bifurcation, illustrating the notation used in the text. Flow direction marked by arrows.

mean flow, with higher velocities near the outside, and oscillatory components, with higher velocities near the inside.

Branched vessels In the analysis of wave reflection at arterial junctions the local flow pattern is unimportant. This means that the detailed geometry of the junction can be ignored, and represented solely by the cross-sectional areas of the arms of the junction. Much more geometrical detail is required, however, for a prediction of the detailed flow patterns. We need to know the angle at which a branch initially comes off its parent vessel, and the angle which it ultimately makes with it (the two are usually different; see **Fig. 12.2**); we need to know the curvature of all the walls at a junction, including the sharpness of the flow divider; we need to know the way in which the area and shape of the parent tube change as the junction is approached. At present, however, very little of this information is available. It can only be obtained by painstaking measurements of casts of arteries, carefully made at physiological values of transmural pressure and length.

Even if we knew all the geometrical details, prediction of the flow patterns would be very difficult, because there have been very few experimental or theoretical studies of flow in branched tubes, steady or unsteady. We can therefore only make very general predictions about flow in arterial branches.

The type of junction about which most is known is a symmetric bifurcation with equal flow rates out of each daughter tube (**Fig. 12.44**), for which a number of

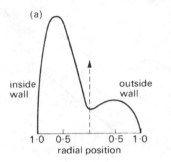
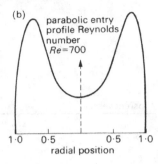

Fig. 12.45. Measured velocity profile two diameters downstream of a symmetric bifurcation, at a parent tube Reynolds number of 700 (Poiseuille flow in the parent tube). (a) Profile in the plane of the junction; (b) profile in the perpendicular plane. (After Schroter and Sudlow (1969). Flow patterns in models of the human bronchial airways. *Resp. Physiol.* **7**, 341.)

experiments have been done with air or water flowing steadily in glass or perspex models. Even with this arrangement, the flow depends on several different quantities. These include the area ratio of the junction (summed area of daughter tubes divided by area of parent), the angle of branching, the radius of curvature of the outside walls of the junction and of the flow divider, the Reynolds number of the flow entering the junction from the parent tube and the velocity profile in the parent tube. Model experiments have been performed on junctions (representative of the airways of the lung) with area ratios of 1.2, branching angles of 70° and Reynolds numbers between 100 and 1000. In large blood vessels there are very few symmetric bifurcations, the only familiar one being the human aortic bifurcation, which normally has an area ratio of 0.8 and a branching angle of about 70°. Nevertheless, the mean Reynolds number there is about 400, and the results of the model experiments will give an idea of the principal features of the flow.

The most obvious feature is that the flow is split into two. Thus, if we suppose the upstream velocity profile to be approximately symmetrical, we see that immediately after the flow has passed the flow divider the greatest velocities will be very close to the inside wall, next to the flow divider itself. The velocity of the fluid in contact with that wall must be zero, so a new boundary layer develops from the flow divider, where the shear rate is very high. The axial velocity profile in the plane of the junction is expected to resemble that shown in **Fig. 12.45**a.

Another obvious feature of the flow is that each of the two streams is made to turn a bend. This means that, as in a uniform curved pipe, secondary motions are set up, as shown in **Fig. 12.46**, and these maintain the highest velocities near the outer wall of the bend; that is, the inner wall of the junction (**Fig. 12.44**). Furthermore, the secondary motions are sufficiently strong to sweep some of the faster moving fluid round to the top and bottom walls, with the result that the velocity profile in the plane

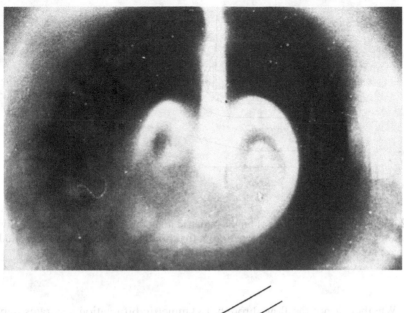

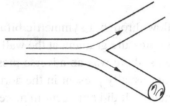

Fig. 12.46. End view of the daughter tube of a symmetric bifurcation (with Poiseuille flow in the parent tube), showing the secondary motions. The flowing fluid was air and the flow patterns were revealed by smoke injected into the outside wall of the daughter tube. (From Schroter and Sudlow (1969). Flow patterns in models of the human bronchial airways. *Resp. Physiol.* **7**, 341.)

perpendicular to the junction has an M-shape, as shown in **Fig. 12.45**b. The secondary motions were visualized by introducing smoke into the flow (**Fig. 12.46**).

The flow patterns change with distance downstream until, at a distance comparable to the usual entrance length (Equation (5.4)), Poiseuille flow is again set up in each branch. However, the initial development of the flow is considerably different from simple axisymmetric entry flow, because the secondary motions help to maintain the peak velocities near the inside wall for a distance of several diameters.

The velocity profiles of **Fig. 12.45** show that the shear rate at the wall remains high on the inside walls of the junction, not only at the flow divider, but for some distance downstream and for some distance round the circumference of the daughter tubes. On the outside wall of the junction, however, the shear rate is very low. Indeed, in many of the experiments, regions of reversed flow are reported on the outside wall just downstream of the bifurcation, indicating flow separation (see Chapter 5,

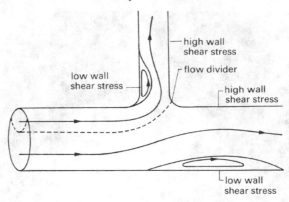

Fig. 12.47. Longitudinal section of an asymmetric junction. Dashed line: surface dividing the fluid which flows down the side branch from that continuing in the main tube; solid lines: streamlines; note also the closed eddies in regions of separated flow.

p. 64). Whether or not the flow through a symmetric bifurcation separates depends on whether there is an adverse pressure gradient at the wall, tending to slow down the flow. In the model experiments, there was an adverse pressure gradient, because of both the expansion in area (not usually present in the aorta) and the sharply curved walls at the outside of the junction. It did not occur in more smoothly curved models.

The curvature of the walls at the aortic bifurcation is not normally very sharp, so the flow is believed to remain attached. However, if the geometry were distorted by disease, separation might occur, and this would be associated with local reversal of flow direction, and reduced wall shear rates. This would increase the likelihood of turbulence developing there, and could influence mass transport across the artery wall; both of these topics are dealt with later in this chapter.

The only type of asymmetric junction in which the flow has received attention is that in which a daughter tube branches at right angles off an otherwise uniform parent (**Fig. 12.47**). This is not a good model, even for those cardiovascular junctions in which the ultimate angle between the parent vessel and the branch is a right angle (e.g. where the intercostal or renal arteries come off the aorta); thus, the flow pattern in such a junction is an oversimplification of the real flow, but it has a limited value because the geometry is precisely known, and the junction is asymmetric.

Experiments with dye in steadily flowing water have shown that the surface dividing the fluid which flows down the branch and that which continues in the main tube is clearly delineated (**Fig. 12.47**). The furthest downstream point of this surface lies somewhere on the part of the junction wall opposite to the oncoming flow, which is analogous to the flow divider in the previous example. The exact location of this point depends on the relative flow rates in the two tubes. It is a stagnation point, where the fluid velocity even just outside the boundary layer is zero, but the boundary layer is very thin there, and a little way downstream from it in either direction is a region where the wall shear stress is very high. The flow is generally observed to separate from the inner wall of the junction, and a region of reversed flow and low shear occurs on this wall just downstream of the junction. Strong secondary motions are set up in the branch. Sometimes, separation occurs also in the main tube, at a point roughly opposite the entrance to the side branch. This is because the sucking of fluid down the branch creates a low-pressure zone in the parent tube, just downstream of which an adverse pressure gradient occurs. Typical paths taken by particles in the plane of symmetry are also shown in **Fig. 12.47**.

Recently, quantitative experiments have been performed in a cast of a dog's aorta, in which the shear stress at various sites on the wall was measured. The measurements confirm the above qualitative description of the flow pattern, although, because the branch comes off initially at a small angle and then curves rapidly, the separated region in the daughter tube may be displaced downstream or may be absent altogether. The results also show clearly that very small perturbations to the geometry of the vessel wall at a junction can have a profound effect on the local flow. In one artery, the flow divider tip (see **Fig. 12.47**) protruded very slightly into the lumen of the tube (by less than 1 mm), but still that was enough to create another small region of separated flow on the wall of the main tube just beyond the flow divider.

It is clear that the patterns of flow in branching arteries are very complicated, involving secondary motions and contiguous regions of high wall shear (where the boundary layer is thin) and of low wall shear (where flow separation may occur). The complicated distribution of wall shear stress may have important implications for the distribution of early atheroma, if that is in any way associated with wall shear (p. 333 *et seq.*).

Instability and turbulence In very viscous or slowly moving fluids (i.e. at low Reynolds numbers), the motion is orderly, with well-defined streamlines and the transfer of material, momentum and energy across the streamlines taking place only by molecular motion (i.e. diffusion). Such flow is termed *laminar*; Poiseuille flow is an example. However, it must be made clear that laminar flow may be steady or unsteady, and may have a very complicated structure because of the presence of vortices or secondary motions. A characteristic of laminar flow is that disturbances within it die out due to viscous action. The hallmark of *turbulent* flow is that it contains disturbances which are *random* in amplitude, frequency and direction and do not die out (Chapter 5).

As the Reynolds number of a laminar flow is increased, the damping of small disturbances is decreased; ultimately, a *critical Reynolds number* is reached, beyond which the flow is no longer stable and at least some disturbances are amplified. When *Re* in a flow exceeds this value, and sufficient time is available, this amplification leads

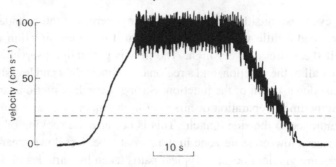

Fig. 12.48. Turbulent flow. Record made with a hot-film probe in a pipe in which the flow rate was slowly increased until turbulence occurred, and later stopped. Peak Reynolds number 9500. The dotted line shows the velocity corresponding to the usual critical Reynolds number of 2300. (After Nerem and Seed (1972). An in vivo study of aortic flow disturbances. *Cardiovasc. Res.* **6**, 1.)

to turbulence. The critical Reynolds number for fully developed pipe flow, based on average velocity and pipe diameter, is usually given as approximately 2300, although it may be considerably higher if great care is taken to avoid introducing disturbances into the flow. However, if an already turbulent flow is slowed down so that *Re* falls, the turbulence will always persist until *Re* is below about 2300. The disturbances that initiate turbulence may come from a number of sources and their magnitude has a considerable influence on the critical Reynolds number. The usual sources are eddying upstream (e.g. in the reservoir) or roughness of the pipe wall.

Because the eddies which represent turbulence in a flow are random in size and velocity, they will therefore be 'seen' (for example, by a velocity sensor at a point in the flow) as fluctuating velocity components superimposed upon the bulk flow velocity. An example of turbulent pipe flow is shown in **Fig. 12.48**. Strictly speaking, a flow can only be characterized as turbulent if such fluctuations can be demonstrated to be present in all three dimensions, but in practice the process has been so well characterized in many fluid dynamic situations that fluctuations in one dimension which have the appropriate characteristics are sufficient identification. Apart from illustrating the high-frequency velocity fluctuations, **Figure 12.48** illustrates a number of other points about the onset and decay of turbulence. First, turbulence does not appear until Reynolds numbers greatly in excess of the usual critical value of 2300 have been reached. This may in part be because accelerating flow is more stable than steady flow and in part because the flow upstream was free of disturbances due to the design of the apparatus. Second, the turbulence persists as the flow is decelerated, even at *Re* below 2300. This is partly because decelerating flow is inherently less stable than steady flow and partly because existing eddies take a finite time to decay.

Since the turbulent eddies cover a wide range of sizes and are carried with the flow, the velocity seen by a stationary probe will contain fluctuating components having a wide range of frequencies, and a *frequency distribution* will exist. This can be characterized in a number of ways (for example, by Fourier analysis, p. 266) and expressed as a graph of velocity (or energy) against frequency. Typically, this *spectrum* is centred on some frequency f_0, and velocities diminish as frequency becomes both very large and very small compared with this value f_0. The value of f_0, the 'centre frequency', is dictated by the size of the largest eddies, and these depend on the geometry of the flow. In a pipe flow, for example, the largest eddies have a diameter d comparable to that of the pipe. Thus, if the flow velocity is u, the value of f_0 is approximately equal to u/d. The precise point in the spectrum where the maximum amplitude of velocity or energy comes is dictated by a number of interacting and often complicated features of the flow, including local and upstream geometry and fluid viscosity.

It is unfortunate from the point of view of arterial flows that most of the engineering studies of transition and turbulence have been carried out when the underlying mean flow is steady, and there have been few studies of the stability of unsteady pipe flows. Those which have been carried out have involved flows in which the oscillatory components were small compared with the mean, so that the mean flow properties dominate the flow; this is unlike the situation in large arteries, where the peak velocity may be four or five times greater than the mean.

Until recently, therefore, discussion about the onset of turbulence in arteries has been largely speculative, and based on comparison with steady flow regimes. Such speculation has suggested that turbulence might occur during systole in the larger arteries, and the few attempts made to examine arterial flow in the living animal, by direct cinematography of injected dye, or by cineradiography, have tended to bear this out. Such flow visualization techniques are, however, difficult to interpret in unsteady flow, since they are observed for short periods at a fixed site, and almost any disturbance (e.g. vortices or secondary motions) may affect the dye stream and look like turbulence. The other classical method of detecting turbulence – demonstrating the breakdown of the Poiseuille pressure–flow relationship – is of course inapplicable in unsteady flow.

As we have seen, the conditions governing transition to turbulence in steady flow are firmly established; the controlling parameter is the Reynolds number, which has a critical value in the region of 2000 (influenced somewhat by the extent of upstream disturbances), and the turbulence has a frequency spectrum with well-defined properties. When the underlying flow is unsteady (as in an artery) the situation is more complicated for a number of reasons. First, it is not obvious how the appropriate *Re* should be chosen, since the mean value is unlikely to be relevant in such highly unsteady flows. In general, the peak Reynolds number has been thought more relevant, since the available evidence has suggested that flow disturbances in the aorta tend to occur at or near peak systolic velocity.

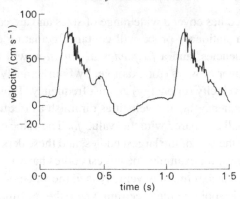

Fig. 12.49. Velocity waveforms from the upper descending aorta of the dog, showing turbulence during the deceleration of systolic flow. (After Seed and Wood (1971). Velocity patterns in the aorta. *Cardiovasc. Res.* **5**, 319.)

However, it is unlikely that peak *Re* will be the only governing parameter. As was previously mentioned, a finite time is needed for flow disturbances to grow to turbulence in supracritical Reynolds number flow, so it is likely that the time course of the aortic velocity wave will be important; at high heart rates, even if the Reynolds number is supracritical, there might be insufficient time available in one cardiac cycle for transition to take place. If we wish to express this in a non-dimensional way, so that it is applicable generally, the appropriate quantity is Womersley's parameter α (see p. 273).

At low values of α, the flow is expected to behave as if it were steady, with a critical Reynolds number of about 2000. As α increases, there is a decrease in the time available in one cycle, compared with the time required for amplification of a disturbance. Thus, the observed critical Reynolds number might be expected to rise with increasing heart rate.

Recently, it has become possible to study the problem directly in arteries, using hot-film anemometry (p. 306), and this technique shows that turbulence can be present in the thoracic aorta. An example is shown in **Fig. 12.49**, where high-frequency fluctuations can be seen superimposed on the underlying velocity waveform as the flow slows down in late systole. Analysis shows that such fluctuations cover a wide frequency range, up to 500 Hz, and down to less than 25 Hz, so that they merge with the high-frequency components of the main velocity waveform. Unlike the latter, however, the velocity components at each frequency show random variation in amplitude from beat to beat, and are thus genuinely turbulent. By varying the conditions of flow (velocity and heart rate) during the experiments with drugs and nervous stimuli, it has also been shown that transition from laminar to turbulent flow does depend on both

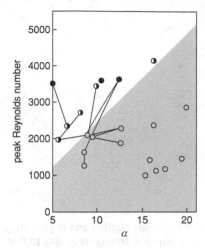

Fig. 12.50. The effect of peak Reynolds number and α on the stability of the flow in the descending aorta of anaesthetized dogs. The lines join observations made in the same animal under different conditions. Open circles: laminar flow; half-filled circles: transitional flow (disturbances only briefly seen at peak velocity); filled circles: turbulent flow. The stippled area indicates the conditions under which flow is stable and laminar. (After Nerem and Seed (1972). An *in vivo* study of aortic flow disturbances. *Cardiovasc. Res.* **6**, 1.)

Reynolds number and α (effectively heart rate) as predicted above; this is shown in **Fig. 12.50**. However, as **Fig. 12.50** shows, the aortic flow was not turbulent in the majority of experiments, and there is as yet no direct evidence about whether aortic flow is normally turbulent in the dog.

Nonetheless, there are good reasons for suspecting that it is; all the experiments shown in **Fig. 12.50** were performed in animals under anaesthesia, and this strongly influences both heart rate and peak velocity of flow in a manner which would displace the points in **Fig. 12.50** downwards and to the right, tending to stabilize the flow. If we examine the data which are available in the literature for normal, conscious animals (**Fig. 12.51**a), there are strong grounds for predicting turbulent flow. In man there are similar grounds, and also some direct evidence from studies with hot films mounted on catheters (**Fig. 12.51**b). Thus, it seems that, under normal resting conditions, flow in the proximal aorta is turbulent in humans and probably also in the dog. There is as yet no clear evidence about what precipitates the turbulence – upstream eddies or local disturbances in the wall region – and little information about how far downstream it may exist.

From a fluid dynamic point of view, there is no reason why turbulence should be undesirable; the increased energy dissipation which occurs in turbulent flow would increase the pressure drop along the large arteries; but as we have seen, this is very small compared with the overall pressure drop in the systemic circulation. There are, however, a number of interesting physiological implications. The first relates to sound

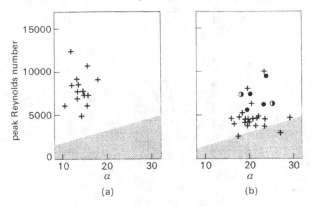

Fig. 12.51. (a) Values of peak Reynolds number and α prevailing in the ascending aorta of the normal, conscious dog. The stippled area corresponds to that in **Fig. 12.50**. Note that in the absence of anaesthetic effects, *Re* tends to be higher; all values lie in the region of the graph where turbulence would be predicted. (Data from Noble, Trenchard and Guz (1966). Left ventricular ejection in conscious dogs: measurement and significance of maximum acceleration of blood from the left ventricle. *Circulation Res.* **19**, 139). (b) Comparable data, from many sources, for normal, conscious humans. Again, turbulence would be predicted in almost all cases. The circles (see caption of **Fig. 12.50**) show results obtained with catheter-tip hot-film probes. They demonstrate the existence of turbulence.

production. It has often been assumed that turbulence is what causes murmurs, and therefore that it is usually absent in the normal circulation; and the view that laminar flow is silent has been widely accepted. Neither of these is necessarily true. There are a number of phenomena associated with laminar flow (for example, vortex shedding behind a body, p. 75) which may generate sound. There is also no a priori reason why turbulent pressure fluctuations should always be detectable at the chest wall; there might well be an audibility 'threshold'.

The presence of turbulence in a flow may have an important influence on the forces exerted by the blood on the artery wall; these may affect both the mechanical properties of the wall and mass transfer between it and the blood (see p. 333 *et seq.*). Turbulence affects these forces both through the high-frequency fluctuations in pressure and shear stress which accompany the random variations in velocity and because the mean velocity profile, and thus the mean shear stress at the wall, is altered. We first consider the mean flow. When the mean flow rate through a straight pipe is steady, the fully developed mean turbulent profile differs from that in Poiseuille flow, being flatter in the core, with a higher mean shear rate at the wall (see Chapter 5). However, in the entrance region of a pipe, the growing boundary layer is thicker in turbulent flow than in laminar flow. This is because the random eddying motion causes fluid elements to be moved about laterally, so that the lateral mixing is much more effective than in laminar flow under the sole action of viscosity. The boundary-layer thickness

δ can still be approximately represented by an equation of the form

$$\delta = k\sqrt{\frac{v^*x}{\bar{u}}}$$

(see Equation (5.3), p. 53), but now the kinematic viscosity v is replaced by an effective 'eddy viscosity' v^*, which incorporates the effect of turbulent mixing. One consequence of this is a shorter entry length in turbulent flow (p. 58); another is smaller mean wall shear rates in the entrance region.

If the underlying *mean*[15] flow is itself oscillatory, with angular frequency ω, then the thickness of the oscillatory boundary layer will be proportional to $\sqrt{v^*/\omega}$, which is again much greater than the corresponding laminar value. Thus, in flows where the laminar oscillatory boundary layer is relatively thin ($\alpha \gtrsim 4$), transition to turbulence will thicken it and the mean wall shear stress will be smaller. This conclusion applies throughout the aorta.

Another physiologically interesting implication of turbulence is in relation to the effect of vibration on the arterial wall. It is a very old clinical observation that the walls of a blood vessel often show localized dilatation just downstream of a partial obstruction to the flow, such as a narrowed (stenosed) aortic valve, for example, or an atheromatous constriction of a vessel. This post-stenotic dilatation occurs regardless of the origin of the obstruction, as long as the latter causes turbulence.

No really long-term studies have been made of the effect of vibration on arterial walls. In the short term (over periods of up to several months), however, it has been shown that post-stenotic dilatation occurs reproducibly downstream of localized stenoses (produced by partially tying off an artery) when a murmur is audible and palpable over the downstream segment. Such segments can be shown to be more distensible than normal, though there is no structural abnormality to be seen microscopically. Furthermore, excised segments of artery have been rendered more distensible by exposure to small-amplitude vibrations in the frequency range 30–400 Hz for periods of several hours. This range corresponds closely to the frequency spectrum of velocity fluctuations in turbulent blood flow. Thus, there is strong evidence to implicate turbulence as the cause of post-stenotic dilatation. It is interesting to speculate further that the dilatation of the aorta which takes place with age might be the result of prolonged exposure to low-amplitude vibration rather than of intrinsic degenerative changes. It has also been shown that experimentally induced turbulence may affect the permeability of the arterial wall to large molecules, and may damage the endothelium (see p. 336).

---

[15] The word 'mean' in the context of turbulent conditions has to be defined carefully. In order to define the mean velocity we suppose that a certain flow is observed a large number of times. The measured velocity at a fixed point and a fixed time after the start of the flow will be different on each occasion because of the random turbulent fluctuations. However, if we take its average over many observations we obtain a well-defined average, called the 'ensemble average', and it is this which we call the mean velocity at that point and time. The averaging process removes all the random fluctuations and leaves a mean which can still vary with time at any point, as, for example, when a wave passes.

Mixing and mass transport in arteries

The main function of the circulation of the blood is to transport materials and heat to and from the tissues of the body. The exchange of material between the blood and the tissues takes place largely in the capillaries, and is therefore discussed in the next chapter. However, the walls of arteries (and veins) are also permeable, and materials are transported across them too. This transport has a nutritive function for the artery wall, but it may also play a role in the genesis of arterial disease such as atherosclerosis. In addition, it has recently been recognized that the forces exerted on the artery wall by the flowing blood may themselves be important factors in determining the permeability of the wall to large molecules and, hence, may be a significant controlling influence in arterial disease. Some of this evidence is reviewed later in this section.

In order for the material which is transported to and from the tissues in the arteries and veins to be evenly distributed among the capillaries of any organ it must be well mixed in the large vessels serving that organ. The system would not function well if a stream of blood emerging from a particular organ remained as a coherent stream, unmixed with the rest of the blood, as it passed round the circulation. Furthermore, a widely used method of measuring cardiac output and regional blood flow, the 'indicator-dilution technique', is also based on the assumption of good mixing between marked and unmarked blood. The extent to which mixing occurs in the heart and large vessels must, therefore, be examined.

Mixing in the heart and large blood vessels The indicator-dilution method of measuring cardiac output was originally suggested in the nineteenth century by the physicist Adolf Fick (who also formulated the laws governing diffusion as described in Chapter 9). He pointed out that if both the difference in oxygen concentration between blood entering and leaving the lung and the total oxygen uptake of the animal per unit time are known, then the rate of blood flow through the lung (i.e. the cardiac output) is simply the ratio of the latter to the former. Much later, when the necessary techniques of sampling and measurement had been developed, this became a standard technique for measuring cardiac output in both animals and humans. So, too, did the derived methods of using boluses or continuous infusions of injected dye (dye dilution), isotopically labelled materials or thermal indicators like cooled blood or saline solution; some of these are used to measure cardiac output, others to measure the blood flow through individual organs or regions of the body. Samples of blood have to be taken at the entrance to and exit from the organ and the concentration of indicator measured in them. It will be obvious that the method can be reliable only if the measured concentrations are actually representative of all the blood passing through the organ. For example, if the arterial blood used in the Fick calculation has an oxygen content lower than the rest of the arterial blood, the cardiac output will be overestimated.

A number of methods have been used to examine the sites, and the completeness, of mixing in the circulation. These include direct flow-visualization in transparent vessels or models following injection of coloured materials, X-ray studies of the behaviour of injected radio-opaque material within the circulation and sampling from multiple sites after indicator injection. It is a unanimous conclusion from all these studies that an indicator passing through the heart is completely mixed with the blood by the time it emerges. On the other hand, mixing elsewhere in the circulation, or within a single chamber, may be incomplete.

Mixing in the heart appears to occur predominantly during passage of the blood through the ventricle. Samples from multiple sites in the right atrium, where streams of venous blood from different parts of the body merge, may show different concentrations of an indicator injected into a single vein, but in the right ventricle or pulmonary artery the concentration is always effectively uniform. The same conclusion holds for the left side of the heart.

The mixing which takes place in the heart is convective mixing; that is, it is a consequence of the flow patterns in the ventricle, which stir the blood and mix it up. The mixing is then much more rapid than it would be by diffusion alone, although diffusion still plays an essential part in the process, as it causes the final transfer of solute from a fluid element rich in solute to an element with little that have been brought together by the stirring. That stirring is much more effective than diffusion at mixing solute with solvent is a familiar experience, for example with a cup of tea.

It was shown in Chapter 9 that the time taken for diffusion to act over a distance Δx is proportional to t_D, where

$$t_D = \frac{(\Delta x)^2}{2D}$$

and D is the diffusivity of the material being diffused in its solvent (Equation (9.3), p. 132). Now the diffusivity of, say, oxygen in blood is about $2 \times 10^{-9}\,\mathrm{m^2\,s^{-1}}$, so the time taken for diffusion over a distance of 1 cm (rather smaller than the distance over which diffusion would have to act in the ventricle if there were no convection) is $2 \times 10^4\,\mathrm{s}$, which is far too long for complete mixing during a single heartbeat. The diffusivity of heat in blood is larger (about $10^{-7}\,\mathrm{m^2\,s^{-1}}$), but still the diffusion time is too long (500 s). Only if the distance over which diffusion has to act has been reduced by convection to 0.06 cm could the mixing time for oxygen be reduced to 1 s.

In the ventricle the stirring motions of the blood are of two types. First, and probably more important, are the eddying motions set up as the blood passes through the atrioventricular valves; this has been observed in animals and in models of the left ventricle (see Chapter 11), and is a powerful mixing influence, though it does not produce complete mixing within one cardiac cycle if material is injected locally within the ventricle. The main eddy is a ring vortex around the valve cusps; but the chordae tendinae and the irregularity of the inner walls of the ventricles must add smaller scale disturbances. These eddies will be present whether the flow is laminar or turbulent, but the second possible source of mixing is turbulence itself. We have no direct experimental evidence of turbulence in the heart, but since the Reynolds number for flow through the mitral valve may reach 8000, and since the flow into the ventricle takes the form of a jet with highly sheared edges, turbulence almost certainly does exist there for at least part of the cycle (see p. 66).

Flow in the ascending aorta is also very effective at mixing the blood, as judged from direct X-ray flow visualization and by injection of a thermal indicator into the ascending aorta. Again, the source of the stirring may lie in eddying motions either generated by the opening of the aortic valve or convected into the aorta from the ventricle (these are not necessarily turbulent), or it may arise from turbulence itself, which probably does occur in the aorta (see above).

In the descending and abdominal aorta, and the smaller arteries and veins, the evidence suggests that only partial mixing occurs. That is, samples from different sites in the vessel may have different concentrations of an indicator injected at a nearby upstream site; also, continuously injected streams of radio-opaque dye remain as coherent streams. The situation here is somewhat different from that in the ascending aorta, since Reynolds numbers are lower, and turbulence does not occur. Therefore, mixing can occur only through secondary motions caused by changes in geometry (bending and branching), and perhaps through the oscillatory motions, both lateral and longitudinal, which are associated with the passage of the pulse wave.

Mixing in small vessels does not appear to have been studied in the dog. In small arteries in the rabbit (0.1 cm) streaming is also seen, but in the aorta (0.3 cm) dye streams rapidly break up and cannot be followed. In humans the evidence is conflicting. It is a very old observation that dye injected into the arch of the aorta enters the two iliac arteries at different concentrations. On the other hand, injection directly into the brachial artery leads to fairly complete mixing as judged by simultaneous sampling from the radial and ulnar arteries downstream. In small veins in the rabbit, flow is largely steady, and dye streaming suggestive of laminar flow is seen, as long as precautions like remote injection are taken to prevent the injection itself from precipitating disturbances (the presence of a needle or catheter can be a very potent source of local eddying and mixing in small vessels). In larger veins (the inferior vena cava, diameter 0.34 cm) flow is unsteady but streaming is still identifiable. In humans, similar streaming can be seen radiologically in leg veins with diameters of the order of 0.5 cm, though the observation must be interpreted with caution because the appreciable density difference between blood and the radio-opaque injection medium may inhibit mixing.

There is a further interaction between flow and diffusion which influences the results of indicator-dilution studies. This is the longitudinal spreading or dispersion of a bolus of indicator as it is carried through the circulation. Some dispersion occurs because different elements of marked blood traverse pathways of different lengths

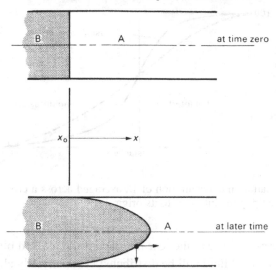

Fig. 12.52. The progressive dispersion of material B in a long tube initially full of fluid A, under Poiseuille flow conditions.

within the organ being studied. However, even along a single vessel (for example, an artery) considerable dispersion is observed, far more than can be explained by pure longitudinal diffusion, with the molecular diffusivity D as described by Equation (9.3), p. 132.

It is easiest to understand this dispersion in terms of fully developed Poiseuille flow in a long straight tube (**Fig. 12.52**). Suppose that at a certain time there is a sharp flat interface between unmarked blood A and marked blood B. As time proceeds, this interface will be pulled out into a paraboloid, the tip moving at twice the average velocity, the fluid in contact with the wall not moving at all because of the no-slip condition. If the two fluids were not able to diffuse into each other at all, there would be a progressive extension of the interface as flow proceeds.

It is possible to compute the concentration of B (averaged across a cross-section) along the tube from the origin x_0 to the tip of the marked blood at $x = L$; the result is shown in **Fig. 12.53** (broken line). However, if diffusion is possible, then a different curve is obtained because material B is capable of diffusing both axially and radially into fluid A. The axial diffusion only causes a minor blurring of the interface, but the radial diffusion is very important because it transports material B from the faster flowing core fluid to regions of lower velocity where it is transported downstream less rapidly. Consequently, the longitudinal extension of the interface (dispersion) is inhibited and the concentration of B varies with distance x in a manner more like the solid curve in **Fig. 12.53**. The concentration profile for a flat interface is also shown.

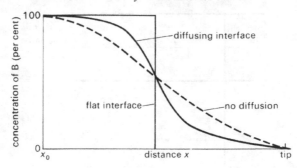

Fig. 12.53. The variation in concentration of B, averaged across a cross-section, with distance x back from the tip of the interface to its origin (x_0).

The same mechanisms operate at the rear interface if the marked blood has the form of a bolus, and the marked fluid will be distributed symmetrically about the midpoint of the bolus.

Figure 12.53 shows how the concentration of marked fluid varies with distance at a given time. In indicator-dilution studies, however, blood is sampled at a fixed position and the concentration measured as it varies with time. It is assumed that there has been sufficient time for complete mixing across the cross-section to occur, so it is important that the measuring points are sufficiently far downstream of the injection site. As the front of the bolus arrives, the measured concentration will rise, and it will fall again at the rear. However, the graph of concentration against time will not be symmetrical about the midpoint of the bolus, whether or not parts of the bolus have traversed pathways of different lengths. The reason is that, by the time the back of the bolus arrives, the mixing process has been under way for significantly longer than when the front arrived, so the tail will be more elongated than the front. It should be remembered that in physiological applications the tail of the bolus may also be strongly modified by recirculation.

In Poiseuille flow, the distribution of the bolus with distance along the pipe can be described as if there were pure diffusion, but with an effective diffusivity K which is very much larger than D.[16] If the flow is not Poiseuille flow, or if the bolus has not travelled far downstream from its origin, the above results are not quantitatively accurate. Nevertheless, the basic mechanism of dispersion *caused* by an axial velocity profile, and *inhibited* by radial diffusion, is still effective. Turbulence in the flow, or

---

[16] This was first demonstrated mathematically by G. I. Taylor in 1953. He showed that K varies with D, with average velocity $\bar{u}$, and with tube diameter d according to the equation

$$K = \frac{(\bar{u}d)^2}{192D} + D.$$

This equation has been so completely verified by experiment that it is now used as the basis of a popular method for measuring D.

secondary motion caused by curvature or branching of the tube, stirs up the fluid and makes the lateral mixing much more vigorous than pure diffusion. This means that the axial dispersion, inhibited by lateral mixing, is significantly less than it would be in Poiseuille flow at the same flow rate. Nevertheless, it is still considerable. It has been shown, again by Taylor, that the effective longitudinal diffusion K_T of any solute in fully developed turbulent flow is independent of the molecular diffusivity D. At a Reynolds number of 2300, K_T takes a value of about $3.5 \times 10^{-3} \, \text{m}^2 \, \text{s}^{-1}$. In order to compare this with the corresponding value in Poiseuille flow (K) we must consider a particular case, for example the dispersion of oxygen in blood, where $D = 2 \times 10^{-9} \, \text{m}^2 \, \text{s}^{-1}$. When $Re = 2300$, the value of K is about 250 and the turbulent value is smaller by a factor of about 10^{-5}; nevertheless, it is still 2 million times greater than the molecular value D. A bolus of dissolved oxygen in an artery at this value of the Reynolds number will spread out to occupy a length $\sqrt{2Kt}$ in time t (Equation (9.3), p. 132, with D replaced by K). If the flow is turbulent and $t = 1 \, \text{s}$, this is a length of 8 cm; if Poiseuille flow is present, it is a length of 20 m; if the flow is laminar, but disturbed by secondary motions, we do not know what the length is, although it will lie between the two and we expect it to be closer to the turbulent value because the mechanism of lateral mixing is similar. It is appropriate to mention here that it has been predicted that the effective diffusivity can be increased even above the Poiseuille flow value if longitudinal pulsations are imposed on the flow; this prediction has not yet been confirmed experimentally.

Mass transport across artery walls The artery wall is metabolically active, so that materials have to be transported to and from it. One route for this transport is through the vasa vasorum, which arise from both adventitial arteries and from the vessel lumen and appear to supply the adventitia and the media of large arteries (p. 248). However, the intima of systemic arteries is devoid of capillaries and appears to be supplied across the endothelium, which is the other major route for mass transport. These conclusions are based on studies of uptake of labelled materials from the blood into the artery wall in vivo. Transport by the vasa vasorum has not been extensively studied because of technical difficulty, and we consider it no further. However, the trans-endothelial route has recently been the subject of numerous studies, because among other reasons it is believed to be implicated in the development of atheroma.

Atheroma nowadays increasingly manifests itself as a disease process in middle rather than old age; however, changes in the large arteries are apparent at post mortem even in young subjects, and many people believe that these are the early stages of atheroma. The earliest change visible under the microscope is the accumulation of lipid in the tunica intima, in so-called *fatty streaks*, which are covered by an intact layer of endothelium and do not deform the vessel wall. At a subsequent stage, raised plaques become visible on the internal surface of the artery. The endothelium overlying these may still be intact, and their microscopic structure is quite variable, depending upon the relative proportions of free fat deposits, fat-laden ('foam') cells and fibrous (scar) tissue. Later, the endothelium may break down, bringing these elements into direct contact with the blood. The endothelium may then regrow, or the plaque may increase in size due to further fat deposition and scarring and the progressive accumulation of layers of *thrombus* (formed through aggregation of platelets, by the mechanism described in Chapter 10). This thrombus may remain in situ, gradually being destroyed by cells which infiltrate it, and covered by a layer of endothelium which grows over it; even if the vessel is completely blocked, it frequently recanalizes in this way. Alternatively, pieces of thrombus may be shed into the vessel to lodge in a smaller branch downstream. Such fragments are known as *emboli* and cause a variable amount of damage depending upon the downstream anatomy, particularly the extent of interconnections – *anastomoses* – between the small vessels which can bypass the block. Interruption of the blood supply to vital organs by atheromatous lesions or their complications is nowadays responsible for about half the deaths that occur in most developed countries. The chief sites are the brain ('stroke') and heart (myocardial infarction).

Atheroma occurs patchily, developing preferentially at certain sites in the cardiovascular system. For example, both early atheromatous plaques and fatty streaks are generally found in the major central arteries and not in the peripheral circulation.[17] Within an artery early atheroma is more often to be found near junctions than elsewhere; the abdominal aorta is more severely affected than the thoracic aorta; the posterior wall of the descending thoracic aorta almost always exhibits fatty streaking, whereas the anterior wall does not; the carotid sinus is usually severely affected. Such a preferential distribution of atheroma can arise in one or both of two ways: either the normal structure of artery walls varies, so that some areas are predisposed to develop atheroma, or they are subjected to different influences (for example, different shear stresses exerted by the blood) so that some physiological or biochemical process is altered. The first possibility is currently being investigated, but so far the second alternative has generally been preferred.

The fatty streaks observed in arteries represent an accumulation of lipid, much of which is in the form of cholesterol esters, in the tunica intima. Lipid deposition is also observed in early atheromatous plaques, together with other changes, including accumulation of protein and some thickening of the fibrous and muscular layers. Some of this lipid is thought to be transported to the site from the blood and 'deposited' there, a process which presumably represents a slight net imbalance between the fat transported into the wall and that which simultaneously is degraded or transported out; mass transport is of course a two-way process, and a state of equilibrium requires a balance between incoming and outgoing molecules of the material. But other lipid is contained within cells, and is believed to be manufactured there. In this case a net

---

[17] We concentrate here on early manifestations of atheroma, because advanced lesions may themselves have a significant influence on the blood flow past them, and hence on the development of atheroma at neighbouring sites.

accumulation of lipid would require that it be more difficult for the fat to escape from the wall and into the blood than for its precursors (whatever they are) to be transported in. As well as small molecules, these precursors probably include plasma lipoproteins, since cholesterol is not soluble in plasma, and must therefore be transported in conjunction with a protein that is. The observations briefly summarized above have led to the hypothesis that atheroma develops in regions of the artery wall where for some reason the rate of efflux of lipids or lipoproteins from the wall is slightly inhibited relative to other wall regions. Support for this hypothesis comes from experiments in which animals are fed high-cholesterol diets. In these animals the blood cholesterol level is much higher than normal, and fatty streaks develop preferentially in those regions of the wall which, from human post mortem studies, we would expect *not* to be affected. This suggests that influx of cholesterol from the blood to the wall, driven by a concentration difference assumed to be the reverse of normal, is enhanced in just those regions where, normally, there are no fatty streaks. (At a later stage in such studies lipid is found to accumulate in the *same* regions as in humans.) The above hypothesis has led to a number of in vitro and in vivo experiments designed to examine the permeability of arterial endothelium to proteins, or lipoproteins and other large molecules, and to investigate the effect of haemodynamic forces on the permeability.

Many such experiments have been performed in vivo with the dye Evans blue, which is bound to albumin in the plasma. Regions in which the rate of albumin transport into the wall is relatively large show up after post mortem dissection because they are stained blue. A blue-stained area is one which is relatively permeable to albumin; the flux of marked albumin is proportional to the concentration gradient of marked albumin and is independent of any other albumin which may be present. The unmarked albumin remains in dynamic balance as it was before injection of the marked albumin; presumably the rate at which unmarked albumin passes into and out of the wall is also increased in blue-stained areas, but this does not affect the equilibrium levels. No inference can be made from Evans blue experiments about the total amount of albumin in the wall. The permeability can be deduced because in sufficiently short-term studies it is proportional to the rate at which the concentration of unmarked albumin in the wall increases.

The obvious implication that a piece of artery with an enhanced permeability to albumin also has an enhanced permeability to other proteins and lipoproteins has been borne out by studies with radioactively marked fibrinogen and cholesterol, the latter presumably bound to a protein. All these studies have indicated that endothelial permeability is greater in those regions of the wall which are normally spared from early atheroma, but which in cholesterol-fed animals initially exhibit lesions. Thus, the hypothesis that early atheroma is normally associated with low permeability of the endothelium, inhibiting the efflux of cholesterol, is borne out. The question now arises as to what controls the permeability of the endothelium.

One of the most popular suggestions is that the differences in the forces exerted on the wall by the flowing blood are responsible for the differences in permeability. These forces consist of the pressure and the viscous shearing stress. At any point of the artery wall both these quantities have well-defined mean values, about which they oscillate every time the pulse wave passes, and they may also include random high-frequency components if there is turbulence present.

One of the first suggestions was that atheroma is associated with damage to the endothelium caused by high wall shear stresses, since it was shown that a shear stress greater than $40\,\mathrm{N\,m^{-2}}$ can damage endothelial cells. However, as we show below, the shear stress in blood vessels is normally considerably smaller than this, and it would require the local geometry of the vessel wall to be already distorted for such high stresses to arise. In any case, it has subsequently been demonstrated in vitro that the main effect of damaging the endothelium is to *increase* its permeability to large molecules, which according to the above ideas would not be expected to lead to atheroma.

A more recent hypothesis has been that atheroma develops most readily in regions where the mean wall shear stress is relatively low. This was based on both the observed distribution of fatty streaks in human arteries and also fluid dynamical predictions of regions where the mean shear stress is low. The observations on fatty streaks included those referred to above, showing that, except along the aorta, fatty streaking develops less readily at more peripheral sites. Also included were data on the local distribution of fatty streaking near junctions (the outer walls of the junction, **Fig. 12.44**, were seen to be affected, but the flow divider tended to be spared), in curved arteries (more fatty streaking on the inside of the bend) and in the carotid sinus. The regions which tend to be spared from fatty streaking are those which are normally found to be more permeable to large molecules.

The fluid dynamic predictions were based on the known distribution of shear stress in tubes when the flow is steady, on the assumption that the mean flow in pulsatile conditions is similar to steady flow with the same average velocity. The shear stress on the wall in Poiseuille flow, with an average velocity equal to $\bar{u}$, is

$$\tau_0 = 8\mu\bar{u}/d, \tag{12.29}$$

where μ is the viscosity of the blood and d the tube diameter (see p. 49). If Poiseuille flow were present in all blood vessels then the wall shear stress would be distributed along them as shown in **Table 12.3**, constructed from the data on dog's vessels given in **Table I**. This shows a marked increase (about tenfold) from the aorta to the arterioles, and supports the contention that the lack of fatty streaking in small arteries is related to a greater mean shear stress there. Of course, the mean flow within an artery is not Poiseuille flow, because there is an entrance length (so that the mean shear in the ascending aorta is much larger than that shown in **Table 12.3**, consistent with the finding of greater permeability there), and there are secondary motions

Table 12.3. *Wall shear stress (N m^{-2}) for various blood vessels, assuming Poiseuille flow*

Ascending aorta	0.43
Abdominal aorta	0.53
Femoral artery	0.80
Arteriole	4.8
Capillary	3.7
Venule	3.0
Inferior vena cava	0.64
Main pulmonary artery	0.28

caused by bends and branches. Greater fatty streaking on the inside of a bend is consistent with the lower shear stress to be expected there (p. 316). Again, the details of steady flow in junctions (p. 318) shows that the shear stress would indeed be high on the flow divider and low on the outer walls, consistent with the observations of fatty streaking.[18]

The 'low mean shear stress' hypothesis is thus a very attractive one. However, it does involve the assumptions that the unsteady components of the shear stress have no qualitative effect on the distribution of permeability, and that the mean shear stress in unsteady flow is distributed in a similar way as the shear stress in steady flow. The second assumption may be approximately justified (because the oscillatory part of the flow is fully developed except very near the entrance of a tube, and can be shown theoretically not to affect the mean) but the first is less easy to justify. This is because the unsteady shear stresses in arteries are expected to be extremely large compared with the mean shear stress. To illustrate this, we consider fully developed sinusoidal flow in a straight pipe, with mean velocity $\bar{u}$, velocity amplitude u_1 and a large value of the frequency parameter α. The mean wall shear stress is τ_0 (Equation (12.29)) and the amplitude of the shear-stress oscillations is $2\alpha\mu u_1/d$. We can apply these results to flow in the human aorta, where d is about 2.5 cm, $\bar{u}$ is about 0.2 m s^{-1} and u_1 about 1 m s^{-1}. Using a value of α of 20, we see that the peak oscillatory shear stress is approximately 6.4 N m^{-2}, which is some 25 times the mean. Furthermore, when the composite velocity waveform contains higher frequency oscillations (as occurs in vivo) the shear-stress amplitude becomes larger still (up to 50 times the mean). It seems unlikely that the permeability of endothelial cells should depend on the low mean shear stress, when the level of shear stress actually experienced is so much greater.

---

[18] This is not as clear cut as it appears, however, because model experiments on flow through a cast of the aorta have shown that the distribution of shear stress near junctions (for example, of the coeliac and mesenteric arteries) is very sensitively dependent on the precise local geometry. If the far lip of a junction (the flow divider in **Fig. 12.47**) protrudes by 0.5 mm into the lumen of the aorta, the shear stress just beyond it is reduced, not enhanced, presumably because of flow separation.

Thus, the 'low mean shear stress' hypothesis cannot be established until the mechanism by which shear stress controls endothelial permeability is understood, and as yet it is not. A number of experiments on perfused excised segments of artery have been carried out, and the uptake of radioactively marked substances measured quantitatively in various flow situations. Several possible mechanisms have been eliminated in this way, but the real one is yet to be discovered.

One suggestion was that the permeability is dictated by the ease with which large molecules travel through the channels between endothelial cells, by diffusion or filtration. However, there is no direct evidence of large molecules (molecular weight greater than 70 000) taking this path.

Another suggestion was that the rate of transport into the wall is controlled by a thin *mass-transfer boundary layer* (Chapter 9, p. 142) across which the concentration of solute varies from its value in the core of the artery to its value at the wall. This mechanism has the virtues that it would be shear dependent and that it can be clearly demonstrated that it is the *mean* shear which is important, not the fluctuations. However, the quantitative in vitro studies have shown that the resistance to mass transport into the wall is about 1000 times greater than could possibly result from a boundary layer in the blood. Furthermore, the longitudinal variation of uptake along an excised segment of artery, predicted because the mass-transfer boundary layer would grow with distance along it (Equation (9.19) or (9.22)), was entirely absent.

Thus, the resistance to transport of large molecules across the endothelium must lie in the cells themselves. Possible mechanisms of transport include so-called 'pinocytosis', in which small vesicles at the cell surface entrap small amounts of plasma, carry them bodily across the cell and release them into the intima. Such a process might be a passive, diffusional one, or it might involve active mechanisms. Other types of active transport might also be involved. No mechanism of this type has yet been proposed which would show how shear stress could affect permeability, although the observation has been made in vitro that an effectively steady shear stress does enhance the permeability to macromolecules. There is also some evidence that the presence of turbulence not only affects the elastic properties of the artery wall (p. 327) but also makes the endothelium more permeable. This could be part of the explanation of greater permeability variations observed in the neighbourhood of junctions where flow separation and, therefore, turbulence might occur.

Appendix: Impedance

Because of their value as an aid to the interpretation of physiological measurements, the mathematical forms of the relationships referred to in the text are widely used by physiologists. We therefore quote them here, although they require the use of complex numbers, which we have elsewhere tried to avoid. In these terms, the equation for the pressure in a wave travelling in the positive x-direction (cf. Equation (12.20)) is

$$p = \Re\left\{p_+ e^{i\omega(t-x/c_1)}\right\}. \tag{12.30}$$

Here, $i = \sqrt{-1}$, e is the familiar exponential function, which has the property that

$$e^{iz} = \cos z + i \sin z,$$

and the symbol $\Re$ means that the real part of the complex quantity on the right-hand side has to be taken. Thus, if p_+ is itself a complex number of the form

$$p_+ = p_a + i p_b,$$

then Equation (12.30) means

$$p = p_a \cos[\omega(t - x/c_1)] - p_b \sin[\omega(t - x/c_1)].$$

Similar expressions can be written for the flow rate and for waves travelling in the opposite direction. It is conventional to omit the symbol $\Re$, so that whenever a complex number is supposed to represent a physical quantity it is assumed that its real part is being used.

With complex numbers it is possible to write down the ratio of the pressure and flow rate at a point, for example the point A in a tube AB (**Fig. 12.28**), by means of a complex quantity called the *effective impedance*, $Z_{1\ eff}$:

$$Q_A = Z_{1\ eff}^{-1} p_A. \tag{12.31}$$

Calculations for the double junction in **Fig. 12.28** shows that $Z_{1\ eff}$ can be expressed in terms of the characteristic impedance of the tube AB, Z_1 (which equals $\rho c_1 / A_1$, where A_1 and c_1 are the area and wave speed of that tube) and of the effective impedance of tubes 3 and 4 at B. The result is

$$Z_1^{-1} = Z_1^{-1} \frac{(Z_{3\ eff}^{-1} + Z_{4\ eff}^{-1}) + i Z_1^{-1} \tan(\omega l/c_1)}{Z_1^{-1} + i(Z_{3\ eff}^{-1} + Z_{4\ eff}^{-1}) \tan(\omega l/c_1)}. \tag{12.32}$$

By repeated use of this equation, it is possible, as described above, to relate the excess pressure and flow rate in the parent tube of a complicated network, given their values at the distal end. It must be remembered that Equation (12.31) refers just to waves of a single frequency ω. In order to relate the pressure and flow rate of a composite wave with many frequency components, the calculation must be repeated for each one. The effective impedance at A, $Z_{1\ eff}$, is often called the *input impedance* of the system distal to A.

We may note the dependence of Equation (12.32) on the ratio between the length l of AB and the quantity c_1/ω (equal to $1/2\pi$ times the wavelength of the wave in AB). If $\omega l/c_1$ is very small, then so is $\tan(\omega l/c_1)$, and

$$Z_{1\ eff}^{-1} \approx Z_{3\ eff}^{-1} + Z_{4\ eff}^{-1}.$$

That is, if l is much smaller than the wavelength, the tube AB forms part of the junction and its properties are irrelevant. A similar result holds if $\omega l/c_1$ is equal to a whole number of multiples of π (i.e. if l is equal to a whole number of half-wavelengths), as predicted above. However, if $\omega l/c_1$ is equal to an odd multiple of $\pi/2$ (i.e. l equal to an odd number of quarter-wavelengths), then $\tan(\omega l/c_1)$ is infinite and

$$Z_{1\,\text{eff}}^{-1} \simeq \frac{Z_1^{-2}}{Z_{3\,\text{eff}}^{-1} + Z_{4\,\text{eff}}^{-1}}. \tag{12.33}$$

In this case, if Z_1^{-1} is less than $Z_{3\,\text{eff}}^{-1} + Z_{4\,\text{eff}}^{-1}$, then $Z_{1\,\text{eff}}^{-1}$ is smaller still, and vice versa. This agrees with the earlier conclusion (p. 289) that the wave eventually transmitted will be large if the reflection at B is positive ($Z_1^{-1} > Z_{3\,\text{eff}}^{-1} + Z_{4\,\text{eff}}^{-1}$) and small if it is negative.

The effective, or input, impedance of a system is a complex number, so it has real and imaginary parts Z_r and Z_i:

$$Z_{\text{eff}} = Z_r + iZ_i.$$

In many analyses of the cardiovascular system this is rewritten in the following, rather different, form:

$$Z_{\text{eff}} = M e^{i\theta}. \tag{12.34}$$

Here, M is the magnitude, or modulus, of the impedance, given by

$$M = \sqrt{Z_r^2 + Z_i^2},$$

and θ is its phase, given by

$$\tan\theta = Z_i/Z_r.$$

The modulus M represents the magnitude of the opposition to flow offered by the system distal to the point of measurement and θ represents the phase lead of pressure over flow rate. The dependence of these quantities on more distal conditions can be seen, by using Equation (12.32) in conjunction with Equation (12.33), to be very complicated.

The idea of impedance is very useful both for predicting the relationship between the pressure and flow-rate waveforms in a system and for interpreting data. However, it is also a concept which has caused considerable confusion among physiologists because of the temptation to use the word without being very clear about its meaning. Part of the trouble lies in the large number of slightly different quantities all called impedance. We list many of them here, not because we recommend their use, but rather in the hope that familiarity with them all will help to avoid confusion. We have already met the *characteristic impedance* Z of a tube: the ratio of pressure to flow rate in that tube when a sinusoidal wave is travelling along it at constant speed in one direction. For elastic tubes containing non-viscous fluid, which we have taken as models of arteries so far, this is a real quantity, indicating that the pressure and flow-rate waveforms are in phase. A related quantity is the *longitudinal impedance* Z_L, which is the ratio of pressure *gradient* to flow rate when the same type of wave is propagating (it therefore has different dimensions from other impedances: $[ML^{-5}T^{-1}]$ not $[ML^{-4}T^{-1}]$). In the tubes considered so far this is a purely imaginary quantity ($Z_L = -i\omega Z/c$), since the pressure gradient and flowrate are 90° out of phase. In a straight rigid tube in which the frequency is so low that Poiseuille flow is always present, the pressure gradient and flowrate are in phase, so Z_L is real and Z is imaginary. Another quantity we have already met is the *effective impedance* Z_{eff} of a system, equal to the ratio of pressure and flow at the entrance to the system. This is also called *input impedance*, especially when it refers to the whole vascular bed, and sometimes even *terminal impedance*, when it refers to that part of the vascular bed distal to the part of current interest. Finally, there is the term *source impedance*, which refers to a flow source or pump; for example the heart. If a pump is short-circuited, so that the flow from its output is pumped directly into its input, the system to which it is connected has negligible impedance. Nevertheless, any real pump will produce only a finite flow rate Q and there will be a non-zero pressure difference Δp across it. Their ratio, $\Delta p/Q$, is therefore not zero, and the circuit as a whole has non-zero impedance, which is all associated with conditions in the pump itself. This 'source impedance' is present whatever system the pump is pumping into.

The source impedance of the left ventricle, at any volume, is thus equal to the difference between the maximum pressure which the muscles could exert at that volume (under isometric contraction) and the actual ventricular pressure, divided by the volume flow rate out of the heart. In early systole, this quantity is about six times the input impedance of the aorta, indicating that the normal myocardium contracts as rapidly as allowed by its own rate-limiting mechanisms (represented by the force–velocity relationship, see Chapter 11) rather than by the load it pumps against. Later in systole, however, the source impedance falls and the aortic load increases in importance.

Further reading

Bergel, D. H. (ed.) (1972). *Cardiovascular Fluid Dynamics*. Chapter 2: Pressure measurement in experimental physiology, by I. T. Gabe; Chapter 3: Measurement of pulsatile flow and flow velocity, by C. J. Mills; Chapter 9: Pressure and flow in large arteries, by D. L. Schultz; Chapter 10: Vascular input impedance, by U. Gessner; Chapter 11: The rheology of large blood vessels, by D. J. Patel and R. N. Vaishnav; Chapter 12: The influence of vascular muscle on the viscoelastic properties of blood vessels, by B. S. Gow; Chapter 13: Post-stenotic dilatation in arteries, by M. R. Roach; Chapter 19; Synthesis of a complete circulation, by R. Skalak. Academic Press, London (2 vols).

Bergel, D. H. and Schultz, D. L. (1971). Arterial elasticity and fluid dynamics. *Prog. Biophys. Molec. Biol.* **22** (eds J. A. V. Butler and D. Noble).

Charm, S. E. and Kurland, G. S. (1974). *Blood Flow and Microcirculation.* John Wiley, New York.

CIBA Foundation Symposium (1969). *Circulatory and Respiratory Mass Transport.* Articles on: 'The optimum elastic properties of arteries', and 'Velocity distribution and transition in the arterial system' J. & A. Churchill Ltd, London.

CIBA Foundation Symposium (1973). New Series, No. 12. *Atherogenesis: Initiating Factors.* Associated Scientific Publishers, Amsterdam.

Fung, Y. C., Perrone, N. and Anliker, M. (eds) (1972). *Biomechanics: its Foundations and Objectives.* Chapter 5: The properties of blood vessels, by D. H. Bergel; Chapter 15: Toward a non-traumatic study of the circulatory system, by M. Anliker; Chapter 16: Flow and pressure in the arteries, by T. Kenner. Prentice-Hall, New Jersey.

Guyton, A. C. and Jones, C. E. (eds) (1974). *Cardiovascular Physiology.* Chapter 1: The mechanics of the circulation, by C. G. Caro, T. J. Pedley and W. A. Seed. MTP International Review of Science. Butterworth & Co. Ltd, London, and University Park Press, Baltimore.

Lighthill, M. J. (1975). *Mathematical Biofluiddynamics.* S.I.A.M., Philadelphia.

McDonald, D. A. (1968). Haemodynamics. *Annu. Rev. Physiol.* **30**, 525–56.

McDonald, D. A. (1974). *Blood Flow in Arteries* (2nd edn). Edward Arnold, London.

Patel, D. J., Vaishnav, R. N., Gow, B. S. and Kot, P. A. (1974). Haemodynamics. *Annu. Rev. Physiol.* **36**, 125–54.

Taylor, M. G. (1973). Haemodynamics. *Annu. Rev. Physiol.* **35**, 87–116.

Whitmore, R. L. (1968). *Rheology of the Circulation.* Pergamon Press, Oxford.

13

The systemic microcirculation

We saw in the last chapter that in the large arteries blood may be treated as a homogeneous fluid and its particulate structure ignored. Furthermore, fluid inertia is a dominant feature of the flow in the larger vessels since the Reynolds numbers are large. The fluid mechanical reasons for treating the circulation in two separate parts, with a division at vessels of 100 μm diameter, were also given in that chapter. In the microcirculation, which comprises the smallest arteries and veins and the capillaries, conditions are very different from those in large arteries and it is appropriate to consider the flow properties within them separately.

First, it is no longer possible to think of the blood as a homogeneous fluid; it is essential to treat it as a suspension of red cells and other formed elements in plasma. As will be seen later in the chapter, this comes about because even the largest vessels of the microcirculation are only approximately 15 red cells in diameter. Second, in all vessels, viscous rather than inertial effects dominate and the Reynolds numbers are very low; typical Reynolds numbers in 100 μm arteries are about 0.5 and in a 10 μm capillary they fall to less than 0.005 (see **Table I**).

In larger arteries, the Womersley parameter α (p. 60) is always considerably greater than unity. In the microcirculation, however, α is very small; in the dog (assuming a heart rate of 2 Hz) it is approximately 0.08 in 100 μm vessels and falls to approximately 0.005 in capillaries. This means that everywhere in these small vessels the flow is in phase with the local pressure gradient and conditions are quasi-steady. The pressure and flow may, as we shall see later (p. 371), still be pulsatile, but the flow pattern at all points and all times will be determined solely by the balance between pressure and viscous forces. Inertial forces, which are those associated with both local and convective accelerations, are negligible (see Equation (4.5), p. 39). Another important consequence of the small diameter of microcirculatory vessels is that, despite the low blood velocity and Reynolds numbers, the wall shear rates tend to be considerably higher (of the order of $1000 \, \mathrm{s}^{-1}$) than in large vessels. This has a significant effect on blood flow, as discussed later.

In addition to the fluid mechanical arguments there are physiological reasons for distinguishing two regions of the circulation. The large vessels provide a supply and drainage system for the microcirculation, which is where most mass transport between blood and tissue occurs.

Little work on the mechanical properties of the microcirculation has been done in the dog; our understanding is based largely on studies of smaller laboratory animals, such as the rat, rabbit and frog. The fact that studies must almost always be performed by microscopy has limited them mainly to flat, almost two-dimensional vascular beds, such as those in the *mesentery* and *omentum* (thin transparent tissues supplying the gut), the *cremaster* muscle (a superficial sheet of muscle in the scrotal wall), the *tenuissimus* muscle (in the leg) and the hamster cheek pouch. The bed of the bat wing has also been used and is of particular interest because, unlike the other preparations, it can be studied without anaesthesia. Three-dimensional beds such as those in skeletal muscle have not yet been studied nearly as extensively, even though considerable effort has been made to develop the necessary experimental techniques.

Direct measurement of microcirculatory flow conditions is extremely difficult for reasons other than simply the small scale at which studies are performed. The vessels are highly reactive to any mechanical stimulus; thus, the surgical procedure, the introduction of pressure-measuring needles and the effects of maintaining tissues on a microscope stage are all potential causes of artefact. Such difficulties have led many workers to study models of the microcirculation rather than handle physiological preparations.

Although our knowledge of the microcirculation is built up from such fragmented sources, the mechanical properties and dimensions of vessels, in particular beds of the various mammalian species, do not show marked differences. It is reasonable at present to base a description of a typical microcirculation on the information available from them.

The organization of a microvascular bed

A schematic representation of a microcirculatory bed, based on observations in the mesentery, is shown in **Fig. 13.1**. Comparison of the numerical data for different beds is extremely difficult because different investigators used different criteria for subdividing the microvasculature. Nevertheless, recent data on various commonly studied microcirculatory beds are presented in **Table 13.1** and **Table 12.2**.

The arteriolar system As can be seen from **Fig. 13.1**, the arterioles divide first into *metarterioles* before the capillary network itself is reached. The separate existence of metarterioles is not acknowledged by some authors, who characterize them simply as the smallest terminal arterioles. This lack of distinction is easy to appreciate, as it is often difficult, at a bifurcation where two daughter vessels are similar in size, to distinguish between the vessels. Furthermore, there is a continuous distribution of vessel sizes, and thus the distinction between various categories of vessel is rather arbitrary. In the model shown, the metarterioles are lateral branches off a feeding arteriole, and

Table 13.1. *Summary of available data on vessel dimensions and flow rates in the microcirculatory vessels of various species. Figures for the bat wing relate to the whole bed, data for the mesentery include the omentum in some cases*

Property	Arteriole Large	Arteriole Small	Capillary Artery end	Capillary Vein end	Venule Small	Venule Large
Diameter (μm)						
Bat wing	19	7	3.7	7.3	21	37
Cat mesentery	70	20	10		20	70
Rat cremaster muscle	80	14	5.5	6.1	24	74
Cat tenuissimus muscle	22	10	4.7	5.9	10	40
Length (μm)						
Bat wing	3500	950	450		1000	
Cat mesentery	380	200	130		130	350
Rat cremaster muscle			615		300	
Cat tenuissimus muscle	300	100	1000		100	
Number						
Bat wing	12	120	1700		350	25
Cat mesentery	2	9	20		10	2
Rat cremaster muscle						
Cat tenuissimus muscle						
Velocity (mm s^{-1})						
Cat mesentery	1.0–31.7		0–1.7		0.5–11.1	
Rat cremaster muscle	0.8–12.9		0.2–1.2		0.4–6.6	
Cat tenuissimus muscle			0–1.5			

Data derived mainly from the following sources:
Schlechta and Fulton (1963). Blood flow-rates in small vessels of the hamster cheek pouch. *Proc. Soc. Exp. Biol. Med.* **112**, 1076.
Smaje, Zweifach and Intaglietta (1970). Micropressures and capillary filtration coefficients in single vessels of the cremaster muscle of the rat. *Microvasc. Res.* **2**, 96–110.
Gaehtgens, Meiselman and Wayland (1970). Erythrocyte flow velocities in mesenteric microvessels of the cat. *Microvasc. Res.* **2**, 151–62.
Intaglietta, Tompkins and Richardson (1970). Velocity measurements in the microvasculature of the cat omentum by on-line methods. *Microvasc. Res.* **2**, 462.
Erikson and Myrhage (1970). Microvascular dimensions and blood flow in skeletal muscle. *Acta Physiol. Scand.* **86**, 211–22.
Wiedeman (1963). Dimensions of blood vessels from distributing artery to collecting vein. *Circulation Res.* **12**, 375–8.

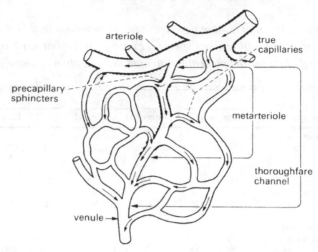

Fig. 13.1. Diagram of the microcirculatory unit. In this illustration the metarteriole does not terminate by dividing into capillaries but continues until it joins the venule as a thoroughfare channel. (From Friedman (1971). Microcirculation. In *Physiology* (ed. Selkurt), Chapter 12, p. 260. Churchill Livingstone, Edinburgh.)

a single arteriole will thus be capable of providing a number of metarterioles along its length; it also thus generates a number of capillary systems which can operate in parallel along its path. The angle at which metarterioles branch off the parent arteriole depends strongly upon the particular vascular bed. In many mesenteric preparations they have been found to branch off at 30–60° from the axis of the parent arteriole, while in skeletal muscle they frequently arise at right angles.

The diameter of the largest arteriole is in the region of 50–100 µm; by progressive bifurcation it is decreased until at the level of the origin of the metarterioles it is about 30 µm. Metarterioles at their origin are in the region of 20 µm in diameter and occur at something like 600 µm intervals along the arteriole. It has been calculated that in the tenuissimus muscle of the cat the typical distance from larger arteriole to the capillary network is 1–2 mm. As the diameter decreases, so does vessel wall thickness. The largest vessels, with diameters in the region of 100 µm, have a wall thickness of about 20 µm, whereas the smallest arterioles, 20 µm in diameter, have a wall thickness of approximately 6 µm. Thus, the ratio of vessel wall thickness to diameter increases from around 0.2 to 0.3 as the calibre decreases over this range.

The capillary system The branching pattern observed in the capillary system varies considerably not only from tissue to tissue, but also within any one bed; again, **Fig. 13.1** gives some idea of the organization. The smooth muscle in the wall of the metarteriole (see p. 353) becomes progressively sparse along its length until the vessel branches into a number of true capillaries, some of which have precapillary *sphincters* at their origins. At this level the metarteriole may either disappear or continue as a somewhat enlarged capillary-like structure without smooth muscle – the *thoroughfare* or *preferential channel.*

The precapillary sphincter consists of one or two smooth muscle cells, wrapped around the arteriolar–capillary junction. The internal diameter of the sphincter when relaxed is the same as that of the capillary itself. Such sphincters are not observed in all microvascular beds, nor are they present at the entrance to every capillary in beds where they have been observed. They have been shown to exist in the frog tongue and mesentery, but have not been demonstrated in skeletal muscle. It has been suggested that, in some beds, sphincter-like activity may reside in the perivascular muscle of the smallest arterioles (see p. 353).

The question of whether or not thoroughfare channels exist has generated much interest because, functioning in association with the precapillary sphincters, they are capable of markedly controlling the blood flow and transcapillary exchange within a unit. Thus, if all the precapillary sphincters were to constrict, perfusion should be shunted predominantly to the low-resistance pathway of the thoroughfare channel and the unit effectively bypassed. This proposed mechanism of bypassing a unit is somewhat different from the older one that arterial-to-venous anastomoses are responsible for perfusion shunts; such anastomoses would have a wall structure similar to small arterioles and venules, containing some smooth muscle. However, such shunts have been identified only in a few microcirculations, namely the cutaneous circulation and the submucosa of the stomach. In general there are numerous artery-to-artery and vein-to-vein anastomoses which permit effective shunting between units, but give no degree of control within a unit. Examples of these can be seen in **Fig. 13.20**.

In many beds the capillary network is tortuous, with multiple cross-connections; the arrangement can be seen in **Fig. 13.20**. However, in skeletal muscle the arrangement is far more uniform. **Figure 13.2** shows the organization of the capillary system in the rat cremaster muscle. The small arteries and veins run parallel to each other, the arterioles branch off at right angles and the capillary networks are arranged so that the capillaries run parallel to the muscle fibres. **Figure 13.2** illustrates two layers of capillaries running parallel to the layers of muscle fibres which lie approximately at right angles to each other. That the capillaries in skeletal muscle run parallel to the muscle fibres can be seen very clearly in **Fig. 13.3**, in which the majority of fibres are at right angles to the plane of the paper and the capillaries can be seen to be parallel to the fibres.

The organization of the capillaries in the rat cremaster muscle is schematically illustrated in **Fig. 13.4**. It will be specially noted that the capillaries are greater in diameter at the venular end than at the arterial end, but always less than the nominal red cell diameter. It is now widely accepted that capillaries are smaller in diameter than the red cells which have to pass through them and also that the arterial end of the capillary is the narrowest part of the circulation, the capillary increasing in diameter

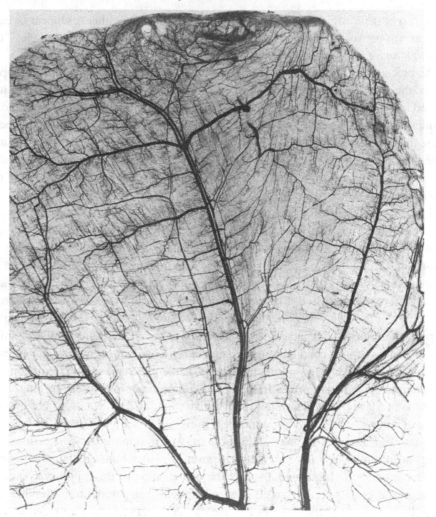

Fig. 13.2. Micrograph of rat cremaster muscle after intravenous injection of carbon to visualize the vascular pattern. The main arteries and veins can be seen running parallel to one another (veins larger). Muscle fibres are arranged in two layers approximately at right angles to one another, with capillaries running parallel to the fibres. (From Smaje, Zweifach and Intaglietta (1970). Micropressures and capillary filtration coefficients in single vessels of the cremaster muscle of the rat. *Microvasc. Res.* **2**, 99.)

along its length. Thus, the cat tenuissimus muscle capillary increases from approximately 4.7 μm to 5.9 μm along its length while the cat red cell is 6.0 μm in diameter. The bat wing capillary has an average internal diameter of 3.7 μm, whilst the bat red cell diameter is 6.4 μm. The capillary length in rat cremaster muscle is approximately 600 μm, whilst it is approximately 1000 μm in the cat tenuissimus muscle. In both

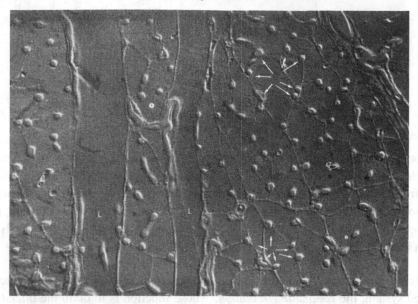

Fig. 13.3. Interference micrograph of a cross-section through skeletal muscle. The dense capillary network surrounding the muscle fibres can be seen (arrowed). Muscles are visible in longitudinal section (*L*) at the left of the photograph, as are parts of associated capillaries. (Reproduced by courtesy of Dr G. Gabella, University College, London.)

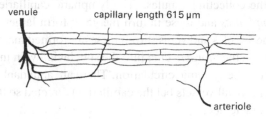

Fig. 13.4. Idealized diagram of the capillary network in the rat cremaster muscle. Capillary diameter at arteriolar end: 5.5 µm; capillary diameter at venular end: 6.1 µm; distance between capillaries: 34 µm; distance between cross-connections between capillaries: 210 µm; capillary density: 1300 mm^{-2}; capillary surface area: 244 cm^2 per cm^3 muscle; average rat red cell diameter: 7.5 µm. (From Smaje, Zweifach and Intaglietta (1970). Micropressures and capillary filtration coefficients in single vessels of the cremaster muscle of the rat. *Microvasc. Res.* **2**, 105.)

muscles there are numerous intercapillary cross-connections occurring about every 200 µm.

The venular system At their downstream end the capillaries merge, usually in pairs to form first the *postcapillary* and then the *collecting venules*. These then progressively merge to form the larger venules and ultimately the veins. This pattern of

Table 13.2. *Approximate composition of blood vessels*

	Vessel type %		
	Arteriole	Capillary	Venule
Endothelium	10	100	20
Elastic tissue	10		
Muscle	60		20
Connective tissue	20		60

dichotomous convergence, whilst very common, is not universal; in the tenuissimus muscle for instance there tends to be a single large venule with smaller ones joining it at frequent intervals. **Figure 13.1** implies that all capillaries originating from one arteriole converge into a single venule; however, this is not necessarily the case, particularly in three-dimensional beds. The dimensions of venules in various beds are given in **Table 13.1**.

The lymphatic system In addition to the blood circulation, there exists a separate but related system, the *lymphatic circulation*, whose function is to drain the *interstitial space* (see p. 363). As explained later (p. 400), fluid passes out of the capillaries at the arterial end of the capillary system and is then partially reabsorbed at the venular end. The excess fluid that is not reabsorbed passes into the *lymphatic capillaries*. These are thin-walled, blind-ended sacs distributed throughout the interstitial space, especially around the collecting venules. The lymphatic capillaries join together to form *collecting lymphatics* and these in turn merge to form larger lymphatic vessels which run beside the venules and veins. From these are formed main lymphatic trunks which largely drain via the *thoracic duct* into the subclavian vein in the neck, thereby returning the lymph to the systemic circulation. There are abundant valves throughout the lymphatic system (in all vessels but the capillaries) directed so that flow can occur only towards the thoracic duct.

Only scant information is available on the distribution of the lymphatic capillaries and collecting vessels, since they are difficult to observe in vivo. They have been visualized principally by injecting dyes and particulate materials like carbon suspensions into the interstitial space, or into the lymphatic capillaries themselves through micropipettes. The lymphatic capillaries in cat and rabbit mesentery have been shown to be flattened sacs some 40–60 μm wide by 5–6 μm deep.

Total lymph flow rates depend strongly upon the level of exercise, food and fluid intakes and many other factors, but in the normal adult man are in the region of 150 ml h^{-1}.

The structure of the vessels of the microcirculation

The overall composition of the three main types of microcirculatory vessel is given in **Table 13.2**. The figures in **Table 13.2** are approximate, since the proportion of each constituent varies with vessel size; the table gives no indication of the organization of the various components. To obtain this we must consider each vessel type separately. All vessel walls are composite in structure and cannot be considered to be isotropic. In all vessels, the innermost cellular lining is a layer of endothelial cells, and smooth muscle cells can be identified in all vessels except capillaries. It has recently been demonstrated by use of special staining techniques that there is a lining material present on the luminal surface of the endothelium. It appears to be a very thin layer (approximately 5–10 nm thick) of a mucopolysaccharide complex, and it is this substance rather than the endothelium which forms the real luminal surface in all blood vessels throughout the circulation.

The arterioles The arterioles can be regarded as very small arteries and have a wall structure similar to that of the artery described in Chapter 12, though they have no vasa vasorum. The layered wall structure is similar in all arterioles, but there are progressive changes in the relative proportions of the components as vessel size decreases. **Figure 13.5** shows a transmission electron micrograph of a transverse section of an arteriole, in which the three principal concentric layers of tissue are seen, just as in an artery.

The intima contains endothelial cells, a basement membrane and some connective tissue. The endothelial cells are arranged in a single layer and are close together (**Fig. 13.6**), separated by narrow *intercell clefts* running from the luminal surface through towards the basement membrane. Near the luminal end of the intercell cleft, adjoining cells form a close contact or *junction*. In arterioles, this junction is said to be 'tight', as explained later (p. 362). The endothelial cell layer appears to be attached directly to the *basement membrane* under normal physiological conditions, but it can become separated by mechanical stress or under circumstances of excessive protein leakage from the plasma. The basement membrane appears to be a mesh of collagen, though its precise form and mechanical properties have yet to be established. If the luminal surface is examined in a preparation fixed under unstretched conditions, it has a highly convoluted form; but when fixed at physiological pressures the surface appears smooth.

The endothelial cells of arterioles of all sizes appear similar and have a thickness of about 0.1–0.3 µm except in the region of a nucleus, where they bulge to 2–4 µm. All the cells contain *vesicles*, the structural characteristics of which are discussed later (p. 363); their function is considered on p. 421. Endothelial cells also contain *microfilaments* within their cytoplasm. These have been shown to contain contractile proteins, similar to those in muscle cells, and there is a possibility that they may have a contractile function.

The media of the arterioles shows considerable variation according to vessel size. The smooth muscle cells in this part of the wall are arranged circumferentially or spirally; as in arteries, individual muscle cells are spindle shaped, being 30–40 µm long by 5 µm wide. In the larger arterioles (50–100 µm) two or three layers of smooth

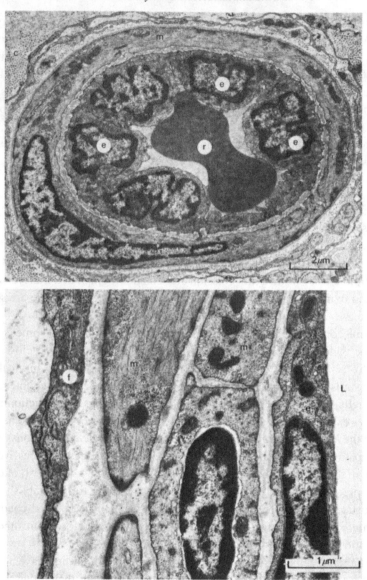

Fig. 13.5. The upper panel shows an electron micrograph of an arteriole in the rat oesophagus. A single smooth muscle cell fully encircles the endothelium. (c: collagen fibrils; e: endothelial cell; m: smooth muscle cell; r: red blood cell.) The lower panel shows an electron micrograph of a section of an arteriole in the rat heart. Two layers of smooth muscle cells can be seen. (e: endothelial cell (intima); f: fibroblast (adventitia); L: lumen; m: smooth muscle cells (media).) (Reproduced by courtesy of Dr G. Gabella, University College, London.)

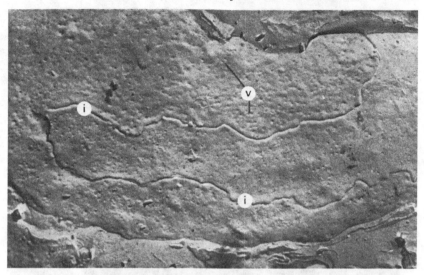

Fig. 13.6. Freeze-fracture electron micrograph of endothelial cells in heart capillary. Inter-cell junctions (i) and surface invaginations (v) formed by vesicles opening onto the surface can be seen. (From Leak (1971). Frozen-fractured images of blood capillaries in heart tissue. *J. Ultrastruct Res.* **35**, 127.)

muscle can be observed, but these are reduced to a single layer in vessels less than approximately 50 μm in diameter. In even smaller arterioles the amount of smooth muscle in the single layer is further reduced, becoming sparse in the metarterioles. The medial smooth muscle layer may extend to form a ring around the origin of the capillaries; when this is clearly identifiable it is described as the *precapillary sphincter*. Collagen fibres tend to be arranged longitudinally among the muscle cells, the number of fibres decreasing with the amount of smooth muscle.

In arterioles larger than 30 μm the basement membrane separates the endothelial and muscle cells, but in smaller vessels the muscle cells often form a tight junction with endothelial cells through gaps in the membrane. This phenomenon is observed with increasing frequency as vessel size is decreased. Small nerve fibres can be seen running parallel to the arterioles and they make occasional junctions with the smooth muscle cells of the media; these are most common in the smallest arterioles.

The adventitia consists mainly of collagen fibres with a few elastic fibres inter-spersed in a loose matrix. The outer limit of this layer is hard to identify, as the vessel wall progressively merges with the loose connective tissue of the interstitial space.

The capillaries As the terminal arterioles divide into the capillary system and thor-oughfare channels, the structure of the wall changes considerably. The capillary wall consists simply of a single layer of endothelial cells surrounded by its basement mem-brane, which splits to enclose occasional cells called *pericytes* (**Figs 13.7** and **13.8**). These are thought to have the potentiality to become smooth muscle cells.

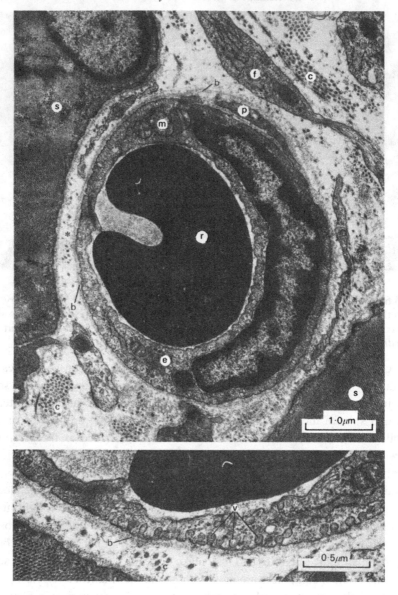

Fig. 13.7. The upper panel is a transmission electron micrograph of a capillary in the rat oesophagus and surrounding striated muscle. The lower panel shows a further magnified view of the wall in the vicinity of the asterisk. (b: basement membrane; c: collagen fibrils; e: endothelial cell with nucleus; f: fibroblast; m: mitochondria; p: pericyte; r: red blood cell; s: striated muscle fibre; v: pinocytic vesicles.) (Reproduced by courtesy of Dr G. Gabella, University College, London.)

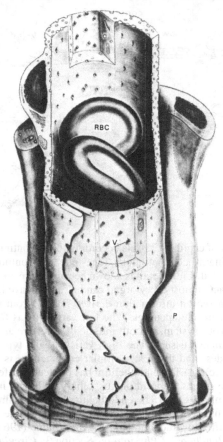

Fig. 13.8. The three-dimensional architecture of the heart muscle capillary with its luminal content of red blood cells (RBC) and wall of endothelium (E), basement membrane (bl) and pericyte (P) illustrated. This diagram was constructed from three-dimensional relief images obtained from replicas of heart capillaries of both fixed and unfixed frozen etched preparations. Both luminal and connective tissue surfaces of the endothelium are populated by vesicles. The topographical association between opposing cell margins to form intercellular junctions (j) is depicted in both surface and cross-views. The basement membrane (bl) is continuous over the endothelial surface as well as the adjacent pericyte (P). (From Leak (1971). Frozen-fractured images of blood capillaries in heart tissue. *J. Ultrastruct. Res.* **35**, 127.)

The endothelial cells of the capillary wall are arranged with their edges closely opposed to one another and with a variable degree of overlap (**Figs 13.6** and **13.7**). There are usually either one or two endothelial cells encircling a capillary at any cross-section. Vesicles are again abundant in the endothelial cells. An idea of the three-dimensional structure of a capillary can be gained from **Fig. 13.8**. The endothelial

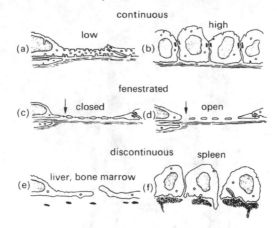

Fig. 13.9. Classification of capillary vessels according to the continuity of the endothelium. Three main types are distinguished (continuous, fenestrated, discontinuous) and, for each, two main varieties are represented. Little detail is shown because there are considerable variations from organ to organ; in fact, almost every organ can be said to have its own type of capillary vessels. The scheme is based on information derived mainly from mammals. (a), (b) The endothelium has no recognizable openings. The low variety (a) is found in striated muscle, myocardium, central nervous system, smooth muscle of digestive and reproductive systems, and subcutaneous and adipose tissue. The high variety (b) is typical of the postcapillary venules of the lymph nodes and thymus; a similar endothelium is found also in the large arteries when contracted. (c), (d) The endothelium has intracellular fenestrae (arrows), either closed (c) as in endocrine glands, choroid plexus, ciliary body and intestinal villus, or open (d) as in the renal glomerulus. (e), (f) The endothelium has intercellular gaps. These vessels are also referred to as 'sinusoids'. They are typical of liver, bone marrow and spleen; in each of these sites they differ in structural detail. (From Majno (1965). Ultrastructure of the vascular membrane. In *Handbook of Physiology*. Section 3: *Circulation*, vol. II. American Physiological Society, Washington, DC.)

lining and overlap between cells can be seen, as can the close association of the basement membrane, including the manner in which it splits to surround the pericyte.

The appearance of the endothelial cell lining of capillaries depends upon which microcirculation is under study. Three major types are distinguished, based upon the continuity of the endothelium; they are shown in **Fig. 13.9**. In *continuous* endothelium, the endothelial cells are usually tightly joined with a considerable length of contact between them (**Fig. 13.7**). In the vessels of striated muscle, the cells may be quite flat with a thin periphery; in others, such as the postcapillary venules, they may be cuboidal and form a thick layer.

In *fenestrated* endothelium the cell is so thin that the opposite surfaces of its membrane are very close together and form small circular areas known as diaphragms or

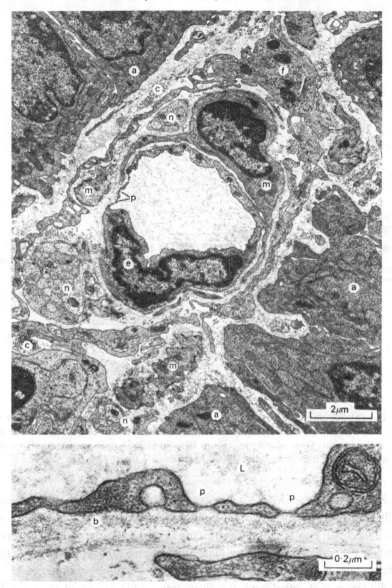

Fig. 13.10. A transmission electron micrograph of a fenestrated capillary within the mucosa of the guinea-pig ileum (upper panel). The details of the fenestrae can be seen more clearly in the lower panel. (a: epithelial absorptive cells; b: basement membrane; c: collagen fibrils; e: endothelial cell; f: fibroblast; L: lumen; m: smooth muscle cell; n: nerve; p: fenestrations.) (Reproduced by courtesy of Dr G. Gabella, University College, London.)

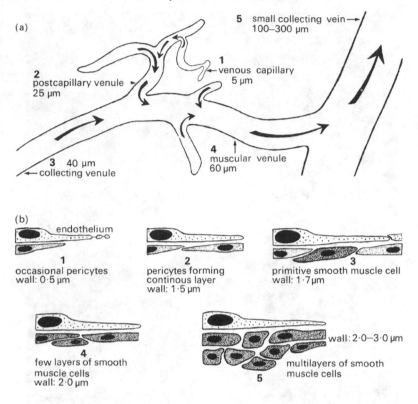

Fig. 13.11. (a) Schematic representation of the venular system indicating location of sections illustrated in (b) and (c). (b) Schematic representation of the endothelium and underlying cells in the various sizes of venule. (From Rhodin (1968). Ultrastructure of mammalian venous capillaries, venules and small collecting veins. *J. Ultrastruct. Res.* **25**, 452.)

fenestrae approximately 25 nm thick and of the order of 0.1 μm across. Adjacent endothelial cells are still usually tightly joined. Such endothelium has been found in the wall of the small intestine of the guinea pig (**Fig. 13.10**), the choroid plexus of the brain and the ciliary body of the eye. Fenestrae are also present in the glomerulus of the kidney, where they are thought to be open.

Yet a third type of endothelium is the *discontinuous* form in which there are distinct intercellular gaps. These occur in the liver, the spleen and bone marrow. In such cases the basement membrane is also discontinuous.

The venules The postcapillary venules, although larger than the capillaries supplying them, have virtually the same structure, the only significant difference being the presence of an almost complete layer of pericytes around the endothelium (**Fig. 13.11**). In collecting venules of diameter 30–50 μm there is a complete layer of

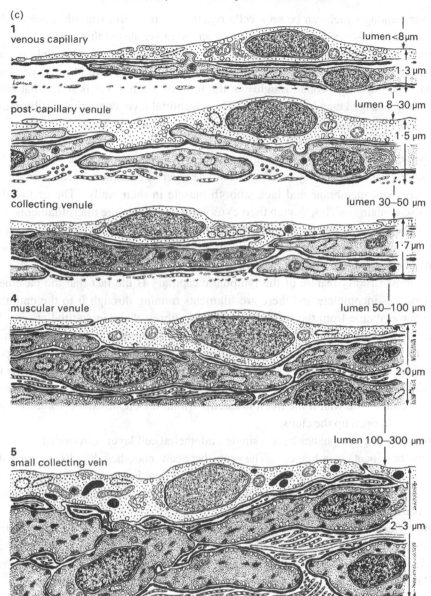

(c)

1
venous capillary

lumen <8 μm

1·3 μm

2
post-capillary venule

lumen 8–30 μm

1·5 μm

3
collecting venule

lumen 30–50 μm

1·7 μm

4
muscular venule

lumen 50–100 μm

2·0 μm

lumen 100–300 μm

5
small collecting vein

2–3 μm

Fig. 13.11. (c) Redrawn electron micrographs of the walls of venules of various sizes showing actual arrangement of cells. (From Rhodin (1968). Ultrastructure of mammalian venous capillaries, venules and small collecting veins. *J. Ultrastruct. Res.* **25**, 452.)

pericytes among which can be seen cells regarded as primitive smooth muscle cells. An outer layer of fibroblasts may also be observed in venules of this size.

In venules 50–100 μm in diameter the endothelium and basement membrane are surrounded by one or two complete layers of muscle cells forming a continuous media. The muscle is wound roughly in the form of a spiral, but is somewhat more irregularly arranged than in an arteriole. The adventitial layer consisting of fibroblasts and collagen becomes increasingly prominent as vessel size increases.

The lymphatics In many ways the structure of lymphatic capillaries is similar to that of systemic capillaries. They consist of a single layer of endothelial cells surrounded by a basement membrane and lack smooth muscle in their walls. The endothelial cells also contain vesicles, though their cytoplasm contains more microfilaments than that of the systemic capillary cells. Adjoining endothelial cells overlap by several micrometres but are separated by a distance which varies from zero at occasional points of close contact to about 0.5 μm.

The most striking feature of the lymphatic capillary is the fact that the basement membrane is incomplete and there are filaments running through it to the endothelial cell membrane from the collagen and elastic fibres of the interstitial space. An impression of the three-dimensional structure of a lymphatic capillary is given in **Fig. 13.12**. It is thought that the anchoring filaments may influence the transport of fluid into the lymphatic capillaries; contraction of the filaments could cause the interendothelial cell clefts to open wider. Alternatively, the anchoring filaments would tend to stabilize the wall if the microfilaments within the endothelial cytoplasm were to contract to open up the clefts.

The collecting lymphatics have a single endothelial cell layer surrounded by a continuous basement membrane. The gap between endothelial cells is virtually eliminated, the cleft being less than 5 nm across. There is a layer of smooth muscle cells surrounding the membrane, and bundles of collagen fibres and a few fibroblasts lie outside the muscle layer.

The collecting lymphatics and all larger vessels contain valves consisting of paired leaflets originating from opposite sides of the lymphatic wall. They contain no smooth muscle, but the cytoplasm of the individual endothelial cells contains numerous microfilaments which may serve a contractile function.

The junctions between vascular endothelial cells The nature of the bonding between adjacent endothelial cells deserves consideration because of its relevance to both the mechanical strength and the permeability of the endothelial layer. The gaps between cells are very narrow and our knowledge of the structure and nature of intercellular junctions awaited the introduction of the electron microscope.

The membranes of adjacent cells are mainly parallel and separated by an intercellular space, the intercell cleft, of approximately 15–20 nm. Whilst the contents of

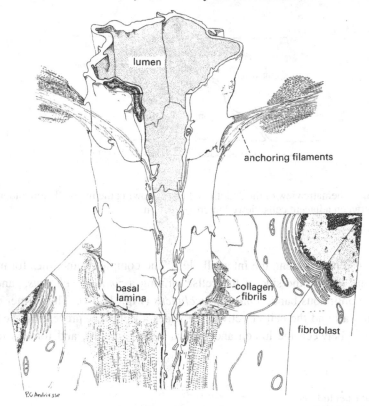

Fig. 13.12. A three-dimensional diagram of a lymphatic capillary, reconstructed from collated electron micrographs. The relationship of the lymphatic capillary to the surrounding interstitial space is illustrated. The anchoring filaments appear to originate from the endothelial cell surface and extend among collagen bundles, elastic fibres and cells of the adjoining tissue area, providing a firm connection between the lymphatic capillary wall and the surrounding connective tissue. An irregular basement membrane and collagen fibres are marked. (From Leak and Burke (1968). Ultrastructure studies on the lymphatic anchoring filaments. *J. Cell Biol.* **36**, 129.)

the gap are not known with certainty, they are thought to consist of mucopolysaccharide. It is also suspected that, because the gap is so uniform in thickness, there is some cohesive force operating between the cells over the entire length of the gap, but whether this is due to the intercellular material or to forces between the membranes is not known. There is in addition a localized site of firmer attachment, the *junctional complex.*

The junctional complex, which consists of a number of specialized structures, is usually positioned near to the luminal surface of adjoining cells, approximately one-third of the way along the intercell cleft (**Fig. 13.13**). At this site the opposing cell membranes come very close together and an impermeable zone is formed; molecules

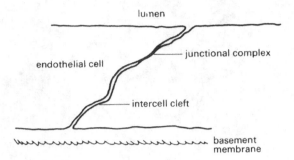

Fig. 13.13. Schematic view of the endothelial layer showing the intercell cleft and junctional complex between adjacent cells viewed in cross-section.

cannot diffuse past it along the intercell cleft. The complex sometimes forms a continuous band around the side of the cells, forming a *zonula occludens*, and sometimes a discontinuous band, a *macula occludens* (**Fig. 13.14**). The former continuous junction around the cell effectively provides a tight seal preventing the passage of molecules between the lumen and cell base via the cleft, and for this reason is

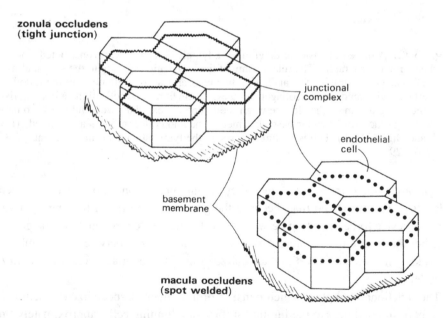

Fig. 13.14. Schematic representation of the arrangement of the junctional complex between endothelial cells.

referred to as a tight junction. The discontinuous or 'spot welded' junction, on the other hand, does leave a diffusion pathway available for molecular transport.

In the past it was generally believed that the tight junction exists in all endothelial layers other than in the capillaries, but this view is no longer held with certainty.

The pinocytic vesicles Reference has already been made to the existence of vesicles in endothelial cells in both the systemic and lymphatic circulations. These are believed to be important in the transport of materials, particularly large molecules, across such cells by the mechanism known as pinocytosis. This will be discussed later; here, we are concerned with a brief description of their structure and distribution. As can be seen in **Fig. 13.7**, vesicles exist in large numbers both free within the cytoplasm of the cells and attached to the cell membrane; in the latter case they may be open to the exterior of the cell. The material comprising the wall of the vesicle appears to be the same as that of the cell membrane. The free vesicles are roughly spherical and typically 60–80 nm in diameter. Those which are attached to the cell membrane and open on to the surface usually do so through a neck about 25 nm in length and 10–25 nm across. There are approximately 500 vesicles per cell, corresponding to around 25% of the total cytoplasmic volume. There are on average about 120 vesicles attached per square micrometre of cell surface.

The interstitial space The interstitial space is the region which surrounds the blood vessels and cells of all tissues. It contains a complex and as yet poorly understood arrangement of intercellular materials such as collagen and elastin bathed in some kind of fluid. It is suggested that the fluid is a heterogeneous colloidal system (p. 133) of two components in equilibrium with one another. One component tends to be more gel-like than the other, whose properties are more like a dilute sol. The gel phase is highly aggregated, consisting of mucopolysaccharides, particularly hyaluronic acid, and it tends to stay more in contact with the structural elements of the space. The water-rich sol is mobile and contains unaggregated solutes; it is more capable of dissolving and, hence, transporting water-soluble materials within the space. The distribution of the two phases is considered to vary continuously and it is not possible to identify particular sites at which the fluid will have any particular composition.

Static mechanical properties of the microcirculatory vessels

The static mechanical properties of the vessels in the microcirculation will naturally depend upon their structure, and thus it is not surprising that experiments have demonstrated widely ranging mechanical properties. However, because of the size and the highly reactive nature of small vessels, measurements are extremely difficult; our understanding is still very limited and partly based on inference. The elastic properties of both arterioles and venules are explained reasonably well on the basis of the composition of their walls, but at present the distensibility of capillaries is not so well understood.

It has already been pointed out in Chapter 7 that arteries exhibit visco-elastic stress–strain properties. This is also true of the vessels of the microcirculation, though because of experimental difficulties little information on the visco-elastic behaviour is available. In this section we shall therefore consider only the elastic behaviour of the small vessels.

Elastic properties of the arterioles In Chapter 7 (p. 102) we considered the statics of an elastic tube and showed that the absolute value of the hoop stress in the wall of most blood vessels is compressive rather than tensile; this resulted from the relatively small excess of internal pressure over external pressure and from the fact that the vessel walls are relatively thick. It was also pointed out that we are not normally concerned with the absolute level of stress, but rather with changes; thus, we are interested to know the change in stress in the wall if the pressure within the vessel is changed by some amount. In order to ascertain the stress required to balance a given distending pressure (relative to atmospheric) we may use the law of Laplace (Equation (7.6)). The amount of extra stress s' in a tube wall of thickness h and internal diameter d produced by a distending pressure p relative to atmospheric is given by

$$p = \frac{2hs'}{d}. \tag{13.1}$$

The product hs' is the total tension in the wall.

We may calculate the total wall tension and stress in a large and small artery to see how they compare; to do so we will consider the canine femoral artery and a $50\,\mu m$ arteriole. The diameter and wall thickness of the artery are approximately 0.4 cm and 0.04 cm and the average internal pressure is about $1.30 \times 10^4\,\mathrm{N\,m^{-2}}$ (100 mm Hg). The wall thickness of the arteriole is approximately $20\,\mu m$, and thus it is a relatively thick-walled vessel ($h/d = 0.4$ compared with 0.1 for the artery); the internal pressure is approximately $0.9 \times 10^4\,\mathrm{N\,m^{-2}}$ (70 mm Hg).

From Equation (13.1) we can see that the total tension in the wall of the femoral artery is $26\,\mathrm{N\,m^{-1}}$ compared with $0.22\,\mathrm{N\,m^{-1}}$ in the arteriole; however, the excess *stresses* in the walls are much closer ($6.5 \times 10^4\,\mathrm{N\,m^{-2}}$ in the artery compared with $1.1 \times 10^4\,\mathrm{N\,m^{-2}}$ in the arteriole). The tension in the wall of the arteriole is probably accommodated by the vessel basement membrane and media, which are known to be stronger than the endothelium and adventitia. Thus, the stress would be doubled to $2.2 \times 10^4\,\mathrm{N\,m^{-2}}$ because the membrane and media comprise half the wall thickness. Furthermore, in such a thick-walled tube the stress would not be uniformly distributed throughout the load-bearing part of the wall. In an isotropic elastic tube the stress at the innermost part of the wall would be approximately 20% greater than the average value, that is about $2.7 \times 10^4 \, \mathrm{N\,m^{-2}}$.

Whilst these calculations can give only approximate indications of the hoop stress within the vessel walls, they do indicate the close similarity between conditions in large and small vessels.

The stress–strain relationship of arterioles has been measured in a few instances, though such experiments have proved extremely difficult technically. The curve of cross-sectional area against distending pressure is nonlinear and has a similar shape to that obtained from arteries. The arteriole becomes stiffer and the effective Young's modulus increases as the strain increases. Studies on the arterioles of the frog mesentery have indicated that when measured at a diameter of 15% above resting diameter at zero transmural pressure the incremental Young's modulus was $6 \times 10^4 \, \mathrm{N\,m^{-2}}$ rising to $12 \times 10^4 \, \mathrm{N\,m^{-2}}$ at 45% distension. The Young's modulus of collagen is about 10 000 times greater. Such findings would suggest that the collagen in the vessel walls is not taking a significant amount of the load applied. However, this statement ignores the fact that collagen only exhibits its great stiffness when the stress is applied along the axis of the fibre; in a vessel where the fibres are aligned predominantly longitudinally only a small percentage will be experiencing a significant axial stress and hence their contribution to the vessel's elasticity is hard to assess.

The effect of active tension (Chapter 11, p. 189) in arteriolar smooth muscle is very important since it is capable of causing marked changes in both the calibre and the stiffness of the vessel. The effects of active tension on the stress–strain properties have recently been studied directly in vivo in the frog mesentery. The arteriole was stretched by various levels of distending pressure and the vessel diameter (d) was measured with the smooth muscle relaxed. The vessel was then stimulated to contract maximally with noradrenalin (norepinephrine) at various levels of initial stretch and the reduction of diameter (Δ) was noted. As can be seen from **Fig. 13.15**, the response of the vessel is extremely dependent upon the initial length. Thus, as the distending pressure increases from low to intermediate levels, the ability of the arteriole to constrict actively increases markedly, but then at higher initial stress the ability to constrict decreases again. It should be noted that the maximum response of the arterioles occurs at the normal physiological level of applied stress, namely $1.2 \times 10^4 \, \mathrm{N\,m^{-2}}$; thus, they are normally close to their optimal diameter and can respond to vasoactive stimulae with calibre changes in the region of 40–50%. The relationship shown in **Fig. 13.15** is similar to that found in many other types of muscle, including that in the larger muscular arteries.

This pattern of active constriction is important in relation to the possibility of active closure of blood vessels. For a number of years it was assumed that a fixed amount of active tension is developed by the smooth muscle for a given stimulus, independent of vessel diameter. Under these conditions a vessel whose internal pressure is slowly reduced would reach a point at which the active tension coupled with the

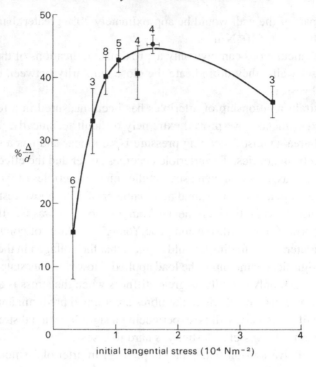

Fig. 13.15. Relation between magnitude of active constriction and initial stress in arterioles of the frog mesentery. This is equivalent to an active length–tension curve for smooth muscle; the behaviour is similar to that of skeletal and heart muscle (**Fig. 11.9**). The bars on the curve indicate ±1 standard error and the numbers refer to the number of vessels studied. (From Gore (1972). Wall stress: a determinant of regional differences in response of frog microvessels to norepinephrine. *Am. J. Physiol.* **222**, 82.)

elastic restoring force in the wall exceeded the distending pressure; the vessel would then close. The pressure at which this was postulated to occur is called the 'critical closing pressure'.   This concept has been useful in developing ideas of how active and pressure forces are interrelated; but on the basis of the evidence described above it must now be re-examined, for as distending pressure is reduced so too is the active tension which opposes the pressure force. This is not to say that vessels cannot close under any circumstances, for clearly spasm does occur.

Mechanical properties of the capillaries When a capillary is viewed under the microscope it does not appear to pulsate in the way that an arteriole does. However, if the venous end of a capillary is occluded, the red cells trapped within the upstream section of the capillary oscillate back and forth in time with the heart beat and can be seen to pack slowly towards the occluded end. The slow movement of the cells is due to filtration of fluid out of the vessel, as will be explained later, but the fast, oscillatory movement is thought to result from a very slight distension of the capillary during systole. On the basis of measurements of this cell motion and of the oscillatory upstream pressure, the stress–strain relationship of the capillary has been studied and an effective Young's modulus calculated to be of the order of $3 \times 10^5 \, \mathrm{N\,m^{-2}}$.

Typical transmural pressures across the capillary wall are in the region of $0.3 \times 10^4 \, \mathrm{N\,m^{-2}}$ (25 mm Hg). Thus, for a capillary of $7 \, \mu\mathrm{m}$ internal diameter, the total tension in the wall is approximately $1 \times 10^{-2} \, \mathrm{N\,m^{-1}}$. If we consider the capillary to be simply a tube floating in the interstitial fluid, then this tension has to be accommodated solely by the endothelium and basement membrane. If we assume that it is supported by the inner and outer membranes of the endothelial cell (each membrane being 10 nm thick), then the effective Young's modulus of the membrane would have to be approximately $1.3 \times 10^7 \, \mathrm{N\,m^{-2}}$. Such a Young's modulus is far higher than that normally expected for cell membranes and, thus, it is unlikely that they are responsible for the support. If the support is by the basement membrane (20 nm thick) then its Young's modulus would also have to be about $1.3 \times 10^7 \, \mathrm{N\,m^{-2}}$. In fact, the basement membrane of renal tubules, which appears very similar to systemic capillary basement membrane (but slightly thicker), has been calculated to have a Young's modulus of 0.7–$1.0 \times 10^7 \, \mathrm{N\,m^{-2}}$. Therefore, it is possible that the basement membrane could be responsible for supporting the tension in the wall. If the total tension were accommodated by the whole wall (approximately $0.7 \, \mu\mathrm{m}$ thick), then, on the assumption that it is a gel-like material, its Young's modulus would be approximately $4 \times 10^5 \, \mathrm{N\,m^{-2}}$.

However, an alternative hypothesis has been developed to explain the mechanical properties of the capillaries. This supposes that the capillary should be considered not simply as a tube but rather as a tunnel within the interstitial tissue and that its elastic properties are derived from those of the tissue. This is known as the 'tunnel in gel' model. A mathematical analysis has been made of the distensibility of such a tunnel. The distensibility D of the capillary (radius R) is defined as $(1/A)(\mathrm{d}A/\mathrm{d}p)$, where A is the cross-sectional area and p the distending pressure (see Equation (12.8)). The analysis shows that the distensibility of the capillary when considered as a tunnel compared with that of a simple tube is approximately given by

$$\frac{D_{\text{tunnel}}}{D_{\text{tube}}} = \frac{h(1+\sigma)E_{\text{t}}}{RE_{\text{h}}},$$

where E_{t} and E_{h} are the Young's moduli of the tube wall and surroundings respectively. The capillary wall thickness is h and σ is the Poisson's ratio of the surrounding material. If it is assumed that the elastic properties of the capillary and surroundings are the same and that the tissue is isotropic and incompressible ($\sigma = 0.5$), then for a capillary of typical dimensions the tunnel is approximately one-third as distensible as the tube.

On the basis of measurements of capillary distensibility, the effective Young's modulus of the gel surrounding such a tunnel would be of the order of $1.5 \times 10^4 \, \mathrm{N\,m^{-2}}$. Thus, the interstitial tissue could be capable of providing significant support for the capillary. Further experiments are needed to distinguish between the 'tunnel in gel' and 'tube' models; it may be that in different tissues the support of the capillaries is provided by these mechanisms in different proportions.

Elastic properties of the venules The postcapillary venules, whose structure is similar to that of capillaries, appear also to be rather stiff vessels. The larger venules, on the other hand, contain smooth muscle and are probably more distensible, but little is known of their detailed stress–strain properties. At low transmural pressures, considerable increases in vessel size can be achieved with only small increments in pressure, but this is probably largely the result of collapsed vessels being opened up. It has been estimated that these vessels contain approximately 25% of the total blood volume and that as a result their calibre greatly affects the volume distribution of blood within the circulation. (The detailed mechanics of the collapse of vessels is considered in Chapter 14.)

Measurements of the distensibility of the venular system (with zero arteriolar pressure) have indicated that when venous perfusing pressure is raised from a resting value of zero to $0.4 \times 10^4 \, \mathrm{N\,m^{-2}}$ (30 mm Hg), the vessel diameter rises more than 90% above its resting value; thereafter, diameter rises little as pressure is raised as high as $0.9 \times 10^4 \, \mathrm{N\,m^{-2}}$ (70 mm Hg). Other experiments, in which venular dimensions were measured while venous pressure was maintained at zero but arteriolar perfusing pressure varied, indicated that vessel calibre was affected but increased linearly to only about 130% of its resting value as arteriolar pressure was raised from zero to $2 \times 10^4 \, \mathrm{N\,m^{-2}}$ (150 mm Hg). The effect of venous pressure on vessel volume is thus very great indeed, since raising venous pressure from zero to $0.13 \times 10^4 \, \mathrm{N\,m^{-2}}$ (10 mm Hg) produced a 200% rise in volume, and elevation of the pressure to $0.4 \times 10^4 \, \mathrm{N\,m^{-2}}$ (30 mm Hg) produced an increase of approximately 350% above resting volume. At higher pressures the volume changes were slight (about 20%) because of the greatly increased vessel stiffness.

Pressure in the microcirculation

The distribution of pressure Many attempts have been made to obtain representative information about pressures within capillaries and the feeding and collecting vessels; numerous isolated measurements have been made in vessels of varying size. However, enormous scatter is evident in the reported data for two main reasons. First, vessels of a given size in different preparations, or even within a single bed, do not necessarily serve precisely the same function. Second, pressures within vessels show periodic fluctuations which can be quite considerable; such variation can result both from local factors and from changes in central arterial and venous pressures. Perhaps one of the clearest impressions gained from any visual study of the microcirculation is that at any instant the blood flow in about a quarter of the vessels is either stopped or sluggish.

Recently, a very extensive survey has been made of the distribution of pressures within the mesentery of the cat, which provides a clear idea of the pressure profiles within a single microcirculatory bed. (It must be appreciated that the mechanical properties of the mesenteric bed may be unusual and, therefore, caution must be exercised in generalizing from these results.) The vessels of the bed may be divided into five main groups on the basis of size. First, there is the rather heterogeneous group of arterioles ranging from about $50\,\mu m$ to $20\,\mu m$; then there are the highly contractile precapillary vessels. These both comprise the feeding system. After the capillaries there are first the non-contractile postcapillary vessels, which are essentially wide capillaries with diameters from 10 to $20\,\mu m$. These are followed by the larger, and contractile, collecting venules (20–$50\,\mu m$).

Measurements in all the groups of vessels indicated that the pressures within two vessels of similar size commonly differ by as much as 0.26–$0.33 \times 10^4\,N\,m^{-2}$ (20–25 mm Hg); these differences were not, however, the consequence of short-term temporal fluctuations such as the pulse. The pressure measured within a given vessel remained virtually constant over periods 2–3 min apart from small cardiac oscillations, and over longer periods would vary by only about 0.04–$0.05 \times 10^4\,N\,m^{-2}$ (3–4 mm Hg) even though there were considerable variations in local blood flow. **Figure 13.16** shows the pressure distribution found in this study for vessels in the mesenteric bed smaller than $60\,\mu m$ in diameter. The range of pressure is somewhat larger in the largest arterioles than it is in the more distal vessels. Whilst the greatest *fall* in pressure is seen to be in the vessels before the capillaries (approximately 70%), the pressure *gradient* through the bed shows a somewhat different pattern. As shown in **Fig. 13.17**, the pressure gradient increases as vessel size is decreased, which is consistent with any viscous flow theory (Chapter 5).

The data presented in **Fig. 13.16** were obtained from cats whose central arterial pressures were in the normal range of 1.31–$1.85 \times 10^4\,N\,m^{-2}$ (101–142 mm Hg). Data were also obtained in other groups of cats which were found to be hypertensive with pressures in the range 1.85–$2.54 \times 10^4\,N\,m^{-2}$ (142–194 mm Hg) and hypotensive with pressures of 0.78–$1.31 \times 10^4\,N\,m^{-2}$ (60–100 mm Hg) and are shown in **Fig. 13.18**. The pressure drop in small arterioles and precapillaries in the hypertensive cats was greater than in the normals, while it was less in the hypotensive animals. Consequently, pressures in the capillaries and postcapillary vessels were similar in the two groups, the difference between the mean pressures being only 0.04–$0.06 \times 10^4\,N\,m^{-2}$ (3–5 mm Hg). Locally applied vasoactive substances were also observed to produce changes of pressure which were greatest in the small arterioles and precapillary vessels, the effects on capillary and venular pressure being

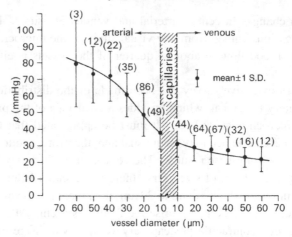

Fig. 13.16. Distribution of pressures in cat mesenteric vessels less than 60 μm diameter. A smoothed line of best fit is shown together with the standard deviation for each of the designated vessel groups. The means were computed for ±5 μm intervals about each abscissa point. The numbers in parentheses indicate the number of vessels sampled for each interval. (From Zweifach (1974). Quantitative studies of microcirculatory structure and function, I. Analysis of pressure distribution in the terminal vascular bed in cat mesentery. *Circulation Res.* **34**, 843–57. By permission of the American Heart Association Inc.)

far less. These observations strongly suggest that alterations in central arterial pressure are only very weakly reflected in the capillaries, and that pressures there are strongly influenced by local factors.

A further observation in these experiments was that pressures in about 10% of the venules were sometimes higher than expected (approximately $0.9 \times 10^4 \, \mathrm{N\,m^{-2}}$ (70 mm Hg) compared with the usual value of $0.4 \times 10^4 \, \mathrm{N\,m^{-2}}$ (30 mm Hg). It might be thought that these pressures reflect the stoppage of flow, for example by large leukocytes, but this is unlikely because such blockage would be expected to occur farther upstream. The finding therefore suggests the probable existence of shunts (such as thoroughfare channels) bypassing the capillary system.

Within the capillaries themselves the pressure distribution showed two particular variations. The average capillary pressure varied by up to $0.13 \times 10^4 \, \mathrm{N\,m^{-2}}$ (10 mm Hg) between different sectors of the mesentery and also varied as much within the same local area. However, the pressure within particular capillaries remained fairly constant over long periods; vessels in which the average pressure was as high as $0.5 \times 10^4 \, \mathrm{N\,m^{-2}}$ (40 mm Hg) or as low as $0.26 \times 10^4 \, \mathrm{N\,m^{-2}}$ (20 mm Hg) maintained these average levels over several minutes. As might be expected, the capillaries which arose from the proximal end of the arterioles had in general higher pressures than those arising more distally.

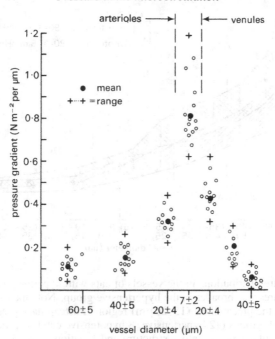

Fig. 13.17. Pressure gradient measured by two microprobes separated by the maximum distance possible between branches. In arterioles, the gradient was measured along a 1500–2500 μm interval; in venules longer segments were usually available, 2000–3000 μm. In the capillary region, the probe separation was usually 200–350 μm. Values cover a diameter range of ±5 μm for each or the categories listed, except for capillaries in which the range is ±2 μm. (From Zweifach (1974). Quantitative studies of microcirculatory structure and function, I. Analysis of pressure distribution in the terminal vascular bed in cat mesentery. *Circulation Res.* **34**, 843–57. By permission of the American Heart Association Inc.)

There were several time scales of pressure variation within a given capillary. First, there were cardiac oscillations, which could be observed throughout the length of the capillary. Their normal amplitude was about $0.01–0.02 \times 10^4 \, \mathrm{N\,m^{-2}}$ (1–2 mm Hg) (see **Fig. 13.19**) but could increase to about twice this level if the precapillary sphincter became dilated. The propagation of these oscillations through the microcirculation is discussed below. The second form of fluctuation was one lasting for 15–20 s and was associated with pressure variation in the region of $0.04–0.07 \times 10^4 \, \mathrm{N\,m^{-2}}$ (3–5 mm Hg); the pattern of such changes was random. The third type of pressure variation was more substantial and lasted longer; it was in the region of $0.13 \times 10^4 \, \mathrm{N\,m^{-2}}$ (10 mm Hg) and occurred over a period of 5–8 min, followed by a return to the steady state condition in about 2–3 min.

Although local pressure does vary with time, an indication of the progressive fall in pressure with distance along a capillary can be obtained from **Fig. 13.19**, where

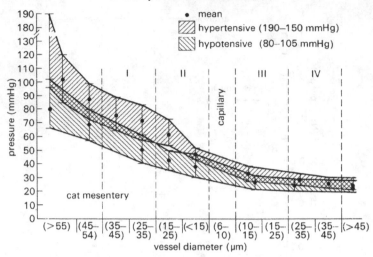

Fig. 13.18. Pressure distributions in the vessels of cats with hypertension and hypotension. The range of pressures was broader in the hypertensive group. Note the trend for pressure to converge rapidly in the precapillary (15–20 μm) region. Postcapillary pressures were essentially the same for all groups (12 hypertensive, 8 hypotensive cats). (From Zweifach (1974). Quantitative studies of microcirculatory structure and function, I. Analysis of pressure distribution in the terminal vascular bed in cat mesentery. *Circulation Res.* **34**, 843–57. By permission of the American Heart Association Inc.)

the pressures at various sites in a capillary are shown. Pressures were obtained by introducing a micropipette into a side branch and occluding it to obtain a local lateral pressure in the capillary without causing any local flow disturbances or blockage.

Attempts have been made to develop mathematical models to predict the distribution of pressures within the microcirculation. However, these have met with only limited success for perhaps two main reasons. First, it is difficult to relate the model predictions to the experimentally measured varying pressure fields. Second, in all models, blood is treated as though it were a homogeneous Newtonian fluid and flow conditions are everywhere Poiseuille in form. From what has been said already in Chapter 10, these assumptions are probably unacceptable, and this is confirmed later in the chapter (p. 384). The results of one such attempt are presented in **Fig. 13.20**, where a typical sector of the cat mesentery is shown in (a); artery and vein pairs lie together dividing the mesentery up into what are considered to be triangular sectors. Within each sector are numerous units bounded by smaller artery–vein pairs (b); such a unit may be schematically represented as shown in (c), where all vessels are assumed to be straight and interconnected as shown. From a knowledge of pressures and flow rates within the boundary vessels and the assumption of Poiseuille flow in all vessels in the network, the distribution of pressure in the system was computed. It is plotted

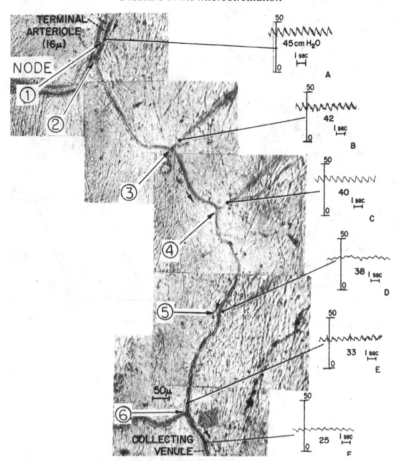

Fig. 13.19. Photographic reconstruction of capillary from arteriole to collecting venule with direct recordings of pressure and mean values taken at the points indicated. The capillaries ranged from 7.5 to 9 μm in width. Note the persistence of the pulse throughout. Each reading was taken at a side branch so as not to interrupt the flow through the feeding capillary. (From Zweifach (1974). Quantitative studies of microcirculatory structure and function, II. Direct measurement of capillary pressure in splanchnic mesenteric vessels. *Circulation Res.* **34**, 858–66. By permission of the American Heart Association Inc.)

against vessel size in (d). The predictions may be compared with the experimentally measured distribution obtained in a large number of such units (cross-hatched region). The experimental distribution is very similar to that shown in **Fig. 13.16**, but the prediction differs significantly. The predicted curve would suggest that there is a very steep fall in pressure at the level of the capillaries, whereas experiment shows most of the pressure drop to be in the vessels just before the capillaries. This discrepancy may result from several factors, but one obvious weakness of the theory is its neglect of the

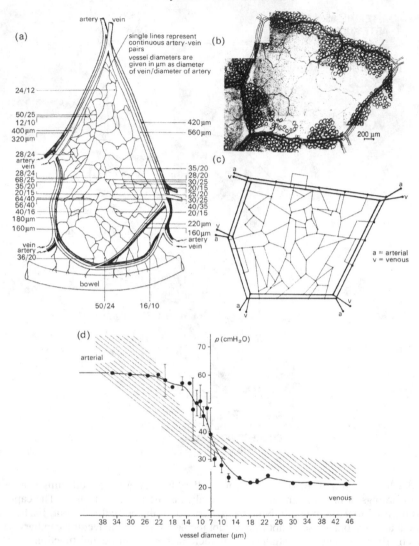

Fig. 13.20. (a) A schematic representation of a typical sector of the mesentery of the cat. Such sectors are bounded by relatively large arteries and veins running in pairs. (b) Micrograph of a typical unit found within a sector in which the capillary network is bounded by pairs of arteries and veins. (c) Schematic representation of the network of vessels depicted in the micrograph of (b). (d) The distribution of pressure within the model unit of (c) (continuous line) computed as described in the text. The cross-hatched region represents the band of pressures measured in vivo in 150 different units. (From Lipowski and Zweifach (1974). Network analysis of microcirculation of cat mesentery. *Microvasc. Res.* **7**, 73.)

anomalous viscous properties of blood flowing in small vessels, which are discussed later in this chapter.

The propagation of cardiac pressure oscillations **Figure 13.19** shows that pressure oscillations of cardiac origin persist all along a microcirculatory pathway, although they are significantly attenuated. These oscillations have been further studied in other experiments in which the pressures were measured simultaneously at the entrance and exit of one capillary. These revealed phase differences which confirmed that the pulse propagates from the arterial end of the pathway towards the venous end. The speed of propagation within the capillary can be deduced from the observed phase changes (25° phase lag between small arteriole and small venule at a heart rate of 2.5 Hz) if the distance between the pressure measuring sites is known. Unfortunately this distance was not reported, but we would expect it to lie somewhere in the region of 2000 μm. Taking this value, one predicts a propagation speed of $7.2 \, \mathrm{cm \, s^{-1}}$. The pulse amplitude is attenuated by about 30% between the two sites, which is equivalent to attenuation of 83% per centimetre if the attenuation is exponential (see p. 120).

The mechanism of pulse propagation in the microcirculation is not the same as in arteries, because of the fact that inertia is negligible. In an artery, the pulse propagates as a wave, in which the elastic restoring force of the wall is balanced by inertia of the blood, as explained on p. 271. This mechanism brings the pulse to the small arteries and large arterioles, where the Womersley parameter α is close to 1. The pressure therefore oscillates at the entrance to the smaller vessels where α is less than 1. In the microcirculation, the elastic pulsations of the wall are resisted by viscous forces, not inertia forces, in the fluid. When the wall of a small vessel has been locally expanded by high pressure, its elasticity pulls it back towards its equilibrium position, forcing fluid out into the low-pressure region downstream. This causes the vessel wall in that region to expand, driving the pressure up so that the process is repeated farther downstream, and so on. Unlike the situation in arteries, however, viscous forces prevent the wall from overshooting its equilibrium position. Thus oscillations are 'overdamped', like the oscillations of a simple pendulum in a highly viscous fluid (see p. 116). Because viscous resistance involves the dissipation of mechanical energy, the amplitude of the pulse is rapidly attenuated.

This process can be analysed, if blood is taken to be a Newtonian fluid, by using the general theory for wave propagation in an elastic tube with a very small value of α (this theory is described in Chapter 12, and the results shown in **Fig. 12.36**, p. 302); the speed of propagation c of a sinusoidal pulse is predicted to be

$$c = \tfrac{1}{4}d\sqrt{\frac{\omega}{\mu D}},$$

where d is the capillary diameter, D is its distensibility, ω is the angular frequency of the pulse and μ is the blood viscosity. A propagation speed of about $10 \, \mathrm{cm \, s^{-1}}$ is predicted for a wave of frequency 2.5 Hz in a 10 μm capillary. Furthermore, the pulse is predicted to be very sharply attenuated, by a factor of about 99.8% per wavelength ($k = 2\pi$: contrast the large α-result given on p. 303), which is equivalent to about 80% per centimetre. The difference between the predicted values and the measurements quoted above is a result of the oversimplifications involved in the theory, including the neglect of tissue visco-elasticity and the non-Newtonian properties of blood.

Pressure in the interstitial space When considering the static mechanical properties of the various blood vessels we showed how diameter and area are related to the effective distending pressure (transmural pressure) of the vessel. However, in practice the internal pressure of the vessels is measured relative to atmospheric, and it is tacitly assumed that extravascular pressure (the pressure within the interstitial space) is atmospheric. This is an important assumption when the transmural pressure is small, and must be considered carefully.

Numerous methods have been used to measure interstitial pressure, with different results. The obvious method, of inserting a fine needle into the tissue, usually gives a reading close to atmospheric pressure and for a long time this was believed to be the true pressure. In fact, this method is inappropriate because it causes tissue damage and distortion.

More recent methods of investigation usually give values which are negative relative to atmospheric pressure. The precise value depends upon the technique, the species studied and the site of measurement; most values lie in the range 0–$0.9 \times 10^3\,\mathrm{N\,m^{-2}}$ (0–$10\,\mathrm{cm\,H_2O}$) less than atmospheric pressure.

Two principal methods of measurement have been developed and it is important for us to consider them briefly in order to appreciate the difficulty of interpreting the experiments. The first method involves implanting into the subcutaneous tissue space a small hollow porous capsule (approximately 1 cm diameter), which is left in the tissue for approximately 3 weeks. During this time the surfaces of the capsule become covered with loose connective tissue and the core is filled with fluid. Then, a fine needle is inserted into the centre of the capsule and the fluid pressure measured; a negative pressure is usually obtained. The technique assumes that the capsule contents and the surrounding interstitial space have come into equilibrium. It further assumes that the osmotic pressure of the fluid within the capsule is the same as that in the tissue space; thus, the measured hydrostatic pressure within the capsule will be the same as that in the surrounding tissue. However, the fluid contents and structure of the material within the capsule do not appear to be the same as in the interstitial space, and therefore the assumption that the osmotic pressures are equal is questionable. Thus, the measurements should be viewed with some reservation.

The second method involves the implantation of a cotton wick into the space. A few strands of cotton are placed inside a narrow polythene cannula with approximately

1 cm of wick exposed at one end, a pressure transducer is fixed to the other end and the whole assembly is filled with isotonic saline. The wick end of the cannula is introduced into the subcutaneous space via a needle which is then removed and the cannula is left undisturbed; fibres of the wick are assumed to maintain hydrostatic communication between the cannula and the interstitial space by providing tiny channels. Within an hour the measured pressure becomes approximately constant and is considered to represent the interstitial fluid pressure; a negative value, similar to that obtained with the capsule technique, is observed.

A possible cause of sub-atmospheric tissue fluid pressure is the partially hydrated gels, made up of very large molecules such as hyaluronic acid, which are present in the interstitial space and tend to imbibe water. The gel phase will tend to swell by extracting water from the liquid phase (sol) and those protein molecules in solution in the sol which are too large to enter the structure of the gel become concentrated. Thus, tissue osmotic pressure is raised. Unless water entering from the vascular space is removed continuously from the interstitial space, the latter will swell until the gel is fully hydrated. Then the sol would equilibrate osmotically with the plasma, and the interstitial pressure would increase to atmospheric or above. Dehydration is maintained by removal of the water via the lymphatics and reabsorption by the venules. It follows that the osmotic pressure in the interstitial space is normally slightly above the plasma osmotic pressure. If a cannula, containing saline of the same colloid osmotic pressure as plasma, is inserted into the space, water will tend to be drawn out of it. On the other hand, the tissue gel cannot enter the cannula, so a negative pressure will be recorded.

Pressures within the lymphatic capillaries have been measured with micropipettes, and found to be slightly positive. It has been argued that these pressures reflect the pressure within the interstitial space, since the lymphatic capillaries are in free hydrostatic communication with their surroundings because their endothelium is discontinuous. However, we have already seen (p. 355) that the discontinuous endothelial cells overlap each other, and the proponents of negative interstitial pressure argue that hydrostatic continuity occurs only when the interstitial space experiences a compression force, due for example to muscle contraction. Interstitial hydrostatic pressure then becomes positive, exceeding that within the lymphatics. It is postulated that these are held open by the anchoring filaments shown in **Fig. 13.12** and, therefore, flow into them can occur.

In summary, it seems that the existing methods all effectively measure the osmotic pressure within the interstitial space and not the true hydrostatic pressure. To measure this would require a probe filled with a fluid with the same colloid osmotic pressure as the tissue sol, and of dimensions less than the tissue fluid spaces. Thus, there are drawbacks to all existing methods, but the bulk of the observations suggest that interstitial pressure is slightly subatmospheric.

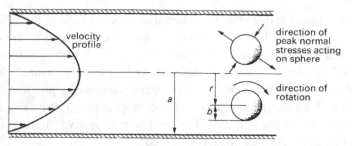

Fig. 13.21. The effects of suspending a neutrally buoyant rigid sphere of radius b in Poiseuille flow in a tube of radius a. The sphere rotates in the direction indicated whilst moving along the tube at a constant radial position r. The sphere is also subjected to a varying normal stress over its surface; the stress is compressive over part of the surface and tensile over the remainder.

Flow in models and in the large vessels of the microcirculation

Any analysis of flow in the microcirculation involves a consideration of the particulate nature of blood, for as we have already seen when considering blood viscosity measurement (Chapter 10), anomalous behaviour occurs when blood flows through gaps comparable in size to the red cell. As with so many other aspects of microcirculatory mechanics, much of our understanding is inferred from studies of particulate suspensions or whole blood flowing in glass tubes. The concentration of red cells in blood is far greater than that of any other formed elements, and the details of the flow are dominated by the red cells. The discussion that follows is thus mainly concerned with the behaviour of red cells in the flow. The motion of blood has been shown to be similar in many ways to that of suspensions of model spheres, discs and rods. We shall therefore look at such flows first, building up in complexity from single to multiple particle suspensions and from tubes much larger than the particle to tubes where there is a tight fit and the particles move in single file. In all cases we are concerned with flows in which the tube or vessel Reynolds number is usually low (that is, less than unity) and in which the wall shear rates can be as high as $1000\,\text{s}^{-1}$.

The motion of single particles at very low flow rates We will first consider the behaviour of single spheres, rods or discs suspended in a Newtonian fluid of the same density, flowing extremely slowly in a long straight tube with the tube Reynolds number less than one. This is the simplest case, because particle interactions and collisions cannot occur. When there are no particles present, the fluid downstream from the entrance region flows with the characteristic parabolic velocity profile of Poiseuille flow, and the velocity gradient or shear rate G at any radial position r in the flow will increase linearly from zero at the axis to a maximum at the wall. (See **Fig. 13.21** and Chapter 5, p. 49.)

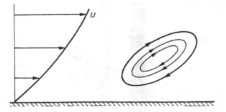

Fig. 13.22. The deformation and rotation of a fluid drop flowing in a shear field such as the Poiseuille flow in **Fig. 13.21**. The internal fluid circulation patterns are also shown.

The motion of spheres If we introduce into the flow a small rigid sphere which is neutrally bouyant we shall observe first that it moves along with the flow and second that it rotates at a constant angular velocity about a transverse axis. The speed of translation along the tube is slightly less than the velocity *u* the fluid would have at the centre of the sphere; the difference (known as the slip velocity) increases as the particle size increases. The rotation results from the distribution of shear rate over the particle surface, and takes place in the sense shown in **Fig. 13.21** and earlier in **Fig. 10.2**. The sphere also experiences a varying stress normal to its surface, as shown in **Fig. 13.21**; thus, part of the surface is subjected to compression while the remainder experiences a tensile stress. At these very low flow rates the rigid sphere is observed to move along the streamlines of the flow and not experience any migration across them.

The behaviour of an immiscible fluid drop (i.e. a flexible sphere) introduced into the same flow is very different from that of a rigid sphere. Because of the variation in normal stress around the surface of the drop, it tends to be deformed into an ellipsoid, as shown in **Fig. 13.22**. At equilibrium the fluid stresses tending to distort the drop are balanced by the interfacial tension forces which tend to restore the spherical shape. Close inspection of the surface of the drop shows that it is rotating in the same direction as a solid sphere rotates; but, in addition, the interior fluid rotates, as shown in **Fig. 13.22**. It is also observed that the drop migrates radially towards the axis of the tube from whatever radial position it originally occupies (**Fig. 13.23**). The radial migration velocity is greatest nearest to the wall and decreases as the drop moves towards the axis. It has been shown theoretically that this radial migration results from the deformability of the drop, the distribution of shear rate in the fluid and the presence of the wall. As the deformability becomes less, so does the migration velocity. The migration velocity increases as the particle size increases relative to the tube radius, and as the flow rate (and thus the gradient of shear rate across the drop) increases. Clearly, if the drop was flowing along in a region of uniform velocity, there would be no shear forces, no deformation and, as a result, no migration.

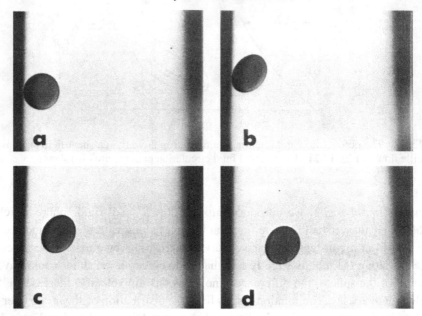

Fig. 13.23. Photomicrographs of the deformation and accompanying radial migration of a suspended drop; the camera was moved along the tube at the speed of the drop. (a) Drop at rest near wall. (b), (c), (d) Successive radial positions of the drop deformed into an ellipsoid as liquid flows upward through the tube. Note that there is migration from the wall, the rate decreasing with the decreasing deformation and shear rate as the drop moves towards the axis. (600 μm drop, tube diameter 0.8 cm (ratio tube/particle diameter = 13).) (From Mason and Goldsmith (1969). The flow behaviour of particulate suspensions. In *Circulatory and Respiratory Mass Transport* (ed. Wolstenholme). J. A. Churchill, London.)

The motion of rods and discs The motion of rigid rods and discs is somewhat more complicated than that of rigid or liquid spheres. Translation of such particles along the tube is still associated with rotation, because even when the rod axis or disc plane is parallel to the flow there is still a velocity gradient across it, as shown in **Fig. 13.24**; but the rotation is no longer at a constant angular velocity. Because the force causing their rotation increases as their angle φ to the flow increases, both rods and discs tumble as they move along the tube. They spend most of the period of rotation aligned with their long axis nearly parallel to the flow and flip rapidly in the remainder of the cycle.

Flexible rods and discs are also subjected to deformation, as shown in **Fig. 13.24**, as a result of the alternating compression and tension on the particle. Flexible particles are also observed to migrate towards the tube axis in a similar way to deformable liquid drops. Rigid rods and discs do not show this migration at low flow rates.

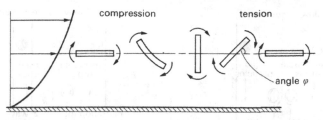

Fig. 13.24. The rotation of rods and discs in a sheared flow. Notice that, if the rod or disc is flexible, it may become deformed, as shown, during the compression part of the cycle.

The motion of single particles at high flow rates At higher Reynolds numbers (greater than unity), where inertial forces become important within the flow, deformable bodies still migrate towards the tube axis. But in this case rigid particles also migrate, though this migration is not always from wall to axis.

The behaviour of rigid particles under these conditions is referred to as the 'tubular pinch' effect or 'Segré–Silberberg' effect after those who first reported it. Particles initially near the wall move towards the axis and particles near the axis move towards the wall; they all move towards an equilibrium radial position approximately 0.6a from the axis, where a is the tube radius. The migration velocity depends upon radial position and decreases as the equilibrium location is approached; the velocity increases with Reynolds number and as the particle size increases relative to tube size. The underlying mechanisms controlling the motion are complex and involve both the inertia of the fluid and the interaction of the particle with the wall.

The types of motion of rigid and deformable particles at very low and somewhat higher flow rates are summarized in **Fig. 13.25**. It is clear that, except for rigid particles at very slow flow rates, there is always a migration of particles away from the tube wall.

The motion of single red blood cells in Poiseuille flow The motion of isolated red cells in plasma reflects the behaviour of both rigid and deformable particles. When the tube flow rate is low enough for the local shear rate G in the vicinity of the cell to be less than approximately $20\,s^{-1}$ the cell is seen to rotate like a rigid disc, retaining its biconcave shape. The flipping and the periodicity of rotation are just as would be predicted for a rigid disc of the same diameter, and the cell also rotates about its minor axis in the same manner as a rigid disc. This behaviour is summarized in **Fig. 13.26**. At these low shear rates the cells spend approximately 50% of the time with their major axes aligned within $\pm20°$ of the direction of flow and do not exhibit radial migration.

At shear rates greater than $20\,s^{-1}$ the behaviour of normal red cells progressively changes from that of rigid discs. The cells spend a greater time with their

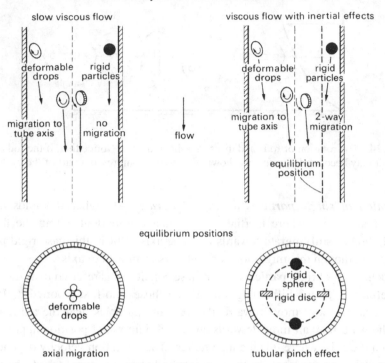

Fig. 13.25. Schematic representation of the differences in the observed migration of rigid and deformable particles in the median plane of a tube, on the left at very low Reynolds numbers and on the right at Re significantly greater than unity. In the lower part, the tube is shown end on with the equilibrium positions reached in migration due to particle deformation (left) and that due to inertia of the fluid – the tubular pinch effect (right). (After Goldsmith (1971). Red cell motions and wall interactions in tube flow. *Fedn. Proc. Fedn. Am. Socs. Exp. Biol.*. **30**, 1578.)

orientation close to that of position 3 in **Fig. 13.26**. At shear rates in excess of $100\,\mathrm{s}^{-1}$ the vast majority of the cells are observed to lie in position 3. Careful study of the cell membrane suggests that it is in motion round the interior. Measurement of the cell dimensions indicates that it has become deformed but that its biconcave shape is retained. At these higher shear rates, radial migration from the wall is also observed, and the migration velocity increases with the rate of shear. At very high shear rates of the order of $5000\,\mathrm{s}^{-1}$, single red cells experience the tubular pinch effect, but with an equilibrium position nearer to the axis than would be obtained with a rigid disc.

At shear rates less than about $50\,\mathrm{s}^{-1}$, rouleaux (Chapter 10, p. 164) can be observed in the flow; these behave just like flexible rods, and exhibit strong radial migration. As the shear rate is increased, the rouleaux become progressively broken down in size by the stretching and shearing forces acting during their rotation.

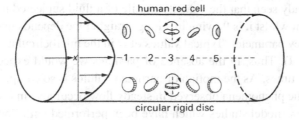

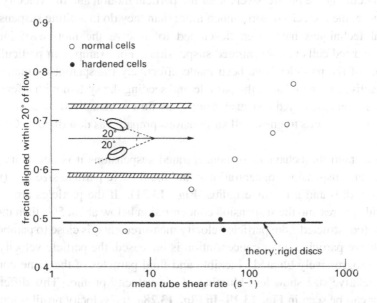

Fig. 13.26. The effect of suspending rigid discs and human red cells in Poiseuille flow at shear rates below $20\,\text{s}^{-1}$. The figure shows tracings from photomicrographs of the rotating particles at various orientations in one half revolution. The short dashed lines show the axes of revolution of the particles whose angular velocities are greatest at positions 1 and 5. (After Goldsmith (1971). Red cell motions and wall interactions in tube flow. *Fedn. Proc. Fedn. Am. Socs. Exp. Biol.* **30**, 1578.)

Fig. 13.27. Fraction of normal and glutaraldehyde-hardened cells lying with their major axes oriented within ±20° of the direction of flow (as illustrated) as a function of the shear rate. The horizontal line showing an orientation distribution independent of shear rate is that calculated for rigid discs whose periods of rotation at a given shear rate are the same as those of human red cells. (After Goldsmith (1971). Red cell motions and wall interactions in tube flow. *Fedn. Proc. Fedn. Am. Socs. Exp. Biol.* **30**, 1578.)

Studies have also been performed with red cells which have been hardened with glutaraldehyde. Such cells behave like rigid discs at all shear rates (**Fig. 13.27**) and do not exhibit radial migrations until shear rates are high enough for the tubular pinch phenomenon to occur.

We have already seen that the velocity profile of a fluid subjected to an oscillatory pressure gradient whilst it is flowing in a long straight pipe depends upon the value of α, the Womersley parameter. Typical values of α in the microcirculation are less than 0.08 (see **Table I**). Thus, in the absence of particles we would expect quasi-steady conditions at all times. As the motion of particulate fluids is so complex, it is not safe to assume that the phenomena observed in steady flow occur also in oscillating flows. However, the few model studies which have been performed with both particles and red cells indicate that the processes described above do occur.

The flow of concentrated suspensions of particles and red cells As the concentration of the flowing suspension in a tube is increased above the very low value of about 0–5%, experiments have shown that particle–particle interactions or near collisions begin to occur. These progressively alter the particle motion and the velocity profile. Furthermore, the red cells deform much more than they do in a dilute suspension.

Special techniques have been developed to observe the motion of individual particles and red cells in concentrated suspensions. For example, in particulate systems most of the particles have been made effectively transparent by matching the fluid refractive index to that of the particle and seeding the system with a few opaque particles. In the case of red cell suspensions, the flow is visualized by adding a small proportion of red cells to ghost cell suspensions prepared as described in Chapter 10 (p. 160).

To understand the behaviour of concentrated suspensions it is necessary to consider both the suspension concentration c and the relative particle size b/a (where b is particle radius and a the tube radius, **Fig. 13.21**). If the particles are relatively small rigid spheres and the suspension concentration below about 5%, then the velocity profile (constructed from particle velocity measurements) is close to parabolic; but as the relative particle size or concentration is increased, the particle velocity profile becomes progressively blunted. Flexible and fluid particles of the same concentration and relative size show less blunting of the velocity profile. This difference in behaviour can be seen in **Fig. 13.28**. In **Fig. 13.28**a, the velocity profiles are plotted for large and small rigid spheres and large liquid drops, all at the same concentration of approximately 30%. It can be seen that at this concentration the particle velocity profile obtained with the large rigid spheres is flat, and the plug flow region, in which all spheres move with uniform axial velocity, extends right across the tube almost to the wall. Smaller rigid spheres do not show such a wide zone of plug flow in the core. When rigid sphere behaviour is compared with that of liquid drops, it can be seen that large liquid drops exhibit less blunting than rigid spheres of half their diameter.

In the case of rigid particles, the velocity profile is determined solely by the suspension concentration and relative particle size and is independent of the flow rate. Liquid drop and flexible particle suspensions, on the other hand, show a dependence

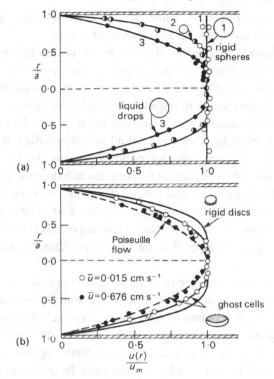

Fig. 13.28. Velocity profiles at very low Reynolds numbers in straight tubes for particle suspensions and ghost red cells. Local velocities $u(r)$ are plotted as ratios of the centre-line velocity u_m; this means that profiles are scaled so that centre-line velocity is unity in all cases. In (a) velocity profiles are shown for the same concentrations of (1) large ($b/a = 0.12$) and (2) small ($b/a = 0.06$) rigid particles and (3) large ($b/a = 0.12$) liquid drops. In (b) velocity profiles are also shown for suspensions of ghost red cells and rigid discs of approximately the same size and concentrations. The profiles for the ghost cells are shown at two different flow rates. (From Goldsmith (1972). The flow of model particles and blood cells and its relation to thrombogenesis. In *Progress in Hemostasis and Thrombosis*. By permission of Grune & Stratton Inc., New York.)

of the velocity profile upon flow rate, in that the degree of blunting decreases as the flow rate increases.

Figure 13.28b contrasts the behaviour of suspensions of ghost red cells with rigid discs of approximately the same size and concentration. The particle velocity profile for the discs contains a wider region of plug flow than is present in the case of the ghost cell suspensions. The cell suspensions also show how increasing the flow rate causes the plug flow region to diminish. At the higher flow rate, the ghost cell velocity profile is close to parabolic.

The detailed mechanism underlying this difference in behaviour is extremely complex, but it clearly results from the deformability of the drops and cells. In concentrated suspensions of red cells, severe deformation of the membrane is evident at low shear rates, in the region of $3\,\mathrm{s}^{-1}$, which is very much less than the rate of about $100\,\mathrm{s}^{-1}$ necessary to deform individually suspended red cells.

It must be emphasized that these velocity profiles do not necessarily reflect the velocity profile within the fluid itself, particularly near the wall. This is obvious in the case of the suspension of rigid spheres in which the particle velocity profile was flat all across the tube; in the vicinity of the wall, the fluid velocity must decrease rapidly to be zero at the wall because of the no-slip condition (p. 38). However, at least for the motion of rigid particle suspensions, the fluid velocity profile is determined by the particle velocity profile and is independent of flow rate. This may be deduced from the fact that the 'viscosity' of rigid particle suspensions is independent of flow rate and determined solely by particle size and concentration. Because the viscosity of these suspensions is independent of the shear rate, they are said to be 'quasi-Newtonian'. This constancy of viscosity is not demonstrated by suspensions of droplets, since their particle velocity profiles depend on flow rate; this is presumably true also of the fluid velocity profile.

There is only qualitative evidence of radial migration in concentrated suspensions; a small depletion in particle concentration near the wall is observed and the tubular pinch effect has been reported. In the case of red cell suspensions of normal haematocrit the situation is more complicated. In flows at reasonably high shear rates, the red cells move as individual cells and there appears to be a thin cell-free layer near the wall. However, at lower shear rates the cells tend to aggregate into rouleaux, and the cell-free layer increases considerably in thickness; such an increase is to be expected, because with larger particles the forces causing radial migration are larger.

The viscosity of whole blood These observations on the motion of concentrated suspensions of particles and red cells provide a qualitative explanation of the observed effect of shear rate on whole blood viscosity (Chapter 10, **Fig. 10.16**). At the lowest shear rates, the majority of the blood cells combine to form rouleaux. These are long, possess a degree of rigidity and provide an interlocking structure which makes the blood behave like a solid. As the shear rate is increased beyond the yield stress of this solid (Chapter 10, p. 175), so the shearing forces bend and break the rouleaux, so that the bridging lattice breaks down. As the shear rate is further increased the rouleaux length is progressively reduced till the suspension consists of single red cells. Observations of the distortion of flexible rods in concentrated and in dilute suspensions indicate that even at low shear rates the rods are much more distorted when in concentrated suspension. Thus, the breakdown process of the rouleaux in whole blood would be expected at very much lower shear rates ($1\,\mathrm{s}^{-1}$) than those required for dilute rouleaux suspensions. The rapidly decreasing viscosity of whole blood at rates of about $1\,\mathrm{s}^{-1}$ may thus be attributed to the breakdown of rouleaux and to the decreasing particle size.

The observed decrease in whole blood viscosity at shear rates between about 10 and $100\,s^{-1}$ is probably the result of the progressive distortion of individual red cells in the suspension. As explained above, they experience considerable distortion in concentrated ghost cell suspensions at shear rates as low as $3\,s^{-1}$. Finally, at very high shear rates (of the order of $5000\,s^{-1}$) single red cells begin to behave as though they were rigid particles (p. 385). However, the fact that there is a plateau in the viscosity–shear rate curve of whole blood at shear rates above about $150\,s^{-1}$ suggests that red cells in concentrated suspension behave like rigid particles at this relatively low shear rate.

Radial dispersion of red cells When concentrated suspensions of red cells flow in a tube there are continuous intercell collisions and individual red cells move along very erratic paths. This behaviour has been studied using red cells in ghost suspensions, and it has been found that the magnitude of a cell's radial fluctuations depends upon its radial position (**Fig. 13.29**).

By measuring the radial displacements it is possible to obtain a diffusion coefficient (see Chapter 9, p. 132) for the red cells. In the example described in **Fig. 13.29**, the diffusion coefficient ranged from $3 \times 10^{-8}\,cm^2\,s^{-1}$ near the tube axis to $1.5 \times 10^{-7}\,cm^2\,s^{-1}$ at a radial position $r = 0.7a$. This is considerably greater than that anticipated on the basis of Brownian motion resulting from the impact of the liquid molecules, which would suggest a diffusion coefficient of the order of $4 \times 10^{-10}\,cm^2\,s^{-1}$. It is interesting to see that the diffusion coefficient of the red cells is comparable to that which we would expect for macromolecules in free diffusion in plasma, where a typical value is $1 \times 10^{-7}\,cm^2\,s^{-1}$ for albumin in plasma at $37\,°C$.

The radial fluctuations of the red cells are associated with displacement of the adjacent plasma and other formed elements. Studies of the radial movements of $2\,\mu m$ rigid spheres (to simulate platelets) have shown that the displacements are even greater in a flowing red cell suspension than those of the red cells themselves (**Fig. 13.29**). If the diffusion coefficient is measured over short time intervals it is found to be approximately the same as for red cells at the same radial position. However, over longer time scales the coefficient is larger, as the microspheres move from the influence of one red cell to that of another. Thus, we can see that the effective mobility of platelets in flowing blood is very much enhanced by the presence of the red cells.

The cell-free layer All the dynamic processes described above imply that when blood is flowing along a tube, either steadily or in pulsatile manner, there is a tendency for cells to move away from the wall and to concentrate to some degree in the core of the tube. The effect of this is to create a layer of suspending fluid very close to the tube wall that is depleted of cells – the cell-free layer. This layer is not entirely free of cells, for the high core concentration and the agitated nature of the flow tend to push cells into this region where they collide with the wall (**Fig. 13.29**).

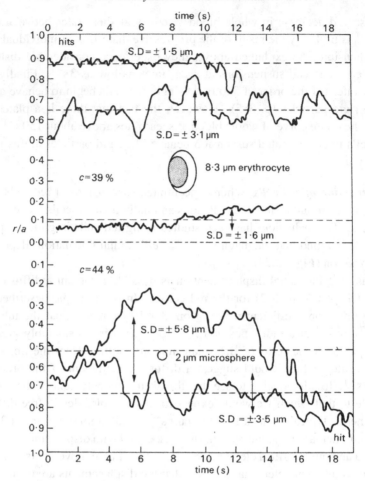

Fig. 13.29. Variation in radial position with time of tracer red cells (upper part) and polyvinyl toluene latex microspheres (lower part) in concentrated ghost cell suspensions flowing at a mean velocity of $0.015\,\mathrm{cm\,s^{-1}}$ through a $76.5\,\mu m$ glass tube. The size of the particles relative to the tube is shown. Points at which the cell and microsphere were observed to touch the wall are marked as 'hits'. The microsphere adhered to the wall after colliding with it. (After Goldsmith (1971). Red cell motions and wall interactions in tube flow. *Fedn. Proc. Fedn. Am. Socs. Exp. Biol.* **30**, 1578.)

However, it is worth remembering that even in the absence of flow a cell-free layer is formed for purely geometrical reasons. Imagine a uniformly packed bed of spheres (**Fig. 13.30**) in which a circle is described to represent the wall of a tube containing the spheres. Clearly, spheres whose centres are closer than one sphere radius to the wall cannot fit in; this implies a reduction in sphere concentration near the wall.

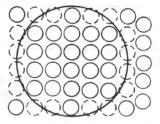

Fig. 13.30. The cell-free layer which results from the impossibility of having spheres within one sphere radius of the tube wall.

In flowing blood, the thickness of the cell-free layer depends upon both flow rate and cell-to-tube radius ratio. It has never been measured directly with any success in vivo because of the technical difficulties of viewing through vessel walls. However, measurements in glass tubes of the order of 100 μm in diameter suggest that it is approximately 2–4 μm in thickness. The effect of this cell-free layer is to create a sheath of low viscosity fluid (plasma) around the suspension which flows in the core of the tube. The physiological consequences of the cell-free layer in the microcirculation may be considerable and must be examined.

The dynamic haematocrit of blood If we measure the haematocrit of whole blood in a reservoir and flowing out through a tube attached to the reservoir, it will be found to be lower in the tube. This effect has been known for a long time and is partly responsible for the variations in blood viscosity in narrow tubes, as described later.

This reduced dynamic haematocrit is a consequence of the cell-free layer, for the suspended red cells move down the central portion of the tube at a relatively fast velocity, whereas plasma also flows in the slower moving region near the wall. This effect will occur regardless of the velocity profile. As a result, the mean time for red cells to traverse a given length of tube is less than that for the plasma; if the dynamic haematocrit were the same as the static value at the entrance, then we would end up with an increased concentration of red cells at the end of the tube! In fact, the dynamic haematocrit measured in any fairly narrow tube is always less than the static haematocrit, so that whilst the transit time per cell is reduced relative to the plasma, the total number of cells passing through the tube in that time is maintained at the appropriate level.

Plasma skimming If we consider a large microcirculatory vessel with a very small side vessel attached (**Fig. 13.31**) then it is likely that the small vessel, supplied relatively slowly from the larger vessel, will contain blood with a greater proportion of plasma than that in the parent tube. This is because the blood supply to this vessel

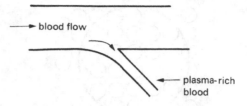

Fig. 13.31. Plasma skimming effect of a small side vessel.

comes mainly from near the wall in the large vessel where the blood is plasma rich. This phenomenon is known as *plasma skimming*.

Fårhaeus–Lindqvist effect Fårhaeus and Lindqvist measured the apparent viscosity of blood in tubes of various diameters from larger than 0.1 cm down to approximately 10 μm. They performed their experiments at shear rates which were sufficiently high for rouleaux formation to be absent and for viscosity to be independent of shear rate (Chapter 10, p. 174).

It was found that in tubes greater than approximately 0.1 cm in diameter the measured apparent viscosity was independent of tube diameter. However, in smaller tubes the apparent viscosity decreased as the tube size became smaller (**Fig. 13.32**); even in vessels of the same diameter as the undeformed red cell, the effect is very strongly apparent.

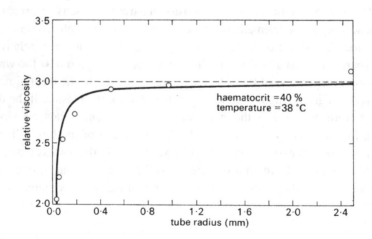

Fig. 13.32. The relationship between the apparent viscosity of blood relative to that of plasma and tube diameter in flow through narrow cylindrical tubes. (From Haynes (1960). Physical basis of the dependence of blood viscosity on tube radius. *Am. J. Physiol.* **198**, 1193.)

The implication of the Fårhaeus–Lindqvist effect is that the shear stress at the wall decreases as tube size is reduced. It will be remembered from Chapter 5 that the pressure gradient causing flow along a tube is balanced by the shear stress exerted by the wall on the fluid. In turn, the shear stress is determined by the product of the velocity gradient within the fluid at the wall and the local fluid viscosity. In a flowing particulate suspension such as blood, the cell-free layer viscosity is reduced, but at the same time the velocity gradient within the layer is greater than it would be in the presence of particles; thus, the velocity gradient at the wall may be elevated. Consequently, it is not possible to be sure whether the overall effect will be an increase or decrease in wall shear stress. In a large-radius tube the relative scale of the cell-free layer is small and in consequence adjustments in velocity profile within the cell-free layer occur over a relatively small distance; in a small tube, however, the scale of the cell-free layer increases and velocity profile adjustments can occur over a relatively greater distance; thus, the velocity gradient at the wall may not be as great as expected. The second factor affecting the shear stress is the dynamic haematocrit, which falls with diminishing tube or vessel size; as a result, the apparent viscosity also falls.

Velocity profiles in vessels The measurement of representative flow rates and velocity profiles in the very small vessels has proved even more difficult than the measurement of pressure. This has been mainly due to the difficulty of observing the movement of individual red cells in vessels larger than capillaries. But in addition, anyone who has observed a microvascular bed in vivo with a microscope will have noted the variability of flow in a given vessel; flow can be steady for a period of time and then suddenly slow down or stop altogether. Such changes in flow rate have not been closely correlated to vascular pressure; this is perhaps to be expected, because in such a network the pressures and flow rates in adjacent vessels are related.

In capillaries, where individual red cells can be observed in the flow, flow rates have been estimated on the basis of red cell velocity, assuming plug flow conditions in the vessel. Typical velocities measured in capillaries are to be found in **Table 13.1**. In somewhat larger vessels, measurements of flow rate have recently become possible using a photometric technique involving illumination of the vessel through two closely separated slits. This technique was at one time used to measure not only average velocities, but also velocity profiles across the lumen. The measurements suggested that velocity profiles were blunt. However, the technique has recently been shown to be faulty and to overestimate the blunting; there is strong evidence from other experiments that the velocity profile in whole blood flowing through narrow tubes is virtually parabolic at all haematocrits. Closely similar results are obtained with ghost cell suspensions, as indicated in **Fig. 13.28**. Thus, it would appear that the red cell is sufficiently flexible to allow an approximately parabolic velocity profile to be established in vitro under all conditions which might reasonably be expected to occur in vivo. Until good direct experimental information is available, it is reasonable, therefore, to assume that velocity profiles are approximately parabolic in all vessels of the microcirculation other than capillaries. However, the cell-free layer very near the wall may have a marked effect on the value of the shear rate at the wall and, hence, it would not be appropriate to compute the pressure drop along a vessel on the basis of the velocity profile in the core of the vessel.

The flow conditions in the vicinity of vessel branches are unknown, but it is unlikely that the profiles will be significantly different from parabolic for a distance of more than a radius from the junction, and thus they will be unimportant in determining the flow. This is because the Reynolds numbers in the small vessels are very much less than unity, and it was shown in Chapter 5 (p. 54) that the entrance length for Newtonian fluids at such Reynolds numbers is of the order of one tube diameter.

Blood flow in capillaries

Mammalian capillaries range in diameter from 10 μm down to 3 μm, so that blood cells, in particular red cells, have to flow down them in single file. In capillaries whose diameter is less than the diameter of a red cell (7–8 μm) the cells must be deformed in order to pass through the vessel. Observation shows that normal red cells take up various configurations within small capillaries; some are bent back in an approximately symmetrical parachute shape (**Fig. 13.33**a), while others may roll up and flow down the vessel end on (**Fig. 13.33**b). As we have seen, the shape of a normal cell permits considerable distortion without change of either volume or surface area (see p. 161). Abnormal cells which are hardened or are almost spherical cannot be deformed so easily and, therefore, cannot pass through the smallest vessels.

Experiments have been performed in which red cells were made to pass through membranes containing pores of different sizes. It is found that normal cells can pass through undamaged when the pore diameter is greater than 3 μm, but that they are damaged when forced through smaller pores (**Fig. 13.34**). Cells which are artificially hardened or are made to swell into a spherical shape by osmotic means, both of which make them less flexible, cannot pass through pores less than 8 μm in diameter without haemolysis.

There is a hereditary disease of humans, called *hereditary spherocytosis*, in which the red cells are spherical and have a diameter of about 6 μm. The patient suffers from excessive haemolysis, so that, although red cell production is increased several fold, it cannot keep pace with cell destruction and anaemia develops. Removing the spleen improves the situation. The reason seems to be that, in flowing through the capillaries in the splenic pulp, the red cells have to squeeze through gaps as little as 3 μm across. Normal cells can do this without excessive haemolysis, in the same way as they can pass through artificial micropores of similar diameter, but the abnormal spherical cells cannot.

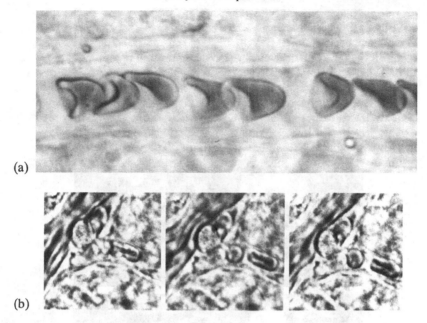

(a)

(b)

Fig. 13.33. Photographs of the configuration of red cells flowing in small capillaries. (a) Parachute configuration. (b) Rolled up, flowing end on. (From Skalak and Branemark (1969). Deformation of red blood cells in capillaries. *Science*, **164**, 717.)

In describing the mechanics of red cell motion in narrow capillaries, we can distinguish two situations according to the ease with which the cells fit into the vessels. First, when the capillary has a diameter noticeably larger than that of the cell, the cells can fit into the tube without distortion (there is *positive clearance*), and suffer little or no distortion as they flow in single file along it. In this case, the pressure in the plasma is comparable everywhere (apart from a more or less uniform pressure gradient driving the flow), and is unlikely to cause significant distortion of the cells. Therefore, the deformability can be neglected in an analysis of the dynamics. The second and more usual case arises when the capillary diameter is smaller than the cell diameter (*negative clearance*). In this case the cell has to be deformed in order to fit into the capillary, and suffers further distortion as it flows. High pressures must be generated locally in the thin layer of fluid round the edge of the cell in order to deform it, and therefore depend on the elastic properties of the cell (and of the capillary wall, although that is considerably stiffer than the cell). The whole flow pattern is dictated by the motion in these thin regions of high pressure, which can be analysed by the methods used in the theory of hydrodynamic lubrication (outlined on p. 397). We shall consider the two cases separately.

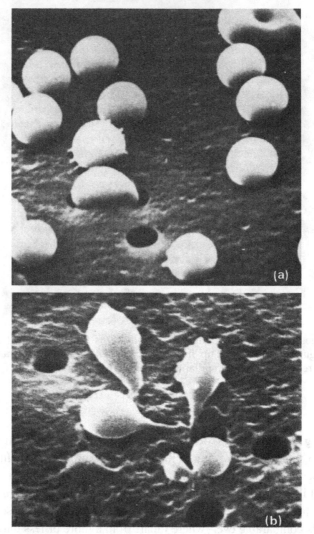

Fig. 13.34. Scanning electron micrographs of red blood cells (a) before and (b) after passing through a polycarbonate sieve of pore diameter 2.2 μm. In (a) many of the cells appear spherical because they have already partially entered the pores. (From Chien, Luse and Bryant (1971). Hemolysis during filtration through micropores: a scanning electron microscopic and hemorheologic correlation. *Microvasc. Res.* **3**, 197.)

Positive clearance In this case the detailed shape and spacing of cells, as well as their elastic properties, are relatively unimportant. The main feature of the flow can be understood by considering a simple model in which the cells are replaced by rigid spheres, flowing steadily and symmetrically in single file down a cylindrical tube (**Fig. 13.35**). A mathematical analysis can be carried out for this simplified geometry,

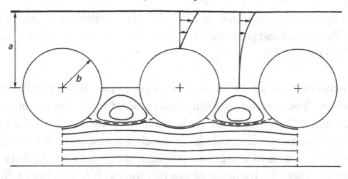

Fig. 13.35. Streamlines and velocity profiles for flow of rigid spheres down a rigid tube. Observer is moving with spheres so that they appear fixed and walls move. (From Wang and Skalak (1969). Viscous flow in a cylindrical tube containing a line of spherical particles. *J. Fluid Mech.* **38**, 88.)

and we shall summarize the numerical results of this analysis, emphasizing their physical basis. The chief parameter governing the motion, given that inertial forces can be neglected, is the ratio of the radius b of the sphere to the radius a of the tube. The way the relative velocity of the spheres and the fluid depends on this ratio can be illustrated by the two limiting cases, in which b/a is very small and in which $b/a = 1$. In the former case, which is physiologically unrealistic, the spheres do not significantly affect the fluid motion, which is Poiseuille flow. Because they are situated on the centre-line, they have a velocity equal to twice the average velocity of the fluid. As b/a is increased, the ratio between the particle velocity and the average velocity of the flow falls, approaching unity as b/a approaches unity. In this extreme case the particles fill the tube and the fluid cannot pass around them, being carried along in 'boluses' between the particles. Except in this extreme case, however, the particles travel faster than the fluid (resulting in a reduced dynamic haematocrit, as discussed on p. 390). This result implies that, relative to the particles, there is a 'leak back' of fluid which drives a circulatory motion in the 'bolus' of fluid carried along between them (**Fig. 13.35**).

Another important result is that the ratio of the pressure drop required to push the suspension of spheres down a narrow tube at a given flow rate to the equivalent pressure drop for the suspending fluid alone increases with increasing radius ratio b/a. For example, when (1) $b/a = 0.9$ and (2) the spheres are touching (which gives the greatest excess pressure drop for a given b/a), the ratio reaches a value of about 2; that is, the total pressure drop for fluid plus spheres is about twice the fluid-only pressure drop. This implies that the apparent viscosity of the suspension is approximately twice that of the suspending fluid.

If we were to apply this result to blood in a capillary, we would predict its viscosity to be about twice that of plasma. In fact, we know that the viscosity of whole blood as measured in a large-scale viscometer in bulk is about three times that of plasma (p. 174), so the Fårhaeus–Lindqvist effect (p. 390) is shown to operate even when b/a is as large as 0.9 (**Fig. 13.32**).

Negative clearance In this case the red cells have to pass down capillaries narrower than themselves. The cell must be squashed in order to fit into the tube whether it is moving or not. When it is at rest, it is jammed up against the vessel wall, or at least against the mucopolysaccharide layer which lines the wall. When the cell starts moving, under the action of a pressure gradient, there must be slip between it and the vessel wall, resisted by frictional or viscous forces. It is a matter of common experience that the frictional forces between two slipping surfaces are greatly reduced, once the motion has begun, by the introduction of a layer of fluid between the surfaces. This is the lubricating layer. The most convincing theoretical model of red cell motion in narrow capillaries is one in which such a lubricating layer is formed round the cell, and we here outline the principal features of this model and report its main physiological predictions.

We have seen that red cells can take up a number of different orientations as they flow along narrow capillaries, but we do not know which arise most commonly. They are all very difficult to analyse, and the only one for which an approximate quantitative theory has been developed is the axisymmetric parachute configuration. The physical principles underlying the motions are the same in each case, but the details of the pressure distribution around the red cell, for example, will be different.

Because the elasticity of the deformed cell tends to squeeze it against the vessel wall, there must be a relatively high pressure in the lubricating layer of plasma in order to keep the two surfaces separate. This is one of three factors which are crucial to a correct mechanical explanation of the motion. The second is that fluid within the layer must move in such a way that the pressure gradient and viscous forces within it are in balance. Third, the net pressure force acting on the cell must be in balance with the total viscous retarding force on it.

In analysing the motion of the cell and of the plasma in the lubricating layer, it is convenient to suppose that we, the 'observers', are travelling at the same speed (U) as the cell (**Fig. 13.36**). The cell then appears to be at rest and the capillary wall moving backwards with speed U. This is convenient because only if we move with the cell do the pressure and layer thickness appear constant in time.

A consideration of the flow within the lubricating layer shows that there must be leak-back of fluid past the cell in this case too. If there were no leak-back, the velocity profile everywhere would be linear (**Fig. 13.36**a); and, since the layer is thinner at X than at Y, the volume flow rate would be smaller there. That is impossible in an incompressible fluid and, hence, the profile cannot be everywhere linear; there must also be another component, as shown in **Fig. 13.36**b, and this in turn implies the existence of a backward (i.e. positive) pressure gradient. If the volume flow rate associated with this departure from the linear profile is Q, and if x is the distance

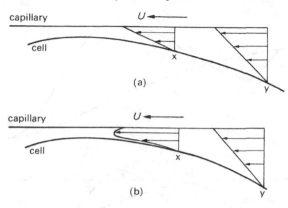

Fig. 13.36. Diagram showing the need for leak-back past a red cell in a capillary (cell taken to be at rest, capillary wall moving): (a) linear profile everywhere leads to non-uniform flow rate, which is not permitted; (b) continuity upheld by the superposition of a parabolic component, requiring a pressure gradient.

measured along the capillary wall from left to right, then the force balance in the layer requires that the pressure gradient (dp/dx) at a point where the layer thickness is h is the sum of two terms, one positive and proportional to Q/h^3 and the other negative, proportional to $-U/h^2$. The pressure gradient is given by

$$\frac{dp}{dx} = \frac{12\mu Q}{h^3} - \frac{6\mu U}{h^2},$$ (13.2)

where μ is the viscosity of plasma. It is the detailed analysis of the force balance leading to this equation which is called 'lubrication theory'.

The nature of the pressure distribution can now be predicted qualitatively as follows, with reference to **Fig. 13.37**. The pressure far upstream of the cell, p_+, is greater than that far downstream, p_-, since the average pressure gradient must be a favourable one (i.e. negative), as in the absence of cells. Far away from the cell the thickness of the lubrication layer becomes large and Equation (13.2) shows that the pressure gradient must be negative (since $1/h^3$ decreases more rapidly than $1/h^2$ as h increases). However, at or near the position where the cell would have maximum diameter if it were not distorted by the tube, the point X in **Fig. 13.37**, there must be a pressure maximum, because it is there that the cell has to be squashed the most. Just upstream of that pressure maximum the pressure gradient is positive, and there must be a minimum of pressure yet further upstream (the point Z) for consistency with the negative pressure gradient far away. Thus, the pressure distribution must be as shown in **Fig. 13.37**a. Equation (13.2) indicates that the minimum layer thickness is associated with the maximum positive pressure gradient, and should therefore occur at the point Y (**Fig. 13.37**b), further upstream than the minimum layer thickness in the absence of flow.

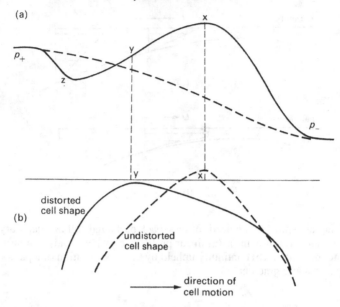

Fig. 13.37. (a) Approximate pressure distribution in the thin lubricating layer of plasma between red cell and capillary wall; (b) approximate shape of the distorted red cell. Broken line in (a) shows approximate pressure distribution in tube in absence of cell.

These qualitative predictions are borne out by calculations which incorporate detailed information about the elastic behaviour of red cells. The calculations also lead to the following conclusions, which can in principle be tested by physiological experiment.

(1) For a given overall pressure gradient, the volume flow rate is much *smaller* than for plasma-filled tubes (in contrast with the case of positive clearance).

(2) The cell velocity depends *nonlinearly* on the overall pressure gradient, decreasing very rapidly (proportional to the square of the pressure gradient) as the pressure gradient tends to zero.

(3) For a given positive clearance, the thickness of the lubricating layer is relatively insensitive to velocity.

(4) When the clearance is negative, the layer thickness decreases as the velocity (or pressure gradient) decreases; it is proportional to the pressure gradient as that tends to zero.

From this last result we conclude that if the pressure gradient or velocity becomes small enough, the lubricating layer of plasma will become so thin (a few molecules across) that it no longer acts like a continuous fluid layer and loses its lubricating properties. The layer may be supposed to 'seize up' and flow to stop. It would require a very much higher pressure gradient to overcome solid friction and restart the motion after seize-up than was required just before.

As yet it has not proved possible to confirm all these predictions by physiological measurement. It is known that the resistance to blood flow through capillaries is larger than predicted for a Newtonian fluid with viscosity equal to the viscosity of whole blood in bulk. This effect is the reverse of the Fårhaeus–Lindqvist effect apparent in larger tubes, and is in agreement with (1) above. It has also been verified in model experiments that there is a nonlinear relationship between pressure gradient and cell velocity, but physiological evidence is still conflicting. In particular, it appears that the prediction of seize-up of the lubrication layer is not borne out in vivo. One possible explanation is non-Newtonian behaviour of the plasma, which may be important because the predicted layer thickness at seize-up is less than 0.5 μm and is therefore comparable to the diameter of the largest plasma protein molecules. (The layer of mucopolysaccharide on the endothelial surface may also be involved.) Another explanation may be that the assumptions made about red cell elasticity are inadequate.

Both the high pressures in the lubrication layer and the circulating motions in the 'boluses' of plasma between cells may have an influence on mass transfer between the blood and the capillary wall.

Mass transport in the microcirculation

For a long time it was considered that the transport of materials from blood to tissue, and in the opposite direction, took place exclusively across the walls of the capillary vessels. However, it is now clear that transport is not thus confined, but occurs mainly in both the capillaries and the postcapillary venules in most microcirculatory beds. Thus, in this section we shall consider transport in and out of both types of vessel, though often the actual vessel being considered will not be defined and the term capillary used. The reader is referred to Chapter 9 for a general discussion of mass transfer.

Transport across the capillary wall does not usually appear to be active in form, because poisons such as potassium cyanide have no direct effects on transport rates. There are two principal mechanisms of transport: bulk fluid movement under the action of a hydrostatic pressure gradient and transcapillary molecular exchange, a diffusional process driven by a concentration difference across the vessel wall. An enormous range of types and sizes of molecules cross the wall, including water, dissolved gases such as oxygen and carbon dioxide, low molecular weight ions such as Na^+ and HCO_3^- and organic molecules whose sizes range from a molecular weight less than 60 (urea) to several millions (certain lipoproteins). A number of different pathways and mechanisms for transport exist and these will be considered in the following sections.

Whilst the two principal mechanisms of bulk and diffusional transport are independent, transport rates are linked because the driving pressure and concentration differences are related through osmotic effects, since the vessel wall acts as a semipermeable membrane.

Filtration and reabsorption of water within single capillaries It was widely appreciated in the nineteenth century that the hydrostatic pressure difference across the capillary wall is capable of transporting water from the blood into the surrounding interstitial space. However, it was another of Starling's great achievements in 1896 to realize that this filtration pressure is counteracted by a resorptive force due to the presence of *protein* in the plasma. Starling considered that the capillary wall was permeable to water, electrolytes and very small molecules within the plasma, but not to protein molecules. Consequently, as the water and electrolytes were filtered out of the plasma, an osmotic imbalance was established and a colloid osmotic pressure difference was set up opposing filtration. Thus, if the hydrostatic pressures within the capillary and tissue were p_c and p_t respectively and the colloid osmotic pressures in the two compartments were Π_p and Π_t, then the net filtration force Δp_F would be given by

$$\Delta p_F = (p_c - p_t) - (\Pi_p - \Pi_t). \tag{13.3}$$

When Δp_F is positive (i.e. when the hydrostatic pressure difference is greater than the osmotic pressure difference) filtration of fluid out of the capillary occurs. When Δp_F is negative, reabsorption of fluid from the interstitial space into the capillary will take place. Starling proposed that the interstitial space could be considered to possess an effectively uniform hydrostatic pressure p_t and colloid osmotic pressure Π_t. Thus, it was considered that at the arterial end of a capillary the capillary hydrostatic pressure was greater than plasma oncotic pressure; as a result, filtration out of the vessel took place (**Fig. 13.38**). At the venous end of the capillary the hydrostatic pressure within the capillary would have fallen and the plasma oncotic pressure would have become elevated above the hydrostatic pressure and as a result reabsorption of fluid would take place. At some point along the capillary there would be neither filtration nor reabsorption, for at that point the two driving forces would be equal. Starling further suggested that, under normal conditions, there was a small excess of filtration over absorption and that this accounted for lymphatic flow.

It was 31 years later, in 1927, that Landis provided experimental evidence to support Starling's hypothesis. He occluded single capillaries of the frog mesentery with a fine glass rod under direct microscopic observation (**Fig. 13.39**). When the capillary was occluded near to the venous end, the red cells upstream of the probe could be seen to oscillate in time with the heartbeat and slowly pack towards the probe. When the occlusion was at the arterial end, the cells downstream of the probe could frequently be seen to move out of the vessel and into the venule. These observations were

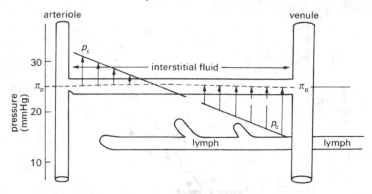

Fig. 13.38. Schematic diagram of a typical capillary to illustrate the filtration–absorption hypothesis of Starling. Capillary pressure p_c falls along the length of the vessel; capillary oncotic pressure Π_p is also shown. (After Landis and Pappenheimer (1963). Exchange of substances through the capillary walls. In *Handbook of Physiology*. Section 2: *Circulation* vol. II (eds Hamilton and Dow), pp. 961–1034. American Physiological Society, Washington, DC.)

interpreted as evidence to support Starling's hypothesis, since when the venous end of the capillary was occluded the whole capillary was exposed to arteriolar pressure and filtration of plasma resulted. When the arterial end was occluded, the capillary was exposed to venular pressure and net absorption of interstitial fluid generally occurred. Landis assumed that the red cells fitted tightly in the capillary with no leak of plasma around the cell margin and that the capillary cross-section was circular; then, from measurement of the speed of movement of the red cells along the capillary he was able to make direct estimates of the filtration and absorption rates per unit area of capillary wall. These he related to the mean capillary hydrostatic pressure obtained by micropuncture of side branches of the vessel under observation. Results from his studies are presented in **Fig. 13.40**. The slope of the line through the data is the

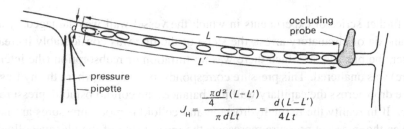

Fig. 13.39. The occlusion method for determining the filtration coefficient (J_H). On occlusion the red cells can be seen to oscillate in time with the heartbeat and slowly pack towards the probe. In time t the end red cell moves by $L - L'$. Vessel internal diameter is d.

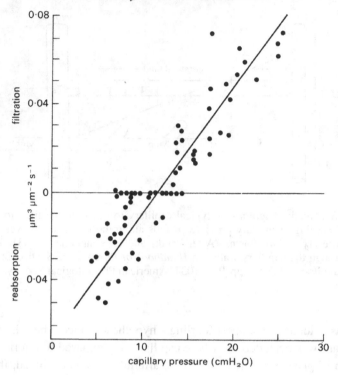

Fig. 13.40. The relation between fluid movement through the capillary wall and capillary pressure as measured in single capillaries of the frog mesentery. The slope of the line indicates the filtration coefficient of fluid filtered or reabsorbed. The intercept on the abscissa indicates the effective plasma colloid osmotic pressure. (From Landis (1927). Micro-injection studies of capillary permeability II: the relation between capillary pressure and the rate at which fluid passes through the walls of single capillaries. *Am. J. Physiol.* **82**, 233.)

filtration coefficient J_H (volume flow rate, per unit time, area and pressure drop; see Chapter 9, p. 133); this was found to be approximately $5.6 \times 10^{-5} \, \mu m \, s^{-1} \, N^{-1} \, m^2$.

In a further series of experiments in which the vessel wall was damaged by alcohol or anoxia or circulatory arrest, the slope of the line was considerably increased. However, the pressure at which there was no filtration or reabsorption (the intercept pressure) was unaltered. This pressure corresponds to the point where the hydrostatic pressure drop across the capillary wall just balances the colloid osmotic pressure difference. If in reality the tissue hydrostatic and colloid osmotic pressures are nearly zero, then the intercept pressure represents the mean value of the plasma colloid osmotic pressure.

The original technique of Landis has been modified to permit its application to mammalian capillaries whose diameters are less than those of the frog. Capillary

Table 13.3. *Capillary filtration coefficients*

Tissue	Coefficient (μm s^{-1} N^{-1} m^2)
Single capillaries	
Frog mesentery	4–16×10^{-5}
Rabbit omentum:	
arterial end	2–8×10^{-5}
venous end	16–25×10^{-5}
Rat cremaster muscle	1×10^{-5}
Cat mesentery[a]	
capillaries	10–30×10^{-5}
collecting venules	30–50×10^{-5}
Capillary beds	
Human forearm	1×10^{-6}
Dog hind limb	2.5×10^{-6}
Cat hind limb	3.5–5.2×10^{-6}
Rabbit heart	8.6×10^{-6}
Dog lung	0.22×10^{-6}

Note that the data for capillary beds have been calculated assuming figures for the capillary surface area per unit weight of tissue.
Data reproduced from Michel (1972). Flows across the capillary wall. In *Cardiovascular Fluid Dynamics* (ed. Bergel), vol. II. Academic Press, London.
[a]From Verrinder (1974). *Pressure and permeability studies in single vessels of mammalian microvasculature.* Ph.D. thesis, University of London.

filtration coefficients have now been measured in a number of tissues in different species; a few are listed in **Table 13.3**. In some cases venular filtration coefficients have also been measured, and it can be seen that they are greater than in the feeding capillaries. It has further been observed, in both skeletal muscle and the mesentery, that filtration out of the plasma occurs throughout the length of the capillary and in the case of the cat mesentery may even do so in venules as large as 20–30 μm in diameter. The balance point between filtration and reabsorption is thus not necessarily in the mid-capillary region but may be within the venule. It may even be non-existent in the mesentery with no reabsorption taking place, all the plasma filtrate being removed via the lymphatics.

The observation that the filtration coefficient varies along the length of the microvascular bed means that the simple form of Starling's hypothesis cannot be used to model the behaviour of the whole exchange bed. The balance point may, however, be computed from Equation (13.3), for at that point

$$p_c = (\Pi_p - \Pi_t) + p_t.$$

The plasma colloid osmotic pressure is approximately 3.0×10^3 N m^{-2} (30 cm H$_2$O) and the sum of tissue hydrostatic and colloid osmotic pressures was shown earlier (p. 376) to be about -0.5×10^3 N m^{-2} (-5 cm H$_2$O). Thus, at the balance

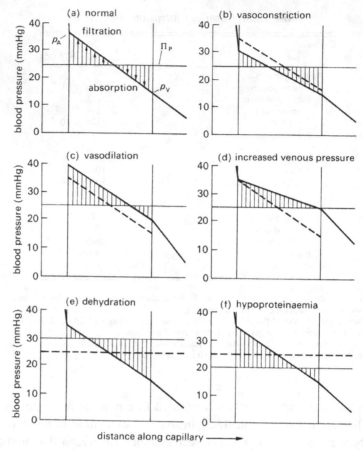

Fig. 13.41. Variations in filtration and absorption produced by changes in arterial and venous pressures, resistance and plasma colloid osmotic pressure, p_A and p_V, are the pressures at the arterial and venous ends of the capillary and Π_p represents plasma colloid osmotic pressure. The broken lines represent normal values taken from (a). (From Friedman (1971). Microcirculation. In *Physiology* (ed. Selkurt), pp. 259–273. Little Brown & Co., Boston.)

point, capillary hydrostatic pressure would be about $2.5 \times 10^3\,\mathrm{N\,m^{-2}}$ ($25\,\mathrm{cm\,H_2O}$). This is less than typical values for capillary pressures reported earlier (p. 370) but is representative of pressures to be expected within the venules.

The effect of changes in capillary hydrostatic pressure on capillary filtration and reabsorption can be seen in **Fig. 13.41**. In (a) the normal situation can be seen. The effect of vasoconstriction (b) is to decrease the capillary pressure; as a result, filtration is reduced and reabsorption from the interstitial space is enhanced. Vasodilatation (c) has the reverse effect. The effect of elevating venous pressure is to increase the capillary pressure, which favours filtration out of the capillary. The effects of dehydration (effective increase of plasma colloid osmotic pressure) and hypoproteinaemia (a reduction in plasma colloid osmotic pressure) are shown in (e) and (f) respectively.

Capillary pressure and filtration of water in whole organ preparations The Landis occlusion technique can be used only for the study of single capillaries in flat beds such as the mesentery and cremaster muscle; alternative methods are required for beds such as limb muscles and organs. To study such beds, the isogravimetric technique has been developed; in this, the isolated organ or limb is perfused and weighed continuously. Under steady-state conditions, the preparation is of a constant weight – the isogravimetric state. There is no net exchange of fluid between blood and tissue. If the blood flow rate is varied it is found that the weight of the preparation tends to change, but that such alterations can be prevented by adjustment of central venous pressure (p_v).

Assuming that the resistance to blood flow R_v between the capillaries and central veins is independent of blood flow rate Q, then the capillary pressure p_c is given by

$$p_c = p_v + R_v Q. \tag{13.4}$$

By plotting the blood flow rate under isogravimetric conditions for various central venous pressures, an estimate may be made of p_c from the intercept for venous pressure at zero flow rate (**Fig. 13.42**). The linearity of the relation between pressure and flow shown in the figure supports the assumption of a constant postcapillary resistance (R_v). Data for two experiments using plasma of differing colloid osmotic pressures are shown and the close correspondence between the estimated mean capillary pressure p_c and the plasma oncotic pressure Π_p, which is predicted by the Starling hypothesis, can be seen.

The isogravimetric technique may also be used to obtain an estimate of the capillary filtration rate for a whole organ bed. Elevation of venous pressure with a constant blood flow rate results in a sudden rise in the weight of the preparation followed by a continued but slower weight gain. The initial rapid increase is considered to result from the filling of the venous capacitance vessels, whilst the slow continuing rise in weight is due to net filtration into the tissues. The rate of filtration per unit weight of tissue is found to be linearly proportional to the increase in capillary pressure above the isogravimetric value calculated on the basis of Equation (13.4). The slope of the line is the filtration rate per unit weight of organ per unit pressure difference – this is not the filtration coefficient, as no estimate of the area over which filtration occurs has been made.

Measurements of the filtration rate have been made in a number of tissues and a few measurements are given in **Table 13.4**. It can be seen that filtration occurs far more readily through fenestrated endothelium, such as is present in the intestine, than

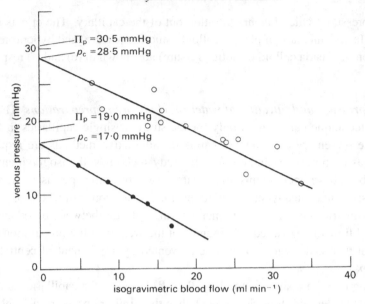

Fig. 13.42. The relation between blood flow rate and venous pressure in the isolated perfused hind limb of the cat. The two sets of data were obtained for different plasma colloid osmotic pressures. (From Pappenheimer and Soto-Rivera (1948). Effective osmotic pressure of the plasma proteins and other quantities associated with the capillary circulation in the hind limb of cats and dogs. *Am. J. Physiol.* **152**, 479.)

through the continuous endothelial wall of skeletal muscle capillaries, even though the density of capillaries per unit weight of intestine is less than that in muscle. In practice, the capillary filtration coefficient for a particular preparation may be calculated if the capillary surface area per unit mass of tissue is known; filtration coefficients for a number of tissue beds have been estimated, some of which are shown in **Table 13.3**.

Table 13.4. *Capillary filtration rates related to endothelium type*

Tissue	Capillary type	Filtration rate $(\mathrm{g\,s^{-1}N^{-1}m^2}$ per kg muscle)
Cat hind limb	Continuous	18×10^{-4}
Rat hind limb	Continuous	40×10^{-4}
Rabbit heart	Continuous	30×10^{-4}
Dog lung	Continuous	2×10^{-4}
Cat intestine	Fenestrated	2×10^{-2}
Cat kidney	Fenestrated	20×10^{-2}

Data derived from sources given in Altman and Dittmer (eds) (1971). *Biological Handbooks, Circulation and Respiration*, pp. 502–3. Federation of American Societies of Experimental Biology.

The filtration coefficients thus obtained are always considerably less than those obtained more directly on single capillaries. However, this is perhaps not too surprising, as morphometric studies of capillary surface area obtained post mortem give a poor indication of the number of capillaries open and functioning under more physiological conditions. Any overestimate of capillary surface area will lead to an underestimate of the filtration coefficient. Furthermore, measurements on single capillaries involve surgical exposure, cannulation and thermal effects from microscope stage lighting, all of which may increase permeability.

The dependence of plasma oncotic pressure on protein concentration As can be seen from Equation (13.3), the rate of filtration of water out of the plasma depends upon the values of both hydrostatic and osmotic pressures in the plasma and tissue space. As water and small molecules are filtered out of the plasma, the protein concentration increases; the resultant increase in plasma oncotic pressure should be predictable by application of van't Hoff's law (Equation (9.14), p. 140), which relates osmotic pressure Π to molar concentration c for dilute solutions:

$$\Pi = cRT,$$

where T is the absolute temperature and R the Universal Gas Constant. However, such predictions underestimate the actual increase in osmotic pressure; as the concentration of protein increases the osmotic pressure increases disproportionately. This effect can be seen from **Fig. 13.43**, in which the plasma oncotic pressure is plotted as a function of plasma protein concentration. Also plotted is the osmotic pressure of an albumin solution as a function of concentration compared with that predicted on the basis of the van't Hoff relationship assuming dilute solutions. The cause of this behaviour is very complex, depending partly on the effects of the large molecules trapping water and also on electrical charge effects on the molecules themselves.

Evidence for the existence of filtration pores in the capillary wall The transport of water and water-soluble small molecules across the capillary wall for a given hydrostatic or osmotic pressure gradient is far greater than it is across the red cell membrane and it has long been postulated that such substances cross the wall via narrow pores. The suggestion that lipid-insoluble materials can only filter or diffuse through a very small fraction of the capillary wall is also highly suggestive of the existence of specialized water-filled pores or slits penetrating the endothelial lining of the vessels.

It was shown earlier (p. 362) that the cleft between endothelial cells is a long slit approximately 15–20 nm wide, except over a narrow zone where adjacent membranes almost fuse together. When the zone of fusion is intermittent (macula occludens), the oval slit or gap between adjacent sites of fusion is thought, from electron microscopic studies, to be approximately 8 nm wide. Thus, this gap would provide the greatest

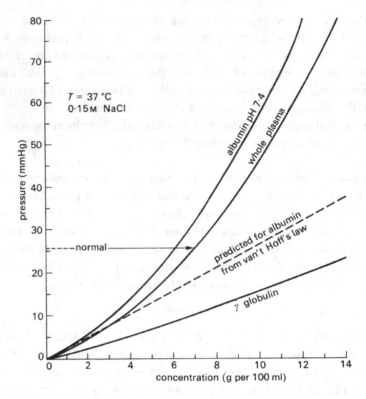

Fig. 13.43. The variation of osmotic pressure with concentration for whole plasma, albumin and gamma-globulin. The dotted line indicates the predicted osmotic pressure for albumin based upon van't Hoff's law. (After Landis and Pappenheimer (1963). Exchange of substances through the capillary walls. In *Handbook of Physiology*. Section 2: *Circulation* vol. II (eds Hamilton and Dow), pp. 961–1034. American Physiological Society, Washington, DC.)

restriction to filtration or diffusion along an intercell cleft. It was also pointed out that fenestrated capillaries may be so designed to provide a good diffusion path via the circular fenestrae and that discontinuous capillaries possessed very large intercell gaps which would clearly provide a very low transport resistance.

It is considered that the intercell cleft is in fact the predominant pathway for the filtration of water and small water-soluble molecules.

Diffusion across the capillary wall The process of filtration and reabsorption can occur only through water-filled passages across the capillary wall. In principle, however, diffusion of molecules across the wall can take place across the membrane of the endothelial cells. Indeed, it does seem to do so for highly lipid-soluble substances

Table 13.5. *Permeabilities of skeletal muscle capillaries to lipid-insoluble molecules in skeletal muscle*

Molecule	M.w.	Radius (nm)	P ($\times 10^{-6}$ cm s^{-1})	P/D ($\times 10^{-3}$ cm^{-1})
H_2O	18	0.15	0.54	17.2
Urea	60	0.26	0.26	13.5
Glucose	180	0.37	0.090	10.0
Sucrose	342	0.48	0.050	6.9
Raffinose	504	0.57	0.039	6.0
Inulin	5 500	1.30	0.005	2.6
Myoglobin	17 000	1.90	0.0004	0.3
Serum albumin	67 000	3.60	$<10^{-5}$	0

Renkin (1959). Capillary permeability and transcapillary exchange in relation to molecular size. In *The Microcirculation: Factors Influencing Exchange of Substances across a Capillary Wall.* University of Illinois Press.

such as the respiratory gases; with such molecules it appears that the area available for diffusion is virtually the total surface area of the membrane.

Larger molecules and lipid-insoluble molecules, however, do not appear to be capable of diffusing across the endothelium with such facility. Nevertheless, the capillary wall possesses a far higher permeability to such substances than does the red cell membrane and the wall also has a low but definite permeability to larger molecular weight molecules such as albumin.

It will be recalled that, in free dilute solution, the diffusional rate of transport J_s of a substance may be described by Fick's law (p. 130):

$$J_s = -D_s A \frac{\Delta c}{\Delta x}, \tag{13.5}$$

where $\Delta c / \Delta x$ is the gradient of concentration of the substance in the direction of transport, A is the cross-sectional area across which transport occurs and D_s is the diffusion coefficient. As was explained on p. 138, the above equation may be conveniently rewritten

$$J_s = P_s A_m \Delta c, \tag{13.6}$$

where $P_s = -D_s / \Delta x$, because the exact distance over which the concentration of the substance is changing cannot really be determined. A_m is the surface area of membrane and P_s is called the 'permeability coefficient' or simply 'permeability'.

Permeability coefficients have been obtained for a number of non-lipid-soluble molecules in mammalian skeletal muscle; some of them are given in **Table 13.5**. It can be seen that as molecular weight and approximate molecular radius increase, the permeability decreases considerably. A decrease in permeability is to be expected because the diffusion coefficient decreases with increasing molecular weight;

however, **Table 13.5** shows that permeability falls relative to the diffusivity as molecular weight increases. This would suggest that the area available for diffusion decreases, or that the diffusion path length increases, or that the resistance to diffusion increases for some other reason. If the path length for diffusion is approximately the thickness of the endothelial cell and independent of molecular size, then the effective area for diffusion of lipid-insoluble materials is far less than the actual surface area of the capillary wall.

If the effective area of membrane available for transport of a particular solute is A_s, then Equation (13.6) should be rewritten as

$$J_s = (P_s A_s)\Delta c$$

or

$$(P_s A_s) = \frac{J_s}{\Delta c}. \tag{13.7}$$

Most studies of permeability are in fact concerned with obtaining the product $(P_s A_s)$ for different molecules from the experimentally derivable quantities J_s and Δc. If the effective area of membrane available for the transport of two molecules of differing sizes is the same, then comparison of the value of $(P_s A_s)$ gives a direct indication of the relative permeability; however, it is not always appropriate to make such an assumption, as will be discussed later.

Methods of measuring permeability coefficients A number of methods have been developed to measure the permeability of the wall. All are very elegant but suffer from various experimental and computational difficulties; measured permeabilities for a given molecule in a particular tissue vary considerably according to the method used. There appear to be significant differences in the value obtained in different tissues, and whilst this may be due to genuine differences between the endothelial cell linings, it is thought also to reflect the shortcomings of the various techniques.

In order to determine the permeability of a vessel, it is necessary to know not only the net solute flow between blood and tissue, but also the concentration gradient across the wall. Because the blood is flowing through the capillary while transport is taking place, the concentration gradient and solute transport will vary with distance along the capillary. Since it is not possible to measure these quantities from point to point, it is necessary to estimate an effective concentration gradient over the whole length of the vessel; in turn, the choice of a suitable form of mean gradient depends upon the complexity of the model describing exchange between the blood and tissue.

It must also be appreciated that the blood flow rate can independently have a significant effect on exchange in certain circumstances, and this should not be confused with an alteration in the permeability of the wall. Consider the situation where an exchangeable molecule is injected into the arterial blood approaching a capillary. As

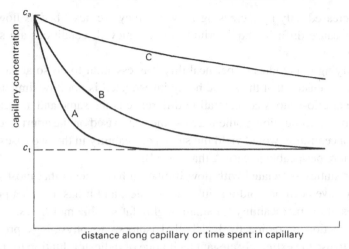

Fig. 13.44. The fall in blood concentration of three substances as they exchange with the tissues during flow through the capillary. The capillary wall is most permeable to substance A and least to C. In the case of A and B, equilibrium between blood and tissues is achieved before the blood leaves the capillary and conditions are 'flow limited'. Case C is not equilibrated and exchange is said to be 'diffusion limited'. (From Michel (1972). Flow across the capillary wall. In *Cardiovascular Fluid Dynamics* (ed. Bergel), vol. II. Academic Press, London.)

the blood passes through the capillary the concentration of the marker molecule will fall progressively from the arterial to venous end, as shown in **Fig. 13.44**. If the wall is highly permeable to the marker but has a low capacity, the concentration will initially fall rapidly (curve A) and then, when equilibrium is achieved, the concentration will fall no further with distance along the capillary. Curve B represents the situation where the molecule achieves equilibrium with the tissue just at the end of the capillary. Curve C illustrates the situation where the capillary is not long enough to allow the marker to equilibrate between blood and tissue. It can be seen that, since the rate of solute uptake is the product of the blood flow rate and the arterio-venous concentration difference ($c_a - c_v$), the uptake in situations A and B is the same, but is less for C.

If the blood flow rate were reduced, the rate of uptake would be reduced in cases A and B in direct proportion to the degree of decrease in blood flow. The uptake of C would also be reduced, but to a much lesser degree, since the increased residence time in the blood would allow a greater time for exchange and an increase in the arterio-venous difference. An increase in the blood flow rate, whilst increasing the rate of uptake in all situations, will have most effect in case A and least in case C. These differences in the dependence of uptake on blood flow rate have led to the terms 'flow-limited' and 'diffusion-limited' uptake. Because uptake in cases A and

B can be increased only by increasing flow rate they are described as flow limited. In case C, because diffusion equilibrium is not achieved, conditions are said to be diffusion limited.

When carrying out studies of permeability it is essential to do so at various flow rates in order to ensure that the solute being investigated is not flow limited. In both A and B the arterio-venous concentration difference is the same and any calculation of the effective concentration gradient across the wall based on the arterio-venous difference in concentration would yield the same permeability in the two cases. Clearly, the wall is more permeable in case A than case B.

The uncertainties associated with flow limitation have meant that most studies of permeability have been on lipid-insoluble molecules, and it has not been possible to obtain values of the permeability for small, highly fat-soluble molecules.

The introduction of radioactively labelled isotopes to physiology provided the opportunity to use the extremely elegant technique of indicator dilution for measuring the product of permeability and effective area. A radioactive isotope of a molecule known to remain in the blood and a differently labelled isotope of a molecule capable of traversing the capillary wall are injected together as a bolus into a small artery. Samples are taken from a vein draining the capillary bed. The relative concentration of the isotopes in the venous samples compared with the relative values when injected indicates the fraction of the exchangeable marker which has crossed the wall. **Fig. 13.45** shows the results of an experiment on the permeability of sucrose, in which albumin was used as the non-exchanging intravascular marker. The sucrose concentration rises more slowly and to a lower peak than the albumin because of its extraction into the extravascular space. At some time after the peak concentrations have been reached, the curves cross. This is because the albumin has been largely washed out of the capillary system but the venous sucrose concentration is maintained relatively high by transport back from the tissues where the sugar concentration has become higher than the falling blood concentration.

If it is assumed that initially the transport out of the capillary at any point is proportional to the local concentration in the blood (the tissue concentration of labelled sucrose is initially zero), then the blood concentration of sucrose will fall exponentially along the capillary. It can then be shown that the mean capillary concentration $\bar{c}$ is given by

$$\bar{c} = (c_a - c_v)/\ln(c_a/c_v),$$

where c_a and c_v are the sucrose concentrations at the arterial and venous ends of the bed. These concentrations may be obtained directly from the heights of the peaks of the curves in **Fig. 13.45**; the height of the albumin peak gives the arterial concentration of sucrose and the sucrose peak gives the venous sucrose concentration. Because the initial tissue concentration of the sucrose isotope is zero, the mean concentration difference across the vessel walls is equal to $\bar{c}$. Also, the rate of uptake of sucrose is

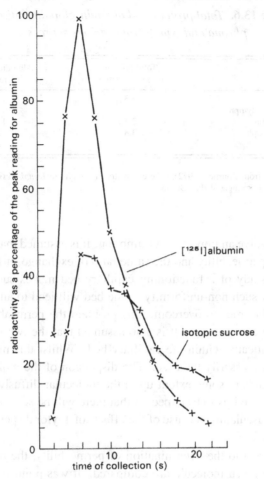

Fig. 13.45. The relative concentrations of radioactively labelled albumin and sucrose as monitored in a vein after simultaneous injection in an artery.

given by

$$J_s = \dot{Q}(c_a - c_v),$$

where $\dot{Q}$ is the blood flow rate through the capillary bed. Thus, substitution into Equation (13.7) gives

$$(P_s A_s) = \frac{\dot{Q}(c_a - c_v)}{(c_a - c_v)/\ln(c_a/c_v)} = \dot{Q}\ln(c_a/c_v).$$

Great advantages of the indicator-dilution technique are that it may be carried out on an intact circulation and it is relatively easy to perform experimentally. However,

Table 13.6. *Total protein and albumin:globulin ratios in plasma and lymph from various sources*

	Total protein (g per 100 cm^3)	Albumin:globulin ratio
Plasma	6.3	1.43
Thoracic duct lymph	4.0	1.50
Muscle lymph	2.3	1.54
Intestinal lymph	3.8	1.50
Liver lymph	5.3	1.48

Data derived from Courtice (1972). The chemistry of lymph. In *Lymph Vessel System* (ed. Meessen). Springer Verlag, Berlin.

the method does involve an important assumption. It is assumed that all vessels in the bed have the same permeability and the same ratio of exchange surface area to blood flow. Any visual study of a functioning capillary bed makes such an assumption seem dubious. Any such non-uniformity in the bed will lead to an underestimate of the permeability. In order to overcome this problem the permeability is measured at as high a flow rate as possible. It is also assumed that the two molecules studied are distributed identically within the capillary bed. Whilst this might at first seem probable, it is not necessarily the case. The dispersion of the molecules within the flowing blood depends to some extent upon the molecular diffusivity, and since the molecules differ in size it is to be expected that there will be some difference in their distribution in the vasculature because of the effects of Taylor dispersion, as discussed in Chapter 12 (p. 332).

In another approach to the determination of permeability, the plasma and lymph concentrations of a given molecule are compared. It was pointed out earlier in the chapter that, under steady-state conditions, while the greater part of the water and small molecules which pass out of the blood at the arterial end of the capillary is returned at the venous end, the remainder leaves the interstitial space via the lymphatic drainage system. Plasma and lymph electrolyte concentrations are very similar, but their protein concentrations are not, being higher in the plasma than in the lymph; furthermore, the relative concentrations of the various proteins themselves are different, as shown in **Table 13.6**. Globulin, a molecule rather larger than albumin, appears to be less capable of crossing the capillary wall of some organs than others.

Molecules of known molecular size are introduced into the circulation intravenously, and the plasma and lymph concentration is monitored. (Direct measurement of the concentration of the marker molecule has been attempted in the interstitial space when radioactive molecules have been used.) When steady-state conditions are achieved, the net flow rate of the solute through the capillary walls is equal to the flow rate of the solute in the lymphatic channels. The flow of the molecule out of the capillaries

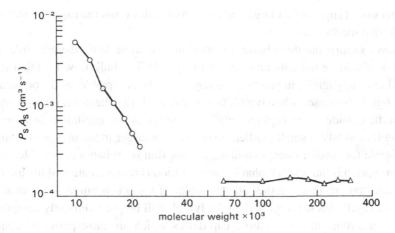

Fig. 13.46. Capillary permeability area product as obtained from plasma:lymph ratios and lymph flow rate for PVP molecules of a range of sizes. Note that permeability falls with increasing molecular weight in the low weight region but becomes constant at high molecular weights. (From Renkin (1964). Transport of large molecules across capillary walls. *Physiologist* **7**, 20.)

will be by both filtration and diffusion. However, when the permeability of relatively large molecules is being studied, the lymph concentration (c_L) is found to be much less than the plasma concentration. Thus, it is reasonable to assume that the net filtration is very small because the resistance to the molecules within the pores is very high, and transport is mainly by diffusion. Then

$$J_L c_L = (P_s A_s)(c_p - c_L),$$

where J_L is the lymphatic flow rate, c_p is the plasma concentration of the molecule and $(P_s A_s)$ is the permeability area product for the molecule as defined in Equation (13.7). The effective concentration difference across the capillary wall as given by $(c_p - c_L)$ is based first on the fact that the net exchange rate is very low and the arterio-venous fall in concentration is very small compared with the absolute level, and second on the assumption that the diffusion resistance is all at the capillary wall and interstitial fluid concentrations are the same as the lymphatic concentration.

The permeability of the wall to a number of substances has been measured in this way, in particular polysaccharides such as dextrans, with a range of molecular weights, and the polymer polyvinyl pyrrolidone (PVP), whose molecular weight ranges from 10 000 to over 200 000. The variation in the permeability area product with molecular weight is found to vary as shown in **Fig. 13.46**. In the lower molecular weight range the permeability falls with increasing molecular weight, but in the higher molecular weight range the permeability becomes independent of molecular size. This would suggest that large and small molecules cross the capillary membrane by a different mechanism.

We now examine the three basic assumptions which underlie the principle of this method of obtaining the capillary permeability. (1) The bulk flow filtration term is assumed to be negligible; in practice this appears to be acceptable, as the permeability is only slightly increased when lymph flow is elevated. (2) The concentration gradient between the outside of the capillary wall and the lymphatic capillaries is assumed to be negligible. Whilst a small gradient may exist for larger molecules, it is thought to be negligible for smaller ones, so that the assumption is probably reasonable. (3) The third assumption is that the lymphatic concentration is representative of the interstitial fluid concentration; this is rather questionable. If there is a range of permeabilities among the capillaries in a region then the lymph will be predominantly composed of the water and molecules from those capillaries which are most permeable, and the local interstitial fluid will be relatively high in protein concentration. Thus, there will be a spectrum of interstitial fluid compositions, and the regional lymph composition will represent a mean of the ultrafiltrates of all capillaries.

One of the earliest methods of determining permeability used the isolated hind limb preparation. The solute molecules were added to the perfusate flowing to the capillary bed; this caused water to be drawn out of the limb tissue into the plasma. To prevent this, capillary pressure was raised by increasing venous pressure until the limb regained and maintained its original weight. The amount by which the capillary pressure was raised equalled the osmotic pressure difference across the capillary walls resulting from the introduction of the solute. The transport rate J_s of the test solute across the wall was estimated from the product of the arterio-venous concentration difference and the blood perfusion rate. The effective concentration difference across the wall was estimated from the osmotic pressure change ($\Delta\Pi$) using van't Hoff's law (Equation (9.14)). Then, on substituting into Equation (13.7), one obtains

$$(P_s A_s) = J_s(\Delta\Pi/RT).$$

Whilst this technique appears very attractive, its major drawback lies in the assumption that the effective concentration difference across the wall can be computed simply from van't Hoff's law. In fact, the law breaks down for many large molecules at all but the most dilute concentrations (see p. 407). Furthermore, the law assumes that the membrane acts as a perfect semi-permeable barrier and that all the solute molecules are prevented from crossing; the fact that the membrane is restricting but not prohibiting solute transfer leads to a reduced osmotic pressure (see Chapter 9, p. 140).

The size of the osmotic reflection coefficient σ (Equation (9.15)) depends upon hydrodynamic and diffusional factors influencing the motion of molecules within the membrane passages, and currently no satisfactory theory predicting its magnitude is available. The estimation of this coefficient is one of the most important areas of membrane physics of today.

The value of σ is believed to depend upon the relative size of the restricted diffusion coefficients of the solute (D_s') and solvent (D_w'); thus,

$$\sigma = 1 - \left(\frac{D_s'}{D_w'}\right). \qquad (13.8)$$

The effective diffusion coefficient under these circumstances may be given by

$$D' = DA_s/A_p, \qquad (13.9)$$

where D is the free diffusion coefficient and A_s and A_p are the apparent and true areas for diffusion. Thus, combining Equations (13.8) and (13.9) we obtain

$$\sigma = \left(1 - \frac{A_s}{A_w}\frac{D_s}{D_w}\right), \qquad (13.10)$$

where A_s and A_w are the apparent areas for diffusion of solute and solvent respectively, assuming that the true areas available for diffusion of solute and solvent are in fact the same. In fact, it is now thought that the area of membrane available for the transport of water and most lipid-insoluble molecules is probably not the same and thus Equation (13.10) is a poor description of the osmotic reflection coefficient.

The diffusion pathway across the capillary wall Two general types of study have been conducted to try to elucidate the diffusion pathway across the capillary wall. In the first, lipid-insoluble molecules of known molecular size, capable of detection by electron microscopy, were introduced into the capillary lumen and allowed to diffuse out through the capillary wall. The tissue was then fixed and prepared for microscopy, and their location and concentration in the wall and interstitial space studied. In the second approach, the permeability studies reported above have been used to predict theoretically the existence of pores of a certain size and then to relate the pore size to structures in the wall that are identifiable with an electron microscope.

In continuous capillaries, it has been shown that molecules of intermediate size (horse-radish peroxidase, m.w. about 40 000) enter the extravascular space beyond the endothelium via the vesicles and are not extensively observed within the intercell clefts. Smaller molecules (cytochrome peroxidase, m.w. about 12 000) were observed to move out of the capillaries more rapidly and could be observed in the intercell clefts in concentrations diminishing with distance from the luminal surface. On the basis of these and similar studies it has been suggested that the intercell cleft is capable of allowing the passage of relatively small molecules, but inhibits the transport of large molecules.

These studies are compatible with the observed dependence of permeability on molecular weight as shown in **Fig. 13.46**. At the high molecular weight end, permeability is largely independent of molecular size; the size at which this happens is in the region of 40 000.

The Pappenheimer equivalent pore theory Throughout this chapter it has been strongly suggested that the filtration and diffusion across the capillary wall occur via pores. This suggestion has long been proposed, but it was in 1951 that Pappenheimer and his colleagues proposed a quantitative model for the diffusion of molecules through pores of dimensions comparable to their size. The theory provides a very convenient description of flow across a membrane and is capable of making very precise predictions about the structure of the membrane.

The theory of Pappenheimer gives an estimate of the restricted diffusion coefficient compared with the free diffusion coefficient in terms of the apparent and true pore areas as defined in Equation (13.9). It suggests that the effective diffusion coefficient is reduced because the effective pore area is diminished as the result of hindrance at the entrance to the pore due to shape differences, and that the increased hydrodynamic drag on a molecule passing through a confined space, compared with that in an unbounded medium, causes a diminution in the diffusion coefficient.

Thus, for a molecule of radius a to enter a pore of radius r without touching its sides, the true target area A_s is $\pi(r-a)^2$ or

$$\frac{A_s}{A_p} = \frac{\pi(r-a)^2}{\pi r^2} = \left(1 - \frac{a}{r}\right)^2. \tag{13.11}$$

The amount by which the drag on a molecule is increased in a narrow pore is very hard to assess as it will depend not only upon the relative dimensions but also upon the shape of the molecule, the charge distribution on its surface, its flexibility and many other factors. Some of these effects (e.g. shape and flexibility) are partially allowed for when defining the molecular radius on the basis of its Stokes–Einstein equivalent (Equation (9.6)). If the frictional resistance in free motion is f and that when passing through a pore is f_R, then a good approximation of the effect has been found to be

$$\frac{f_R}{f} = 1 - 2.10\left(\frac{a}{r}\right) + 2.09\left(\frac{a}{r}\right)^3 - 0.95\left(\frac{a}{r}\right)^5. \tag{13.12}$$

On the basis of the Nernst equation (Equation (9.4), p. 132),

$$\frac{D_R}{D} = \frac{f}{f_R}. \tag{13.13}$$

Substituting for Equations (13.11), (13.12) and (13.13) in Equation (13.5)

$$J_s = -D_s\frac{\Delta c}{\Delta x}\left(1 - \frac{a}{r}\right)^2\left[1 - 2.10\left(\frac{a}{r}\right) + 2.09\left(\frac{a}{r}\right)^3 - 0.95\left(\frac{a}{r}\right)^5\right]. \tag{13.14}$$

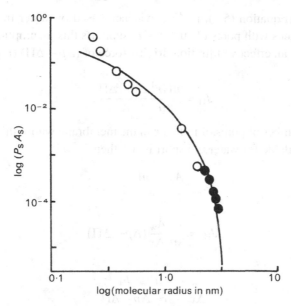

Fig. 13.47. The permeability of cat and dog hind limbs to molecules of different molecular radius. The curve is that predicted on the basis of restricted diffusion through pores of 4 nm radius. (From Michel (1972). Flow across the capillary wall. In *Cardiovascular Fluid Dynamics* (ed. Bergel), vol. II, Academic Press, London.)

Thus, if the permeability area product (Equation (13.7)) is measured in a given preparation for a range of molecular sizes, it is possible (using Equation (13.14)) to obtain an estimate of the equivalent size of the pores through which diffusion takes place. Comparison of the pore theory with experimental data obtained in the hind limb preparation is shown in **Fig. 13.47**. It can be seen that, as molecular size is increased, the permeability falls in a manner very close to that predicted from Equation (13.14), assuming the pores to be circular and of 4 nm radius.

Such agreement between experiment and theory is very attractive, particularly when electron microscopic studies of the intercell cleft indicate that the gap is of this order of size. However, such close agreement is not always obtained and many further studies are needed to draw firmer conclusions about the structure of the diffusion pathway in a wide range of tissues.

The pathway for water transport across the capillary wall Using the Pappenheimer pore theory it is also possible to obtain an estimate of the size of the pores in the capillary walls from a knowledge of the water filtration rate. If the pores are of the order of 4 nm in radius, they are considerably larger than the radius of the water molecule and it is therefore reasonable to assume that the water flow may be described using

Poiseuille's law (Equation (5.1), p. 47); evidence based on water transport through artificial membranes with pores of such a size supports this assumption. If the filtration rate is J_H for an effective filtration driving force of $(\Delta p - \Delta \Pi)$ (Equation (13.3)), then

$$J_H = \frac{n \Pi r^4}{8 \mu} \frac{\Delta p - \Delta \Pi}{\Delta x},$$

where n is the number of pores of radius r in the membrane whose thickness is Δx. If the pore area available for water transport is A_w, then

$$A_w = n \Pi r^2$$

and

$$J_H = \frac{r^2}{8 \mu} \frac{A_w}{\Delta x} (\Delta p - \Delta \Pi)$$

or

$$\frac{A_w}{\Delta x} = \frac{8 \mu}{r^2} \frac{J_H}{\Delta p - \Delta \Pi}. \tag{13.15}$$

Thus, we may obtain the pore area per unit path length for water if we already know the pore radius from measurements of the filtration coefficient as obtained on p. 406 using the isogravimetric preparation. It is also possible to obtain the effective pore area per unit path length $(A_s/\Delta x)$ for the diffusion of molecules of various radii using the same preparation:

$$\frac{A_s}{\Delta x} = \frac{J_s RT}{D_s \Delta \Pi}. \tag{13.16}$$

If we assume that the pore area A_w available for water transport is the true pore area A_p for solute diffusion, then dividing Equation (13.15) by Equation (13.16) and equating the result with Equation (13.14) provides an explicit equation for pore radius in terms of experimentally derivable quantities. Solution of this equation gives an effective pore radius of 14 nm, which is considerably larger than the value obtained by other methods and the discrepancy cannot be reconciled on the basis of the inaccuracies associated with the experiments.

However, the discrepancy can be accounted for if it is not assumed that water is constrained to pass only through the pores available to lipid-insoluble molecules. If the pores are taken to be 4 nm radius, then it can be calculated that only half the water passes through the pores available to the solute, the remainder using other pathways through the wall.

The transport of large molecules For a long time it was believed that the capillary wall was impermeable to molecules larger than albumin. However, studies such as those in **Fig. 13.46** have shown that the wall has a very low but definite permeability to large molecules and that it is virtually independent of molecular size. There have been a number of electron microscope studies aimed at identifying large pores in the capillary wall capable of allowing the passage of such molecules; the fenestrae in some capillary walls may provide such a pathway. The cytoplasmic vesicles present in all capillary walls are also considered to be responsible for the transport of the large molecules. Experimental studies have shown that plastic microspheres of diameter less than about 30–70 nm are capable of crossing the hind limb capillary walls.

If the transport of such large molecules and particles is via large pores then we would expect the permeability to vary in proportion to the molecular diffusivity. Studies in the hepatic and intestinal beds with dextran molecules, whose diffusivity decreases with the square root of molecular weight, have shown a consistent linear decrease in permeability with the square root of molecular weight; indeed, these are fenestrated capillary systems. Similar studies carried out on skeletal muscle beds, which contain continuous endothelium, do not show such a clear relation. It is possible that permeability is much less dependent upon molecular diffusivity and that transport is via the vesicles rather than unidentified pores.

It was Palade in 1953 who first proposed that the vesicles acted as a transport system across the endothelium, and since that time some effort has been devoted to assessing their importance and elucidating the mechanism. As was explained earlier, the membrane bounding vesicles is, as far as can be determined, the same as the membrane bounding the cell, and vesicles can be found either free within the cytoplasm of the cell or attached to the cell wall. Because of their small size, their behaviour is extremely difficult to study and great reliance must be placed on the use of the electron microscope.

Three important processes associated with the vesicles have been identified. These are the movement of vesicles within the body of the cell, their attachment and detachment at the cell surfaces and the exchange of their contents with either the blood or interstitial fluid.

The movement of vesicles within the cell is thought to occur through Brownian motion as a result of collisions with the constituent molecules of the cytoplasm. However, it is hard to predict the effective diffusivity of the vesicles, even if their density is assumed to be the same as that of the cell, because the effective viscosity of cytoplasm is unknown. Furthermore, the long-range vesicle displacements cannot be the same as in an unbounded medium; the gap between opposing cell walls is in the region of only 0.3–0.4 μm or four to five vesicle diameters (approximately 70 nm). It is possible, however, to predict (using elementary kinetic theory) that the impact of surrounding molecules could rapidly accelerate a vesicle to a velocity of about $30 \, \text{cm} \, \text{s}^{-1}$. The viscous drag on such a small particle is very large and as a result it will be brought back to rest in a small fraction of a second.

The processes associated with attachment and detachment of vesicles to and from the cell wall are even less well understood. If a vesicle moving around in the body of

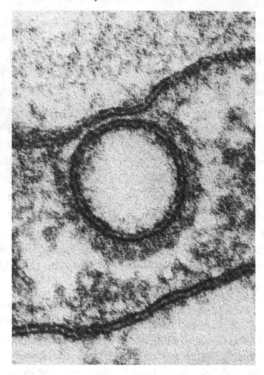

Fig. 13.48. The close approach of a vesicle to the endothelial cell surface showing the deflection of the surface of the cell membrane. (From Palade and Bruns (1968). Structural modulations of plasmalemmal vesicles. *J. Cell. Biol.* **37**, 633.)

the cell comes sufficiently close to the wall (say a distance of 2–3 nm), then it enters a region in which the van der Waals' attraction force will pull it towards the wall. For this to happen the cytoplasm must be squeezed out of the gap between the vesicle and the wall; as the gap narrows, the force required to exclude the fluid increases, but so does the van der Waals' force. When the vesicle approaches very close to the wall, the latter becomes distorted, as shown in **Fig. 13.48**. The two membranes then fuse to form a diaphragm, which finally splits to give the vesicle an opening to the external environment of the cell (**Fig. 13.49**). The remnants of the diaphragm can be clearly seen in **Fig. 13.50**. The mechanism of detachment is obscure; however, once it has occurred the vesicle is again free to move under the influence of Brownian motion. It can either be attracted back to the wall and reattach at or near the original point, or move to some remote site possibly on the opposite wall. The typical lengths of time a vesicle spends free within a cell or attached at a wall have been estimated very tentatively to be about 5 s and 3 s respectively.

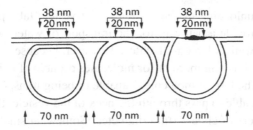

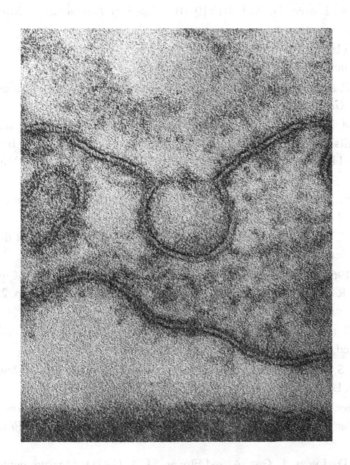

Fig. 13.49. Schematic illustration of the progressive fusion and rupture of the membrane between a vesicle and the cell wall. (From Palade and Bruns (1968). Structural modulations of plasmalemmal vesicles. *J. Cell. Biol.* **37**, 633.)

Fig. 13.50. A vesicle shortly after the diaphragm across the surface has shattered. The central knob of the diaphragm can be clearly seen. (From Palade and Bruns (1968). Structural modulations of plasmalemmal vesicles. *J. Cell. Biol.* **37**, 633.)

The loading and unloading of the contents of a vesicle take place by diffusion. Thus, the vesicles provide a pathway for the transport of water and small molecule weight materials as well as large molecules. However, because of the relative slowness of this pathway, it is unimportant for molecules capable of being transported via the intercell clefts. The largest molecules capable of being transported by the vesicle system are those just able to pass through the neck of the vesicle; that is, lipoproteins of molecular diameters of about 50 nm diameter. Somewhat smaller molecules will enter the vesicle but experience restricted diffusion passing through the neck. The permeability of the wall to molecules transported via the vesicles thus may depend to some extent upon the diffusion coefficient of the species.

Further reading

Crone, C. and Lassen, N. A. (eds) (1970). *Capillary Permeability*. Munksgaard, Copenhagen.

Fung, Y. C. (1968). Biomechanics, its scope, history, and some problems of continuum mechanics in physiology. *Appl. Mech. Rev.* **21**, 1–20.

Gaehtgens, P. and Uekermann, U. (1971). The distensibility of venous microvessels. *Pflug. Arch. Ges. Physiol.* **330**, 206–16.

Landis, E. M. and Pappenheimer, J. R. (1963). Exchange of substances through the capillary walls. In *Handbook of Physiology*. Section 2: *Circulation*, Vol. II (eds W. F. Hamilton and P. Dow), pp. 961–1034. American Physiological Society, Washington, DC.

Lightfoot, E. N. (1974). *Transport Phenomena and Living Systems*. Wiley–Interscience, London.

Johnson, P. C. (1973). The microcirculation and local humoral control of the circulation. In *Cardiovascular Physiology* (ed. A. C. Guyton). MPT, London.

Majno, G. (1965). Ultrastructure of the vascular membrane. In *Handbook of Physiology* (eds W. R. Hamilton and P. Dow), Section 2: *Circulation*, Vol. II, pp. 2293–375. American Physiological Society, Washington, DC.

Michel, C. C. (1972). Flows across the capillary wall. *Cardiovascular Fluid Dynamics*, Vol. II (ed. D. H. Bergel), Chapter 17. Academic Press, New York.

Middleman, S. (1972). *Transport Phenomena in the Cardiovascular System*. Wiley–Interscience, London.

Pappenheimer, J. R., Renkin, E. M. and Borrero, L. M. (1951). Filtration, diffusion and molecular sieving through peripheral capillary membranes. *Am. J. Physiol.* **167**, 13–46.

Snashall, P. D., Lucas, J., Guz, A. and Floyer, M. A. (1971). Measurements of interstitial 'fluid' pressure by means of a cotton wick in man and animals: an analysis of the origin of the pressure. *Clin. Sci.* **41**, 35–53.

Weinbaum, S. and Caro, C. G. (1976). A macromolecule transport model for the arterial wall and endothelium. *J. Fluid. Mech.* **74**, 611–40.

Wolstenholme, G. E. W. (ed.) (1969). *Circulatory and Respiratory Mass Transport.* Churchill, London.

Zweifach, B. W. (1974). Quantitative studies of microcirculatory structure and function. *Circulation Res.* **34**, 834–66.

14

The systemic veins

The study of the mechanics of blood flow in veins has been far less extensive than that of blood flow in arteries. However, virtually all the blood ejected by the left ventricle must return to the right atrium through the veins;[1] they normally contain almost 80% of the total volume of blood in the systemic vascular system and have an important controlling influence on cardiac output. It is therefore important to understand their mechanics.

The venous system resembles the arterial system, in that it consists of a tree-like network of branching vessels; the main trunks are the venae cavae, which come together and lead into the heart. However, it is fundamentally different from the arterial system in several respects:

(1) As can be seen from **Fig. 12.11**, p. 257, the pressure in a vein is normally much lower than that in an artery at the same level, and may be less than atmospheric (for example in veins above the level of the heart).

(2) The vessels have thinner walls and their distensibility varies over a much wider range than that of arteries at physiological pressures.

(3) The blood flows *from* the periphery *towards* the heart, and the flow rate into a vein is determined by the arterio-venous pressure difference and the resistance of the intervening microcirculation.

(4) Many veins contain valves which prevent backflow.

The consequence of these differences is that the distribution of pressure and flow rate in the venous system is quite different from that in the arteries, although many of the physical factors determining them, for example the transmission and reflection of waves, are qualitatively similar.

This chapter begins by describing the anatomy and arrangement of veins and goes on to discuss the transmural pressures to which veins are subjected and their elastic response to changes in transmural pressure. Then we discuss the dynamic pressure

---

[1] Some arterial blood supplied to the lungs by the bronchial circulation returns to the left atrium; this is normally a very small fraction (less than 5%) of the cardiac output.

426

and flow-rate fluctuations which are observed in veins, describing the physical factors which determine them. A separate section is devoted to the interesting phenomena which may occur when blood flows through a vessel which can collapse; and finally we examine some aspects of the mechanics of the systemic venous system as a whole.

Anatomy

The systemic veins begin at the venules, where the capillaries join together, and end where the venae cavae enter the right atrium of the heart. There are one or two exceptions to this pattern, because in some regions blood flows successively through two microcirculations in series (for example the portal circulation[2] or the kidneys), and in others there are arteriovenous anastomoses, which bypass the capillaries altogether. Nevertheless, the system can be represented by a network of vessels which come together and increase in diameter the nearer they are to the heart, while the total cross-sectional area decreases. The venous system has a much greater capacity than the arterial system, both because the number of vessels is greater, and because at a particular level the veins draining a vascular bed tend to be larger than the corresponding arteries supplying it. The greater variability of the capacity of the veins is governed by their greater distensibility together with the mode of action of the muscles in and around their walls.

There is some disagreement about exactly how the blood volume is distributed within the veins, because they are difficult to study. As we have seen in Chapter 13, the vascular beds which have been most widely studied are those which are two-dimensional and run through transparent tissues. An example is the bat's wing; the dimensions and cross-sectional areas of different blood vessels in this circulation are given in **Table 12.1** (p. 243). It shows that the area increases between the capillaries and the venules, and that the area of the venules exceeds that of the small veins by a factor of 4–6 and of the large veins by a factor of 26. The area of the venules also exceeds that of the arterioles by a factor of about 25. There is of course a less marked difference between the capacities of the venules, small veins and large veins than between their cross-sectional areas, because the smallest vessels are also the shortest.

Rather different results have been obtained, however, from measurements in the dog omentum. Here, the total cross-sectional area of the venules is reported to be only about four to five times that of the arterioles. It would be desirable to confirm these findings by another method, for example by measuring the blood velocity in the different vessels (as the total cross-sectional area increases, the blood velocity should fall in proportion), but as far as we know no such measurements have been made in either the bat's wing or the dog omentum. However, measurements of blood velocity in the cat omentum are consistent with the relative cross-sectional areas in the dog.

---

[2] Here, blood from the intestines, spleen, stomach, pancreas and gall bladder drains into the portal vein, from which it flows directly into the liver, which has its own microcirculation. From here it enters the hepatic vein, and thence passes straight into the inferior vena cava.

The most probable reason for the discrepancy in the area measurements between bat and dog is that the relative dimensions of the venous and arterial systems vary from place to place in different animals.

The walls of veins are much thinner than those of arteries. For example, the ratio of wall thickness to diameter in a vein at a physiological transmural pressure is typically 0.01–0.02, while the corresponding ratio in a large systemic artery is 0.06–0.08 (**Table I**). Too little is known about the venous bed for us to be able to describe in detail the taper of individual veins, if any, or the area ratios and angles of branching at even the major junctions.

On the basis of their size and of microscopic studies of their wall structure, the veins are classified into five groups: postcapillary venules, venules, small veins, medium veins and large veins, the last two sometimes being lumped together in a single category. The wall structure of postcapillary venules (diameter 8–30 μm) and of venules (diameter 100–200 μm) was described in detail in Chapter 13 and is briefly summarized here. The transition from capillary to vein is a gradual one, with the connective tissue elements appearing very near the capillary end and smooth muscle cells slightly more proximally. When several capillaries join together, they form the postcapillary venules, about 20 μm in diameter, which consist of a layer of endothelium surrounded by a thin layer of longitudinally directed collagen fibres and fibroblasts. In venules of 40–50 μm diameter, isolated circularly arranged muscle fibres appear between the endothelium and connective tissue. In veins of 200–300 μm diameter, the circular smooth muscle forms a continuous layer and the adventitia becomes thicker, having scattered fibres in addition to the longitudinally disposed collagen fibres. In small veins (up to 0.2 cm diameter), the smooth muscle becomes multilayered and the individual layers of muscle are separated by loose collagenous tissue. It is at this level that the three characteristic layers of blood vessel walls (intima, media and adventitia, see p. 244) are first apparent; thereafter the structure of veins is similar to that of arteries. The main differences are that the media is very much thinner and the adventitia, composed principally of collagen fibres, comprises the bulk of the wall. Also, the internal elastic lamina is poorly defined. Veins in parts of the body normally below the heart, such as the lower limbs, tend to have relatively thick media, though whether this is because they experience a relatively high transmural pressure is unknown.

The walls of veins with a calibre greater than about 0.1 cm have vasa vasorum. These originate from nearby small systemic arteries, and run in the adventitia, forming a profuse capillary network which extends through the thickness of the wall almost to the intima, and connects with venules in the adventitia. Veins are well supplied with nerves, largely derived from the sympathetic system. The smooth muscle in the walls of veins is normally in a state of active tension (i.e. of tone). Cutting of the nerves, or the inhibition of nerve activity with drugs, causes veins to dilate, because the tension in the wall is reduced. This dilation may considerably increase the cross-sectional area of a vein.

Many medium-sized veins, especially those in the limbs, are provided with valves, which act to prevent blood from flowing backwards. (In humans, there are no valves in the veins which drain the abdominal viscera, except the portal vein, or in the venae cavae.) The valves usually have two cusps which consist of connective tissue membranes containing elastin fibres and covered with endothelium. Adjacent to each cusp on the downstream side (the side nearer to the heart) there is usually a sinus; that is, an outpouching of the wall. This presumably has the same mechanical function, of ensuring stable opening and smooth closing of the valve, as the sinus of Valsalva behind each cusp of the aortic valve, whose mechanics were described on p. 231. Detailed studies of the operation of venous valves have not been made.

Transmural pressure and static elastic properties

At the level of the heart in a supine subject the mean pressure within veins, relative to atmospheric, varies from about $2.0 \times 10^3 \,\mathrm{N\,m^{-2}}$ ($20\,\mathrm{cm\,H_2O}$) in the venules to about one-third of that value in the venae cavae, although pressures within the chest can be sub-atmospheric because of respiratory pressure swings (see **Fig. 14.20**). The difference between these pressures is responsible for driving the mean flow through the veins back to the heart. Since the pressure outside a vein is normally about atmospheric, these are also the values of the transmural pressure tending to distend the veins. If the subject becomes upright, the excess pressure above hydrostatic (p_e, see p. 259, Chapter 12) which drives the mean flow remains virtually the same, but the transmural pressure remains the same only at the level of the heart. The veins of the foot will fill from the microcirculation, until the internal pressure has increased by about $10^4 \,\mathrm{N\,m^{-2}}$ (the hydrostatic pressure due to a 1 m column of blood), while in those of the neck or a raised arm it is significantly reduced, and becomes sub-atmospheric. As we shall see, it is possible to reduce the distending transmural pressure of veins in the dependent limbs by contracting the skeletal muscle around them and thereby increasing the external pressure; this has the effect of reducing their cross-sectional area and hence their capacity, the blood being displaced towards the heart. The veins above the heart are commonly collapsed (see p. 258, Chapter 12), and cannot be opened by muscular action.

Thus, the physiological range of cross-sectional area and shape of a vein is far greater than that of an artery, whose circular cross-section changes its area by a relatively small amount when subjected to the same variation in transmural pressure. The possibility of collapse has a number of important mechanical consequences, which are explored in the next three sections. Here, we describe, as far as they are known, the changes in cross-sectional area, shape and distensibility of a vein which follow gradual changes of transmural pressure (in the physiological range). All the quoted measurements have been made on excised segments of large veins held at their physiological length but without tone. In vivo the response will be quantitatively rather

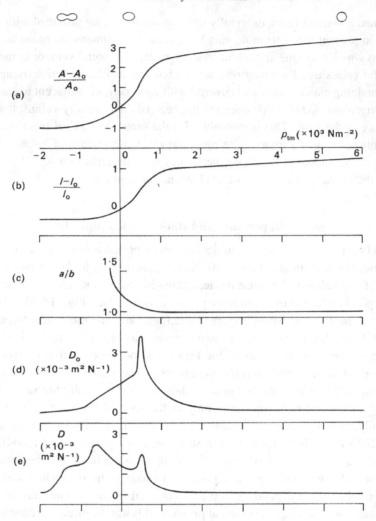

Fig. 14.1. A sequence of graphs illustrating the elastic properties of a segment of vein. The abscissa in each case is the transmural pressure p_{tm}, measured in units of $10^3\,\mathrm{N\,m^{-2}}$ (each unit $= 10\,\mathrm{cm\,H_2O}$). The ordinates are: (a) the change in cross-sectional area $(A - A_0)$ measured as a fraction of the cross-sectional area at zero transmural pressure A_0; (b) the change in perimeter $(l - l_0)$ measured as a fraction of the perimeter at zero transmural pressure l_0; (c) the ratio between the lengths of the major and minor axes of the cross-section, a/b, defined only while it remains circular or elliptical; (d) the 'pseudo-distensibility' D_0, defined by Equation (14.2) and equal to the slope of graph (a); (e) the actual distensibility D, defined by Equations (14.1); the maximum at a transmural pressure of $+0.5 \times 10^3\,\mathrm{N\,m^{-2}}$ is associated with the maximum in the slope of curves (a); the maximum at a negative transmural pressure is a consequence of the very small values of A. Above curve (a) are sketches to illustrate the shape of the cross-section at values of transmural pressure corresponding to the position of the sketches. (Parts of data from Moreno, Katz, Gold and Reddy (1970). Mechanics of distension of dog veins and other very thin-walled tubular structures. *Circulation Res.* **27**, 1069, by permission of the American Heart Association Inc.; and from Attinger (1969). Wall properties of veins. *Trans. Biomed. Eng.* **BME-16**, 253, Institute of Electrical and Electronics Engineers Inc., New York.)

different because contraction of the smooth muscle renders a vein less distensible, although it does not greatly alter its resistance to collapse because the walls are so thin.

The data are best presented by means of **Fig. 14.1**, which has been constructed both from measurements and from inferences which can be made from them. This figure summarizes what is known of the static behaviour of veins, and is of central importance in the rest of the chapter. The curves are not direct recordings of measurements from a single segment, and should be regarded as a schematic representation. A number of quantities are plotted against transmural pressure p_{tm}. From top to bottom these are as follows.

(a) The change in cross-sectional area $(A - A_0)$ measured as a fraction of the cross-sectional area at zero transmural pressure A_0; this is a dimensionless measure of the area change and can be used to compare results from different vessels. For transmural pressures between $\pm 2 \times 10^3 \, \text{Nm}^{-2}$, this curve was taken from direct measurements on the vena cava of a dog.

(b) A similar ratio for the perimeter l of the cross-section. The central section of the curve (between $\pm 0.5 \times 10^3 \, \text{Nm}^{-2}$) was plotted from the data of **Fig. 14.2**(b) (p. 434) and (a) above; the section for transmural pressures greater than $1.5 \times 10^3 \, \text{Nm}^{-2}$ was taken from **Fig. 14.3**a (p. 435); the rest was drawn in to give a smooth fit.

(c) The ratio between the lengths of the major and minor axes of the cross-section, a/b; the cross-section is observed to become elliptical when it first ceases to be circular, but during the final stages of collapse it is not elliptical and, therefore, this ratio is not defined. This curve was taken from a number of measurements.

(d, e) Two quantities, D_0 and D, each of which is a measure of the distensibility of the vessel. On p. 273 the distensibility was defined as the relative change in area $(\Delta A / A)$ resulting from a small change in transmural pressure (Δp_{tm}); that is,

$$D = \frac{1}{A} \frac{\Delta A}{\Delta p_{tm}}. \tag{14.1}$$

The quantity D measures the ease with which a change in transmural pressure causes a change in cross-sectional area: a vessel with a low distensibility is very stiff, and vice versa. It is this definition of distensibility which is used in the equation for the speed of propagation of pressure waves (Equation (12.9), p. 273). In the case of arteries, the cross-sectional area does not change very much as the transmural pressure varies over the physiological range. Thus, only a small error is introduced if the area A in the denominator of Equation (14.1) is replaced by a fixed area, for example that at a distending pressure of $13.3 \times 10^3 \,\mathrm{N\,m^{-2}}$ (100 mm Hg); this is commonly done because it is more convenient. In the case of veins, however, a considerable error is introduced if A is replaced by a fixed area, because the area varies greatly over the physiological range of pressures. Nevertheless, some workers use the fixed area at zero transmural pressure A_0 in place of A, and thus record a pseudo-distensibility D_0, defined by

$$D_0 = \frac{1}{A_0} \frac{\Delta A}{\Delta p_{\mathrm{tm}}}. \tag{14.2}$$

To show the considerable difference between the two distensibilities, we plot D_0 and D in **Fig. 14.1**d and e respectively. Both curves are plotted from measured slopes of the area curve in **Fig. 14.1**a. It is the lower distensibility curve, **Fig. 14.1**e, which should be regarded as the more useful.

Consider what happens to the vein as the transmural pressure is gradually reduced from a high value of more than $7 \times 10^3 \,\mathrm{N\,m^{-2}}$ (70 cm H$_2$O). At this pressure the vein is circular ($a/b = 1$) and relatively stiff, since it has a small distensibility and a large Young's modulus. This is presumably because the collagen in the wall dominates its response to changing pressures. The vessel remains circular, with little change in distensibility, as the transmural pressure is reduced to about $1.5 \times 10^3 \,\mathrm{N\,m^{-2}}$; the area falls slowly and the perimeter more slowly still (because while the vessel is circular the perimeter is proportional to the square root of area). When the transmural pressure falls from $1.5 \times 10^3 \,\mathrm{N\,m^{-2}}$ to $1.0 \times 10^3 \,\mathrm{N\,m^{-2}}$ the vessel still remains circular, but becomes somewhat more distensible, presumably because the elastin in the wall takes over from the collagen (as in arteries; see p. 96). When the transmural pressure falls below about $10^3 \,\mathrm{N\,m^{-2}}$ (but is greater than $0.5 \times 10^3 \,\mathrm{N\,m^{-2}}$), the vessel becomes slightly elliptical, the area falls more rapidly and the distensibility continues to rise. The perimeter also continues to fall, although it is no longer exactly proportional to the square root of area because the cross-section is not circular, and it is this reduction in perimeter, not the change in cross-sectional shape, which makes the larger contribution to the reduction in area and hence to the distensibility. As the transmural pressure falls to $0.5 \times 10^3 \,\mathrm{N\,m^{-2}}$ and below, the cross-section becomes markedly elliptical, and now the change in shape makes an increasing contribution to the area change and distensibility. The distensibility has a maximum at a transmural pressure of $0.5 \times 10^3 \,\mathrm{N\,m^{-2}}$, associated with the rapid change in cross-sectional area. The slope of the area curve, and hence the pseudo-distensibility D_0, continues to fall thereafter, because it becomes increasingly difficult to bend the vessel wall at its point of maximum curvature (see below). Such bending has to occur before the final collapsed state is achieved at a negative transmural pressure between $-1 \times 10^3 \,\mathrm{N\,m^{-2}}$ and $-2 \times 10^3 \,\mathrm{N\,m^{-2}}$ (-10 to -20 cm H$_2$O). Because the cross-sectional area becomes extremely small as the vessel collapses, the actual distensibility D has another maximum at a transmural pressure of about $-5 \times 10^3 \,\mathrm{N\,m^{-2}}$, before falling in an irregular way to a very low value again in the final collapsed state. (Because data are so sparse, some of the irregularities in this curve may reflect inaccurate measurements of the small area and of the slope of the area curve.)

During the last stages of collapse, the perimeter scarcely changes at all. After the final stage is reached, the cross-section has a dumb-bell shape, with two small, almost circular channels separated by a completely flattened portion of tube. There is some histological evidence that the inner surfaces of vessels become corrugated as they collapse, and the small side channels may become completely blocked. It is important to bear in mind that the whole of the collapse process for veins takes place at values of transmural pressure between about $-10^3 \mathrm{N\,m^{-2}}$ and $+10^3 \mathrm{N\,m^{-2}}$, which are in the middle of the physiological range.

Measurements of the elastic properties of blood vessels are often expressed in terms of an effective incremental Young's modulus E for circumferential stretch. When vessels are circular, E is related to the distensibility and the wall thickness-to-diameter ratio h/d through the equation

$$\frac{1}{D} = \frac{Eh}{d} \tag{14.3}$$

(cf. Equation (7.3b), p. 96). This shows how a vessel with a large Young's modulus can be very distensible if its walls are sufficiently thin. Use of the Young's modulus in this way is inappropriate when the vessel cross-section is not circular, because it varies around the cross-section. However, the Young's modulus is still important in determining the ease with which a vessel may collapse, as we shall see; and if two tubes of similar dimensions have different Young's moduli, their behaviour during collapse is different. For example, it is important not to generalize about the behaviour of a vein from observations on a rubber tube of similar diameter and wall thickness. Rubber has a Young's modulus of about $2.1 \times 10^6 \mathrm{N\,m^{-2}}$, while a typical value for a vein when its cross-section is circular (derived from Equation (14.3)) is about $5 \times 10^4 \mathrm{N\,m^{-2}}$.

The difference in behaviour between the two types of tube, as the transmural pressure is reduced from more than $+2 \times 10^3 \mathrm{N\,m^{-2}}$ to less than $-10^3 \mathrm{N\,m^{-2}}$, is illustrated in **Fig. 14.2**. **Figure 14.2**a shows the pressure–area relationships of the two types of tube and **Fig. 14.2**b shows the relationship between area and perimeter. The maximum area of the latex tube at high values of transmural pressure is much smaller than that of the vein, and the latex tube is much less distensible. The latex tube remains circular until the transmural pressure falls almost to zero, and then it starts to become elliptical and then dumb-bell shaped, just like a vein. However, during collapse, the perimeter of the latex tube remains almost constant and the area change is associated solely with change of shape. Furthermore, the maximum rate of change of area of the latex tube occurs at a negative value of p_{tm} (about $-0.25 \times 10^3 \mathrm{N\,m^{-2}}$), whereas that of a vein occurs at a positive value (about $+0.5 \times 10^3 \mathrm{N\,m^{-2}}$). It can

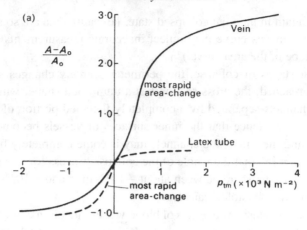

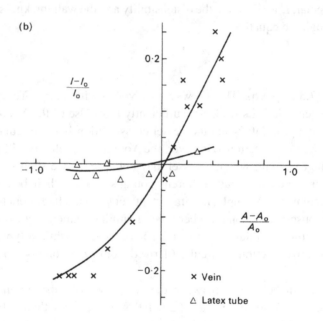

Fig. 14.2. (a) Pressure–area relationships for a segment of vein and a thin-walled latex tube: area expressed as a fractional area change as in **Fig. 14.1**a; pressure expressed as transmural pressure p_{tm}. Each vessel has an internal diameter of 1.2 cm and a wall thickness to diameter ratio h/d of approximately 0.04, but the latex tube had a Young's modulus 40 times that of the segment of vein. (b) Fractional change in perimeter plotted against fractional change in area for the vein and the latex tube. (After Moreno, Katz, Gold and Reddy (1970). Mechanics of distension of dog veins and other very thin-walled tubular structures. *Circulation Res.* **27**, 1069–1080, by permission of the American Heart Association Inc.)

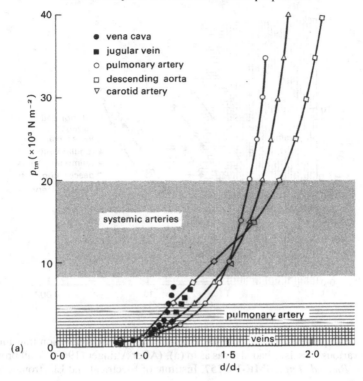

Fig. 14.3. (a) Relationship between diameter and transmural pressure p_{tm} for different vessels of a dog (diameter d is scaled relative to the diameter d_1 at a transmural pressure of $10^3 \, \text{N m}^{-2} = 10 \, \text{cm} \, \text{H}_2\text{O}$). The shaded areas indicate the normal physiological range of transmural pressures for systemic arteries, pulmonary arteries and veins at the level of the heart.

also be seen that the slope of the graph in **Fig. 14.2**a (i.e. the pseudo-distensibility) changes much less abruptly for a vein than for a latex tube. This is because the change in perimeter in a vein continues to contribute to the area change even after the shape has begun to alter.

We shall see in the next chapter that pulmonary arteries and veins, which are thin-walled, flexible vessels, like systemic veins, exhibit behaviour similar to that of veins (and not latex tubes) at small but physiological values of transmural pressure. It is of interest to compare the elastic behaviour of the three main types of large blood vessel which have been studied: a systemic vein, a pulmonary artery and a systemic artery. Such a comparison is provided for the vessels of a dog in **Fig. 14.3**a and b, which show measurements made in vessels in a pressure range where their cross-sections remained circular. **Figure 14.3**a shows the diameter of the vessel, relative to the diameter at a transmural pressure of $10^3 \, \text{N m}^{-2}$ (10 cm H$_2$O), plotted against distending pressure. The slopes of these curves, converted into values of an effective incremental Young's modulus of the vessel wall (related to distensibility by Equation (14.3)), are plotted

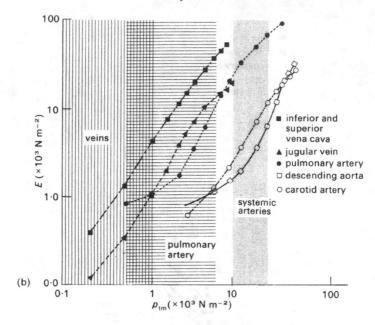

Fig. 14.3. (b) Dependence of effective incremental Young's modulus E on transmural pressure p_{tm} for various vessels. Shaded areas as in (a). (After Attinger (1969). Wall properties of veins. *Trans. Biomed. Eng.* **BME-16**, 257, Institute of Electrical and Electronics Engineers Inc., New York.)

in **Fig. 14.3**b. The shaded areas of the two graphs indicate the physiological range of transmural pressure to which the vessels are normally subjected. If these results are representative of such vessels in general, two deductions can be made:

(1) In their normal pressure range, pulmonary arteries have a similar Young's modulus to systemic arteries, although they are about seven times as distensible because of their smaller thickness to diameter ratio (**Table I**); systemic veins are about three times as distensible as pulmonary arteries, because the Young's modulus is about one-third of that of the pulmonary artery, while h/d is similar.

(2) However, at equal values of transmural pressure, throughout the whole range of studies, veins are stiffer than pulmonary arteries, which in turn are stiffer than systemic arteries. These results may reflect the fact that the sharp increase in stiffness associated with the increasing importance of collagen at high transmural pressures comes into effect at about the upper end of the normal operating range. We have already seen in the discussion of arteries (p. 258) that progressively greater area changes will follow equal increments of transmural pressure unless the vessel becomes progressively stiffer in this

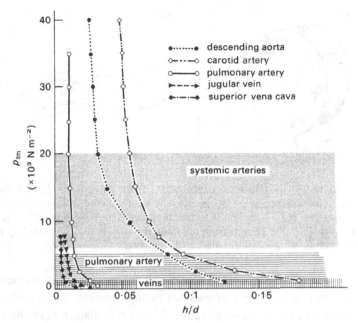

Fig. 14.4. Relationship between wall thickness: diameter ratio h/d, and transmural pressure p_{tm} for various vessels. Note the large changes in this ratio for low distending pressures. Shaded areas as in **Fig. 14.3**. (After Attinger (1969). Wall properties of veins. *Trans. Biomed. Eng.* **BME-16**, 257, Institute of Electrical and Electronics Engineers Inc., New York.)

way. The pressure required to burst a vein is very much greater than the normal transmural pressure, being about $5 \times 10^5 \, \mathrm{N\,m^{-2}}$ (5 atm), as was shown as early as 1733 by the English clergyman Stephen Hales (1677–1761) during one of his many experiments on the systemic circulation of animals.

Since the wall is incompressible, it must become thinner as the vessel is distended. Quantitative data for medium-sized and large veins and other vessels are given in **Fig. 14.4**. This shows that the ratio of wall thickness to vessel diameter h/d falls towards a constant value of about 0.01–0.02, which is a small fraction of the comparable value in a systemic artery.

All the above information was obtained from excised segments of large veins; factors which might affect the relationship between distensibility and transmural pressure in vivo are tone and the presence of surrounding tissue. However, very little information is available because it is very difficult to measure the transmural pressure and shape of veins. We know that large systemic veins can become non-circular in cross-section under physiological conditions; this has been shown with bi-plane X-rays of a vein in the calf muscles of erect human subjects. The X-rays were taken after the local intravenous injection of a radio-opaque dye, first with the leg immobile and a venous

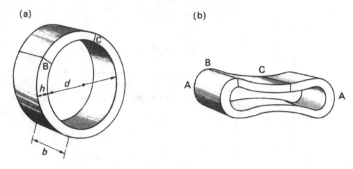

Fig. 14.5. (a) A ring-shaped slice of a tube wall. Internal diameter d, wall thickness h, breadth of slice in axial direction b. BC is an element of the ring which can be regarded as approximately straight, but whose length is much greater than b or h. (b) Illustration of the cross-section of the slice after compression has caused it to collapse. The bending moment is greatest and the radius of curvature smallest at the ends A.

transmural pressure of about $7 \times 10^3 \,\mathrm{N\,m^{-2}}$ and second with this pressure greatly reduced by a contraction of the calf muscles (which squeezes out the blood, see p. 461). In the former case the average values for the two diameters of the vein were 0.61 cm and 0.47 cm and in the latter 0.15 cm and 0.06 cm. It has not been established whether small systemic veins become significantly non-circular at low values of transmural pressure.

The resistance to bending of a tube wall It is clear that the bending of the wall is an important feature of the behaviour of blood vessels at low values of transmural pressure, and we have seen that there are significant differences in behaviour between veins and rubber tubes of similar dimensions. For an explanation of these differences we require an understanding of the mechanics of the bending process. To achieve this, we may suppose that the tube bends in the same manner throughout its length, so that it is only necessary to consider a single slice of the tube, forming a thin ring-shaped element (**Fig. 14.5**a). The thickness of the ring in the radial direction is h; we let the breadth of the slice in the axial direction be b. When the ring begins to buckle under a compressive transmural pressure, it will take up a configuration like that shown in **Fig. 14.5**b; the greatest curvature occurs at the ends A.

Once the ring ceases to be circular, it becomes very complicated to analyse exactly. However, it is possible to treat it approximately by considering an element of it (BC) that is short enough to be regarded as almost straight, but still long enough that the dimensions of its cross-section (h and b) are small compared with its length. Such an element resembles a beam of uniform cross-section, and we can apply to it a simple theory for the bending of a uniform beam, supported at one end (**Fig. 14.6**a). The compressive forces on the tube wall can be represented by a downwards force distributed uniformly along the beam, causing it to bend. An even simpler (but

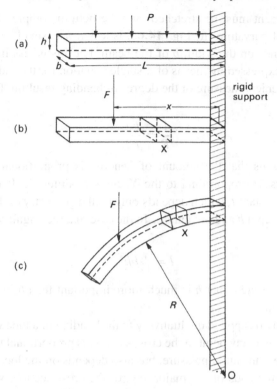

Fig. 14.6. A straight beam of length L, thickness h and breadth b supported at one end and caused to bend by forces acting down on it. (a) Beam shown before bending. A force P distributed uniformly along its upper surface; the bending moment at the support is $\frac{1}{2}PL^2$. (b) Beam shown before bending. A point force F exerted at a distance x from the support; bending moment is Fx. Note the undistorted element X. (c) Beam shown after bending. The element X has been distorted, being stretched at the top and compressed at the bottom. Radius of curvature R.

equivalent) situation is that of a point force acting on the beam (**Fig. 14.6**b). The amount of bending will depend on both the magnitude of the force F and the distance x of its point of application from the point of support. In fact, it depends on the moment of the force about the point of support, Fx, called the bending moment. (The bending moment generated at the support by a distributed force of magnitude P per unit length is $\frac{1}{2}PL^2$, where L is the length of the beam.)

Consider a small element of the beam, marked X in **Fig. 14.6**. As the beam bends, the top of the element is stretched relative to the centre-line and the bottom is compressed (**Fig. 14.6**c). Thus, the ease of bending depends on the ease with which the material of the beam can be stretched and compressed, as characterized by the Young's modulus E (p. 89). Furthermore, for a given value of E, the beam will bend more easily for smaller values of the thickness h; this is because the amount by which the top of the element must be stretched, and the bottom compressed, to achieve a particular radius of curvature (R, **Fig. 14.6**c) is less if h is small. The resistance to bending also depends on the width b of the beam, but this is less important. These factors can all be expressed by means of a single equation for the radius of curvature R, which is a convenient measure of the degree of bending resulting from a particular bending moment:

$$\frac{1}{R} = \frac{Fx}{EI}. \tag{14.4}$$

This equation tells us that the amount of bending is proportional to the bending moment and inversely proportional to the Young's modulus E. It is also inversely proportional to a constant I, which depends only on the geometry of the beam's cross-section (i.e. on its depth h and breadth b). In the case of a rectangular beam, I is given by

$$I = \tfrac{1}{12}bh^3, \tag{14.5}$$

which shows, as expected, that h is much more important than b in determining the resistance to bending.

These results may be applied qualitatively to the bending of a tube wall. In this case the bending moment at any point in the cross-section is proportional to the magnitude of the compressive transmural pressure, but also depends on the local curvature; it is in fact greatest at the ends of the major axis of the cross-section, where the radius of curvature is smallest (the points A in **Fig. 14.5**b). The difference in behaviour between the vein and the latex tube, illustrated in **Fig. 14.2**, is not surprising in view of Equation (14.4). The Young's modulus of the latex tube was 40 times that of the vein ($2.1 \times 10^6 \, \mathrm{N\,m^{-2}}$ compared with $5 \times 10^4 \, \mathrm{N\,m^{-2}}$), while the wall thickness and the undisturbed diameter were the same. Thus, the product EI was 40 times greater in the latex tube. This explains both its greater resistance to bending, exemplified by the fact that it remains circular until the transmural pressure has become negative, and its smaller distensibility while the cross-section is circular. It is also responsible for the fact that, once bending has begun, the perimeter of the latex tube does not change, while that of the vein continues to fall. A detailed explanation of the final development of the dumb-bell configuration is extremely complicated. One simple conclusion which can be drawn is that the tube cannot close off completely, since that would imply zero radius of curvature at the ends of the cross-section and hence an infinite bending moment (Equation (14.4)). In practice they may be effectively closed by corrugation of the inner surface.

Dynamics of blood flow in large veins

As in an artery, the pressure and flow rate in a vein vary with time and, since veins are distensible, pressure waves will propagate along them. We have already dealt at length with the propagation of pressure waves through elastic vessels (Chapter 12), and in this section we extend the discussion to take account of the special properties of veins: their thin walls, great distensibility, and valves. In order to confirm theoretical predictions of wave propagation along veins, and to predict flow patterns within them, it would be desirable to have experimental measurements of how the pressure distribution varies throughout the venous system. We have already seen examples of the pressure waveform in a venule (**Fig. 13.19**, p. 373), and that in the vena cava (**Fig. 11.29**, p. 226), so that we can calculate the total pressure drop across the venous system, but there are very few measurements in between. There is even less information on the flow-rate waveform. This limited amount of data is presented below, but is not sufficient for as complete a description of the unsteady pressures and flows in veins as that given for arteries in Chapter 12. Furthermore, only a brief discussion of the velocity patterns to be expected in veins is feasible.

There are several possible sources of unsteadiness of the pressure and flow in veins. These include: (i) the arterial pulse, transmitted to the venules through the arterioles and capillaries, where it is highly attenuated (p. 376); (ii) contractions of the right heart, transmitted peripherally against the direction of the flow in the large veins; (iii) respiratory manoeuvres, which can influence venous flow drastically because of the swings in extravascular pressure within the thorax; (iv) the action of muscles near which the vein passes. Very large amplitude oscillations may be developed with abnormal conditions of the heart, such as incompetence of the tricuspid valve and heart-block (p. 443). Self-excited oscillations may occur also; they are discussed separately on p. 456 *et seq.* because of their unusual mechanical properties.

Observed pressure and flow-rate waveforms The pressure waveforms measured at four stations between the right atrium and the brachial vein in humans are shown in **Fig. 14.7**. The measurements were all made within a short space of time, so it may be assumed that they correctly represent the way the shape of a particular wave changes with distance along this venous pathway. The oscillations in the superior vena cava are very marked, with an amplitude of about $700\,\mathrm{N\,m^{-2}}$, and the waveform is almost identical to that in the right atrium; there can be no doubt that they are derived from the pulsations of the right heart. A somewhat modified and considerably attenuated waveform, with an amplitude of about $250\,\mathrm{N\,m^{-2}}$, is found in the subclavian vein. Here, a definite phase lag can be seen because the peak pressures occur later than in the vena cava. This indicates that a pressure wave is propagating peripherally (its speed has not been calculated because the distances involved are not known). A further change in shape is found in the axillary vein, where some of the high-frequency oscillations are absent. A dramatic change occurs as the pressure-measuring catheter is withdrawn along this vein, which is almost certainly related to the presence of a valve (see below). In the brachial vein the oscillations have only a very small amplitude, less than $100\,\mathrm{N\,m^{-2}}$. However, we have already seen (**Fig. 13.19**, p. 373) that

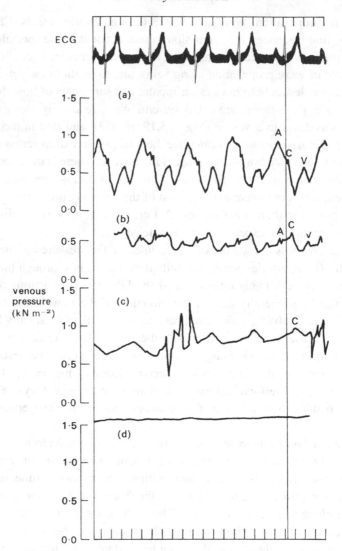

Fig. 14.7. The pressure waveform measured sequentially with a single catheter at four successive sites in the human venous system. These sites are, from top to bottom: the superior vena cava, the subclavian vein, the axillary vein and the brachial vein. The pressures were measured relative to atmospheric pressure; the subject was supine, so that all sites were at approximately the same level. Note the difference in mean pressure between the four sites. The four waveforms are aligned with the ECG record (top trace), so that the phase difference between the vena cava and the subclavian vein can be seen. The 'a', 'c' and 'v' pressure waves are also shown in **Fig. 14.8** and are discussed in the text in relation to it. (By courtesy of Dr G. Miller, Brompton Hospital, London.)

there are pressure oscillations in the venules of the cat mesentery, with an amplitude of about $400\,\text{N}\,\text{m}^{-2}$, which is about 30 times less than that in a systemic artery. It was demonstrated in Chapter 13 (p. 375) that these are almost certainly the remains of the arterial pulse after attenuation in the arterioles and capillaries. Larger oscillations of this sort can be found in small peripheral veins under certain conditions; for example, when the arterial pulse has a particularly large amplitude, as in patients with severe aortic valve incompetence; the attenuation in the microcirculation is then not sufficient to suppress the pulse so completely.

It should be emphasized once more that the waveforms plotted in **Fig. 14.7** were recorded at different sites along one particular pathway. The distributions of pressures along different pathways are likely to vary much more than in arteries. There are two reasons for this: one is the distribution of valves along the venous pathways, which as we shall see are expected to have a strong influence on the pressure pulse, and the other is the fact that the veins along some pathways are likely to be in a collapsed state, depending upon their transmural pressure. These veins will have a different distensibility, and hence wave speed, from those which remain fully open, and will also be a potent cause of wave reflection.

Data on the velocity waveform in the venae cavae of a man are presented in **Fig. 14.8**, which shows that in each cardiac cycle there are two main oscillations of flow velocity, completely out of phase with the pressure oscillations. At the time of atrial contraction, corresponding to the 'a' wave in the right atrial pressure trace, the velocity of blood in the vena cava is greatly reduced and there may even be momentary flow reversal. Thereafter, vena caval blood velocity rises, maintained throughout the 'c' wave and ventricular contraction. The peak velocity occurs during ventricular systole. A second increase in velocity occurs early in ventricular diastole, during the 'y' descent of the atrial pressure, which follows the 'v' wave when the atrioventricular valves open. Even larger oscillations can be generated if there is leakage of the tricuspid valve, or if the impulse conduction in the heart is defective (so-called heart-block) so that the atrium contracts during ventricular systole, while the atrioventricular valves are closed. Velocity waveforms have not yet been measured in more peripheral veins.

Wave propagation in veins Very few experiments have been performed in which the pressure pulse has been recorded simultaneously at two sites in the large veins. Thus, there are few observations of the speed of propagation and rate of attenuation of the pulse along the veins. Those observations which do exist show a wave speed of between 0.5 and $3.0\,\text{m}\,\text{s}^{-1}$ in the superior vena cava of a dog; almost as much variation was found within single animals, according to which part of the pulse was recorded. This variation was presumably related either to the variation in venous pressure as the wave passed (see p. 298) or to the presence of reflections, which alter the apparent wave speed (see p. 276). We have only two further sources of information:

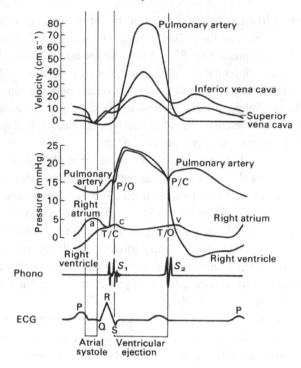

Fig. 14.8. A composite diagram of pressure and velocity of flow into and out of the right heart, constructed from a number of different measurements. The top curves show the velocities in the venae cavae and the pulmonary artery, then come the pressures in the right heart and pulmonary artery, and then the phonocardiogram and ECG. These are included to serve as timing references. T/C: tricuspid valve closed; P/O: pulmonary valve open; P/C: pulmonary valve closed; T/O: tricuspid valve open. (From Wexler, Bergel, Gabe, Makin and Mills (1968). Velocity of blood flow in normal human venae cavae. *Circulation Res.* **23**, 349–359. By permission of the American Heart Association Inc.)

(i) prediction, based on the knowledge of the mechanics of wave propagation as derived from a study of arteries; and (ii) a single series of experiments on the abdominal vena cava of a dog in which a short train of artificial high-frequency sinusoidal pressure waves was generated at one location in the vessel and the pressure recorded at another. As in the similar experiments in arteries (p. 297), measurements were made before the first reflected waves had time to return to the point of observation from the nearest reflection site.

The approximate theory outlined in Chapter 12, pp. 271–274, predicts that small-amplitude pressure waves in an elastic tube should be propagated with speed c given by Equation (12.9); that is,

$$c = \frac{1}{\sqrt{\rho D}}, \tag{14.6}$$

where ρ is the density of blood and D is the distensibility of the vessel, given by Equation (14.1). Now the distensibility of the thoracic vena cava of a dog has maximum values of $2\text{--}2.5 \times 10^{-3}\,\text{m}^2\,\text{N}^{-1}$ at distending pressures of $\pm 0.5 \times 10^3\,\text{N}\,\text{m}^{-2}$ (**Fig. 14.1e**), with a much smaller value at large values of transmural pressure (the distensibility is about $4 \times 10^{-5}\,\text{m}^2\,\text{N}^{-1}$ when $p_{tm} = 3 \times 10^3\,\text{N}\,\text{m}^{-2}$, and falls as p_{tm} rises further). Also, at very negative values of transmural pressure the distensibility becomes very small, equal to $10^{-4}\,\text{m}^2\,\text{N}^{-1}$ when $p_{tm} = -2 \times 10^3\,\text{N}\,\text{m}^{-2}$. The density of blood is about $10^3\,\text{kg}\,\text{m}^{-3}$, so the predicted value of the wave speed varies from $5\,\text{m}\,\text{s}^{-1}$ or more at high transmural pressures, to a minimum of about $0.6\,\text{m}\,\text{s}^{-1}$ when the vein is collapsing, and increases again to about $3\,\text{m}\,\text{s}^{-1}$ when it is fully collapsed.

Another prediction can be obtained from the data of **Figs 14.3b** and **14.4**. The values of Young's modulus E shown in **Fig. 14.3b**, combined with a value of the wall thickness to diameter ratio of 0.01 (**Fig. 14.4**), can be substituted into Equation (14.3) to estimate the distensibility, and the wave speed can again be predicted from Equation (14.6). The values of the wave speed so obtained vary continuously from about $1\,\text{m}\,\text{s}^{-1}$ at a transmural pressure of $0.5 \times 10^3\,\text{N}\,\text{m}^{-2}$ to about $9\,\text{m}\,\text{s}^{-1}$ at a transmural pressure of $10^4\,\text{N}\,\text{m}^{-2}$.

These predictions indicate that the wave speed in veins is less than that in large arteries, as long as the transmural pressure remains small (less than about 3×10^3 $\text{N}\,\text{m}^{-2}$). However, when the transmural pressure is as large as $10^4\,\text{N}\,\text{m}^{-2}$ (as in the foot of an upright human), the wave speed in a vein may be as great as or greater than that in the corresponding artery, despite the fact that the pressure in the artery still exceeds that in the vein by about $10^4\,\text{N}\,\text{m}^{-2}$. (This may be somewhat academic, since, as we have seen, pulsations are not usually transmitted to the veins in the foot.)

The measurements of small, high-frequency sinusoidal waves were made in the abdominal venae cavae of dogs at values of transmural pressure between $0.5 \times 10^3\,\text{N}\,\text{m}^{-2}$ and $2.5 \times 10^3\,\text{N}\,\text{m}^{-2}$. The results for one dog, with different frequencies, are shown in **Fig. 14.9**, where it can be seen that the wave speed varies from 2 to $6\,\text{m}\,\text{s}^{-1}$ over the applied pressure range. The results for different dogs showed a wide scatter, the wave speed at a transmural pressure of $10^3\,\text{N}\,\text{m}^{-2}$ varying from less than $1\,\text{m}\,\text{s}^{-1}$ to nearly $3\,\text{m}\,\text{s}^{-1}$. These values cover a similar range as the predicted values at the same transmural pressures, despite the fact that the applied frequencies were unphysiologically high (20 Hz or above). The measured wave speed falls slightly as the frequency drops. No experiments were performed on vessels whose cross-section had become markedly non-circular, so no measurements are available to check the prediction that the wave speed varies rapidly with transmural pressure when that is very low.

These experiments used short trains of high-frequency waves expressly to avoid distortions due to wave reflection. However, reflections are sure to arise in vivo at every site where the vessel properties change. Such sites will include junctions with other veins, valves and regions where the transmural pressure and, hence, cross-sectional area and distensibility change; for example, where a vein enters the thorax

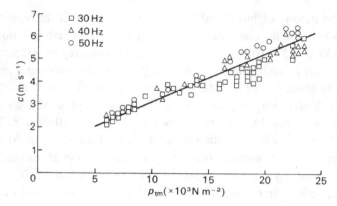

Fig. 14.9. Wave speed c of imposed small-amplitude sinusoidal oscillations in the vena cava of a dog, plotted against transmural pressure p_{tm}. The different symbols represent different frequencies of the waves. (After Aniiker, Wells and Ogden (1969). The transmission characteristics of large and small pressure waves in the abdominal vena cava. *Trans. Biomed. Eng.* **BME-16**, 271, Institute of Electrical and Electronics Engineers Inc., New York.)

and may collapse. Little experimental work has been done to determine how the pulse is reflected and transmitted at any of these sites, and too little is known about the venous bed for us to make any detailed predictions. However, it is observed that the pressure is virtually steady in a medium-sized vein in the arm (**Fig. 14.7**); this is true even when the central venous pressure has transient fluctuations of up to $1.3 \times 10^3 \, \text{N} \, \text{m}^{-2}$ (100 mm Hg) superimposed on it by coughing. This indicates that almost all the wave is reflected before it reaches this site (as in arteries, viscous and visco-elastic effects could not cause such complete attenuation of the wave in the distance available; see p. 304). The complete reflection is probably associated with the valves, because in patients with incompetent valves (who may also have varicose veins, p. 463) transient pulses generated by raising intrathoracic and intra-abdominal pressure (as in a cough) can often be detected in peripheral veins.

We can speculate about some of the mechanisms involved in wave reflection at a venous valve. Model experiments have shown that if a valve is already closed when a pressure pulse arrives from the direction of the heart, then the valve cusps bulge backwards elastically, but a negligible amount of backflow through it occurs. Thus, a little of the energy of the wave is transmitted, but most is reflected as it would be at a closed end (see p. 284). If the valve is open, and the flow forwards through it when the pulse arrives, the situation is more complicated. Initially the pressure gradient will be reversed, so the valve will start to close, presumably by the same mechanism as operates to close the aortic valve (see p. 231). The time during which the pressure gradient is reversed may or may not be long enough for the valve to close completely before the pressure gradient is once more directed towards the heart. Thus, the details of what reflection takes place, and what the flow is like on either side of the valve, will depend on the time course of the pulse, and hence on frequency; this is unlike reflection at a junction (p. 278).

In the experiments to measure the wave speed in veins, the amplitude of the sinusoidal pressure oscillations was less than $0.2 \times 10^3 \,\mathrm{Nm}^{-2}$; in most cases this was a small fraction of the mean transmural pressure, and the waves were observed to remain sinusoidal. The amplitude of the normal pulse may, however, be twice as great as this, and even higher in some abnormal conditions. Furthermore, the mean value of transmural pressure will in some veins lie between $\pm 0.5 \times 10^3 \,\mathrm{Nm}^{-2}$, where the veins are expected to be particularly distensible, and can change their cross-sectional area and shape by a large amount when the pressure changes by $0.4 \times 10^3 \,\mathrm{Nm}^{-2}$. At these levels the passage of the pulse will cause large changes in the vessel cross-section; at the same time, the peak blood velocity of about $0.25 \,\mathrm{m\,s}^{-1}$ (**Fig. 14.8**) will be almost half the predicted wave speed ($0.6 \,\mathrm{m\,s}^{-1}$ – see p. 445). For both these reasons nonlinear effects may be important, as we have seen in the discussion of pressure waves in arteries (Chapter 12, pp. 297–299).

The most striking effect of nonlinearity will be a steepening of the front of the pressure wave. In certain circumstances this may develop into a sharp pressure jump, analogous to a shock wave, and be accompanied by a rapid longitudinal variation in vessel cross-section. Direct evidence of 'shocks' in the venous system comes from the observation of venous 'pistol shot' sounds, which may be heard in the femoral and jugular veins of patients with leakage of the tricuspid valve. The sounds originate with the contraction of the right ventricle, which in such patients causes considerable regurgitation of blood into the right atrium and the venae cavae, leading to a very large amplitude venous pulse (cf. the arterial pistol shot, described on p. 299). The steepening of the front of the waves in the veins has been investigated, again by introducing artificial high-frequency waves into a dog's vena cava and measuring the wave at two stations 4 cm apart. The imposed amplitude was varied between 0.46×10^3 and $1.3 \times 10^3 \,\mathrm{Nm}^{-2}$ (3.5–10 mm Hg); the imposed frequency was 7.5 Hz; the results are shown in **Fig. 14.10**. The non-sinusoidal nature of the waves is evident, as is the progressive steepening of the front as it proceeds down the tube, despite the fact that there is considerable damping, as witnessed by the smaller amplitude at the distal station. Another possible effect of nonlinearity is that the different frequency components in the waveforms of the pressure gradient, velocity profile and flow rate may no longer be independent and may interact to affect the mean quantities as well (see p. 126). This means that inferring the mean flow from the mean pressure gradient may be inappropriate. However, no experimental evidence is available on this question; in arteries it is probably unimportant.

Finally we turn to the attenuation of pressure waves in veins. As with every other property of the waves, this has been examined only with artificial high-frequency waves, and, as can be seen from **Fig. 14.10**, these are considerably attenuated.

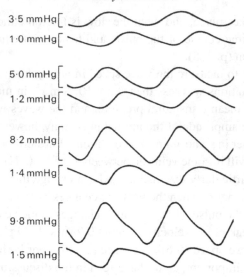

Fig. 14.10. Changes in shape of an imposed pressure wave of frequency 7.5 Hz as it travels along the abdominal vena cava of a dog. Each pair of traces illustrates a single amplitude of the imposed wave, and in each case the lower trace was recorded 4 cm distal to the upper, on an expanded scale. The amplitude of the wave in each case is shown beside the trace; note that the wave is increasingly distorted as its amplitude is increased. (After Aniiker, Wells and Ogden (1969). The transmission characteristics of large and small pressure waves in the abdominal vena cava. *Trans. Biomed. Eng.* **BME-16**, 268, Institute of Electrical and Electronics Engineers Inc., New York.)

Measurements of the amplitude of small sinusoidal pressure waves at pairs of stations separated by different distances show that, as in arteries (p. 303), the amplitude falls off exponentially with distance. By analysing the waves at different frequencies it is found that the amount of attenuation per wavelength is independent of frequency, for frequencies greater than 20 Hz (see **Fig. 14.11**). That is, the amplitude a of the wave is equal to

$$a_0 e^{-kx/\lambda},$$

where x is a distance measured along the vessel, a_0 is the amplitude at $x = 0$, λ is the wavelength and k is a constant. The value of k derived from the venae cavae of dogs was found to lie in the range 1.0–2.5 (attenuation of 63–92% per wavelength), compared with the range 0.7–1.0 (50–63% per wavelength), for the aorta (see **Fig. 12.37**, p. 303). Thus, the rate of attenuation of high-frequency waves is greater in the vena cava than in the aorta. At a heart rate of 2 Hz, a venous wave with a speed of 2 m s^{-1} would be attenuated by 20–30% in a distance of 30 cm, the length of the dog's vena cava. This damping cannot all be due to the viscosity of blood because, as

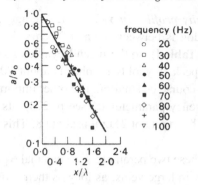

Fig. 14.11. The attenuation of small-amplitude pressure waves in the vena cava of a dog. The ordinate is the ratio of wave amplitude a to the amplitude a_0 at a fixed site. The ordinate (x/λ) is the distance distal to the fixed site x divided by the wavelength of the wave λ. The straight line in this semi-logarithmic plot is given by the equation $a/a_0 = e^{-kx/\lambda}$ with $k = 1.05$. The different symbols refer to different wave frequencies. The transmural pressure was in the range 1.0–$1.5 \times 10^3 \, \mathrm{N\,m}^{-2}$. (After Aniiker, Wells and Ogden (1969). The transmission characteristics of large and small pressure waves in the abdominal vena cava. *Trans. Biomed. Eng.* **BME-16**, 271, Institute of Electrical and Electronics Engineers Inc., New York.)

in arteries, the value of a is large in these experiments (see p. 273) and the energy lost in the viscous boundary layers is relatively small. The actual attenuation must, therefore, be predominantly due to the fact that the vessel walls have visco-elastic, rather than purely elastic, properties. Veins also exhibit significant visco-elastic behaviour at somewhat lower frequencies. For example, hysteresis (p. 97) was found when the volume of a segment of vein was varied sinusoidally at a frequency of 5 Hz (the value of α still being large, about 14); the pressure was higher during inflation than at the same volume during deflation (**Fig. 14.12**).

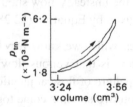

Fig. 14.12. Evidence of hysteresis in a vein. Pressure–volume curve of a segment of the inferior vena cava of a dog, while the volume was subjected to sinusoidal changes at a frequency of 5 Hz. The period of increasing volume is associated with the larger pressures, as indicated by the arrows. (The volume is plotted instead of the cross-sectional area because the length of the segment is unknown.) (After Attinger (1969). Wall properties of veins. *Trans. Biomed. Eng.* **BME-16**, 256, Institute of Electrical and Electronics Engineers Inc., New York.)

Flow patterns and velocity profiles in veins The average velocity of blood in the inferior vena cava of a dog is between 10 and $20\,\mathrm{cm\,s^{-1}}$, and the diameter of the vessel is about 1 cm (see **Table I**), so the mean Reynolds number lies between about 250 and about 500; the peak Reynolds number is about 700. In smaller veins the Reynolds number will of course be smaller. The other important dimensionless number is Womersley's frequency parameter α (see p. 273); its highest value occurs in the vena cava, where at a heart rate of 2 Hz it is about 8. This too is smaller in smaller veins.

From a knowledge of these two parameters we can make predictions of the general characteristics of the flow in large veins, as long as their cross-section remains more or less circular:

(1) We expect the flow to be laminar, because even the peak Reynolds number is far less than the critical value for steady flow of about 2300.

(2) There will be a long entrance length for the mean flow in the largest veins because the mean Reynolds number is quite large (assuming that the mean flow is independent of the oscillatory components, which may not be true – see p. 315). However, it will not be longer than the vessel, as it is in the aorta. In the inferior vena cava, the entrance length, as predicted by Equation (5.4), is 7.5– 15 cm, although the presence of secondary motions in the flow entering a vein may cause these values to be decreased somewhat because the lateral mixing of slower and faster moving fluid is enhanced, so that the flow can develop more quickly. Thus, even if there are thin boundary layers where the flow enters the vessel from subsidiary veins, the mean flow should be approximately fully developed before it reaches the heart, 30 cm from the entrance.

(3) Because the value of α is quite large, the oscillatory boundary layers on the vessel wall should remain thin (see p. 59) and the oscillatory flow in the core should have a flat velocity profile, in contrast with the curved profile of the fully developed mean flow.

(4) The entrance length for the unsteady flow should be short, about 4–5 cm, according to the formula given by Equation (12.28) on p. 315.

Apart from the effect of valves, which we know very little about, a major difference between venous and arterial flow is that venous flow is directed from the smaller vessels to the larger. Thus, the pattern of flow at junctions will be different from that in arteries (p. 318). At a junction two streams come together, so that just beyond the flow divider the velocity profile in the plane of the junction will have a dip in the centre because the peak velocity in each tube upstream is off the axis of the parent tube. Each stream is forced to turn, so that secondary motions are set up (see p. 67), generating two pairs of secondary vortices. This has been verified by observing smoke patterns in a model bifurcation, as shown in **Fig. 14.13**. Measurements of the velocity profiles in the parent tube of the model, downstream of the flow divider,

(a)

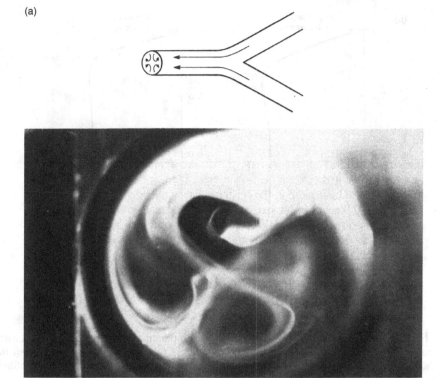

Fig. 14.13. (a) End view of the parent tube of a model symmetrical bifurcation when air flows steadily from the daughter tubes to the parent. Flow patterns visualized with smoke. Two pairs of secondary vortices can be seen. Parent tube Reynolds number is 700. (From Schroter and Sudlow (1969). Flow patterns in models of the human bronchial airways. *Respiration Physiol.* **7**, 341. North-Holland Publishing Co., Amsterdam.)

show that the boundary layers remain thin and that, except near the flow divider, the velocity profile is fairly flat (**Fig. 14.13**b). Presumably this is because of the way the secondary motions redistribute the longitudinal flow. Unsteady flow in a junction of this nature has not been studied. We know of no systematic measurements of velocity profiles in even the largest veins, but secondary motions have been clearly seen in the inferior vena cava, downstream of the confluence of the iliac veins.

Flow in collapsible tubes

In circumstances when a vessel may collapse and its cross-section become markedly non-circular, a number of interesting phenomena can occur, and the relationship between pressure gradient and flow rate becomes very complicated. For this reason a separate section is devoted to flow in collapsible tubes.

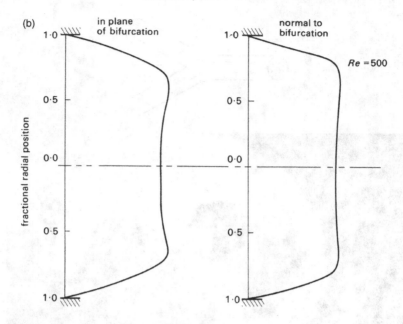

Fig. 14.13. (b) Velocity profiles in two planes in the parent tube of a model symmetrical bifurcating system consisting of four generations of the tube. Air flows steadily from the daughter tubes to the parent. Parent tube Reynolds number is 500. Note how flat the profile is and how thin the boundary layer is.

Curious behaviour of blood flow in a vein has been observed when the transmural pressure is very low. Reduction of the internal pressure at the downstream end is found to have little or no effect on the rate of flow, provided that it is less than the pressure outside the vessel. Moreover, the same result can be obtained with water flowing in a thin-walled rubber tube, in a model system. There has been quite extensive investigation of this phenomenon, most of the work being done with thin-walled latex tubes rather than blood vessels. Caution should be exercised in applying the results to actual vessels, because of the difference between the mechanical properties (p. 433) of the two types of tube. We shall first describe such a model experiment and then suggest possible mechanical explanations for the results. Finally, we shall indicate where in the circulation such phenomena might appear and whether they are observed to do so.

Model experiments The experimental arrangement is depicted schematically in **Fig. 14.14.** A segment of collapsible tubing is attached at its ends to two rigid tubes of the same diameter and surrounded by a chamber whose pressure p_c is controlled. Water flows through the tube and the static pressures at points just upstream and downstream of the collapsible segment are measured; they may be denoted by p_1 and p_2

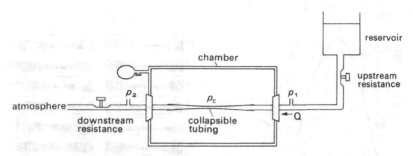

Fig. 14.14. Sketch of the experimental system for studying the flow of a liquid in a collapsible tube. The collapsible tube is attached to rigid tubes of the same diameter and enclosed in a chamber. Chamber pressure p_c can be independently varied; tube pressures upstream and downstream of the collapsible segment (p_1 and p_2) can be measured. The flow rate Q through the system can be altered by varying either the downstream resistance leading to atmosphere or the upstream resistance leading from the supply reservoir. Flow from right to left.

respectively. Further downstream of the downstream pressure tap the water flows to atmosphere through a fixed resistance, so that p_2 exceeds atmospheric pressure by an amount proportional to the flow rate Q through the tube. This flow rate can be varied by adjusting either the height of the upstream reservoir which supplies the water or the value of an upstream resistance between the reservoir and the upstream pressure tap. It can also be varied by adjusting the downstream resistance; the consequences are then somewhat different, and are described later.

The results described first were obtained with the chamber pressure held fixed and the flow rate reduced from a large value to a small one by an increase in the upstream resistance. The results are expressed as a graph of the pressure drop ($p_1 - p_2$) against flow rate Q, and are plotted in **Fig. 14.15**a. The numbers on the curve refer to the photographs in **Fig. 14.15**b, which show a sequence of side views of the collapsible segment of the tube at successive stages in the experiment (flow is from right to left). Note that whenever there is flow, p_1 must be greater than p_2 in order to overcome any viscous resistance in the collapsible segment. The results fall into three distinct categories:

(1) When the flow rate is sufficiently large for downstream pressure p_2 to exceed chamber pressure p_c, the pressure everywhere in the collapsible segment exceeds p_c and the tube remains almost circular (panels 1–6 of **Fig. 14.15**b.) The flow in the tube will everywhere be Poiseuille flow, so the pressure drop is proportional to flow rate, with an almost constant resistance, as given by Poiseuille's law, Equation (5.1), p. 47. This is illustrated by the right-hand segment of the curve in **Fig. 14.15**a, marked I. The resistance is not exactly constant because, as the flow rate is reduced, the diameter of the tube decreases slightly since the tube has a small distensibility (**Fig. 14.2**a).

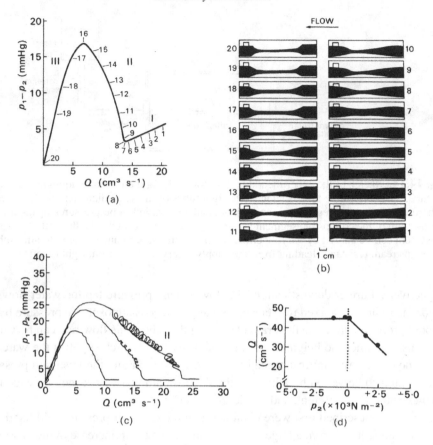

Fig. 14.15. (a) Graph of pressure drop $(p_1 - p_2)$ against flow rate Q in a model such as that shown in **Fig. 14.14**. Q was varied by varying the upstream resistance; the downstream resistance and the chamber pressure $(p_c = 3.3 \times 10^3 \, \text{N m}^{-2})$ were kept fixed. The numbers refer to the photographs in (b). The three flow regimes, I, II and III, are explained in the text. (b) Side views of the collapsible segment of tubing at different stages of the experiment; numbers correspond to the numbered positions on the graph in (a). Flow from right to left. ((a) and (b) from Conrad (1969). Pressure–flow relationships in collapsible tubes. *Trans. Biomed. Eng.* **BME-16**, 284–95, Institute of Electrical and Electronic Engineers Inc., New York.) (c) Three curves, like that in (a), derived with three different values of the downstream resistance, the chamber pressure being kept fixed ($= 3.9 \times 10^3 \, \text{N m}^{-2}$). Each trace is a continuous recording, made while the flow rate was gradually increased, and the self-excited oscillations which occur in phase II of the experiment in some cases can be seen. Similar curves are obtained if the downstream resistance remains fixed and different values of the chamber pressure are taken. ((c) from Katz, Chen and Moreno (1969). Flow through a collapsible tube. *Biophys. J.* **9**, 1261–79.) (d) Graph of flow rate against downstream pressure p_2, measured relative to chamber pressure p_c. When p_2 is less than p_c the flow rate is independent of p_2. ((d) from Holt (1969). Flow through collapsible tubes and through in situ veins. *Trans. Biomed. Eng.* **BME-16**, 274, Institute of Electrical and Electronic Engineers Inc., New York.)

(2) When the flow rate is reduced below a certain critical value, the downstream pressure becomes smaller than the chamber pressure. Thus, for a very small further decrease in flow rate, the transmural pressure at the downstream end of the collapsible segment becomes negative and the cross-section begins to change shape and to collapse. The cross-sectional area falls rapidly as the transmural pressure falls, because of the large distensibility (**Fig. 14.2**a). As the tube collapses in this way, the flow resistance rises rapidly and the pressure drop required to maintain the (gradually falling) flow rate also rises dramatically. This is illustrated by the central section (II) of **Fig. 14.15**a. The progressive collapse of the tube can be seen from panels 7–16 of **Fig. 14.15**b. At first (panels 7 and 8) the tube begins to taper towards its downstream end (where the internal pressure is least, close to p_c), and partially collapses there. As the flow rate is reduced further, the region of collapse extends further upstream (panels 9–16), until eventually the whole of the flexible segment is collapsed (panel 16). At this stage the upstream pressure p_1 has become approximately equal to the chamber pressure p_c. If, during this phase, the downstream resistance is changed so that p_2 varies independently of p_1, very little change in flow rate is observed. However, if the chamber pressure is varied instead, the degree to which the tube is collapsed also varies, and the flow rate changes accordingly. In fact, during such manoeuvres, the flow rate Q is approximately proportional to $p_1 - p_c$, not $p_1 - p_2$.

(3) Finally, when the whole segment is collapsed (upstream pressure less than chamber pressure), its cross-section has the rather rigid dumb-bell configuration already described (p. 433). Thus, as the flow rate is reduced still further, no further change in cross-section occurs (panels 17–20) and the resistance to flow once more becomes constant, the pressure drop falling in proportion to the flow rate (region III in **Fig. 14.15**a). The value of the resistance is 10–100 times higher than before collapse, however, because of the very narrow channels through which the fluid has to pass.

If at any stage the flow is stopped completely by closing the downstream resistance, the pressure everywhere in the tube will become uniform, taking a value determined by the level of the upstream reservoir. If this is such that p_1 is less than the chamber pressure then the tube will remain collapsed, but if p_1 is greater than the chamber pressure then the tube will open again.

A qualitatively similar sequence of results to that outlined above is obtained whatever the Reynolds number of the flow through the tube. A further interesting phenomenon is observed in experiments in which the Reynolds number is quite large. This is that self-excited oscillations develop for a range of conditions in which the chamber pressure, the upstream reservoir and the downstream resistance are held fixed. A steady value of the flow rate in such circumstances is thus expected; but when this value lies in region II (**Fig. 14.15**a), oscillations in tube cross-section, pressure drop and flow rate are seen to develop spontaneously. (When such oscillations are present, mean values of flow rate and pressure drop are plotted in **Fig. 14.15**a.) As the upstream resistance is increased, so that the mean flow rate is reduced, the frequency of the oscillations increases. In the series of experiments from which **Fig. 14.15** was taken the oscillation frequency was between 0.1 and 1 Hz. What governs this frequency is not well understood; possible mechanisms causing the oscillations are discussed below.

It should again be emphasized that the results presented in **Fig. 14.15**a were all obtained with fixed values of the downstream resistance and of the chamber pressure. A different curve is obtained if either of these quantities is changed; all such curves have a similar shape, and all come together in region I when the tube is fully open. However, if either the chamber pressure is reduced or the downstream resistance increased, the slope of the curve in regions II and III is reduced, as is the maximum pressure drop (**Fig. 14.15**c). This makes it clear that the value of the pressure drop required to maintain a given mean flow rate depends on the chamber pressure and the downstream resistance. If the upstream and downstream resistances are held fixed and the chamber pressure is increased from atmospheric, then both the flow rate and the pressure drop $p_1 - p_2$ remain unaltered until the chamber pressure exceeds the downstream pressure. Then the flow rate starts to fall, in proportion to $p_1 - p_c$, as stated above, until the tube is completely collapsed, when it remains at a more or less constant, small value. Similarly, if the downstream resistance is decreased, with chamber pressure fixed above atmospheric, and with the upstream resistance fixed, the downstream pressure will fall initially, and $p_1 - p_2$ and hence the flow rate will increase. However, when the downstream pressure becomes less than the chamber pressure, collapse will begin and the flow-rate will stop falling, ultimately taking a value determined by the value of $p_1 - p_e$ (**Fig. 14.15**d).

Mechanisms We can easily see why the resistance of the flexible tube increases as it collapses. The cross-sectional area decreases, so for a fixed flow rate the average velocity increases and the viscous shear stress on the wall, which retards the fluid, correspondingly increases. The surface area over which this shear stress is operating does not decrease because the perimeter remains almost constant. This increase is enhanced by the flattening of the cross-section, which requires high velocity gradients across the shortest diameter. The flow rate can thus be maintained only by increasing the pressure drop. An indication of the amount by which the pressure drop has to rise can be obtained from the theory of fully developed viscous flow in a straight elliptical tube of major and minor axes $2a$ and $2b$. The ratio between pressure gradient and

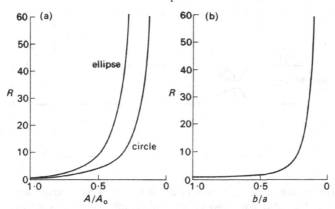

Fig. 14.16. The resistance of a straight elliptical tube of constant perimeter plotted (a) against cross-sectional area and (b) against the ratio of minor to major axis (b/a). The dimensionless parameter R is the resistance of the tube divided by the resistance when the tube was circular; A/A_0 is the cross-sectional area of the tube divided by the area when the tube was circular. Thus, for a circular tube, $A/A_0 = 1$, $b/a = 1$ and $R = 1$. In (a) the curve is compared with that for a circular tube having the same cross-sectional area (but not the same perimeter, of course). The resistance of each tube becomes infinite as the area goes to zero, but the elliptical tube has more than double the resistance of the circular tube when the area has fallen below about $0.5A_0$ (b/a below about 0.25).

flow rate (the resistance) is in this case given by

$$\frac{\Delta p/l}{Q} = 8\mu \frac{a^2 + b^2}{2a^3 b^3},$$

(14.7)

where μ is the viscosity of the fluid. The factor $(a^2 + b^2)/2a^3 b^3$ is equal to $1/a^4$ when the cross-section is circular (so that $a = b$), and in that case Equation (14.7) reduces to Poiseuille's law (Equation (5.1)). The effect on this quantity of reducing the area of the ellipse while keeping its perimeter constant is shown in **Fig. 14.16**a. The graph of resistance against area is compared with the same curve for a circular tube whose radius is decreased; the increasing ellipticity of the tube can be seen to contribute an increasingly large amount to the resistance as the area decreases. Note that the ratio of minor to major axis, b/a, falls to a value of 0.75 with only a 10% change in resistance (**Fig. 14.16**b).

These theoretical results are based on the assumption that the elliptic tube has parallel walls. When there is a localized constriction, other factors come into play, at least if the Reynolds number is large (about 100 or above), and the results may become inaccurate. In that case, the flow, accelerated through the constriction, may separate from the tube walls when the area begins to increase again, forming a jet-like flow downstream (see p. 64). Such separated flows are always associated with large energy losses and pressure drops, over and above the direct viscous losses already

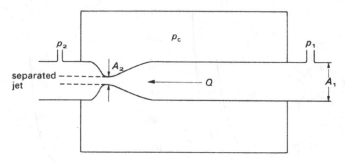

Fig. 14.17. Illustration of the separation of the flow from a constriction when the Reynolds number is large. The upstream, downstream and chamber pressures are p_1, p_2 and p_c respectively; the volume flow rate is Q; the upstream area is A_1 and that at the constriction is A_2.

considered, and may in part be responsible for the rise in resistance observed in phase II of the experiment.

A complete theoretical analysis of the pressure–flow relations at the large Reynolds numbers present in most of the experiments (and appropriate for the study of large veins) is not available. A very approximate analysis can be performed by assuming that the flow does separate downstream of the constriction and that, therefore, no pressure recovery takes place (**Fig. 14.17**). Then the pressure at the constriction is equal to the downstream pressure p_2, whereas the cross-sectional area (A_2) there is much smaller than that at the upstream pressure tap (A_1). Thus, the average fluid velocity at the constriction (Q/A_2) is much greater than that upstream (Q/A_1). Neglecting the direct viscous pressure loss between the upstream pressure tap and the constriction, we can apply Bernoulli's equation (Equation (4.6), p. 43) to obtain

$$p_1 - p_2 = \tfrac{1}{2}\rho Q^2 \left(\frac{1}{A_2^2} - \frac{1}{A_1^2} \right), \tag{14.8}$$

where ρ is the fluid density. All quantities except the area of the constriction A_2 are measured in the experiment; this can be estimated either from photographs like those of **Fig. 14.15**b or by linking it to the known transmural pressure $p_2 - p_c$ through independently measured distensibility relations (**Fig. 14.1** or **14.2**). Then Equation (14.8) can be tested experimentally. One such test has been performed, and shows reasonably good agreement with theory.

The oscillations which develop are even more difficult to explain. One possible mechanism also uses Bernoulli's equation and can be summarized as follows. Suppose that the pressure at the downstream end of the tube is slightly less than that in the chamber, so that the tube begins to collapse. As the constriction develops, the flow rate does not at first fall because of the fluid's inertia. Thus, the velocity in the constriction increases, and so, from Bernoulli's equation, the pressure falls. This fall increases the pressure difference between the chamber and the tube, and the collapse is therefore accelerated. When the cross-sectional area becomes very small, the resistance becomes very high, both through direct viscous action in the constriction and because of separation of the flow and turbulence downstream of it. Therefore, the flow rate falls and the pressure just upstream of the constriction rises. This high pressure causes the collapsed section to reopen, after which the upstream pressure falls again and the cycle is repeated.

This explanation of the oscillations is very plausible, but cannot be entirely correct because it ignores the inertia associated with local acceleration and deceleration of the fluid every cycle. The observed frequency of the oscillations is such that α is large (between 4 and 13), and the inertia will not be negligible. No theory which incorporates it has yet been developed; when one is, it may involve the presence of 'shock' waves in the tube (p. 299). The energy dissipation associated with wall viscoelasticity is likely also to be important.

Physiological evidence: Korotkoff sounds It has already been stated that veins collapse when the transmural pressure is negative, one example being the veins in the neck of an upright subject. However, there appear to be no experiments in which the relationships between transmural pressure, cross-sectional area and flow rate have been measured at the same time in a vein at or near a point of collapse, nor any observations of self-excited oscillations in veins. There is nonetheless a familiar situation in which high-frequency oscillations develop in systemic arteries, and they are probably similar to those described in the model experiments. These are Korotkoff sounds.

When arterial blood pressure is measured with a cuff inflated around a limb, the cuff pressure is usually raised to a level well above the peak (systolic) arterial value, so that the artery collapses and blood flow stops. Then, while the observer listens for sounds with a stethoscope over the artery downstream, the cuff pressure is gradually lowered. At a certain stage, Korotkoff sounds are heard, in time with the heartbeat. It has been shown by angiographic and ultrasonic techniques that the first appearance of the sounds coincides with the onset of blood flow through the collapsed artery segment. With progressive lowering of the cuff pressure, the duration of the Korotkoff sounds during each cardiac pulse lengthens, because the compressed segment is open for longer. However, as the cuff pressure approaches diastolic pressure in the artery, the sounds become muffled and then disappear. It is not certain whether it is the muffling or the disappearance of the sounds which coincides with the cuff pressure being at the diastolic level, but, by internationally agreed convention, it is the pressure at which muffling occurs that is taken as diastolic pressure. If the amplitudes of the different frequency components of the sounds are plotted, as in **Fig. 14.18**, it can be seen that prior to muffling there is a significant contribution to the sounds from components with frequencies up to 180 Hz. With the onset of muffling, however, there is a considerable reduction in all frequencies between 60 and 180 Hz.

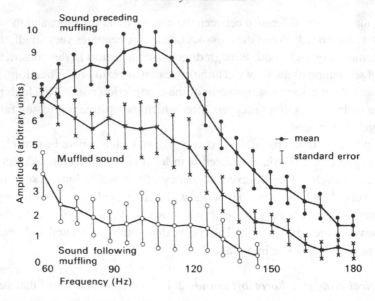

Fig. 14.18. Graphs of amplitude against frequency for different high-frequency components of Korotkoff sounds. The upper trace was obtained at a cuff pressure between systolic and diastolic pressure; the middle trace was obtained when the sounds were muffled, with cuff pressure approximately equal to diastolic pressure; the lower trace was obtained at a cuff pressure below diastolic pressure. (After McCutcheon and Rushmer (1967). Korotkoff sounds: an experimental critique. *Circulation Res.* **20**, 154. By permission of the American Heart Association Inc.)

Observations on casts, made by injecting substances into the arteries of a limb at arterial pressure while the limb was compressed by a cuff, show that the cross-sections of the arteries in this situation are distinctly non-circular. Both the area and the circumference of the vessels were greatly reduced, recalling the behaviour of excised veins under compression, and longitudinal corrugations were observed on the inner surface. This suggests that systemic arteries, like veins, collapse when the transmural pressure is close to zero or negative. It is therefore plausible that Korotkoff sounds are a manifestation of the self-excited oscillations described above.

Mechanics of venous beds

From our study of the properties of veins we turn now to an examination of complete venous beds, and the effect of external influences such as the contraction of skeletal muscle and respiration on the flow of blood through them. Because of the low transmural pressure and the consequent great distensibility of veins, the volume of the system is very sensitive to manoeuvres which change the transmural pressure. Also, the nonlinear mechanical properties of veins mean that the corresponding changes in flow rate are difficult to predict. We can only describe qualitatively the expected response and indicate to what extent it is consistent with physiological observation.

Elevation of a venous bed above the level of the heart When a venous bed (e.g. in the arm) is situated roughly at the level of the heart, the mean venous pressure is everywhere about $0.5 \times 10^3 \mathrm{N\,m^{-2}}$, and none of the vessels is expected to be collapsed. As the arm is raised, the transmural pressure is reduced and the veins collapse. For this to happen the volume of blood contained in them must be reduced. (This will be effected both by a transient surge in the outflow from the bed and by a slight transient decrease in inflow because the cross-sectional area of the veins is reduced and hence the impedance of the venous system is increased; see p. 338 for a discussion of impedance. The former is likely to be much more important, because the impedance of the large veins and heart remains considerably less than that of the microcirculation.) The mean flow rate through the venous bed will remain approximately constant because the mean flow rate through the arteries will be effectively unaltered by the manoeuvre, since the input impedance of the arterial system, determined by the visco-elastic properties of the arteries and the resistance of the microcirculation, is itself virtually unaltered. The average blood velocity and shear rate in the veins will both, therefore, be increased. The velocity and shear rate will transiently be increased by a greater amount in the veins near the heart than in those further away, because the blood displaced by the vein walls while they are collapsing will augment the velocity most in the downstream segments.

Contraction of skeletal muscle If there is mild, sustained contraction of the skeletal muscles in a limb at or below the level of the heart, the transmural pressure of the veins embedded in the muscle can be expected to be reduced, just as it is by elevation of the limb, and with similar effects on the venous bed. However, a sustained powerful contraction of the skeletal muscles in a limb can cause the blood flow in the muscles to cease, presumably by compressing and obstructing first the veins and then perhaps the arteries.

Intermittent contraction of the skeletal muscle in a limb has a more complicated effect on the local venous circulation than a sustained mild contraction. The response depends on a number of factors, including the strength and frequency of the contraction, the volume of blood in the bed, the presence of the venous valves and the time required for the bed to refill after being emptied. The effects can be demonstrated dramatically in the leg veins of a normal subject by measuring the pressure relative to atmospheric pressure (approximately equal to the transmural pressure) when the leg is immobile, and when the leg muscles are contracted intermittently.

If the subject lies horizontally then the mean pressure in an ankle vein is about $0.7 \times 10^3 \mathrm{N\,m^{-2}}$ (5 mmHg). If the subject is then tilted into the upright position and remains motionless the pressure gradually rises to a steady value of about

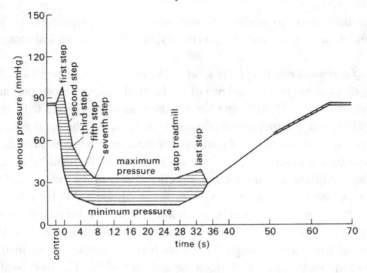

Fig. 14.19. Maximum and minimum pressures in an ankle vein of an erect subject. The subject is initially motionless, then walks and finally again stands still. (From Pollack and Wood (1949). Venous pressure in the saphenous vein at the ankle in man during exercise and changes in posture. *J. Appl. Physiol.* **1**, 649–662. Am. Physical Soc., used with permission)

$11.9 \times 10^3 \, \mathrm{N \, m^{-2}}$ (85 mm Hg). The rise is gradual because the venous valves prevent backflow and the bed fills solely from the microcirculation.

If the leg muscles are then contracted repeatedly, as if walking (**Fig. 14.19**), the pressure in the vein rises during the first step and then falls as the muscle relaxes. The pressure continues to fall with successive steps, and eventually comes to lie in the range $2\text{–}4 \times 10^3 \, \mathrm{N \, m^{-2}}$. If the contractions cease, the pressure in the vein gradually rises to its previous level. Measurements of the total volume of the limb show that intermittent contractions of the muscle cause the volume of the limb to decrease because the venous blood volume is reduced. It follows that muscular contractions will be ineffective in lowering the local venous hydrostatic pressure (or blood volume) if the interval between contractions considerably exceeds the time needed for the bed to refill; the latter varies from about 50 s in a resting limb to about 5 s in a strongly exercising limb.

Contractions occurring too frequently for the bed to refill completely have other important effects on the mechanics of the local venous circulation. If the contractions are gentle there will be no significant active dilatation of small vessels in the muscles and, therefore, the peripheral resistance will be effectively constant. Since the rate of inflow of the blood into the bed is set by the arteriovenous pressure difference and the peripheral resistance, the lowering of the time-average venous pressure will increase the inflow and venous return from the bed in proportion to the extent by which that pressure difference is increased. In normal circumstances, however, venous pressure is so much less than arterial pressure that no significant increase of inflow results. It will be appreciated that though the excess pressure in the ankle vein of an erect immobile person is $12 \times 10^3 \mathrm{Nm}^{-2}$, the excess pressure in the adjacent artery is about $24 \times 10^3 \mathrm{Nm}^{-2}$, so that the arteriovenous pressure difference is still about $12 \times 10^3 \mathrm{Nm}^{-2}$. What such gentle contractions do is to cause the blood flow in the veins to fluctuate and perhaps affect the distribution of the flow, because some veins may be compressed and others dilated. At certain frequencies of contraction outflow from the compressed veins to those downstream will occur only during the period of contraction, because the veins are refilling during the period of relaxation. However, since the time-average volume flow rate of the bed is unchanged and the time-average venous blood volume is reduced, the average velocity of blood in the veins must be increased and its mean residence time (volume of bed/volume flow rate) decreased.

There are several implications of these results. Subjects who stand motionless for long periods will undoubtedly have higher hydrostatic pressure in their veins and more distension of these vessels than those who intermittently contract their leg muscles, as by walking. Subjects who have incompetent venous valves will not be able to reduce the transmural pressure or the diameter of their leg veins as effectively by muscle contraction as normal subjects; incompetence of venous valves is a common sequel to thrombosis within the veins. It has been suggested that sustained distension of veins predisposes them to becoming abnormally dilated (varicose).

It was mentioned previously (Chapter 10) that thrombosis is prone to occur in the leg veins of persons who are confined to bed. No explanation is available for this, though the incidence is reduced if, for example, they exercise their legs, or their legs are subjected to rhythmically applied external compression (without a change in time average volume flow rate).[3] This suggests that the process may be associated with decreased pulsatility of venous blood flow, reduced time-average values for venous blood velocity and shear rate, and an increased residence time of the blood in the leg veins. However, no causal relationship between these mechanical factors and the occurrence of thrombosis has been established.

It is important to appreciate that the intermittent contraction of the muscles in a limb will have a qualitatively similar effect on venous blood pressure and flow whether the venous valves are competent, incompetent or even absent. To illustrate this, consider a horizontal liquid-filled flexible tube whose central portion is compressed suddenly. The instantaneous and time-average outflows from the two ends will be the same only if the impedances at the two ends are identical. Thus, if there is a competent or partially competent valve upstream and only a very flexible chamber (such as the right atrium) downstream, most of the flow will be directed towards the chamber,

---

[3] Experimental measurements have shown that intermittent external compression of a limb, applied for 5 s every minute at a rate of increase of pressure of $10^3 \mathrm{Nm}^{-2}\mathrm{s}^{-1}$, had no effect on the time-average volume flow rate, though it markedly reduced the incidence of deep-vein thrombosis.

which has the lower impedance. In veins, the result will be similar even if the valves are totally incompetent or absent, because the impedance of the upstream vessels far exceeds that of the veins and right atrium downstream.

So far we have dealt with small segments of the venous bed, and have been justi-fied in considering their behaviour in isolation because changes within them have an insignificant effect on the circulation overall. If, however, such changes affect a large part of the venous system then they may significantly alter the venous return to the heart, and this, as discussed in Chapter 11, may affect cardiac output. This in turn affects the flow into the venous bed itself. Many respiratory manoeuvres fall into this category.

Respiratory manoeuvres There is very little established information on the effect of respiratory manoeuvres on the venous system, and some disagreement still exists between different investigators. We therefore deal with the subject briefly, and almost exclusively confine ourselves to reporting experimental measurements made in the vena cava. It must be borne in mind that these manoeuvres can also have a strong influence on conditions in systemic arteries, by modifying both venous return to the heart and the mechanics of intrathoracic arteries. Through reflex action they may also affect the heart rate and the peripheral vascular bed.

There has been interest for over 200 years in the effect of breathing on the veins and venous return. The distinguished Swiss anatomist and physiologist Albrecht von Haller (1703–88) proposed that the enlargement of the thorax during inspiration, and the associated reduction of the intra-thoracic pressure, would cause blood to be drawn from the extra-thoracic veins into the intra-thoracic vessels and heart. Expiration would have the opposite effect, the increased thoracic pressure causing both a reduc-tion in the volume of intra-thoracic veins and an increase in their pressure, so that the flow rate into them from the unchanged extra-thoracic veins would be reduced. This is what happens during quiet breathing, but other respiratory manoeuvres may have more complicated consequences.

The events accompanying a deep sustained inspiration against a closed glottis (a Mueller manoeuvre) are shown in **Fig. 14.20**. The alveolar and intrapleural pressures are reduced and, therefore, so too are the pressures in the right atrium and the thoracic vena cava. The abdominal pressure, however, rises steeply, because of the contraction of the abdominal muscles and the lowering of the diaphragm, and this rise is trans-mitted to the abdominal segment of the inferior vena cava. Thus, there is a greatly enhanced pressure difference between the abdominal and thoracic segments of the vena cava, so the blood in the abdominal vena cava is sharply accelerated. There is only a transient increase in blood velocity, however, and subsequently it remains only slightly elevated. This is because the pressure in the abdominal cavity exceeds that in the abdominal vena cava at its downstream end (i.e. just below the diaphragm) so that the vessel collapses there. The mechanisms described on p. 451 are then operative

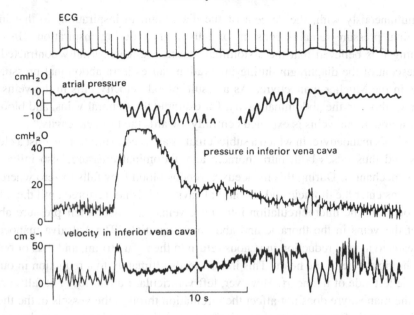

Fig. 14.20. Changes in right atrial pressure and in inferior vena caval (IVC) pressure and velocity during a Mueller manoeuvre. Top trace is the ECG. The probes measuring IVC pressure and velocity were situated below the diaphragm, in the abdominal vena cava. The vertical line is drawn to show that, during the manoeuvre, peak velocity occurred at the same time as peak atrial pressure. (From Wexler, Bergel, Gabe, Makin and Mills (1969). Velocity of blood flow in normal human venae cavae. *Circulation Res.* **23**, 356. By permission of the American Heart Association Inc.)

(although no self-excited oscillations have been observed), and the flow through the abdominal vena cava is independent of the pressure in the right atrium, as long as that remains below the pressure in the abdominal cavity. The record of blood velocity in the abdominal vena cava shown in **Fig. 14.20** reveals that the timing (or phase) of cardiac oscillations is different from normal, with peak velocity (towards the heart) occurring during atrial systole rather than ventricular systole. This is a further illustration of the fact that right atrial pressure has no effect on flow in the abdominal vena cava; the origin of the oscillations in this case is not clear. At a later stage in the manoeuvre, the pressure in the atrium can be seen to rise, that in the abdominal vena cava to fall and the blood velocity to increase, and eventually the normal cardiac pulsations are restored. This was presumably because the subject was unable to maintain the high pressure in the abdominal cavity, or the low pressure in the thorax, so that the vein did not remain collapsed.

Little is known about the variations of intra-abdominal pressure during breathing in general. In quiet breathing the abdominal muscles are believed to relax before, or simultaneously with, the descent of the diaphragm in inspiration, so that intra-abdominal pressure remains constant or falls slightly during inspiration. In deep breathing it is believed that the abdominal muscles remain partially contracted, so that descent of the diaphragm during inspiration causes intra-abdominal pressure to rise as in the Mueller manoeuvre. As a result, blood velocity falls in the veins upstream of those in the abdominal cavity, for example the femoral veins, and blood in the intra-abdominal veins is expelled centrally towards the thoracic cavity.[4]

Valsalva's manoeuvre, in which a subject makes a forced expiration against a closed glottis and thus raises both intra-thoracic and abdominal pressure, also influences venous mechanics. During this manoeuvre venous blood flow falls to zero where the large veins enter the thoracic and abdominal cavities; it is not restored until the inflow of blood from the microcirculation into these veins has raised their pressure above that of the veins in the thoracic and abdominal cavities. The Valsalva manoeuvre thus leads to a great reduction in venous return to the right atrium, and thus of output from the right side of the heart. This in turn leads ultimately to a reduction in output from the left side of the heart. However, left ventricular output does not fall to zero, since the manoeuvre does not affect the circulation through the vessels in the thorax and abdomen. Sufficient cerebral circulation is usually maintained to keep the subject conscious, by reflex peripheral vasoconstriction, although there is a fall in systemic arterial pressure.

---

[4] In patients with certain diseases, such as constrictive pericarditis or conditions when there is excess fluid within the pericardial cavity, the heart is prevented from expanding freely. When such a patient inspires deeply, central venous (i.e. right atrial) pressure rises and the neck veins become distended. Normally, an elevation of right atrial pressure produces an increase in the output of the right ventricle which can compensate for the increased flow into the thorax, but in such subjects this may be impossible. The clinical name for this phenomenon is 'Kussmaul's sign'.

15

The pulmonary circulation

The pulmonary circulation conveys the entire output of the right ventricle via the pulmonary arteries to the alveolar capillaries and returns the blood, via the pulmonary veins, to the left atrium. The lung has a second, though far smaller, circulation, the bronchial circulation. This arises from the thoracic aorta, supplies systemic arterial blood to the lung, has some interconnections (anastomoses) with the pulmonary microcirculation and drains into the systemic venous system.

The pulmonary circulation differs from the systemic circulation in several important respects. For example, it is a low-pressure, low-resistance system; the time-average excess pressure in the pulmonary arteries is only about $2 \times 10^3\,\mathrm{N\,m^{-2}}$ (15 mm Hg or $20\,\mathrm{cm\,H_2O}$), or approximately one-sixth of that in the systemic arteries, while the total blood flow rate through the lungs is the same as that through the systemic circulation. Further differences are that the pulmonary arteries have much thinner walls than the systemic arteries, and the pulmonary vascular bed is apparently not regionally specialized. In addition, vasomotor control in the pulmonary vessels is believed to be relatively unimportant under normal conditions; unlike the systemic arteries and veins, the vessels do not undergo large active changes in their dimensions.

The main function of the lungs is the exchange of oxygen and carbon dioxide between the air and the blood. However, any gas for which there is a difference in partial pressure between pulmonary capillary blood and alveolar gas will diffuse across the alveolar capillary membrane. Important gases include the anaesthetics, for example nitrous oxide; the rate of uptake of respired nitrous oxide is actually used as a means of measuring the instantaneous pulmonary capillary blood flow rate (**Fig. 15.17**). The uptake of carbon monoxide, in tracer quantities, is used for measuring both the diffusing capacity of the lung and the pulmonary capillary blood volume (p. 485).

In addition, the pulmonary circulation performs certain mechanical functions, by virtue of the flexibility of its vessels. We have seen that the return of blood to the right side of the heart from the extra-thoracic and extra-abdominal systemic veins is transiently interrupted by straining, coughing, etc. (p. 465), when the intra-thoracic and intra-abdominal pressures are raised. During such periods the flow of blood to the left atrium and ventricle, and hence to the systemic arteries, is temporarily maintained by blood draining from the pulmonary blood vessels; since there is a reduction of the rate of inflow of blood to the pulmonary circulation, the volume of blood they contain is reduced. The pulmonary blood vessels are effective, furthermore, in attenuating pressure and flow oscillations, generated by opposing contractions of the right and left sides of the heart, which could interfere with the flow of blood in the pulmonary circulation (p. 489 and **Fig. 15.14**).

Two other important functions of the pulmonary circulation may be mentioned here. First, it acts as a sieve to remove abnormal particulate material circulating in the blood. Thus, an embolus resulting from the detachment of a thrombus which has formed in a systemic vein, or the right side of the heart, will lodge in the pulmonary vascular bed. That may not seem desirable, but provided it is relatively small no serious disturbance may ensue, and fibrinolytic mechanisms (p. 167) may lead to its dissolution. This is quite different from the state of affairs when a similar embolus lodges in the circulation of a small highly specialized organ such as the brain. Second, the pulmonary circulation appears to have an important role in influencing the chemical composition of the blood by either removing or adding substances to it, as it flows through the lungs; for example, during a single passage of the blood there is a substantial conversion of the vasoconstrictor substance angiotensin I to its active form angiotensin II, almost complete inactivation of the vasodilator substance bradykinin and removal of 5-hydroxytryptamine (5HT), an amine important in platelet and central nervous function.

The mechanics of the pulmonary circulation are, as we shall see, affected to a major extent by the mechanics of the lung. It is for this reason that an entire chapter is devoted to the pulmonary circulation in which considerable attention is given to the influence of lung mechanics.

Anatomy

Pulmonary circulation The pulmonary arterial system starts at the pulmonary valves. The main pulmonary artery (pulmonary trunk), like the aorta, has sinuses behind the valve cusps. In humans the trunk is 5–6cm in length and 2.5–3.0cm in diameter, and it passes upwards and backwards, as shown in **Fig. 15.1**, to lie beneath the aortic arch, where it divides into the right and left main pulmonary arteries. (In the dog the anatomy is similar, but the vessel is shorter (3–4cm) and narrower (1.4–1.7cm diameter).) These pass towards the corresponding lungs, but before entering them divide into several branches, which supply the different lobes. The arteries, as they enter the lungs at their *hila* (points of attachment) and as they branch farther, are surrounded by sheaths of connective tissue. These are extensions of the *pleura*, the connective tissue membrane which envelops the lung and also surrounds the bronchi. The pulmonary

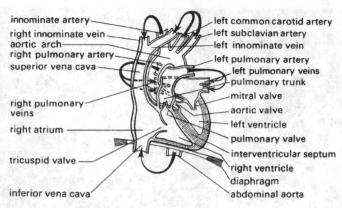

Fig. 15.1. Diagram of the heart and great vessels in a human seen from the front. The arrows denote the direction of blood flow. (After Pauchet and Dupret (1937). *Pocket Atlas of Anatomy* (3rd edn). Oxford University Press.) A simpler illustration of the circulation through the heart is shown in **Fig. 11.1**.

arteries and bronchi do not penetrate these sheaths to enter the substance of the lung until the level of the terminal air spaces (see **Fig. 15.3** and **Table 15.1**).

Figure 15.2 shows a resin cast prepared from a human pulmonary arterial tree; vessels of diameter less than 0.08 cm have been pruned away to reveal the larger arteries.

Table 15.1. *Numbers and dimensions of vessels in human pulmonary arterial system[a]*

Order	Number of branches	Diameter d (cm)	Length l (cm)	l/d	Number of end branches
17	1.00	3.000	9.05	3.00	3.00×10^8
16	3.00	1.483	3.20	2.16	1.00×10^8
15	8.00	0.806	1.09	1.35	3.02×10^7
14	2.00×10^1	0.582	2.07	3.56	1.38×10^7
13	6.60×10^1	0.365	1.79	4.90	3.98×10^6
12	2.03×10^2	0.209	1.05	5.02	1.16×10^6
11	6.75×10^2	0.133	0.66	4.96	3.47×10^5
10	2.29×10^3	0.085	0.47	5.54	8.92×10^4
9	5.86×10^3	0.053	0.32	6.05	4.81×10^4
8	1.76×10^4	0.035	0.21	6.00	1.60×10^4
7	5.26×10^4	0.022	0.14	6.36	5.36×10^3
6	1.57×10^5	0.014	0.09	6.43	1.79×10^3
5	4.71×10^5	0.009	0.07	7.78	5.98×10^2
4	1.41×10^6	0.005	0.04	8.00	2.00×10^2
3	4.23×10^6	0.003	0.03	10.00	6.66×10^1
2	1.27×10^7	0.002	0.02	10.00	2.37×10^1
1	3.00×10^8	0.001	0.01	10.00	1.00

[a]After Singhal *et al.* (1973). Morphometry of the human pulmonary tree. *Circulation Res.* **33**, 190.

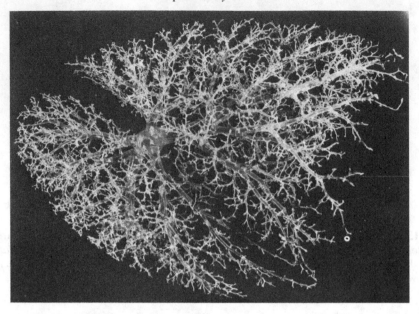

Fig. 15.2. The pulmonary circulation of both lungs with branches broken off at diameters of 0.08 cm, showing the general pattern of branching. (From Cumming, Henderson, Horsfield and Singhal (1969). The functional morphology of the pulmonary circulation. In *The Pulmonary Circulation and Interstitial Space* (eds Fishman and Hecht). University of Chicago Press.)

The branching is not dissimilar from that in the systemic arteries and veins, in that flow dividers are relatively sharp and branching occurs in different ways at different locations. For example, whereas the pulmonary trunk divides into two roughly equal main pulmonary arteries, seen in the centre of the cast, much of the remainder of the branching occurs by small arteries coming off larger ones. The pattern is seen also in **Fig. 15.3**, which shows a fragment from the periphery of the cast; the vessel in the spine of the fragment has a diameter of about 0.07 cm.

It is not a simple task to describe such a branching system. One method is to place the branches in generations, counted downwards from the most central vessel, and another is to place them in orders, counted upwards from the most peripheral vessels. In the pulmonary circulation, where there is considerable asymmetry of branching, vessels of similar size may arise at different generations and the first system cannot give an accurate representation. The second system, discussed in detail in the reference given in **Table 15.1**, though more difficult to apply, does at least class vessels of similar size in the same order. On application of this system to measurements from the human cast shown in **Figs 15.2** and **15.3**, some average values have been computed to describe the tree (**Tables 15.1** and **15.2**). It is seen that the

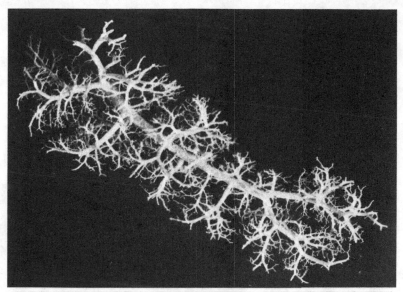

Fig. 15.3. A small branch of the pulmonary arterial tree showing the pattern of branching. (From Cumming, Henderson, Horsfield and Singhal (1969). The functional morphology of the pulmonary circulation. In *The Pulmonary Circulation and Interstitial Space* (eds Fishman and Hecht). University of Chicago Press.)

length-to-diameter ratio of the larger arteries lies between 1.35 and 3.0 and that the ratio is greater in the smaller arteries, reaching a value of 6–10. The total length of an average pathway, from the beginning of the pulmonary trunk to the capillaries, is about 20 cm.

As seen in **Table 15.2**, the total cross-sectional area of the arterial bed actually decreases slightly initially, so that the mean blood velocity must increase for the first two branches away from the pulmonary trunk, but thereafter blood velocity falls. The mean Reynolds number must fall all along the tree, because the product of velocity and diameter falls. The area ratio (ratio of combined cross-sectional area of daughter vessels to that of parent vessel) at the bifurcation of the pulmonary trunk into the main pulmonary arteries is about 0.8, both in humans and in the dog.

In the pre-capillary vessels of the pulmonary circulation the branching pattern is different from that described above: a single pre-capillary vessel 15–25 μm in diameter gives rise to a number of capillaries which lie in the alveolar septa and course over the surface of the alveoli. It seems that one pre-capillary vessel supplies capillaries to several adjacent alveoli, though whether in series or in parallel is not known.

A conventional view of the pulmonary capillaries is that, like those in the systemic microcirculation, they are cylindrical tubes. However, measurements have shown that their lengths are comparable to, and in some cases smaller than, their diameters, and

Table 15.2. *Values computed from the data in* **Table 15.1**[a]

Order	Cross-sectional area (cm²)	Total cross-sectional area (cm³)	Total volume (cm³)	Cumulative volume (cm³)	Velocity (cm s⁻¹)
17	7.07	7.07	63.97	63.97	11.3
16	1.73	5.18	16.58	80.55	15.4
15	5.10×10^{-1}	4.08	4.45	85.00	15.8
14	2.66×10^{-1}	5.32	11.01	96.01	13.8
13	1.05×10^{-1}	6.91	12.36	108.37	10.2
12	3.43×10^{-2}	6.96	7.31	115.68	9.0
11	1.39×10^{-2}	9.38	6.19	121.87	6.7
10	5.67×10^{-3}	12.99	6.09	127.96	4.2
9	2.17×10^{-3}	12.69	4.01	131.97	5.9
8	9.68×10^{-4}	16.99	3.57	135.54	4.5
7	3.94×10^{-4}	20.71	2.86	138.40	3.6
6	1.50×10^{-4}	23.54	2.14	140.54	3.2
5	5.81×10^{-5}	27.38	1.78	142.32	2.7
4	2.29×10^{-5}	32.31	1.42	143.74	2.3
3	9.08×10^{-6}	38.37	1.11	144.85	2.0
2	3.46×10^{-6}	43.85	0.88	145.73	1.8
1	1.33×10^{-6}	398.20	5.18	150.91	0.2

[a]Values of cross-sectional area are for a single branch. Total cross-sectional area is the summed area of all the branches in each order. Total volume is the summed volume of all the branches in each order. Cumulative volume is the total volume added cumulatively from higher to lower orders. Velocity is mean velocity of blood flow if pulmonary blood flow rate is 80 ml s⁻¹. (After Singhal *et al.* (1973). Morphometry of the human pulmonary tree. *Circulation Res.* **33**, 190.)

that they are commonly flattened rather than circular. It is more appropriate, therefore, to regard the blood in the alveolar septum as a sheet of fluid flowing between almost parallel alveolar membranes which are held apart by frequently occurring posts of connective tissue (**Fig. 15.4**). In the cat lung, which has been studied in detail, these posts typically occupy about 10% of the area of the sheet.

The pulmonary arteries have far thinner walls than do systemic arteries of comparable size. The ratio of wall thickness to diameter (h/d) takes a value of about 0.01 for the pulmonary trunk of a dog, comparable to the value for a systemic vein rather than a systemic artery; the value for the aorta, for example, is about 0.07 (**Table I**). Histologically there are also striking differences. Pulmonary arteries, from the pulmonary trunk down to vessels of 0.1 cm diameter, are classified as elastic arteries, because in them the media consists predominantly of elastic tissue with little smooth muscle and collagen. The transition to muscular pulmonary arteries occurs gradually at this level (orders 10–11, **Table 15.1**) and vessels of diameter 0.1–0.01 cm are classified as muscular, because the media consists predominantly of muscle. In the systemic circulation this transition occurs in large arteries; for example, the coronary and femoral arteries, which are several millimetres in diameter and arise respectively from the aorta and common iliac arteries, are muscular.

The pulmonary vascular bed is also distinctive in that it possesses no counterpart to the small muscular arteries and arterioles of the systemic circulation. For example, a

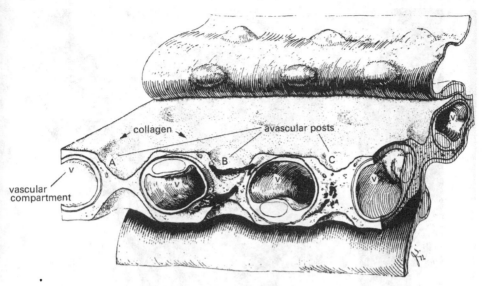

Fig. 15.4. A composite drawing of the pulmonary interalveolar wall of the dog showing the interalveolar microvascular sheet composed of a vascular compartment (the capillary bed) and the intercapillary posts. The alveolar epithelium has been pulled back to show the connective tissue matrix of the wall. Collagen converges on the post from the surrounding capillary wall. (From Rosenquist, Berrick, Sobin and Fung (1973). The structure of the pulmonary interalveolar microvascular sheet. *Microvasc. Res.* **5**, 199.)

60 μm diameter pulmonary artery has a wall thickness of only 8 μm ($h/d = 0.13$) and possesses only a thin media. In the corresponding systemic arteries, h/d takes a value of 0.25–0.4 and the media is thick. Moreover, a 30 μm diameter pulmonary 'arteriole' has no detectable smooth muscle, whereas the muscle in the media is obvious in a systemic arteriole. It is difficult to imagine that the pulmonary arterioles, so lacking in muscle, are capable of intense vasoconstriction; indeed, it seems probable that such vasomotion as occurs in the pulmonary arterial system is mainly effected by 0.01–0.1 cm diameter vessels, which have a well-formed media.

The tissue barrier which separates blood and air in the alveoli is extremely thin, with a thickness generally of about 0.2 μm (**Fig. 15.5**). It consists of the cells which line the alveoli (alveolar epithelium) and the capillaries (vascular endothelium) separated by a narrow interstitial space. In the neighbourhood of cell nuclei the barrier may be as thick as 10 μm. The endothelium is continuous with that of the entire circulation. On the alveolar side of the barrier there is a lining layer on the surface of the alveolar epithelium, composed of lecithins possibly linked to protein and known as the pulmonary surfactant. This has considerable importance in lung mechanics and influences the transmural pressure of the pulmonary capillaries (p. 479).

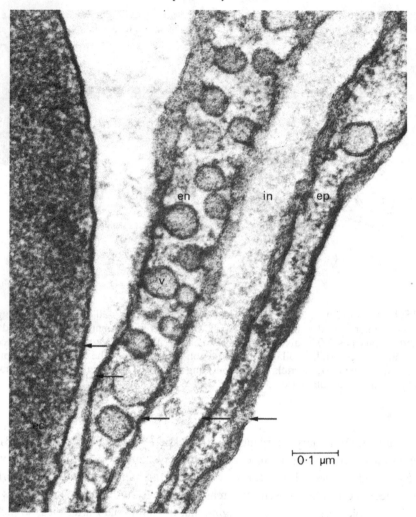

Fig. 15.5. Electron micrograph of the alveolar–capillary barrier where it is very thin. On the left is an erythrocyte (ec). The arrows indicate the five cell membranes, that of the erythrocyte, and those of the capillary endothelial (en) and alveolar epithelial (ep) cells. Between the endothelial and epithelial cells is a narrow interstitial space. Pinocytic vesicles (v) are also shown. (From Weibel (1969). The ultrastructure of the alveolar–capillary membrane or barrier. In *The Pulmonary Circulation and Interstitial Space* (eds Fishman and Hecht). University of Chicago Press.)

The blood from the alveolar capillaries drains into postcapillary vessels and then into pulmonary venules and veins.[1] The pulmonary veins run in the septa between

---

[1] There appear to be some direct vascular communications between the pulmonary arteries and veins other than through the alveolar capillaries; these anastomoses are reported to be present in the neighbourhood of the terminal airspaces and to be up to 500 μm in diameter. Their physiological role has not been established.

pulmonary lobules, and, unlike the pulmonary arteries, are remote from the bronchi in much of their course. They leave the lung at the hilum and enter the left atrium, usually as two veins on each side (**Fig. 15.1**). The pulmonary veins have no valves, but like the systemic veins are thin walled (**Table I**). Small pulmonary veins (diameter 0.01–0.1 cm) contain irregular elastic fibres and connective tissue in their walls and possess very little smooth muscle. Intermediate-sized veins contain some muscle and elastin; and the largest pulmonary veins, while they have a media of smooth muscle, are mainly composed of a thick collagenous adventitia.

There have been only limited studies of the vasa vasorum of the pulmonary arteries. Some of these arise from the bronchial arteries, though in the rabbit, at least, additional branches come from the coronary arteries. The larger vasa vasorum (50–100 μm diameter) runs in the adventitia, whilst only smaller ones occur in the media. It is not known whether the capillaries extend into the intima. The veins are parallel to the arteries in their course and drain into nearby systemic veins. Lymphatic channels surround the pulmonary arteries and veins and the bronchi; they also drain lymph from the interstitial spaces around the terminal airways.

Bronchial circulation Most commonly there are two bronchial arteries supplying each lung. They arise from the aorta and run into the lung with the main bronchus. Throughout their course they communicate with each other repeatedly, giving off branches to the pleura and finally dividing to supply all the bronchi. The bronchial circulation extends as far as the terminal bronchioles and provides blood to the vasa vasorum of the pulmonary arteries (down to a diameter of 0.1 cm) and to the peribronchial nerves. The bronchial veins drain into the superior vena cava, and, on a small scale, into the pulmonary veins.

The existence in normal lungs of pre-capillary anastomoses between the bronchial and pulmonary circulations has been established by injection studies. The vessels vary in diameter from 20 to 200 μm and are found principally in the region of the small bronchi. Their physiological role is not clear, though it has been suggested that they maintain a blood supply for small areas of lung which are temporarily not ventilated or are collapsed.

Transmural pressure and static elastic properties of vessels

In this section we examine the static elastic properties of the pulmonary vessels and the forces acting upon them. The effective transmural pressure acting on pulmonary arteries in situ is difficult to determine for two reasons. One is that the internal pressure varies with position in the lung, as a result of hydrostatic effects. The other is that the effective pressure acting on the outside of the vessels depends upon the mechanical properties of the lung, the pressure in the thorax and the degree of lung inflation.

It is necessary, therefore, to examine in detail the factors which determine both these pressures, as well as the elastic properties of the vessel walls.

Intravascular pressure The pulmonary circulation is a low-pressure system: the mean pressure above atmospheric (mean excess pressure) in the right ventricle and the large pulmonary arteries is normally about $2 \times 10^3 \, \mathrm{N\,m^{-2}}$ (15 mm Hg or 20 cm H_2O), although it can rise to about $4 \times 10^3 \, \mathrm{N\,m^{-2}}$ in exercise (**Fig. 15.11**). The mean excess pressure in the pulmonary veins is about half this value ($10^3 \, \mathrm{N\,m^{-2}}$), indicating that the pressure drop across the pulmonary microcirculation is very much smaller than that across the systemic microcirculation. In a human when upright the lung has a height of about 30 cm and the pulmonary arteries enter the lungs about half-way up. The hydrostatic gradient of pressure in the pulmonary arteries is about $0.1 \times 10^3 \, \mathrm{N\,m^{-2}}$ per centimetre vertical distance, so that the pressure at the top of the lung is about $1.5 \times 10^3 \, \mathrm{N\,m^{-2}}$ (15 cm H_2O) lower than the mean excess pulmonary artery pressure. Similarly, it is $1.5 \times 10^3 \, \mathrm{N\,m^{-2}}$ (15 cm H_2O) higher at the bottom of the lung. In the presence of blood flow there is a pressure drop of about $0.7 \times 10^3 \, \mathrm{N\,m^{-2}}$ between the main pulmonary artery and the capillaries; thus, the mean pressure is normally sub-atmospheric in the microcirculation of the upper part of the lungs.

Perivascular pressure Pulmonary blood vessels can be separated into two groups:

(1) The small vessels which lie in and around the alveolar septa (including the 'sheet' of capillaries), called *alveolar vessels*.
(2) The larger *extra-alveolar* vessels, which can again be subdivided into intra- and extra-parenchymal vessels; that is, vessels lying within and outside the lung substance (actually, although we use these terms, the larger vessels lie within connective sheaths and are not strictly within the lung parenchyma, as noted on p. 468).

The distinction between (1) and (2) is made on functional, not anatomical, grounds, in that the two groups of vessels experience different values of external pressure. This can be demonstrated in an excised lung by filling the pulmonary arteries and veins with a fluid such as kerosene, which cannot enter the smaller vessels, because of interfacial tension. The level in the vasculature reached by the kerosene is determined by a balance between the hydrostatic pressure difference across the interface between the kerosene and the fluid it is displacing and the interfacial tension. Calculations indicate that a kerosene–water interface will be within 50 μm vessels when the hydrostatic pressure difference is $2 \times 10^3 \, \mathrm{N\,m^{-2}}$. When the lung is inflated by increasing the air pressure in the alveoli (without changing the pressure at the surface of the lung, or that of the kerosene reservoirs – 'positive pressure inflation') the volume of the kerosene-filled vessels increases. However, when the whole vascular bed is filled with an aqueous solution of dextran, a similar manoeuvre causes the volume to rise by a smaller amount or to decrease. This can be accounted for only by a fall in the volume of the small vessels. In order to understand the mechanism we must examine what determines the pressure on the outside of the different-sized blood vessels in the lung.

Consider first the extra-alveolar, *intra-parenchymal* vessels (**Fig. 15.6**a). The pressure exerted on the outside of these vessels by the material in the perivascular space or sheath, which we shall call p_{pv}, is not equal to alveolar gas pressure p_{alv} for a number of reasons. The first is that if the neighbouring alveoli are at all inflated, the alveolar membranes pull outwards on the membrane surrounding the perivascular space, and this pulling is equivalent to a reduction of p_{alv} by an amount which we may call the 'elastic stretching pressure' p_{es}. This effect, however, is reduced by the elastic recoil pressure of the perivascular membrane, less the effective outwards pressure exerted on the vessel wall by any elastic connections which may join the vessel wall to the perivascular membrane, which we call p_{er}. Thus

$$p_{pv} = p_{alv} - p_{es} + p_{er}. \tag{15.1}$$

If the lung were uniform, all alveoli would be equally inflated, and the same p_{es} would be exerted on every surface of lung parenchyma, including the outer pleural surface (**Fig. 15.6**b). Under static conditions at any volume p_{alv} is uniform, because all air spaces are in communication with each other, and it also follows that p_{pl}, the pressure in the pleural space, the extremely thin space between the visceral pleura which covers the lung and the parietal pleura which lines the thorax, differs from p_{pv} only if the elastic recoil pressure of the perivascular membrane is different from the net recoil pressure of the pleural membranes. In the case of extraparenchymal vessels, p_{pv} is equal to p_{pl}, because the perivascular space is the pleural cavity.

If the lung is inflated by positive pressure, p_{alv} is increased and p_{es} increases even further, because of the nonlinear elastic properties of the alveolar membranes, which become stiffer as they are stretched. The net effect of lung inflation is that p_{pv} falls, so that, if the pressure inside the vessel has remained constant, the transmural pressure rises and the volume of the vessel also rises. This explains the observations in the kerosene experiments.

The above conclusions are not altered by the fact that, with the lungs inside the thorax, their mechanical properties are not uniform. For example, alveolar volume increases with height, because the pressure at the pleural surface (lung surface pressure) varies with height, in a manner determined primarily by the way in which the weight of the lungs is supported in the thorax. The vertical gradient of lung surface pressure is about $25\,\mathrm{N\,m^{-2}}$ ($0.25\,\mathrm{cm\,H_2O}$) per centimetre; this is only a quarter as large as the hydrostatic gradient which would be expected if the fluid in all parts of the pleural cavity could communicate freely, so clearly it does not. We expect a similar vertical gradient in p_{pv}, but even over a height of 30 cm (as in the lung in an upright human being), the total variation in perivascular pressure is no more than $0.75 \times 10^3\,\mathrm{N\,m^{-2}}$

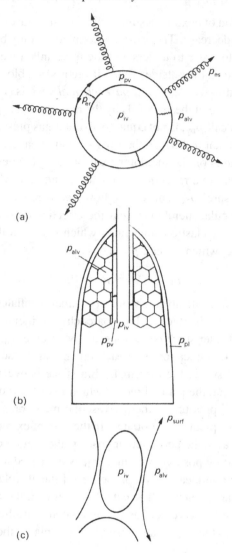

(a)

(b)

(c)

Fig. 15.6. Schematic diagrams illustrating the different factors determining intravascular pressure p_{iv}. (a) An extra-alveolar, intraparenchymal vessel surrounded by a perivascular space (pressure p_{pv}) which is bounded by a membrane. This membrane is pulled outwards by the alveolar membranes (effective outward pressure p_{es}) and pushed inwards both by alveolar pressure (p_{alv}) and by its own recoil pressure and that of the occasional connections between the membrane and the vessel wall (p_{er}). (b) Indicating that, apart from the occasional connections, the perivascular space is continuous with the pleural space, where the pressure p_{pl} is close to perivascular pressure p_{pv} (see text). (c) An alveolar vessel: perivascular pressure is approximately equal to alveolar pressure, apart from the small outwards pull of the alveolar membrane and the surfactant layer (p_{surf}) on the vessel.

(7.5 cm H$_2$O), compared with a 3×10^3 N m^{-2} (30 cm H$_2$O) variation in intravascular pressure.

Now consider the *alveolar* vessels (**Fig. 15.6**c). The blood can be visualized as flowing between two opposing layers of capillary endothelium, held apart in places by endothelium-covered posts (**Fig. 15.4**). The capillary endothelium is covered in turn by a thin region of connective tissue, the alveolar epithelium, and an extremely thin layer of surfactant, which lines the alveoli. Perivascular pressure for these vessels is almost equal to alveolar pressure, the only differences being caused by the outward pull of the alveolar membrane and the surfactant (p_{surf}) at the liquid–air interface. This conclusion has been verified by experiments in which the flow through the pulmonary vascular bed of isolated liquid-filled, liquid-submerged, lungs was observed to cease where arterial pressure was equal to alveolar pressure; under these conditions interfacial tension in the alveoli is essentially abolished, as are the vertical gradients of vascular and transmural pressure. In contrast, in similar, but air-inflated, lungs, suspended in air, the flow ceased only when the arterial and venous pressures were equal, even though venous pressure was slightly less than alveolar pressure. The magnitude of the contribution of the surfactant in these conditions has not, however, been established.

If the lung is inflated very slowly in situ by expansion of the thorax, p_{alv} remains essentially equal to atmospheric pressure, while p_{pl}, and hence also p_{pv}, decreases. In this case it is the fall in p_{pv} which, in Equation (15.1), balances the rise in p_{es} associated with the stretching of the alveolar membranes. The fall in p_{pv} means that the transmural pressure of the extra-alveolar vessels is increased. Since, however, p_{alv} does not change, there is little or no change in the transmural pressure of alveolar vessels, which therefore do not collapse as they may do during positive pressure inflation. They do, however, experience a reduction in thickness, because the alveolar septum is stretched (p. 481).

Consider what happens if the intravascular pressure of a blood vessel is reduced, other pressures being held constant. The reduction will cause a fall in the transmural pressure; if it becomes negative, as may happen normally towards the apex of the lung (see p. 476), this may lead the vessel to collapse, if it is an alveolar vessel. In an extra-alveolar vessel, a reduction of cross-sectional area would occur and that in itself would cause some local alveolar membranes to be stretched. Thus, p_{es} would locally be raised above its value elsewhere in the lung, so that p_{pv} would be reduced (Equation (15.1)), and the tendency to collapse resisted. This phenomenon, known as interdependence, means that the presence of lung parenchyma in a state of inflation effectively reduces the distensibility of blood vessels (or airways) embedded in it. The magnitude of the reduction has not as yet been investigated quantitatively for pulmonary arteries. This limits the value of any predictions one may make of, for example, the pulse wave speed in pulmonary arteries at different degrees of lung inflation.

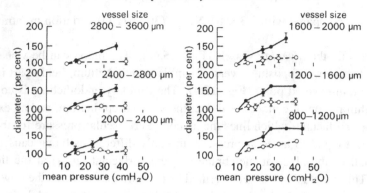

Fig. 15.7. Pressure–diameter relations for dog pulmonary arteries (filled circles) and veins (open circles) grouped according to size. Vessel diameters are expressed as a percentage of the value at an intravascular pressure of 8–10 cm H_2O. Lung inflating pressure was 10 cm H_2O. (From Maloney, Rooholamini and Wexler (1970). Pressure–diameter relations of small blood vessels in isolated dog lung. *Microvasc. Res.* **2**, 1.)

Elastic properties The properties of excised pulmonary arteries and veins can best be described with reference to the discussion of systemic veins in Chapter 14, especially **Fig. 14.1** (p. 430). Like veins, pulmonary arteries are circular at transmural pressures of 2.0–2.5 × 10^3 N m^{-2} (20–25 cm H_2O) but are markedly non-circular when the transmural pressure is as low as 0.5 × 10^3 N m^{-2} (5 cm H_2O). This is also true in the excised but whole rabbit lung; we can expect, therefore, that the vessels become non-circular in vivo. The elastic properties of a large excised dog pulmonary artery, over the range in which it remains approximately circular, are compared with those of systemic veins and systemic arteries, also from the dog, in **Fig. 14.3**, p. 435. At physiological transmural pressures its Young's modulus has a similar value to that of the systemic arteries, but it is more distensible, because it has a much thinner wall than a systemic artery of comparable size; on the other hand, it is less distensible than a systemic vein, because, although its wall thickness is comparable, the Young's modulus is greater. However, at equal values of transmural pressure the order is reversed, with the systemic arteries being the most distensible.

No similar study has been made of large pulmonary veins. However, the elastic properties of medium and small pulmonary arteries and veins in the dog (diameters 0.08–0.36 cm) have been determined. The lungs were excised and perfused and vessel diameters were measured radiologically. The average pressure–diameter results are shown in **Fig. 15.7**. Over the range of intravascular pressure 1.0–3.5 × 10^3 N m^{-2} (10–35 cm H_2O), when the vessels can be expected to be almost circular, arteries of diameter greater than 0.16 cm are seen to increase in diameter almost linearly with pressure, while smaller arteries increase in diameter linearly at first and then more slowly. In the case of the veins the diameter change for unit pressure change is considerably smaller and also, unlike the arteries, it is the medium and larger vessels (diameter greater than 0.12 cm) which tend to reach a diameter plateau as pressure is increased.

Average values of distensibility D (Equation (14.1)) have been calculated from some of the results shown in **Fig. 15.7**. Over the intravascular pressure range $1.0–2.25 \times 10^3\,\mathrm{N\,m^{-2}}$ $(10–22.5\,\mathrm{cm\,H_2O})$ the distensibility of the arteries ranges from $0.35 \times 10^{-3}\,\mathrm{m^2\,N^{-1}}$ for the largest vessels to $0.88 \times 10^{-3}\,\mathrm{m^2\,N^{-1}}$ for the smallest. The distensibility of the veins over the same pressure range varies from $0.17 \times 10^{-3}\,\mathrm{m^2\,N^{-1}}$ for the largest vessels to $0.35 \times 10^{-3}\,\mathrm{m^2\,N^{-1}}$ for the smallest. A distensibility of $0.35 \times 10^{-3}\,\mathrm{m^2\,N^{-1}}$ leads (from Equation (14.6)) to a predicted value for the pulmonary arterial pulse wave velocity of about $1.7\,\mathrm{m\,s^{-1}}$, compared with the measured value (in dogs) of about $2.5\,\mathrm{m\,s^{-1}}$ (**Table I**).

Wall elastic properties have also been studied in the rabbit lung by measuring the apparent velocity of imposed small-amplitude pressure waves, averaged over pathways of different length between the pulmonary trunk and other points in the pulmonary arterial tree. The values found were: interval 2 cm, velocity $0.55\,\mathrm{m\,s^{-1}}$; interval 3.2 cm, velocity $0.62\,\mathrm{m\,s^{-1}}$; interval 5.5 cm, velocity $0.80\,\mathrm{m\,s^{-1}}$. As in the systemic arterial tree, the apparent wave speed clearly increases towards the periphery.

The above results apply to the pulmonary arteries and veins. We should also consider the elastic properties of the alveolar capillaries which comprise the sheet in the alveolar septum. It has been shown that the alveolar capillaries are distensible, by relating the number of red blood cells seen within them in rapidly frozen specimens of isolated perfused lung to the vascular perfusing pressure immediately before freezing (**Fig. 15.8**).

At a lung inflating pressure (*transpulmonary* pressure) of $1.0 \times 10^3\,\mathrm{N\,m^{-2}}$ $(10\,\mathrm{cm\,H_2O})$ the number of red cells increased almost linearly from about 0.4 per $10\,\mu\mathrm{m}$ to 1.5 per $10\,\mu\mathrm{m}$ length of septum, when the perfusing pressure was raised from about 0 to $3.5 \times 10^3\,\mathrm{N\,m^{-2}}$. The increase was much slower when the transpulmonary pressure was $2.5 \times 10^3\,\mathrm{N\,m^{-2}}$.

It has been observed that lung volume independently influences the dimensions of the alveolar capillaries, and this is illustrated in **Fig. 15.9**. The histological sections are from rapidly frozen specimens of lung; in both cases the specimens are taken from sites 32–35 cm below the level where the venous pressure just balances the alveolar pressure so that the capillary transmural pressure is the same, but alveolar pressure is $1.0 \times 10^3\,\mathrm{N\,m^{-2}}$ $(10\,\mathrm{cm\,H_2O})$ in (a) and $2.5 \times 10^3\,\mathrm{N\,m^{-2}}$ $(25\,\mathrm{cm\,H_2O})$ in (b). In (b), consistent with the higher transpulmonary pressure, the alveoli are seen to be larger and there is a striking reduction in the number of red blood cells in the septum. These changes are probably brought about by the greater tautness of the alveolar walls at the higher lung volume. This is analogous to the situation in sheets of elastic material, where stretching in two directions (in the plane of the sheet) results in shrinking in the third (cf. p. 89).

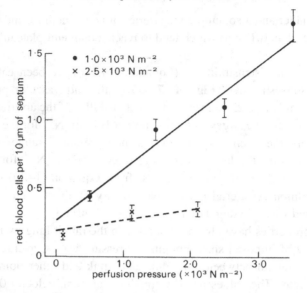

Fig. 15.8. Number of red blood cells per 10 μm of alveolar septum plotted against perfusion pressure. Since arterial pressure rises at the rate of about 1 cm of water ($10^2\,\mathrm{N\,m^{-2}}$) per centimetre distance, the perfusion pressure is equivalent to distance down the lung, with zero representing the lung apex. Venous pressure was low. Measurements are shown at transpulmonary pressures of $1.0 \times 10^3\,\mathrm{N\,m^{-2}}$ and $2.5 \times 10^3\,\mathrm{N\,m^{-2}}$ water. Note the rapid increase in the number of red cells down the lung when the transpulmonary pressure is $1.0 \times 10^3\,\mathrm{N\,m^{-2}}$ and the much slower increase when the lung is inflated to a larger volume (transpulmonary pressure $2.5 \times 10^3\,\mathrm{N\,m^{-2}}$). (From West, Glazier, Hughes and Maloney (1969). Pulmonary capillary flow, diffusion, ventilation and gas exchange. In *Ciba Foundation Symposium on Circulatory and Respiratory Mass Transport* pp. 256–72. J. and A. Churchill, Edinburgh.)

The thickness of the alveolar sheet has been measured directly by perfusing the pulmonary circulation with a low-viscosity silicone rubber at different pressures and allowing the material to set. The results show that sheet thickness increases with perfusing pressure in an approximately linear way, with a slope (proportional to the distensibility) which decreases as the lung volume increases (**Fig. 15.10**). We have already seen that alveolar size varies with position down the lung in situ (p. 477) and we may, therefore, expect corresponding differences in the sizes of the alveolar capillaries.

Pulmonary blood volume One aspect of the static mechanical behaviour of the pulmonary circulation which is of considerable physiological and clinical importance is the volume of blood within it and its distribution between the different types of vessel.

The volume of blood within the pulmonary vascular bed and its constituent arteries, capillaries and veins has been estimated in various ways. The method most commonly employed to measure the total pulmonary blood volume is to inject an indicator into

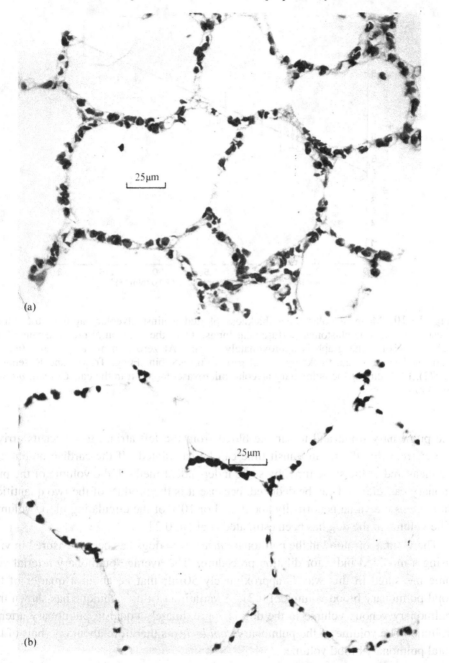

Fig. 15.9. Histological sections from perfused rapidly frozen dog lungs. The effect of transpulmonary pressure on the alveolar capillaries independent of capillary transmural pressure is illustrated – see text for description of study. (From Glazier, Hughes, Maloney and West (1969). Measurements of capillary dimensions and blood volume in rapidly frozen lungs. *J. Appl. Physiol.* **26**, 65.)

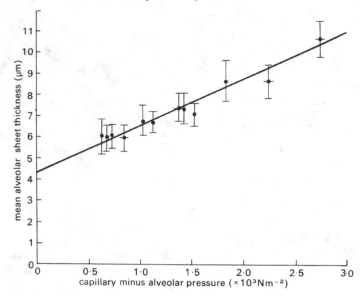

Fig. 15.10. Mean alveolar sheet thickness plotted against alveolar capillary transmural pressure in silicone-elastomer-perfused cat lungs. Over the transmural pressure range 0.6–$2.7 \times 10^3 \, \mathrm{N \, m^{-2}}$ the graph is approximately linear. At zero transmural pressure (an extrapolated value) sheet thickness is 4.28 µm. (From Sobin, Fung, Trewer and Rosenquist (1972). Elasticity of the pulmonary alveolar microvascular sheet in the cat. *Circulation Res.* **30**, 440.)

the pulmonary trunk and to sample blood from the left atrium to detect its arrival there; from this the mean transit time can be calculated. If the cardiac output can be measured at the same time, by some independent method, the volume of the pulmonary vascular bed can be deduced, because it is the product of the two quantities. In humans the value is normally about 0.5 l or 10% of the circulating blood volume. The volume in the dog has been estimated at about 0.2 l.

The volume of blood in the pulmonary *arteries* in dogs has been measured in vivo using a modified indicator dilution procedure. The average pulmonary arterial volume measured in this way is approximately 50 ml; that is, about a quarter of the total pulmonary blood volume of 0.2 l. A variation of this technique has shown that pulmonary venous volume in the dog is approximately equal to pulmonary arterial volume. The volume of the pulmonary *capillaries* is therefore about one-half of the total pulmonary blood volume.

The volume of a cast of the human pulmonary arterial tree has been found to be about 150 ml, of which about 80% is within a few hundred vessels whose diameter is greater than 0.13 cm. The remainder is contained within the millions of smaller vessels. A note of caution should be sounded concerning these results, however, because the measurements on small vessels are particularly likely to be inaccurate. No similar data are known for the pulmonary veins.

The pulmonary capillary blood volume has been determined in vivo by a method involving measurement of the carbon monoxide diffusing capacity of the lung. The value depends on posture (see below), but in a normal supine human it is about 100 ml. Assuming that these two results are correct, the pulmonary veins in humans contain approximately half the total pulmonary blood volume. Note too that, assuming a cardiac output of $5 \, l \, min^{-1}$, the transit time of a red cell through the pulmonary capillaries is only about 1 s. In severe exercise the transit time has been calculated to be less than 0.5 s.

The volume of blood in the pulmonary circulation may vary according to physiological conditions. Measurements in a supine human show that there is an increase of cardiac output, pulmonary arterial and venous pressure and total pulmonary blood volume when the legs are raised. If a supine passive subject is tilted head-up there is a fall in cardiac output, a slight decrease in pulmonary arterial and left atrial pressures and a reduction of total pulmonary blood volume by 50–100 ml. However, as shown in **Fig. 15.11**, when a supine subject stands up, pulmonary artery pressure does not change appreciably; this suggests that the tone in the veins in dependent parts of the body has increased sufficiently to prevent the pooling of blood.

Figure 15.11 also shows the effect of exercise on pulmonary artery pressure. The average pulmonary artery pressure was 15 mm Hg ($2.0 \times 10^3 \, N \, m^{-2}$) when the subjects were standing at rest and approximately 30 mm Hg ($4.0 \times 10^3 \, N \, m^{-2}$) when they were walking at $4.8 \, km \, h^{-1}$ on a treadmill with an incline of 16:100. They then had an average O_2 consumption of $2.35 \, l \, min^{-1}$, compared with the resting value of $0.35 \, l \, min^{-1}$, and a cardiac output of $17.5 \, l \, min^{-1}$, compared with the resting value of $5.1 \, l \, min^{-1}$. Neither pulmonary venous pressure nor pulmonary blood volume was measured in the study. Other studies have shown, however, that exercise causes scarcely any change in left atrial pressure, though pulmonary blood volume is increased. For example, in normal adult subjects exercising and increasing their O_2 consumption per square metre of body surface area from $0.14 \, l \, min^{-1}$ to $0.49 \, l \, min^{-1}$ (less severe exercise than that above), pulmonary artery pressure increased from an average of 13.9 mm Hg to 17.3 mm Hg (1.9–$2.4 \times 10^3 \, N \, m^{-2}$) and pulmonary blood volume per square metre of body surface area from an average of $0.28 \, l$ to $0.35 \, l$.

Studies in dog lungs have confirmed that it is elevation of the pulmonary artery pressure, when the flow rate is increased, which is primarily responsible for the increase of pulmonary blood volume. Thus, when lungs were perfused *in situ* by a pump, in such a way that the pulmonary artery and left atrial pressures could be controlled independently, total pulmonary blood volume did not change when the flow was decreased by increasing the left atrial pressure, while holding pulmonary artery

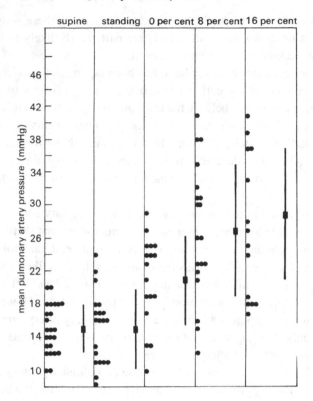

Fig. 15.11. Mean pulmonary artery pressure ($10\,\mathrm{mmHg} = 1.36 \times 10^3\,\mathrm{Nm^{-2}}$) in 24 normal subjects. The columns show individual values and average values $\pm 1\,\mathrm{SD}$, when the subjects are supine, standing at rest and walking at $4.8\,\mathrm{kmh^{-1}}$ on a treadmill which is horizontal, inclined at 8:100 and inclined at 16:100. (From Damato, Galante and Smith (1966). Haemodynamic response to treadmill exercise in normal subjects. *J. Appl. Physiol.* **21**, 959.)

pressure constant. However, as shown in **Fig. 15.12**, if left atrial pressure was held constant and the pulmonary artery pressure was raised by increasing the flow rate, the total pulmonary blood volume increased. The rate of increase of volume with pulmonary artery pressure was $12\,\mathrm{ml}$ per $\mathrm{cmH_2O}$ when the left atrial pressure was $5\,\mathrm{cmH_2O}$, and alveolar pressure exceeded venous pressure throughout the lung (zone II, see p. 499). It was $7\,\mathrm{ml}$ per $\mathrm{cmH_2O}$ when the left atrial pressure was $33\,\mathrm{cmH_2O}$ and venous pressure exceeded alveolar pressure throughout the lung (zone III, see p. 499). It has been suggested to explain this difference that there is a greater recruitment of previously closed channels with increase of pressure in the former case. The subject of recruitment is discussed further below.

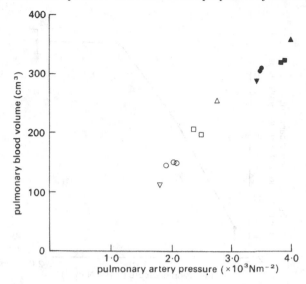

Fig. 15.12. Total pulmonary blood volume in perfused dog lungs, when pulmonary artery pressure was increased by increasing the blood flow rate. Left atrial pressure was held constant at $0.5 \times 10^3 \, \text{N m}^{-2}$ (open symbols) and $3.3 \times 10^3 \, \text{N m}^{-2}$ (filled symbols). (From Maseri, Caldini, Harward, Joshi, Permutt and Zierler (1972). Determinants of pulmonary vascular volume. *Circulation Res.* **31**, 218.)

It is clearly necessary to bear in mind the possibility that active changes in the tone of the smooth muscle of the pulmonary vessels could modify their mechanical response in studies of this type. There is no evidence that exercise is associated with an active change of the distensibility of the pulmonary arteries. However, it has been observed, in dog lungs perfused with a pulsatile flow from a pump, that stimulation of the sympathetic nerves to the lungs causes a reduction of pulmonary arterial distensibility, without altering pulmonary vascular resistance.

Pulmonary capillary blood volume also changes with physiological conditions. For example, manoeuvres which shift blood into the intrathoracic vessels, such as lying down or inflating a pressure suit on the legs and abdomen, increase the capillary blood volume, on average, from 70 ml to 100 ml. It should be appreciated, however, that the pulmonary capillaries have a limited capacity for distension. For example, further tilting the subjects, mentioned above, to a 60° head down position, though it increases the transmural pressure of their pulmonary vessels, caused no additional enlargement of their pulmonary capillary blood volume. Exercise also increased pulmonary capillary blood volume, which may more than double during maximal exercise. As in the experiments recorded in **Fig. 15.12**, this is associated with an increase in pulmonary artery pressure, not pulmonary venous pressure.

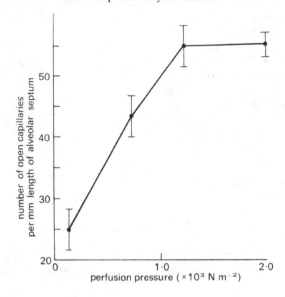

Fig. 15.13. Number of open capillaries per unit length of alveolar septum plotted against pulmonary arterial pressure: rapidly frozen dog lungs. (From Warrell, Evans, Clarke, Kingaby and West (1972). Pattern of filling in the pulmonary capillary bed. *J. Appl. Physiol.* **32,** 346.)

The lung volume also has an effect on the pulmonary capillary blood volume; in normal subjects it was slightly smaller, on average, when measured with the lungs fully expanded than when measured at resting lung volume. This is presumably related to the above observation in an isolated lung, that an increase of transpulmonary pressure reduces the density of the red cells in the alveolar septum.

There are two ways in which pulmonary capillary blood volume can be increased by an increase of pulmonary artery pressure. One is by distension of already open capillaries and the other is by recruitment of capillaries which were previously collapsed. It is known that at the apex of the lung in an upright subject at rest the capillaries are normally collapsed, with little flow through them, but that during exercise, when the pulmonary artery pressure is raised, flow occurs. Evidence from histological measurements of rapidly frozen dog lungs suggests that recruitment plays a very important role in this region when alveolar pressure is intermediate between pulmonary artery and pulmonary venous pressure (p. 499).

As shown in **Fig. 15.13**, the number of open capillaries per unit length of alveolar septum increases rapidly with pulmonary arterial pressure from about $25\,mm^{-1}$ at $1\,cm\,H_2O$ pressure to $55\,mm^{-1}$ at $12\,cm\,H_2O$ and thereafter there is little change. Though not shown here, results from the same study indicate that the length of alveolar septum occupied by each capillary increases steeply as perfusion pressure is increased above $12\,\text{cm}\,\text{H}_2\text{O}$, indicating that there is then capillary distension, but little change in the number of open capillaries. These conditions reflect the situation near the apex of the upright human lung, where there is a reserve of unopened capillaries that can be called on when pulmonary artery pressure rises, as in exercise.

In order to determine the relative contributions of recruitment and distension to an increase in pulmonary capillary volume, it is necessary to know the mechanism by which vessels are reopened after being closed, the pressure required to open them and their distensibility when open at different values of transmural pressure. It has been suggested that closure occurs in the pulmonary arterioles and that as pulmonary arterial pressure rises these vessels open. In fact, the available evidence suggests that it is closure of the capillaries, not the arterioles, which is significant. Because of varying path lengths and resistances upstream of sites of closed capillaries the arterial pressure may need to be very high before all vessels open, even though the pressure required to open individual capillary segments is very small. This subject has been explored by computer simulation.

Dynamics of blood flow in large pulmonary vessels

In this section we describe the waveforms of pressure and flow rate in large pulmonary arteries and veins, and discuss the propagation, attenuation and reflection of pressure waves in these vessels, with reference to their elastic properties which we have already described. We also outline the little that is known about flow patterns and velocity profiles in them.

Waveforms Typical waveforms of blood pressure and flow rate, recorded in the main pulmonary artery of a dog, are shown in **Fig. 15.14**. Also shown are the blood flow-rate waveform in a large pulmonary vein and the blood pressure waveform in the left atrium. The pulmonary arterial pressure has a time-average value of about $2.0 \times 10^3\,\text{N}\,\text{m}^{-2}$ ($15\,\text{mm}\,\text{Hg}$) and an amplitude of about $1.0 \times 10^3\,\text{N}\,\text{m}^{-2}$, representing a 50% variation about the mean, a considerably higher percentage than in a systemic artery (p. 261). The shape of the pressure waveform does not differ greatly from that in the aorta. However, in slightly smaller arteries there is a difference, because the waveform in systemic arteries changes markedly with distance, while there is no obvious change along the large pulmonary arteries. This might be because the pulmonary arteries are so short that the pressure pulse occurs almost simultaneously in all of them; however, this assumption would lead to significant errors in predicting the pulmonary arterial input impedance as a function of frequency (see Chapter 12 and p. 492).

There is a disagreement as to the origin of the pressure and flow-rate waveforms in the pulmonary veins (**Fig. 15.14**). The question is whether they derive mainly from events in the right ventricle, having been propagated through the pulmonary

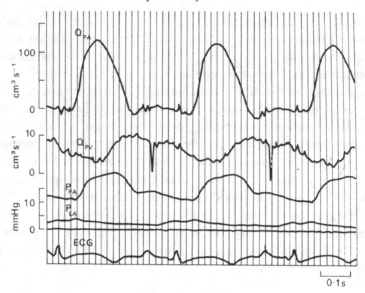

Fig. 15.14. Experimental records of pulmonary pressures and flows in a conscious dog at rest. Symbols: Q, blood flow rate; P, pressure; PA, pulmonary artery; PV, pulmonary vein, within 2 cm of the left atrium; LA, left atrium; ECG, electrocardiogram, lead I. Sharp downward spikes in the pulmonary venous flow tracing and smaller spikes in the pulmonary arterial flow during diastole are electrocardiographic signals. Timing lines appear at intervals of 0–2 s. (From Milnor (1972). Pulmonary haemodynamics. In *Cardiovascular Fluid Dynamics*, (ed. Bergel), vol. 2, pp. 299–340. Academic Press, New York.)

microcirculation, or whether they are caused by retrograde transmission from the left atrium and ventricle. Evidence obtained when a lung is perfused experimentally, with a steady flow, and its normal venous connections to the left atrium are maintained, supports the idea that the venous pressure waves normally originate in the left heart; normal left atrial-type pressure waveforms are seen in the pulmonary veins. However, if the veins are disconnected from the left atrium, the flow that emerges from them, in a lung which is perfused via its artery with a pulsatile flow, is pulsatile. Furthermore, pressure oscillations which resemble those in the pulmonary artery, but are delayed in time, are seen in a pulmonary vein, in vivo, if it is obstructed and the pressure is measured upstream of the obstruction. The time-average pressure has of course risen then to the pulmonary arterial level and the distensibility of the vein is consequently greatly reduced. These findings imply that the flow emerging from the pulmonary microcirculation into the veins, though it is normally pulsatile, generates little pressure oscillation in the veins, because they have a high distensibility at the physiological low values of transmural pressure. Microscopic observations on the pleural surface of the lung have revealed that blood flow in the superficial vessels of the microcirculation, including the precapillary vessels and alveolar capillaries, is pulsatile.

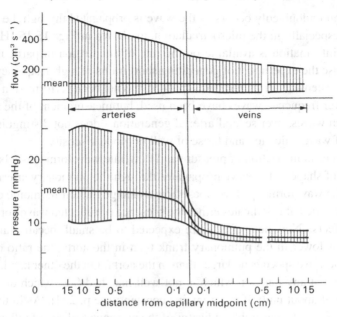

Fig. 15.15. Diagram of mean and pulsatile pressure ($10\,\text{mmHg} = 1.36 \times 10^3\,\text{Nm}^{-2}$) and flow along an average path through the human pulmonary vascular bed. Note changes of scale at 0.5 cm. (From Milnor (1972). Pulmonary haemodynamics. In *Cardiovascular Fluid Dynamics* (ed. Bergel), vol. 2, pp. 299–340. Academic Press, New York.)

Studies of the uptake of the highly soluble gas nitrous oxide by the lung confirm this finding. **Figure 15.15** is a schematic diagram of the variation of mean and pulsatile pressure and flow rate through the pulmonary circulation; we give a description later (**Fig. 15.17**) of the pulsatile blood flow in the pulmonary capillaries.

The mean blood velocity in the pulmonary trunk is approximately the same as in the aorta, because the two vessels have similar diameters and the entire cardiac output passes through them both. The peak velocity, however, is about half as high in the pulmonary trunk as in the aorta (about $0.7\,\text{m s}^{-1}$ rather than $1.2\,\text{m s}^{-1}$).

Wave propagation The speed of propagation of the pulse wave in pulmonary arteries has been measured in a number of species, including humans, but the total amount of data in any one species is rather sparse. A typical value in humans is about $1.75\,\text{m s}^{-1}$, while the corresponding value in a dog (**Table I**) is about $2.5\,\text{m s}^{-1}$. As already remarked (p. 481), the dog result corresponds fairly well with the values predicted from static distensibility measurements. Pulse wave velocities of 4–$8\,\text{m s}^{-1}$ have been measured in patients with pulmonary hypertension in whom there is pathological thickening and stiffening of the vessel walls.

Attenuation undoubtedly occurs as the wave is propagated through the pulmonary circulation, especially in the microcirculation, as shown in **Fig. 15.15**. However, no quantitative information is available on the rate of attenuation in particular vessels. This is because the experiment of generating short trains of high-frequency waves, as was done in systemic arteries and veins (p. 445), has not been performed in pulmonary vessels. Lower frequency waves have been used, but measurement of the attenuation factor of such waves, over several arterial generations, does not distinguish between the effects of wave reflection and those of wall or blood viscosity.

Although visual inspection of pressure and flow-rate waveforms reveals no significant change of shape as the wave propagates through the pulmonary arteries, Fourier analysis of the waveforms (p. 126) does show a change in their harmonic content. As in systemic arteries, this indicates either the presence of reflected waves or significant nonlinear effects. Nonlinear effects are expected to be small, because although the wave speed is lower in the pulmonary trunk than in the aorta, the ratio of the peak velocity to the wave speed is no larger than in the aorta. On the other hand, reflections are anticipated at any junction which is not well matched (i.e. which does not have an area ratio of about unity if h/d remains constant; see p. 284). As in the systemic arteries, a useful and measurable indicator of the presence and site of reflections is the input impedance of the pulmonary circulation, defined for each frequency component as the ratio of the (complex) pressure and flow rate measured at the same site in the pulmonary trunk (see p. 338 *et seq.* for a fuller definition of input impedance and its modulus and phase).

Figure 15.16 shows calculated values of pulmonary arterial input impedance, derived from measurements in anaesthetized, open-chested dogs; the modulus is depicted above and the phase below. We see that the modulus falls from a high value at low frequency to a minimum at 3–4 Hz, when the phase becomes zero. At this frequency the pulmonary trunk is the site of a node (i.e. a minimum of pressure amplitude and a maximum of flow-rate amplitude), a result suggesting that there is partial, closed-end-type reflection, at a distance of one-quarter of a wavelength from the pulmonary trunk. If the wave speed is $2.5\,\mathrm{m\,s^{-1}}$, with a frequency of 4 Hz, a quarter wavelength is about 16 cm, which is approximately the average path length to the microcirculation in a large dog. That there should be such a reflection site in the microcirculation is reasonable. The presence of the second maximum of the impedance modulus at about twice the frequency of the minimum supports the idea of reflection sites in the microcirculation, since the pulmonary trunk would be the site of an antinode for waves of twice the frequency (half the wavelength) of those for which it is the site of a node.

The difference between the pattern shown in **Fig. 15.16** and that found in the aorta (**Fig. 12.29**, p. 292) is probably explained by the relative symmetry of the pulmonary arterial tree, compared with that of the systemic arterial tree, which has shorter path lengths to the upper half of the body than to the lower (**Fig. 12.31**, p. 294).

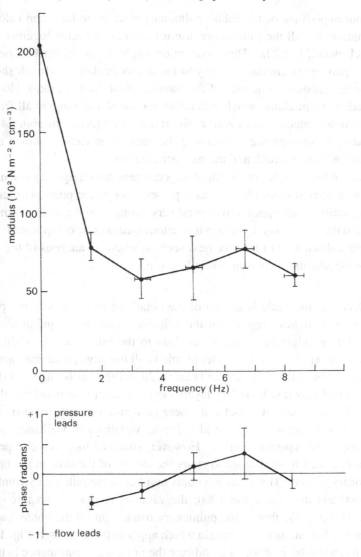

Fig. 15.16. Pulmonary arterial input impedance in anaesthetized, open-chest dogs. The points represent mean values in 29 animals; the bars represent ±1 standard error of the mean. The modulus falls to a minimum at 3–4 Hz and at about the same frequency the phase crosses the zero line. At about 6 Hz there is a second maximum of the modulus. (From Milnor, Bergel and Bargainer (1966). Hydraulic power associated with pulmonary blood flow and its relation to heart rate. *Circulation Res.* **19**, 467.)

The input impedance of the rabbit pulmonary circulation has been calculated on the assumption that all the pulmonary arteries expand and relax together (as in the Windkessel model, p. 270). That assumption might seem reasonable, because the length of a pulmonary arterial pathway to the microcirculation is much shorter than the wavelength (about one-quarter of the wavelength at the heart rate). However, the results predicted a modulus which was below the measured value at all frequencies, with a minimum value close to zero at the resonant frequency of about 4 Hz. This is presumably a consequence of ignoring the multiple reflections which take place between the pulmonary trunk and the microcirculation.

The value of the modulus of the input impedance at zero frequency is a measure of the resistance offered to steady flow as it passes through the pulmonary circulation. This is a quantity which appears to depend very strongly on the degree of inflation of the lung, and the mechanism by which it is determined is quite complicated. We therefore postpone discussion of it to the next section, where an analysis of the resistance offered by the alveolar sheet is also briefly given.

Flow patterns Rather little is known of the details of blood flow in the pulmonary arteries. Recent studies suggest that the velocity profile in the pulmonary trunk is relatively flat, as might be expected so close to the inlet from the right ventricle. Indeed, we can expect a fairly flat inlet profile in all the large pulmonary arteries (and veins), because the Reynolds number is large, the value of α is large and the vessels are short so that new thin boundary layers begin on each flow divider and never fill the tubes. The curvature of vessel walls near bifurcations seems to be small enough to prevent flow separation. We can also expect variations of wall shear stress near bifurcations, as in systemic arteries. However, smaller blood velocity probes than are currently available are needed to study the details of the flow in the branches of the pulmonary vessels. The mean Reynolds number in the pulmonary trunk in a dog is about 600 and the peak about 3000; the value of α (for a heart rate of 2 Hz) is about 15. This puts the flow in the pulmonary trunk right on the borderline between laminar and turbulent, using the criteria which apply to the aorta (see **Fig. 12.50**). No evidence has so far been obtained to indicate the presence of turbulence in the normal pulmonary trunk.

Pulmonary vascular resistance

Flow in the alveolar sheet Blood flow in alveolar capillaries differs from that in systemic capillaries in two important respects. The first is that the capillaries are arranged as a sheet with flexible walls, which are held apart by the 'posts' (**Fig. 15.4**), and the second is that the flow is markedly pulsatile (**Fig. 15.17**). However, there are two fundamental respects in which the flows in the two types of capillary are the same. These are that: (a) the Reynolds number is small (if a typical blood velocity

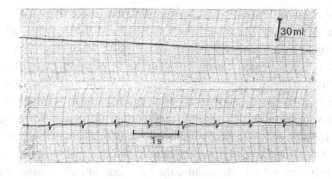

(a)

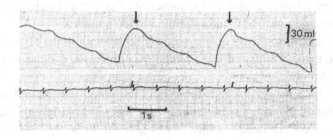

(b)

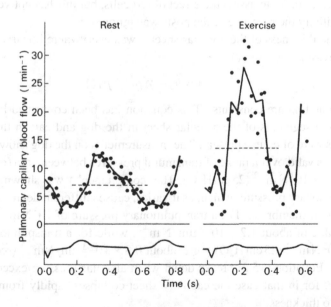

(c)

Fig. 15.17. Records demonstrating that pulmonary capillary blood flow is pulsatile. In (a) and (b) is shown the pressure in a sealed rigid box (whole-body plethysmograph) in which a seated subject is breathholding at resting lung volume with the glottis open. The ECG is also shown. The subject has been breathing (a) air and (b) a mixture of O_2 and the highly *(cont.)*

in the sheet is $0.005 \, \mathrm{m \, s^{-1}}$ and a representative sheet thickness at zero transmural pressure h_0 is $4 \, \mu\mathrm{m}$, the Re is about 0.005); and (b) the value of α is small, even for the highest significant harmonics of the pulse wave ($\alpha = 0.0002$ for a frequency of $10 \, \mathrm{Hz}$). This means that (i) fluid inertia is negligible and the flow everywhere represents a balance between the pressure gradient and the viscous forces and (ii) the flow is quasi-steady, so that the flow-rate waveform is in phase with the pressure gradient waveform at all parts of the sheet, at all times. For a Newtonian fluid in a rigid tube or sheet, these conditions would mean that the flow rate through the system was always directly proportional to the pressure drop across it. In an elastic system that does not follow, because the local pressure determines the sheet thickness, which determines the resistance to flow, so that the resistance is not independent of the driving pressure gradient. The presence of suspended elements in the blood also affects the resistance, as in the systemic microcirculation.

An approximate theory of quasi-steady blood flow in the alveolar sheet has been developed with the help of some model experiments and some simplifying assumptions, as follows.

(1) In both the theory and the models the blood was treated as a homogenous Newtonian fluid, with viscosity about four times that of water, and the presence of red cells in concentrated suspension was ignored (an analysis similar to the lubrication theory given in Chapter 13 for systemic capillaries, p. 396, could presumably be developed in order to assess the effect of red cells, but this has not yet been done).

(2) Initially the presence of the posts was ignored.

(3) The thickness of the alveolar sheet h was everywhere linearly related to transmural pressure ($p_{\mathrm{iv}} - p_{\mathrm{alv}}$):

$$h = h_0 + \gamma(p_{\mathrm{iv}} - p_{\mathrm{alv}}), \tag{15.2}$$

where γ and h_0 are constants. This equation has been confirmed by measurements on excised segments of the alveolar sheet in the dog and cat; in these preparations the posts were of course present. The measurements on the dog showed that Equation (15.2) was valid over a range of transmural pressures between zero (or just above) and about $2.5 \times 10^3 \, \mathrm{N \, m^{-2}}$ ($25 \, \mathrm{cm \, H_2O}$). The 'compliance' γ was shown to depend on the transpulmonary pressure, which, as it varies, causes different degrees of distension of the alveolar membrane. For a transpulmonary pressure of $10^3 \, \mathrm{N \, m^{-2}}$ ($10 \, \mathrm{cm \, H_2O}$), γ has a value of about $1.2 \times 10^{-3} \, \mu\mathrm{m/N \, m^{-2}}$, while for a transpulmonary pressure of $2.5 \times 10^3 \, \mathrm{N \, m^{-2}}$ ($25 \, \mathrm{cm \, H_2O}$) γ was about $0.8 \times 10^{-3} \, \mu\mathrm{m/N \, m^{-2}}$ (see also **Fig. 15.8**, p. 482). Equation (15.2) breaks down when alveolar pressure exceeds intravascular pressure, for in that case the capillary sheet collapses rapidly from thickness h_0 to near zero thickness.

The theoretical analysis of the flow was developed from Equation (15.2) together with the two basic principles governing fluid flow in a narrow space. These are that the pressure gradient is everywhere balanced by viscous forces and that at any

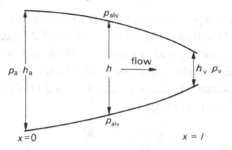

Fig. 15.18. Illustration of sheet thickness and pressure used in the model of sheet flow analysed in the text.

particular time the volume flow rate must be the same everywhere in the sheet, if the fluid is incompressible and the flow is quasi-steady, since fluid is not destroyed or created anywhere. These are the same principles as those used in the lubrication analysis of red cell motion in systemic capillaries (p. 396). The results can best be described in terms of flow in a two-dimensional sheet, as illustrated schematically in **Fig. 15.18**; there is no variation in the thickness of the sheet or in the flow in a direction perpendicular to the plane of the diagram. The pressure in the sheet p_{iv} is assumed to be given at the arterial and venous ends of the sheet (p_a and p_v respectively). The pressure outside is the alveolar pressure p_{alv}. The thickness h of the sheet varies with distance x along it and has the value h_a at the arterial end. The main results of the theory are as follows.

(1) If alveolar pressure is less than pulmonary venous pressure ($p_{alv} < p_v < p_a$), the sheet remains open and h varies with x according to the following equation:

$$h^4 = h_a^4 - \beta x, \qquad (15.3)$$

Fig. 15.17. (*cont.*) soluble gas N_2O. In (a) there is a slight steady rise of pressure due to warming of the air (the record is inverted and pressure changes have been converted to volume changes by calibration). In addition, there are small pressure oscillations, which are due to such factors as compression of gas by the beating of the heart. In (b) plethysmograph pressure falls in an oscillatory way, with the period being that of the heart cycle. (The arrows indicate where, by venting the plethysmograph, the pressure was restored to atmospheric.) The instantaneous pulmonary capillary blood flow was calculated by point-to-point subtraction of the control from the test record, to obtain the instantaneous rate of uptake of N_2O, and by scaling this value by the solubility of N_2O in blood at the measured alveolar partial pressure for the gas. Shown in (c) are values of pulmonary capillary blood flow obtained in a normal subject at rest and after exercise. The dotted line represents the mean flow rate and the ECG is shown. (From Lee and DuBois (1955). Pulmonary capillary blood flow in man. *J. Clin. Invest.* **34**, 1380.)

where β is a term which depends on the values of p_a and p_v and on the length l of the sheet in the direction of the flow. This, with Equation (15.2), means that an expression involving the fourth power of p_{iv} also varies linearly with x, which is quite different from the case in a straight, rigid tube or sheet, in which it is the pressure itself which varies linearly with distance. In this case the flow rate Q is directly proportional to βl, and hence (from Equation (15.3))

$$Q = C(h_a^4 - h_v^4), \tag{15.4}$$

where h_v is the thickness of the sheet at the venous end and C is a constant independent of sheet thickness and of pressure.

(2) When the sheet remains open everywhere, the resistance to flow through it, equal to the pressure drop divided by the flow rate, and denoted by R, is given by

$$\frac{1}{R} = C\gamma\frac{h_a^4 - h_v^4}{h_a - h_v}, \tag{15.5}$$

where γ and C are the same constants as in Equations (15.2) and (15.4). This result will clearly be affected by the presence of the posts. One way of assessing their effect is to find the factor by which the resistance in a parallel-sided, rigid sheet is increased by the presence of posts and to apply the same factor to the case of the elastic sheet. In both model experiments and a theoretical study this factor was found to be about 3. In the absence of confirmation in the real alveolar sheet, or in an accurate elastic model of it, this factor should be applied to the resistance determined from Equation (15.5).

(3) The variation of h with x, described by Equation (15.3), is most rapid at the venous end of the sheet. This is because the rate of change of h with x, dh/dx, is equal to $-\beta/4h^3$, which is greatest for the smallest values of h; that is, for the largest values of x. The variation is particularly rapid, at the venous end, if the alveolar pressure is only just smaller than the venous pressure p_v, for in that case h_v is as small as h_0. For only a very small further increase in alveolar pressure, the venous end of the tube will collapse completely and the flow-rate Q will become equal to Ch_a^4, independent of conditions at the venous end. As alveolar pressure increases further, the thickness of the open portion of the sheet decreases and the flow rate falls, approximately in proportion to $(p_a - p_{alv})^4$, from Equation (15.2). Finally, if alveolar pressure exceeds arterial pressure, the flow ceases altogether. This process is very similar to that described in the section on collapsing veins (p. 429 *et seq.*), only this time inertia plays no part in the mechanism, and self-excited oscillations are never set up.

It is virtually impossible to measure alveolar sheet thickness in vivo, while blood is flowing, so many of the details of this theory cannot be checked directly. However, the predictions of resistance can be checked at least qualitatively, and the volume of blood in the capillaries can be assessed. This is important, because the prediction that sheet thickness varies most rapidly near the venous end (based on Equation (15.3)) implies that capillary blood volume should be proportional to $p_a - p_{alv}$, and be virtually independent of the venous pressure. This is observed experimentally, as mentioned earlier (p. 485).

The above analysis predicts that the state of a capillary sheet and the flow through it takes one of three forms, according to the relative magnitudes of p_a, p_v and p_{alv}, as we shall now see.

Zonal distribution of blood flow It has been recognized for some time that the distribution of blood flow within the lung depends on the relative levels of arterial (p_a), venous (p_v) and alveolar pressures (p_{alv}), and some of the evidence will be mentioned later. We have already seen that flow through collapsible vessels, such as veins, is strongly influenced by the local transmural pressure (Chapter 14, p. 451). In the pulmonary capillaries, as in veins, we can distinguish three regimes or zones:

I. If the alveolar pressure exceeds both arterial and venous pressures the bed is closed and there is no flow.

II. If the alveolar pressure is intermediate between the arterial and venous pressures, the bed is partially collapsed and the flow rate depends on $p_a - p_{alv}$.

III. If the venous pressure is intermediate between the arterial and alveolar pressures the sheet is open everywhere and the flow rate depends on $p_a - p_v$.

In the normal, upright resting human, all three zones may be found, because there is a hydrostatic gradient of intravascular pressure down the lung, as already described, while alveolar pressure is roughly uniform and close to atmospheric. Thus, near the top of the lung, blood pressure is low and the first condition applies; this is termed zone I. In the middle of the lung, the second condition applies, and the region is called zone II. Finally, in the lower part of the lung, zone III, venous pressure exceeds alveolar pressure.[2]

Clearly, the levels at which the zones merge into one another will vary with alveolar pressure. During positive pressure inflation, more of the lung will be in zone I, while during a rapid voluntary inspiration p_{alv} is less than atmospheric pressure and zone II is extended. Even more striking changes are to be expected if pulmonary arterial pressure is raised, for example during exercise, or in pulmonary hypertension; then zone II may occupy the whole lung.

The existence of these zones has been confirmed in a number of ways, for example, by observing the number of red cells in the alveolar capillaries in histological sections of dog lungs, frozen rapidly while being perfused at a series of vascular and alveolar pressures (**Fig. 15.8**). Furthermore, it has been confirmed in isolated perfused lungs and in studies on intact humans.

---

[2] Experimental results show that it is actually necessary to distinguish a fourth region, zone IV, which is adjacent to the bottom of the lung. Since flow in this region is greatly affected by lung volume, discussion of it is deferred to p. 501.

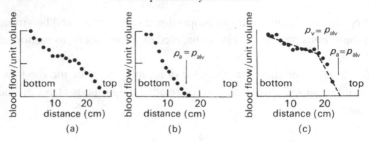

Fig. 15.19. Measurements of the distribution of blood flow in isolated perfused dog lungs by [133]Xe (see text). (From West, Dollery and Naimark (1963). Distribution of blood flow in isolated lung: relation to vascular and alveolar pressures. *J. Appl. Physiol.* **19**, 713.)

Figure 15.19 shows the effect on the vertical distribution of the blood flow rate in an isolated perfused dog lung of varying arterial, alveolar and venous pressures. The blood flow rate has been measured by using radioactive xenon (^{133}Xe), which is dissolved in saline and injected into the pulmonary artery. When it reaches the pulmonary capillaries it is largely transferred from the blood into the alveolar gas because of the large partial pressure difference and its low solubility in blood. Ventilation is stopped while this is occurring and the lungs are scanned with counters; the amount of radioactivity measured at any station after appropriate calibration reflects the local blood flow rate.

Figure 15.19a shows the distribution of blood flow when alveolar pressure is atmospheric and the lung is inflated by a negative pressure of $1 \times 10^3 \, \text{N} \, \text{m}^{-2}$ (10 cmH_2O). Arterial pressure is $3.2 \times 10^3 \, \text{N} \, \text{m}^{-2}$ (32 cmH_2O) at the bottom of the lung, which is 26 cm tall, and the venous pressure reservoir is below the bottom of the lung, so that the venous pressure is sub-atmospheric throughout the lung. There is an approximately linear increase in blood flow down the lung, with the flow rate being almost zero at the top. In **Fig. 15.19**b the effect on the flow distribution of reducing the pulmonary artery pressure to $1.6 \times 10^3 \, \text{N} \, \text{m}^{-2}$ (16 cmH_2O) is shown. Arterial pressure then equals alveolar pressure 16 cm above the bottom of the lung and the flow rate falls to zero at about this level. In **Fig. 15.19**c arterial pressure is $2.55 \times 10^3 \, \text{N} \, \text{m}^{-2}$ (25.5 cmH_2O), whilst the venous reservoir has been raised 19 cm above the bottom of the lung. In the region where venous pressure exceeds alveolar pressure there is only a gradual increase of blood flow down the lung, whereas in the upper part of the lung, where alveolar pressure exceeds venous pressure, the gradient is much steeper. From the arguments given above we may see that there is: (i) no flow in zone I (**Fig. 15.19**b); (ii) a steep increase in flow with distance down the lung in zone II (**Fig. 15.19**b and c); (iii) a slight increase of flow with distance down the lung in zone III (**Fig. 15.19**c). Similar results have been obtained in humans.

In order to understand why the blood flow rate increases with distance down the lung in zones II and III and why the rate of increase is different in the two zones, we must consider what pressure difference determines the blood flow rate and what factors determine the pulmonary vascular resistance within each zone.

In zone II the flow rate is determined by $p_a - p_{alv}$ (p. 499) and since p_{alv} is essentially independent of height in the lung, whereas p_a increases by about $1\,cm\,H_2O$ $(0.1 \times 10^3\,N\,m^{-2})$ per centimetre distance down the lung, the pressure difference causing flow increases by that amount (see footnote 3 below). The pulmonary vascular resistance in this zone is determined by the pressure in the pulmonary arteries and capillaries, venous pressure being unimportant. Because the arterial pressure increases down the lung, the arteries can be expected to be progressively distended, according to the local transmural pressure and the vessel elastic properties. Thus, the resistance offered by the arteries will also fall with distance down the lung, though the relationship between vessel diameter and resistance cannot be predicted exactly, because of the complicated nature of the flow. The transmural pressure of the alveolar capillaries will also increase with distance down the lung and, as a result, there will be both an increase in the number of open capillaries and a progressive distension of those that are patent. Both these changes will cause capillary resistance to fall with distance down the lung, and all the vascular changes, coupled with the increase of the pressure determining flow, will cause the blood flow rate to increase down the lung.

In zone III the flow rate is determined by $p_a - p_v$ (p. 499), and since both p_a and p_v increase by the same amount (about $1\,cm\,H_2O$ per centimetre distance down the lung) the pressure difference which determines the flow rate will be independent of height in the lung.[3] The pulmonary vascular resistance in this zone is determined by the arteries, capillaries and veins. The comments made concerning the dimensions and resistances of the arteries in zone II also apply to the arteries and veins in zone III, so that the resistance to flow through these vessels and the capillaries can be expected to fall with distance down the lung. In the case of the capillaries the predominant change with distance down the lung is thought to be distension, although so far it has not proved possible to distinguish experimentally between distension and recruitment.

These differences in mechanics between zones II and III offer a consistent explanation for the greater increase of flow rate with distance down the lung in the former zone. We shall discuss in the next section the effect of lung volume on the relative resistances of the extra-alveolar and alveolar vessels.

Effect of lung mechanics We have seen above the influence of pleural pressure and lung volume on the mechanics of the extra-alveolar vessels (p. 477) and of alveolar pressure and lung volume on the mechanics of the alveolar capillaries (p. 479). Now

---

[3] This is not precisely true because the distribution of pressures along the vessels of the lung depends on both the arterial and venous resistances, as well as on capillary resistance. Thus, the distribution of pressures and the distribution of flow rates are interdependent.

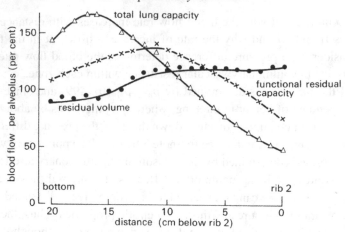

Fig. 15.20. Blood flow per alveolus as a percentage of that expected had all alveoli been perfused equally, plotted against distance down the lung at three lung volumes – total lung capacity, functional residual capacity and residual volume. Points represent the mean values of the right and left lungs of all subjects studied at that volume. Note the reduction of basal blood flow which is more marked at lower lung volumes. (From Hughes, Glazier, Maloney and West (1968). Effect of lung volume on the distribution of pulmonary blood flow in man. *Respiration Physiol.* **4**, 58.)

we examine the effect of lung mechanics on the distribution of pulmonary blood flow and pulmonary vascular resistance.

Experimental studies in erect humans, which confirm the existence of the zones, were made at fairly high lung volumes. However, when similar studies were made with a somewhat refined technique over the whole range of lung volumes, hitherto unrecognized behaviour of pulmonary blood flow was found, as shown in **Fig. 15.20**. At total lung capacity a small region is seen near the bottom of the lung where blood flow actually falls with distance down the lung; this becomes larger as lung volume is reduced to functional residual capacity and at residual volume the flow in the apical region exceeds that near the base, thus reversing the distribution seen at total lung capacity. It is impossible to explain this behaviour in terms of the simple model described (p. 499). It is believed that it is explained by the weight of the lung causing it to be less expanded near the base (p. 477) and the raised perivascular pressure of the basal extra-alveolar vessels causing them to be narrowed and to have a raised resistance. Consistent with this explanation is the finding that alveolar volume is relatively uniform in the fully expanded isolated lung.

Consider next the results obtained when an isolated perfused lung is inflated by an external negative pressure, while maintaining atmospheric alveolar pressure. The differences between arterial and venous pressure and between arterial and alveolar pressure are controlled independently of chamber pressure and lung volume. The

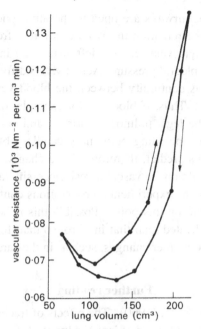

Fig. 15.21. Vascular resistance versus lung volume. All points were determined under static conditions. Arrows indicate direction of volume change. (From Thomas, Griffo and Roos (1961). Effect of negative-pressure inflation of the lung on pulmonary vascular resistance. *J. Appl. Physiol.* **16**, 451.)

blood flow rate was measured at different values of lung volume and the pulmonary vascular resistance was calculated. As shown in **Fig. 15.21**, as lung volume was increased, pulmonary vascular resistance fell slightly, taking a minimum value when the lung was moderately inflated (100–150 ml). With further inflation of the lung, the pulmonary vascular resistance rose to about twice the minimum value.

We have noted above that inflation of a lung, under these conditions, leads to enlargement of the extra-alveolar vessels. It is believed that the observed fall of pulmonary vascular resistance with increase of lung volume is explained in this way. We have noted, moreover, that an increase of lung volume under these conditions causes compression of the alveolar capillaries, because of the increased tension in the alveolar walls, and it is believed that such a change explains the rise of pulmonary vascular resistance with further increase of lung volume. The implication is that the extra-alveolar vessels make the principal contribution to the pulmonary vascular resistance at small lung volume and the alveolar vessels at larger lung volume.

We must ask whether these experimental studies are relevant to conditions in vivo. Certainly, in vivo, lung volume is changed by change of pleural pressure, and, just as with the perfused excised lung, alveolar pressure is atmospheric at any static lung volume, provided that the airways are open to the atmosphere. However, the mechanics of the pulmonary circulation in vivo are different from those in the experiment, because both the right ventricle and left atrium are inside the pleural cavity, and are hence exposed to pleural pressure. Actually, this difference is apparent rather than real, because there is continuity between the blood vessels of the pulmonary and systemic circulations. Thus, if blood is, for example, displaced from systemic vessels, say by raising the legs, pulmonary arterial and venous pressures rise and pulmonary blood volume, including pulmonary capillary blood volume, increases (p. 488). These arguments predict, therefore, that a change in lung volume in vivo will cause a change in pulmonary vascular resistance similar to that seen with the isolated perfused lung. Some experimental observations confirm that prediction, but there is not firm agreement on this point. Possibly this is because the arrangement in vivo is far more complicated than that in vitro, so that, for example, a change of posture can also bring about other changes, such as in the cardiac output.

Further reading

Banister, J. and Torrance, R. W. (1960). The effects of tracheal pressure upon flow: pressure relations in the vascular bed of isolated lungs. *J. Exptl. Physiol.* **45**, 352–67.

Bates, D. V., Macklem, P. T. and Christie, R. V. (1971). *Respiratory Function in Disease*. W. B. Saunders, Philadelphia.

Benjamin, J. J., Murtagh, P. S., Proctor, D. F, Menkes, H. A. and Permutt, S. (1974). Pulmonary vascular interdependence in excised dog lobes. *J. Appl. Physiol.* **37**, 887–94.

Caro, C. G. (1966). Mechanics of the pulmonary circulation. In *Advances in Respiratory Physiology* (ed. C. G. Caro). Edward Arnold, London.

Cumming, G. (1974). The pulmonary circulation. In *Cardiovascular Physiology* (eds A. C. Guyton and C. E. Jones). Physiology Series I, Vol. I, pp. 93–122. MTP International Review of Science. Butterworths, London/University Park Press, Baltimore.

Fishman, A. P. (1963). Dynamics of the pulmonary circulation. In *Handbook of Physiology*. Section 2: Circulation. Vol. II (eds W. F. Hamilton and P. Dow). American Physiological Society, Washington, DC.

Fishman, A. P. and Hecht, H. H. (ed.) (1969). *The Pulmonary Circulation and Interstitial Space*. University of Chicago Press, Chicago.

Fung, Y. C. and Sobin, S. S. (1972). Pulmonary alveolar blood flow. *Circulation Res.* **30**, 470–90.

Fung, Y. C. and Sobin, S. S. (1969). Theory of sheet flow in lung alveoli. *J. Appl. Physiol.* **26**, 472–88.

Harris, P. and Heath, D. (1962). *The Human Pulmonary Circulation: its Form and Function in Health and Disease*. E. & S. Livingstone, Edinburgh.

Howell, J. B. L., Permutt, S., Proctor, D. F. and Riley, R. L. (1961). Effect of inflation of the lung on different parts of pulmonary vascular bed. *J. Appl. Physiol.* **16**, 71–6.

Permutt, S., Howell, J. B. L., Proctor, D. F. and Riley, R. L. (1961). Effect of lung inflation on static pressure–volume characteristics of pulmonary vessels *J. Appl. Physiol.* **16**, 64–70.

Spencer, H. (1968). *Pathology of the Lung*. Pergamon Press, Oxford.

West, J. B. (1974). Blood flow to the lung and gas exchange. *Anesthesiology* **41**, 124–38.

Index

Table I. Normal values for canine cardiovascular parameters. An approximate average value, and then the range, is given where possible

Site		Aorta			Femoral	Carotid artery	Arteriole	Capillary	Venule	Inferior vena cava	Main pulmonary artery
		Ascending	Descending	Abdominal							
Internal diameter d_i	cm	1.5, 1.0–2.4	1.3, 0.8–1.8	0.9, 0.5–1.2	0.4, 0.2–0.8	0.5, 0.2–0.8	0.005, 0.001–0.008	0.0006, 0.0004–0.0008	0.004, 0.001–0.0075	1.0, 0.6–1.5	1.7, 1.0–2.0
Wall thickness h	cm	0.065, 0.05–0.08		0.05, 0.04–0.06	0.04, 0.02–0.06	0.03, 0.02–0.04	0.002	0.0001	0.0002	0.015, 0.01–0.02	0.02, 0.01–0.03
h/d_i		0.07, 0.055–0.084		0.06, 0.04–0.09	0.07, 0.055–0.11	0.08, 0.053–0.095	0.4	0.17	0.05	0.015	0.01
Length	cm	5	20	15	10	15, 10–20	0.15, 0.1–0.2	0.06, 0.02–0.1	0.15, 0.1–0.2	30, 20–40	3.5, 3–4
Approximate cross-sectional area	cm^2	2	1.3	0.6	0.2	0.2	2×10^{-5}	3×10^{-7}	2×10^{-5}	0.8	2.3
Total vascular cross-sectional area at each level	cm^2	2	2	2	3	3	125	600	570	3.0	2.3
Peak blood velocity	cm s^{-1}	120, 40–290	105, 25–250	55, 50–60	100		0.75, 0.5–1.0	0.07, 0.02–0.17	0.35, 0.2–0.5	25, 15–40	70
Mean blood velocity	cm s^{-1}	20, 10–40	20, 10–40	15, 8–20	10, 10–15		{		}		15, 6–28
Reynolds number (peak)		4500	3400	1250	1000		0.09	0.001	0.035	700	3000
α (heart rate 2 Hz)		13.2	11.5	8	3.5	4.4	0.04	0.005	0.035	8.8	15
Calculated wave speed c_0	cm s^{-1}	580	500	770	840	850				100	350
Measured wave speed c	cm s^{-1}	500, 400–600		700, 600–750	900, 800–1030	800, 600–1100				400, 100–700	250, 200–330
Young's modulus E	N m$^{-2} \times 10^5$	4.8, 3–6		10, 9–11	10, 9–12	9, 7–11				0.7, 0.4–1.0	6, 2–10

From C. G. Caro, T. J. Pedley and W. A. Seed (1974). Mechanics of the circulation. Chapter 1 of *Cardiovascular Physiology* (ed. A. C. Guyton). Medical and Technical Publishers, London.

Printed in the United States
by Baker & Taylor Publisher Services